Advanced Classical
Electrodynamics

Green Functions, Regularizations,
Multipole Decompositions

Advanced Classical Electrodynamics

Green Functions, Regularizations, Multipole Decompositions

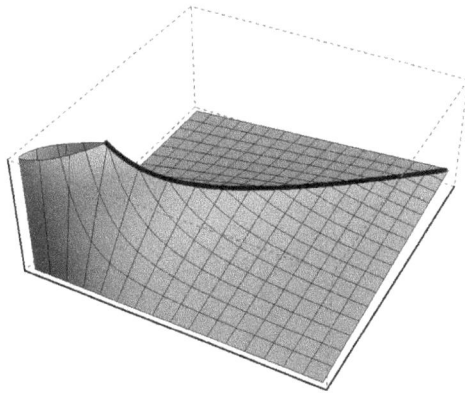

Ulrich D Jentschura

Missouri University of Science and Technology, USA

World Scientific

NEW JERSEY · LONDON · SINGAPORE · BEIJING · SHANGHAI · HONG KONG · TAIPEI · CHENNAI · TOKYO

Published by

World Scientific Publishing Co. Pte. Ltd.

5 Toh Tuck Link, Singapore 596224

USA office: 27 Warren Street, Suite 401-402, Hackensack, NJ 07601

UK office: 57 Shelton Street, Covent Garden, London WC2H 9HE

Library of Congress Cataloging-in-Publication Data

Names: Jentschura, Ulrich D., author.

Title: Advanced classical electrodynamics : Green functions, regularizations, multipole decompositions /
by Ulrich D. Jentschura (Missouri University of Science and Technology, USA).

Description: Singapore ; Hackensack, NJ : World Scientific, [2017] |
Includes bibliographical references and index.

Identifiers: LCCN 2017011376| ISBN 9789813222847 (hardcover ; alk. paper) |
ISBN 9813222840 (hardcover ; alk. paper) | ISBN 9789813222854 (pbk. ; alk. paper) |
ISBN 9813222859 (pbk. ; alk. paper)

Subjects: LCSH: Electrodynamics--Textbooks.

Classification: LCC QC631 .J46 2017 | DDC 537.601/51--dc23

LC record available at https://lccn.loc.gov/2017011376

British Library Cataloguing-in-Publication Data

A catalogue record for this book is available from the British Library.

Dedicated to All Who Seek Beauty in Nature

Preface

Classical electrodynamics is a vast and open field. It would be hard to summarize all that is known in field within a few hundred pages. Still, classical electrodynamics has been the first theory to incorporate vector calculus into the realm of theoretical physics, and notably, the first theory that was (later) shown to be Lorentz covariant, upon a proper identification and grouping of the physical quantities into so-called four-vectors. When Stokes set his theorem as an inspired problem in the Cambridge mathematical tripos examinations, he had no idea how instrumental vector calculus would later be in the development of physics. We shall attempt to summarize the most important aspects of classical electrodynamics, with an emphasis on those concepts that will be useful for further studies in quantum electrodynamics and field theory. Emphasis will be laid on Green functions.

A general survey of the Maxwell equations is given in Chap. 1, with an emphasis on the physical meaning of gauge transformations of the scalar and vector potentials. The Green functions of electrostatics and the concomitant multipole decompositions, in various coordinate systems, are discussed in Chap. 2. In particular, analytic (multipole) decompositions are contrasted with eigenfunction expansions of the Green functions of operators. The considerations are generalized to the Green functions of electrodynamics in Chap. 4. In particular, the resolution of the action-at-a-distance paradox is discussed, which otherwise plagues the theory of electromagnetic interactions in the Coulomb gauge.

In Chap. 3, we discuss some paradigmatic calculations in electrostatics, with an emphasis on the variational principle, and on series expansions of potentials. The Laplace and Poisson equations are treated. The Green function of the Helmholtz equations and the theory of harmonically oscillating sources are discussed in Chap. 5. The multipole decomposition of the radiation into electric and magnetic multipoles, and longitudinal components of the fields, is discussed on the basis of vector spherical harmonics, separating the fields into transverse and longitudinal vector multipole components. We also discuss the calculation of the Liénard–Wiechert potentials generated by a moving point charge, in Lorenz and Coulomb gauges.

Electrodynamics in media should not be ignored in any textbook on classical electrodynamics (see Chap. 6). In view of their illustrative power and technical relevance, waveguides and the resonant eigenmodes of cavities (both rectangular as well as cylindrical) are discussed in Chap. 7.

Finally, in Chap. 8, we discuss special relativity and quantum field theory, and the connection to electrodynamics. Classical electrodynamics can be augmented in two ways, first, supplying a manifestly relativistic interpretation, and second, adding a quantum-field theoretical aspect. These concepts are introduced in Chap. 8. Among other things, the magnetic force $q\vec{v} \times \vec{B}$ is derived purely on the basis of the relativistic Lorentz contraction of charge distributions. Also, the Lorentz transformation of electromagnetic fields is discussed on the basis of explicit calculations. It is shown, e.g., that Gauss's law in the laboratory frame transforms into a superposition of the Gauss's law, expressed in terms of the coordinates of Lorentz-boosted frame, and of the Ampere–Maxwell law. The quantum field-theoretical zero-point energy is introduced in the calculation of the Casimir force between conducting plates. Finally, the connection of electrodynamics and general relativity is highlighted on the basis of covariant derivatives.

The intended audience of this book is threefold. *(i)* First, the book is intended for the advanced student who wishes to enhance his or her background knowledge on an interesting subfield of theoretical physics, possibly supplementing a course on the subject taught at a University. *(ii)* Second, we aim to reach the academic teacher who wishes to give a one-semester course on electrodynamics, in a modern style. Exercises represent challenges and serve as benchmarks for one's own understanding of the subject, but are designed not to represent insurmountable difficulties. The use of SI mksA units in all derivations may enhance the applicability to different subfields of physics where often, different unit systems are used. *(iii)* Last, but not least, the book is also intended as a partial encyclopedia for the active researcher who would like to look up the treatment of, e.g., angular momentum decompositions. A relatively new development in the latter context is the solution of the problem of calculating the Liénard–Wiechert potentials for a moving charge in Coulomb gauge, where, as opposed to Lorentz gauge, the seemingly instantaneous character of the Coulomb interaction has been an obstacle for a deeper understanding of the mechanism of retardation in the past.

In writing this book, we acknowledge helpful conversations with Benedikt J. Wundt and Barbara N. Hale, of Missouri University of Science and Technology, and help in the proofreading of the manuscript by Amanda Wetzel and Chandra M. Adhikari. This book project has been supported by the National Science Foundation (Grant PHY–1403973).

Rolla, Missouri, January 2017

Ulrich D. Jentschura

Missouri University of Science and Technology

Department of Physics, 1315 North Pine Street

Rolla, Missouri 65409-0640, USA, ulj@mst.edu

Contents

Chapter 1

Maxwell Equations

1.1 Basics

1.1.1 Integral and Differential Forms of the Maxwell Equations

There are T-shirts being sold which carry the enigmatic and deep writing: *And God said*

$$\vec{\nabla} \cdot \vec{E} = \frac{1}{\epsilon_0} \rho \,, \tag{1.1a}$$

$$\vec{\nabla} \cdot \vec{B} = 0 \,, \tag{1.1b}$$

$$\vec{\nabla} \times \vec{E} = -\frac{\partial}{\partial t} \vec{B} \,, \tag{1.1c}$$

$$\vec{\nabla} \times \vec{B} = \mu_0 \vec{J} + \frac{1}{c^2} \frac{\partial}{\partial t} \vec{E} \qquad (\text{``T-shirt equations''}) \,, \tag{1.1d}$$

and there was light. God has spoken, but it generally takes a student a little while to figure out and to interpret what He said. Still, it is instructive to consider the "writing on the T-shirt" at the very beginning of this course. Please be reassured that *(i)* the book is written while knowing the hardships of understanding electrodynamics very well, and that *(ii)* everyone needs to dwell on related questions for quite some time before developing full mastery of the subject (hopefully). Classical electrodynamics has a reputation for being a harder course than other courses in the physics curriculum. The reason for this observation is that electrodynamics combines, in quite a unique fashion, the intricacies of physics with those of mathematics. Differential calculus in three and four dimensions, and mastery of GREEN functions[1] are prerequisites for an understanding of electrodynamics. In contrast to analytic mechanics, quantum mechanics, relativistic quantum mechanics, and thermodynamics, electrodynamics is a field theory (a classical field theory perhaps, but still, a field theory).

We will not dwell on the philosophical question here regarding the success of the description of nature by mathematics. Four observations, though, can be noted:

---

[1] George Green (1793–1841)

(i) Augustinus said that everyone who wishes to contribute to the understanding
of nature needs to be willing to overcome considerable personal and professional
obstacles to pursue his or her career; that was true then and it is still true today.
(ii) The discovery of laws of nature by means of mathematics was alternatingly seen
as an act of recognizing and interpreting the works of the almighty in this world
(especially, if science was carried through by pious monks); or the work of the devil
who is taking apart the mysteries of nature by evil means (especially, if science was
carried through by doubtful characters). The truth probably lies in the middle;
there is good science and bad science, in regards to both the methods used to gain
insight into natural phenomena as well as the consequences for mankind. We cannot
really separate the two. We may refer to the famous saga of the goat driven over
the newly constructed bridge at the Gotthard pass, his soul being sacrificed as the
first living being passing the newly established bridge. It takes some invention to
use the fruit of one's quest of knowledge or achievement without facing the evil that
always surrounds us in the current world, whether we like it or not. *(iii)* Still, there
is nothing that has kept the curious characters of every generation from pursuing
science and the quest for knowledge, each one using their means. There is nothing
more gratifying and more adventurous than seeing the Laws of Nature unfold around
us. When these Laws are being explored by various techniques, such as high-
precision laser spectroscopy, or other means, we are happy to see the tiny wheels of
our imagination explain what we observe in the experiments. *(iv)* Finally, the self-
discipline required of any scientist serves, in the best possible sense of the word, as
a guide toward a deeper moral principle in all of us, and toward a factual approach
to the questions surrounding our life; it often solves problems that would otherwise
seem intractable.

It is useful, thus, to ponder the "T-shirt equations" (otherwise known as the
"Maxwell equations") at the beginning of this book for two reasons: *(i)* in order to
clearly state that the subject of classical electrodynamics is nothing but the science
of solving the system of partial differential equations given on the above mentioned
T-shirt, and *(ii)* in order to illustrate right at the beginning how much physics
can be summarized by a set of compactly written differential equations. Let us
remember that electrodynamics covers the physics underlying the telephone as well
as personal computers, the servo-steering of cars and planes, the navigation using
GPS and radar, as well as radio and television. Only in the microworld, where
quantum fluctuations play a role, or in curved space-times, do we need to depart
from Eq. (1.1).

Indeed, electrodynamics is one of the subfields of physics where nothing can be
learned by any means except by constant practice. A famous proverb, probably
invented by some Englishman, states that "understanding is equal to getting used
to something", but understanding electrodynamics is impossible without getting
used to it, and without constant practice. We recall the equations on the T-shirt,

which are the MAXWELL[2] equations,

$$\vec{\nabla} \cdot \vec{E}(\vec{r},t) = \frac{1}{\epsilon_0} \rho(\vec{r},t) \,, \tag{1.2a}$$

$$\vec{\nabla} \cdot \vec{B}(\vec{r},t) = 0 \,, \tag{1.2b}$$

$$\vec{\nabla} \times \vec{E}(\vec{r},t) = -\frac{\partial}{\partial t} \vec{B}(\vec{r},t) \,, \tag{1.2c}$$

$$\vec{\nabla} \times \vec{B}(\vec{r},t) = \mu_0 \, \vec{J}(\vec{r},t) + \frac{1}{c^2} \frac{\partial}{\partial t} \vec{E}(\vec{r},t) \,. \tag{1.2d}$$

These equations are commonly referred to as Gauss's law (1.2a), the law of the absence of magnetic monopoles (1.2b), Faraday's law (1.2c), and the Ampere–Maxwell law (1.2d) (we will consistently suppress the *accent grave* on the name *Ampère* in this book). Here, metric (SI mksA) units are employed, $\vec{E}$ is the electric field, $\vec{B}$ is the magnetic induction field, μ_0 is the vacuum permeability, ϵ_0 is the vacuum permittivity, ρ is the charge density which measures the electric charge per test volume, and $\vec{j}$ is the current density. In the SI mksA unit system, the vacuum permittivity ϵ_0 and the vacuum permeability μ_0 assume the numerical values

$$\mu_0 = 4\pi \times 10^{-7} \, \frac{\mathrm{V\,s}}{\mathrm{A\,m}} \,, \qquad \epsilon_0 = \frac{1}{\mu_0 \, c^2} \,. \tag{1.3}$$

For the Maxwell equations, the use of SI mksA units implies that ϵ_0 and μ_0 are universally kept in all formulas, and that the mechanical units [meter (m), kilogram (kg), and second (s)], are supplemented by the Ampere (A) as the unit of the current, which in turn fixed the unit of charge to be $1\,\mathrm{A} \times 1\mathrm{s} = 1\,\mathrm{C}$, with the Coulomb being denoted as C. Often, in different subfields of physics, different unit systems are used. E.g., in atomic units one would use $\epsilon_0 = 1/(4\pi)$, $\hbar = 1$, and $e^2 = 1/c = \alpha$, where α is the fine-structure constant, ϵ_0 is the vacuum permittivity, $\hbar$ is the reduced Planck constant, and c is the speed of light. However, in natural units, one has $\hbar = c = \epsilon_0 = 1$, and $e^2 = 4\pi\alpha$, where e is the elementary charge. Keeping all factors of c, ϵ_0 (and $\hbar$, where applicable) in the formulas, we hope to arrive at a more general treatment. Finally, the symbol $\vec{\nabla}$ in Eq. (1.2) is the gradient operator. The quantity $\vec{\nabla} \cdot \vec{E}$ is the divergence of the electric field, and the quantity $\vec{\nabla} \times \vec{E}$ is the curl of the electric field. Also, $\cdot$ denotes the scalar product and

$$\vec{\nabla} = \hat{e}_x \frac{\partial}{\partial x} + \hat{e}_y \frac{\partial}{\partial y} + \hat{e}_z \frac{\partial}{\partial z} \tag{1.4}$$

is the so-called gradient operator, where $\hat{e}_x$, $\hat{e}_y$, and $\hat{e}_z$ are the unit vectors in the x, y, and z directions. In writing Eq. (1.2), we have carefully observed that all physical quantities in the Maxwell equations are functions of space and time. Instead of Eq. (1.2), a reader may be more familiar with the integral form of the

---

[2] James Clerk Maxwell (1831–1879)

Maxwell equations, which are familiar from undergraduate physics. These read

$$\int_{\partial V} \vec{E}(\vec{r}, t) \cdot \mathrm{d}\vec{A} = \frac{1}{\epsilon_0} \int_V \rho(\vec{r}, t) \, \mathrm{d}^3 r, \tag{1.5a}$$

$$\int_{\partial V} \vec{B}(\vec{r}, t) \cdot \mathrm{d}\vec{A} = 0, \tag{1.5b}$$

$$\oint_{\partial A} \vec{E}(\vec{s}, t) \cdot \mathrm{d}\vec{s} = -\frac{\partial}{\partial t} \int_A \vec{B}(\vec{r}, t) \cdot \mathrm{d}\vec{A}, \tag{1.5c}$$

$$\oint_{\partial A} \vec{B}(\vec{s}, t) \cdot \mathrm{d}\vec{s} = \mu_0 \int_A \vec{J}(\vec{r}, t) \cdot \mathrm{d}\vec{A} + \frac{1}{c^2} \frac{\partial}{\partial t} \int_A \vec{E}(\vec{r}, t) \cdot \mathrm{d}\vec{A}, \tag{1.5d}$$

with the same identifications as before (Gauss's law, law of the absence of magnetic monopoles, Faraday's law, and the Ampere–Maxwell law). In these equations, V is an arbitrary test volume, and ∂V is its surface, with the surface normal being oriented outward. Also, A is an arbitrary (possibly curved) surface (a two-dimensional manifold embedded in three-dimensional space), and the "surface" of the test area A is denoted as ∂A. Specifically, ∂A is the closed loop (the curve) that constitutes the outer edge of the area A and therefore is a one-dimensional manifold (line). The sign convention is such that the following three vectors constitute a right-handed system: *(i)* the line elements $\mathrm{d}\vec{s}$ which are tangent to ∂A (vector 1), *(ii)* the vector pointing from a line element to the inner region of the area (vector 2), *(iii)* and the surface normal $\mathrm{d}\vec{A}$ (vector 3). The quantity

$$\int_V \rho(\vec{r}, t) \, \mathrm{d}^3 r = q_{\mathrm{encl}}(t) \tag{1.6}$$

is the enclosed charge $q_{\mathrm{encl}}(t)$ in the volume V, still a function of time, and the quantity

$$\int_A \vec{J}(\vec{r}, t) \cdot \mathrm{d}\vec{A} = I_{\mathrm{area}}(t) \tag{1.7}$$

is the vector-summed total current $I_{\mathrm{area}}(t)$ penetrating the area A as a function of time.

1.1.2   *Relation of the Differential to the Integral Form*

We now investigate how to reconcile the differential form of the Maxwell equations with the integral form, and we start with the first Maxwell equation. Indeed, we start from the integral form Eq. (1.5a) of the first Maxwell equation,

$$\int_{\partial V} \vec{E}(\vec{r}, t) \cdot \mathrm{d}\vec{A} = \frac{1}{\epsilon_0} \int_V \rho(\vec{r}, t) \, \mathrm{d}^3 r. \tag{1.8}$$

We consider a small, but not infinitesimally small, test volume $V \to V_\delta$, where V_δ is the cubic volume extending over the intervals $(x, x + \delta x)$, $(y, y + \delta y)$, and $(z, z + \delta y)$, where we consider δx, δy, and δz as small quantities (see also Fig. 1.1). The consideration of small test volume with side lengths δx, δy, and δz enables us to perform a Taylor expansion, and to discard terms of higher order in the displacements δx, δy, and δz.

Let us consider the yz faces ∂V_δ^{yz} of the test volume V_δ in Fig. 1.1. The outward surface normals are $+\hat{e}_x$ and $-\hat{e}_x$. Sampling the electric field in the middle of the test surface, the surface integral is approximated as

$$I \equiv \int_{\partial V_\delta^{yz}} \vec{E}(\vec{r},t) \cdot \mathrm{d}\vec{A} \approx \left[E_x(x+\delta x, y+\tfrac{1}{2}\delta y, z+\tfrac{1}{2}\delta z, t) \right.$$
$$\left. - E_x(x, y+\tfrac{1}{2}\delta y, z+\tfrac{1}{2}\delta z, t) \right] \delta y\, \delta z \,. \tag{1.9}$$

To first order in δx, we obtain

$$I \approx \frac{\partial}{\partial x} E_x(x, y+\tfrac{1}{2}\delta y, z+\tfrac{1}{2}\delta z, t)\, \delta x\, \delta y\, \delta z \,. \tag{1.10}$$

If we are interested only in the term proportional to $\delta x\, \delta y\, \delta z$, then we can replace $y+\tfrac{1}{2}\delta y \to y$, and $z+\tfrac{1}{2}\delta z \to z$ in the argument of the x component of the electric field. Higher-order terms do not contribute to the order we are interested in. Thus,

$$I = \frac{\partial}{\partial x} E_x(x, y, z, t)\, \delta x\, \delta y\, \delta z + \text{higher-order terms} \,. \tag{1.11}$$

With this notion in mind, the left- and right-hand sides of Eq. (1.8) can then be written as follows,

$$\int_{\partial V_\delta} \vec{E}(\vec{r},t) \cdot \mathrm{d}\vec{A}$$
$$\approx \left[E_x(x+\delta x, y+\tfrac{1}{2}\delta y, z+\tfrac{1}{2}\delta z, t) - E_x(x, y+\tfrac{1}{2}\delta y, z+\tfrac{1}{2}\delta z, t) \right] \delta y\, \delta z$$
$$+ \left[E_y(x+\tfrac{1}{2}\delta x, y+\delta y, z+\tfrac{1}{2}\delta z, t) - E_y(x+\tfrac{1}{2}\delta x, y, z+\tfrac{1}{2}\delta z, t) \right] \delta x\, \delta z$$
$$+ \left[E_z(x+\tfrac{1}{2}\delta x, y+\tfrac{1}{2}\delta y, z+\delta z, t) - E_z(x+\tfrac{1}{2}\delta x, y+\tfrac{1}{2}\delta y, z, t) \right] \delta x\, \delta y \,, \tag{1.12a}$$

and

$$\frac{1}{\epsilon_0} \int_{V_\delta} \rho(\vec{r},t)\, \mathrm{d}^3 r \approx \frac{1}{\epsilon_0} \rho(x+\tfrac{1}{2}\delta x, y+\tfrac{1}{2}\delta y, z+\tfrac{1}{2}\delta z, t)\, \delta x\, \delta y\, \delta z \,. \tag{1.12b}$$

We now approximate as follows,

$$\left[E_x(x+\delta x, y+\tfrac{1}{2}\delta y, z+\tfrac{1}{2}\delta z, t) - E_x(x, y+\tfrac{1}{2}\delta y, z+\tfrac{1}{2}\delta z, t) \right]$$
$$\approx \left[E_x(x+\delta x, y, z, t) - E_x(x, y, z, t) \right] \approx \frac{\partial}{\partial x} E_x(x, y, z, t)\, \delta x \,, \tag{1.13a}$$
$$\left[E_z(x+\tfrac{1}{2}\delta x, y+\tfrac{1}{2}\delta y, z+\delta z, t) - E_z(x+\tfrac{1}{2}\delta x, y+\tfrac{1}{2}\delta y, z, t) \right]$$
$$\approx \left[E_y(x, y+\delta y, z, t) - E_y(x, y, z, t) \right] \approx \frac{\partial}{\partial y} E_y(x, y, z, t)\, \delta y \,, \tag{1.13b}$$
$$\left[E_z(x+\tfrac{1}{2}\delta x, y+\tfrac{1}{2}\delta y, z+\delta z, t) - E_z(x+\tfrac{1}{2}\delta x, y+\tfrac{1}{2}\delta y, z, t) \right]$$
$$\approx \left[E_z(x, y, z+\delta z, t) - E_z(x, y, z, t) \right] \approx \frac{\partial}{\partial z} E_z(x, y, z, t)\, \delta z \,. \tag{1.13c}$$

Here, we ignore terms of higher order in the small quantities δx, δy and δz, which would otherwise lead to terms quadratic in the small expansion parameters, such as $\delta x^2\, \delta y\, \delta z$. Analogous approximations are made in Eq. (1.12b).

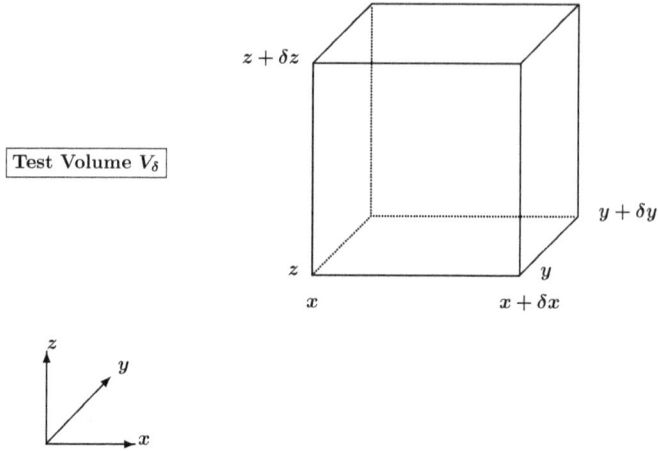

Fig. 1.1   Test volume for the transformation of the first and second Maxwell equations from integral to differential form.

Inserting Eq. (1.13) into (1.12), we see that the volume $\delta x \, \delta y \, \delta z$ cancels, and so

$$\frac{\partial}{\partial x} E_x(x,y,z,t) + \frac{\partial}{\partial y} E_y(x,y,z,t) + \frac{\partial}{\partial z} E_z(x,y,z,t) = \frac{1}{\epsilon_0} \rho(x,y,z,t) . \qquad (1.14)$$

This equation now needs to be transformed into vector form. For the position vector, we write in this book

$$\vec{r} = x\,\hat{e}_x + y\,\hat{e}_y + z\,\hat{e}_z = \sum_{i=1}^{3} x_i \, \hat{e}_i , \qquad (1.15)$$

identifying the components $1, 2, 3$ with the x, y, z directions, respectively. The gradient operator is a vector and reads, in Cartesian coordinates,

$$\vec{\nabla} = \hat{e}_x \frac{\partial}{\partial x} + \hat{e}_y \frac{\partial}{\partial y} + \hat{e}_z \frac{\partial}{\partial z} = \sum_{i=1}^{3} \hat{e}_i \frac{\partial}{\partial x_i} . \qquad (1.16)$$

We can thus write Eq. (1.14) as

$$\vec{\nabla} \cdot \vec{E}(\vec{r},t) = \frac{1}{\epsilon_0} \rho(\vec{r},t) , \qquad (1.17)$$

which is the differential form of the first Maxwell equation (1.2a). The transformation of the integral form of the second Maxwell equation (1.5b) into the differential form (1.2b) is analogous to the above derivation for the first Maxwell equation.

We now turn our attention to the discussion of the third Maxwell equation (1.5c), which involves the curl of the electric field. Its integral form is given as

$$\oint_{\partial A} \vec{E}(\vec{s},t) \cdot \mathrm{d}\vec{s} = -\frac{\partial}{\partial t} \Phi_B = -\frac{\partial}{\partial t} \int_A \vec{B}(\vec{r},t) \cdot \mathrm{d}\vec{A} , \qquad (1.18)$$

where Φ_B is the magnetic flux. The closed-loop integral $\oint_{\partial A} \vec{E}(\vec{s},t) \cdot d\vec{s}$ involves the electric field $\vec{E}(\vec{s},t)$ at the position $\vec{s}$ on the line integral. We assume that the test area A_δ^z is oriented in the xy plane so that its surface normal is oriented along the positive z axis. The test area extends over the interval $(x, x+\delta x)$ in the x-direction and over the interval $(y, y+\delta y)$ in the y-direction, at an arbitrary z coordinate (see Fig. 1.2).

We encircle the line integral about the infinitesimal test area $A_\delta^z \to \partial A_\delta^z$ in the counterclockwise direction and pick up the contributions. It is permissible to approximate the right-hand side of Eq. (1.18) as follows,

$$\oint_{\partial A_\delta^z} \vec{E}(\vec{s},t) \cdot d\vec{s} \approx E_x(x + \tfrac{1}{2}\delta x, y, z, t)\, \delta x + E_y(x + \delta x, y + \tfrac{1}{2}\delta y, z, t)\, \delta y$$

$$- E_x(x + \tfrac{1}{2}\delta x, y + \delta y, z, t)\, \delta x - E_y(x, y + \tfrac{1}{2}\delta y, z, t)\, \delta y\,, \tag{1.19a}$$

$$-\frac{\partial}{\partial t}\int_{A_\delta} \vec{B}(\vec{r},t) \cdot d\vec{A} \approx -\frac{\partial}{\partial t} B_z(x + \tfrac{1}{2}\delta x, y + \tfrac{1}{2}\delta y, z, t)\, \delta x\, \delta y\,. \tag{1.19b}$$

By a Taylor expansion, keeping all terms up to second order in the small quantities δx, δy, and δz, we obtain

$$-\frac{\partial}{\partial y} E_x(x, y, z, t)\, \delta x\, \delta y + \frac{\partial}{\partial x} E_y(x, y, z, t)\, \delta x\, \delta y \approx -\frac{\partial}{\partial t} B_z(x, y, z, t)\, \delta x\, \delta y\,. \tag{1.20}$$

Third-order terms have been neglected on both sides of the equation. One may now cancel the test area $\delta x\, \delta y$ and write

$$\frac{\partial}{\partial x} E_y - \frac{\partial}{\partial y} E_x = (\vec{\nabla} \times \vec{E})_z = -\frac{\partial}{\partial t} B_z\,, \tag{1.21}$$

where we suppress the space-time arguments of the fields. We identify $\frac{\partial}{\partial x} E_y - \frac{\partial}{\partial y} E_x$ as the z component of the curl of the electric field, thus obtaining the z-component of Eq. (1.18). In general, the curl of a vector field is defined as

$$\vec{\nabla} \times \vec{E} = \det \begin{pmatrix} \hat{e}_x & \hat{e}_y & \hat{e}_z \\ \dfrac{\partial}{\partial x} & \dfrac{\partial}{\partial y} & \dfrac{\partial}{\partial z} \\ E_x & E_y & E_z \end{pmatrix}. \tag{1.22}$$

With reference to Eq. (1.15) and (1.16), its components are given as

$$(\vec{\nabla} \times \vec{E})_i = \sum_{j=1}^{3}\sum_{k=1}^{3} \epsilon_{ijk} \frac{\partial}{\partial x_j} E_k\,, \qquad i,j,k = 1,2,3\,. \tag{1.23}$$

Here, ϵ_{ijk} is the totally antisymmetric Levi–Cività tensor, i.e., $\epsilon_{123} = 1$ and ϵ_{ijk} is equal to the sign of the permutation that transforms the triple $(1,2,3)$ into (i,j,k). If an even number of exchanges is required in order to transform $(1,2,3)$ into (i,j,k), then the sign of the permutation is 1, otherwise it is -1. So, $\epsilon_{231} = 1$, but $\epsilon_{213} = -1$.

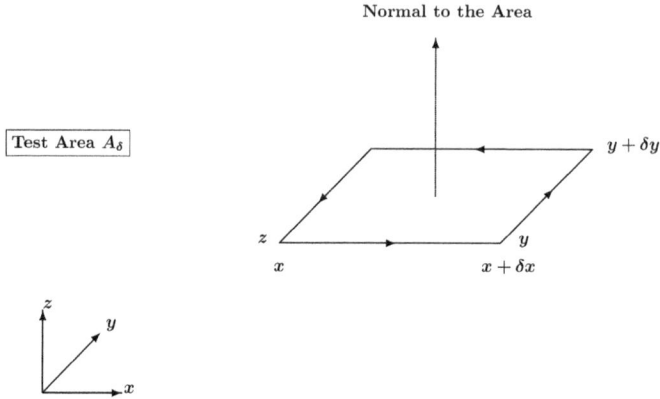

Fig. 1.2   Test area for the transformation of the third and fourth Maxwell equation from integral to differential form.

If two of the indices are equal, such as in $\epsilon_{122} = 0$, the result is zero. We can also write

$$\vec{\nabla} \times \vec{E} = \sum_{i=1}^{3} \sum_{j=1}^{3} \sum_{k=1}^{3} \hat{e}_i \, \epsilon_{ijk} \frac{\partial}{\partial x_j} E_k \, . \tag{1.24}$$

In the above derivation leading to (1.21), we have assumed the test area ∂A is oriented along the positive z axis. If we change the orientation of the test area A_δ, then we recover the x and y components of Faraday's law

$$\vec{\nabla} \times \vec{E} = -\frac{\partial}{\partial t} \vec{B} \, , \tag{1.25}$$

where we suppress the space-time arguments of the electric and magnetic fields. The transformation of the fourth Maxwell equation from integral to differential form is analogous.

For a macroscopic volume V or area A, the validity of the integral and differential forms of the Maxwell equations can be verified based on a partition of the macroscopic volumes and areas into small test volumes and areas, "glued" together along their boundaries (colloquially speaking, interpreting the small test volumes as "bricks in Legoland" from which the entire "Legoland" is being assembled).

One of the most paradigmatic applications of the formalism outlined above involves the so-called continuity equation. The continuity equation is a simple statement that says that the charge leaving a test volume V through currents can be obtained in two alternative ways: *(i)* as the current density $\vec{j}$ summed over the surface area of the test volume ∂V, and *(ii)* as the time derivative of the charge density, summed/integrated over the test volume V. The integral form of the continuity equations is expressed as

$$\int_{\partial V} \vec{J}(\vec{r},t) \cdot d\vec{A} = -\frac{\partial}{\partial t} \int_{V} \rho(\vec{r},t) \, d^3 r \, . \tag{1.26}$$

Using the same test volume as given in Fig. 1.1, we obtain

$$
\begin{aligned}
\Big[J_x\big(x + \delta x, y + \tfrac{1}{2}\delta y, z + \tfrac{1}{2}\delta z, t\big) - J_x\big(x, y + \tfrac{1}{2}\delta y, z + \tfrac{1}{2}\delta z, t\big) \Big]\, \delta y\, \delta z & \\
+ \Big[J_y\big(x + \tfrac{1}{2}\delta x, y + \delta y, z + \tfrac{1}{2}\delta z, t\big) - J_y\big(x + \tfrac{1}{2}\delta x, y, z + \tfrac{1}{2}\delta z, t\big) \Big]\, \delta x\, \delta z & \\
+ \Big[J_z\big(x + \tfrac{1}{2}\delta x, y + \tfrac{1}{2}\delta y, z + \delta z, t\big) - J_z\big(x + \tfrac{1}{2}\delta x, y + \tfrac{1}{2}\delta y, z, t\big) \Big]\, \delta x\, \delta y & \\
= -\frac{\partial}{\partial t}\rho\big(x + \tfrac{1}{2}\delta x, y + \tfrac{1}{2}\delta y, z + \tfrac{1}{2}\delta z, t\big)\, \delta x\, \delta y\, \delta z\,. &
\end{aligned} \tag{1.27}
$$

We have taken the differences at the midpoint of the edges of the test volume. Observing that, to leading order, the midpoint can be replaced by the "lower corner" of the test volume, one can replace the first term in Eq. (1.27) by $J_x(x+\delta x, y, z, t) - J_x(x, y, z, t)$. Then, expanding into a Taylor series and canceling the test volume $\delta x\, \delta y\, \delta z$, we obtain

$$
\frac{\partial}{\partial x}J_x(x, y, z, t) + \frac{\partial}{\partial y}J_y(x, y, z, t) + \frac{\partial}{\partial z}J_z(x, y, z, t) = -\frac{\partial}{\partial t}\rho(x, y, z, t)\,. \tag{1.28}
$$

Finally, the differential form of the continuity equation is obtained,

$$
\vec{\nabla}\cdot\vec{J} + \frac{\partial}{\partial t}\rho = 0\,, \tag{1.29}
$$

where again we suppress the space-time arguments of the current and charge density.

1.1.3 *Dirac δ Function*

Some readers of this book may be familiar with the Dirac-δ function based on other sources, but it is still instructive to recall some basic facts. One may define the (one-dimensional) Dirac delta function by the condition

$$
\int_a^b f(x')\, \delta(x - x')\, \mathrm{d}x' =
\begin{cases}
f(x) & (a < x < b) \\
0 & (x < a \quad \text{or} \quad x > b)
\end{cases}. \tag{1.30}
$$

Essentially, the Dirac-δ function extracts the value of the test function f at the point x. If $x = a$ or $x = b$, the integral is not well defined, but a common definition is

$$
\int_a^b f(x')\, \delta(x - x')\, \mathrm{d}x' = \tfrac{1}{2} f(x) \qquad \text{for} \qquad x = a \qquad \text{or} \qquad x = b\,. \tag{1.31}
$$

This definition is inspired by the fact that any representation of the Dirac-δ function which is symmetric about the center yields the result (1.31), as explained in the following.

The Dirac-δ function (or, more precisely, the Dirac-δ distribution) is very useful in physics. A possible representation of the Dirac-δ function is

$$
\delta(\omega) \longrightarrow \Delta(\eta, \omega) = \frac{1}{\pi}\frac{\eta}{\omega^2 + \eta^2}\,, \qquad \eta \to 0^+\,. \tag{1.32}
$$

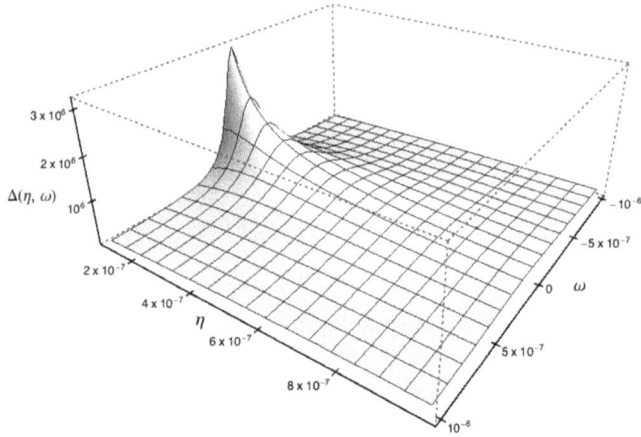

Fig. 1.3 Plot of the function $\Delta(\eta, \omega)$ defined in Eq. (1.32), for small η and ω. The emergence of the peak near $\omega = 0$ is clearly discernible.

The limit is $\eta \to 0$ is approached nonuniformly. The Δ function has the properties

$$\int_{-\infty}^{\infty} d\omega \, \Delta(\eta, \omega) = 1 \,, \tag{1.33a}$$

$$\lim_{\eta \to 0^+} \Delta(\eta, \omega) = 0 \quad (\text{for } \omega \neq 0), \tag{1.33b}$$

$$\lim_{\eta \to 0^+} \Delta(\eta, \omega) = \infty \quad (\text{for } \omega = 0), \tag{1.33c}$$

$$\int_{-\infty}^{\infty} d\omega \left[\lim_{\eta \to 0^+} \Delta(\eta, \omega) \right] f(\omega) = 0 \,, \qquad \lim_{\eta \to 0^+} \int_{-\infty}^{\infty} d\omega \, \Delta(\eta, \omega) \, f(\omega) = f(0) \,. \tag{1.33d}$$

The last two properties illustrate that one cannot take the limit $\eta \to 0$ too early, in the sense of non-uniform convergence. A graphical illustration can be found in Fig. 1.3. In the limit $\eta \to 0^+$, then, $\Delta(\eta, \omega) \to \delta(\omega)$, but the limit needs to be taken after all integrals have been performed.

If the argument of a Dirac delta function is a function of x,

$$\delta(f(x)) = \sum_{i=1}^{n} \left(\left| \frac{df}{dx} \right|_{x=x_i} \right)^{-1} \delta(x - x_i) \,, \tag{1.34}$$

where the x_i ($i = 1, \ldots, n$) with $g(x_i) = 0$ denote simple zeros of $g(x)$. The three-dimensional delta function is equal to the product of three one-dimensional Dirac delta functions,

$$\delta^{(3)}(\vec{r} - \vec{r}') = \delta(x - x') \, \delta(y - y') \, \delta(z - z') \,. \tag{1.35}$$

Thus,

$$\int_V dx' dy' dz' \, f(\vec{r}') \, \delta^{(3)}(\vec{r} - \vec{r}') = f(\vec{r}) \tag{1.36}$$

if $\vec{r}$ is interior to the volume V and 0 otherwise. For multiple integrals such as $\int_V$ (over the three-dimensional volume V), we here follow the convention that the integration domain is being indicated by a subscript of the integral sign. In particular, the integration domain then automatically defines the dimensionality of the integral.

1.1.4 Green Function of the Laplacian

The Dirac-δ function enters the defining equation of the Green function corresponding to the Laplacian operator,

$$\vec{\nabla}^2 g(\vec{r}, \vec{r}') = \vec{\nabla}^2 g(\vec{r} - \vec{r}') = \delta^{(3)}(\vec{r} - \vec{r}'), \tag{1.37}$$

where $\vec{\nabla}^2$ is the Laplace operator, which acts on $\vec{r}$ (not $\vec{r}'$) unless stated otherwise. The desired solution is

$$g(\vec{r}, \vec{r}') = g(\vec{r} - \vec{r}') = -\frac{1}{4\pi |\vec{r} - \vec{r}'|}. \tag{1.38}$$

This result may be shown as follows. We recall Gauss's theorem

$$\int_{\partial V} \vec{V}(\vec{r}) \cdot d\vec{A} = \int_V \vec{\nabla} \cdot \vec{V}(\vec{r}) \, d^3r, \tag{1.39}$$

where V is an arbitrary volume and ∂V is its surface. We apply the theorem to the vector field

$$\vec{V}(\vec{r}) = \vec{\nabla} \frac{1}{|\vec{r} - \vec{r}'|} = -\frac{\vec{r} - \vec{r}'}{|\vec{r} - \vec{r}'|^3}, \tag{1.40}$$

and choose V as a sphere S_ε of radius ε about the point $\vec{r}'$. The surface of S_ε is denoted ∂S_ε. Then,

$$\int_{S_\varepsilon} \vec{\nabla}^2 \left(\frac{1}{|\vec{r} - \vec{r}'|}\right) d^3r = \int_{S_\varepsilon} \vec{\nabla} \cdot \vec{V}(\vec{r}) \, d^3r = \int_{\partial S_\varepsilon} \vec{V}(\vec{r}) \cdot d\vec{A}$$

$$= \int_{S_\varepsilon} \left(-\frac{\vec{r} - \vec{r}'}{|\vec{r} - \vec{r}'|^3}\right) \cdot \frac{\vec{r} - \vec{r}'}{|\vec{r} - \vec{r}'|} |\vec{r} - \vec{r}'|^2 \, d\Omega = -4\pi. \tag{1.41}$$

Because we can choose ε to be arbitrarily small, we conclude that

$$\vec{\nabla}^2 \left(\frac{1}{|\vec{r} - \vec{r}'|}\right) = -4\pi \delta^{(3)}(\vec{r} - \vec{r}'). \tag{1.42}$$

and hence Eq. (1.37) is fulfilled. Because the FOURIER[3] transform (see Chaps. 4 and 6) of the Dirac-δ distribution is just a constant, the function G can be interpreted as the inverse of the Laplacian operator.

1.2 Potentials and Gauges

1.2.1 Transverse and Longitudinal Components of a Vector Field

At any given point in time t, the electric and magnetic fields $\vec{E}(\vec{r}, t)$ and $\vec{B}(\vec{r}, t)$ represent vector fields. Central to the upcoming discussion is the fact that any vector field can be separated into transverse and longitudinal components; we here derive explicit formulas for this separation. Thus, let $\vec{J}(\vec{r}, t)$ be a vector field which

---

[3] Jean-Baptiste Joseph Fourier (1768–1830)

we would like to separate into transverse and longitudinal components. We seek a separation

$$\vec{V}(\vec{r},t) = \vec{V}_\perp(\vec{r},t) + \vec{V}_\parallel(\vec{r},t) , \qquad \vec{\nabla} \cdot \vec{V}_\perp(\vec{r},t) = 0 , \qquad \vec{\nabla} \times \vec{V}_\parallel(\vec{r},t) = 0 . \qquad (1.43)$$

It is not obvious how to do this. We first write a seemingly innocent vector identity,

$$\vec{\nabla} \times \left(\vec{\nabla} \times \vec{V}(\vec{r},t) \right) = \vec{\nabla} \left(\vec{\nabla} \cdot \vec{V}(\vec{r},t) \right) - \vec{\nabla}^2 \vec{V}(\vec{r},t) . \qquad (1.44)$$

One may derive this identity on the basis of the Levi-Cività tensor ϵ_{ijk} and the identity

$$\epsilon_{ijk} \, \epsilon_{k\ell m} = \delta_{i\ell} \, \delta_{jm} - \delta_{im} \, \delta_{j\ell} . \qquad (1.45)$$

Here, the EINSTEIN[4] summation convention for repeated subscripts has been used. (This convention implies that $a_i \, a_i \equiv \sum_{i=1}^{n} a_i \, a_i$ for repeated superscripts or subscripts, with an applicable upper limit n for the summation which is derived from the context in which the equation is written. In the current case, for three-dimensional space, i, j, k, ℓ and m may assume values of $1, 2, 3$; because k is being repeated, the sum $\sum_{k=1}^{3}$ is implicitly understood in Eq. (1.45).)

We now rewrite Eq. (1.44),

$$\vec{\nabla}^2 \vec{V}(\vec{r},t) = \vec{\nabla} \left(\vec{\nabla} \cdot \vec{V}(\vec{r},t) \right) - \vec{\nabla} \times \left(\vec{\nabla} \times \vec{V}(\vec{r},t) \right) . \qquad (1.46)$$

We now appeal to Eqs. (1.36) and (1.37) for the definition of the Green function $g = g(\vec{r} - \vec{r}')$. If we assume that boundary contributions at infinity will vanish, then we obtain (the second step involves a double partial integration)

$$
\begin{aligned}
\vec{V}(\vec{r},t) = \ & = \int d^3 r' \, \vec{\nabla}^2 g(\vec{r},\vec{r}') \, \vec{V}(\vec{r}',t) = \int d^3 r' \, g(\vec{r},\vec{r}') \, \vec{\nabla}^2 \vec{V}(\vec{r}',t) \\
& = -\frac{1}{4\pi} \int d^3 r' \, \frac{1}{|\vec{r} - \vec{r}'|} \left[\vec{\nabla}' \left(\vec{\nabla}' \cdot \vec{V}(\vec{r}',t) \right) - \vec{\nabla}' \times \left(\vec{\nabla}' \times \vec{V}(\vec{r}',t) \right) \right] \\
& = -\frac{1}{4\pi} \int d^3 r' \, \frac{\vec{\nabla}' \left(\vec{\nabla}' \cdot \vec{V}(\vec{r}',t) \right)}{|\vec{r} - \vec{r}'|} + \frac{1}{4\pi} \int d^3 r' \, \frac{\vec{\nabla}' \times \left(\vec{\nabla}' \times \vec{V}(\vec{r}',t) \right)}{|\vec{r} - \vec{r}'|} \\
& = \frac{1}{4\pi} \int d^3 r' \left(\vec{\nabla}' \cdot \vec{V}(\vec{r}',t) \right) \vec{\nabla}' \frac{1}{|\vec{r} - \vec{r}'|} \\
& \quad - \frac{1}{4\pi} \int d^3 r' \vec{\nabla}' \frac{1}{|\vec{r} - \vec{r}'|} \times \left(\vec{\nabla}' \times \vec{V}(\vec{r}',t) \right) \\
& = -\frac{1}{4\pi} \vec{\nabla} \int \frac{\vec{\nabla}' \cdot \vec{V}(\vec{r}',t)}{|\vec{r} - \vec{r}'|} d^3 r' + \frac{1}{4\pi} \vec{\nabla} \times \int \frac{\vec{\nabla}' \times \vec{V}(\vec{r}',t)}{|\vec{r} - \vec{r}'|} d^3 r' . \qquad (1.47)
\end{aligned}
$$

Thus, the longitudinal and transverse parts of the current density, denoted as $\vec{V}_\parallel$ and $\vec{V}_\perp$, respectively, are given as

$$\vec{V}_\parallel(\vec{r},t) = -\frac{1}{4\pi} \vec{\nabla} \int d^3 r' \, \frac{\vec{\nabla}' \cdot \vec{V}(\vec{r}',t)}{|\vec{r} - \vec{r}'|} , \qquad (1.48a)$$

$$\vec{V}_\perp(\vec{r},t) = \frac{1}{4\pi} \vec{\nabla} \times \int d^3 r' \, \frac{\vec{\nabla}' \times \vec{V}(\vec{r}',t)}{|\vec{r} - \vec{r}'|} . \qquad (1.48b)$$

----

[4] Albert Einstein (1879–1955)

These are highly non-local functions of the full current density $\vec{V}(\vec{r},t)$, in the sense that in order to define $\vec{V}_{\parallel}(\vec{r},t)$ and $\vec{V}_{\perp}(\vec{r},t)$, we have to sample values of the original vector potential $\vec{V}(\vec{r},t)$ over wide areas of space.

Furthermore, if we define the scalar potential Ψ and the vector potential $\vec{A}$ as

$$\Psi(\vec{r},t) = -\frac{1}{4\pi}\int d^3r' \frac{\vec{\nabla}'\cdot\vec{V}(\vec{r}',t)}{|\vec{r}-\vec{r}'|}, \tag{1.49a}$$

$$\vec{A}(\vec{r},t) = \frac{1}{4\pi}\int d^3r' \frac{\vec{\nabla}'\times\vec{V}(\vec{r}',t)}{|\vec{r}-\vec{r}'|}, \tag{1.49b}$$

then

$$\vec{V}_{\parallel}(\vec{r},t) = \vec{\nabla}\Psi(\vec{r},t), \qquad \vec{V}_{\perp}(\vec{r},t) = \vec{\nabla}\times\vec{A}(\vec{r},t). \tag{1.50}$$

Given a vector potential $\vec{V}$, we thus have the explicit formulas that tell us how to calculate the scalar potential Ψ and the vector potential $\vec{A}$. These potentials enable us to write the longitudinal component of the vector field $\vec{V}_{\parallel}$ as the gradient of the scalar potential Ψ and the transverse component of the vector field $\vec{V}_{\perp}$ as the curl of a vector potential $\vec{A}$.

1.2.2 *Vector and Scalar Potentials*

The contents of the current discussion are merely restricted to the definition of the scalar and vector potentials, but their significance reaches far beyond, namely, to quantum electrodynamics, which is a gauge theory, and even to the construction of the standard model of particle interactions. Without the scalar and vector potentials, and the concept of a covariant derivation, we would not have the tools at our hand to analyze the gauge theories that define the standard model.

A detour with an illustrative example, regarding the action of the covariant derivative operator on a quantum mechanical wave function, is thus in order. We restrict the discussion to a time-independent configuration. Let us define a covariant derivative operator $\vec{\Pi}$ as follows,

$$\vec{\Pi} \equiv \vec{p} - e\vec{A}(\vec{r}), \qquad \vec{p} = -i\hbar\vec{\nabla}, \tag{1.51}$$

where e is the charge of the particle in question, $\vec{A}(\vec{r})$ is the vector potential (assumed here to be time-independent), $\hbar$ is Planck's constant, and $\vec{p}$ is the momentum operator (in quantum mechanics), or the mechanical momentum of the particle (in classical mechanics). The precise meaning of $\vec{A}$ will be clarified below. The decisive property of the gauge potential is this: Let us assume that the covariant derivative operator $\vec{\Pi}$ acts on a wave function $\psi(\vec{r})$. The addition of a gradient

$$\vec{A}(\vec{r}) \rightarrow \vec{A}(\vec{r}) + \vec{\nabla}\Lambda(\vec{r}) \tag{1.52}$$

is called a gauge transformation. Gauge covariance now means the following: If we simultaneously transform the wave function and the vector potential as follows,

$$\psi(\vec{r}) \rightarrow \psi'(\vec{r}) \equiv \exp\left(\frac{\mathrm{i}e}{\hbar} \int_{\vec{0}}^{\vec{r}} \vec{\Lambda}(\vec{s}) \cdot \mathrm{d}s\right) \psi(\vec{r}) = U(\vec{r})\, \psi(\vec{r}), \tag{1.53a}$$

$$\vec{A}(\vec{r}) \rightarrow \vec{A}'(\vec{r}) \equiv \vec{A}(\vec{r}) + \vec{\nabla}\Lambda(\vec{r}), \tag{1.53b}$$

then it is easy to show that

$$\exp\left(\frac{\mathrm{i}e}{\hbar} \int_{\vec{0}}^{\vec{r}} \vec{\Lambda}(\vec{s}) \cdot \mathrm{d}s\right) \left(\vec{p} - e\,\vec{A}(\vec{r})\right) \psi(\vec{r})$$

$$= \left(\vec{p} - e\,\vec{A}(\vec{r}) - e\vec{\nabla}\Lambda(\vec{r})\right) \exp\left(\frac{\mathrm{i}e}{\hbar} \int_{\vec{0}}^{\vec{r}} \vec{\Lambda}(\vec{s}) \cdot \mathrm{d}s\right) \psi(\vec{r}). \tag{1.54}$$

When we define the prime to indicate the gauge transformed quantity, then this identity takes a particularly sensible form,

$$U(\vec{r}) \left(\vec{p} - e\,\vec{A}(\vec{r})\right) \psi(\vec{r}) = \left(\vec{p} - e\,\vec{A}'(\vec{r})\right) U(\vec{r})\, \psi(\vec{r}), \tag{1.55a}$$

$$\left(\left(\vec{p} - e\,\vec{A}(\vec{r})\right) \psi(\vec{r})\right)' = \left(\vec{p} - e\,\vec{A}'(\vec{r})\right) \psi'(\vec{r}), \tag{1.55b}$$

and simply says that the gauge transform of the covariant derivative is equal to the covariant derivative of the gauge transformed wave function. We shall see that different vector potentials $\vec{A}$ produce equivalent observable electromagnetic field configurations. In differential geometry [1–4], the vector potential $\vec{A}$ has the meaning of a connection 1-form. Roughly speaking it measures how space is "bent" for the charged particle in the presence of the vector potential, just like a light trajectory is "bent" alongside a geodesic in the theory of general relativity when traveling near heavy stars. Or, in differential geometry, the "bending" of the trajectory is due to the curvature of the embedding manifold on which the particle is moving. There is thus a deep philosophical notion behind the definition of the covariant derivative: namely, that we can view the minimal coupling of a charged particle wave function to the field as a consequence of the "bending" of space in the presence of vector fields, for a particle that happens to carry charge, in such a way that gauge freedom exists. In differential geometry, a gauge transformation simply corresponds to a change in the coordinate system that we are using to describe the curved manifold. In electrodynamics, a gauge transformation corresponds to a local phase change of the wave function, leaving the probability density $|\psi(\vec{r})|^2$ unchanged, under a simultaneous change of the scalar and vector potentials, which leave the physically observable fields unchanged. A gauge transformation thus is a "coordinate transformation" involving the phases of the wave function, and the scalar and vector potentials. The gauge freedom corresponds to the freedom of choice of the coordinate system in differential geometry. Incidentally, the gauge freedom is central to the construction of the current standard model of particle interactions. In non-Abelian gauge theories, one replaces $\vec{A}(\vec{r})$ by a field that carries an additional ("color") index, leading to a matrix-valued gauge transform factor.

These remarks are intended to motivate the central character of gauge transformations, and their significance within theoretical physics; their understanding is not required for the following derivation.

Having illustrated the significance of scalar and vector potentials for quantum theory and theoretical physics in general, let us finish the detour and return to the case of classical electrodynamics. The gauge freedom allows us to define vector and scalar potentials, and to choose a particular gauge that makes the calculation easy for a particular source configuration (current and charges). Still, a lot can be learned from gauge transformations on the classical level, and we shall briefly dwell on related questions. In the following, we thus restrict our investigation to the microscopic form of Maxwell's equations. We start with the two source free equations. First, we consider Gauss' law for the magnetic induction, $\vec{\nabla} \cdot \vec{B} = 0$. In this case, according to the discussion in the Sec. 1.1.3, a vector potential $\vec{A}$ can be defined so that the magnetic induction vector is given by

$$\vec{B}(\vec{r},t) = \vec{\nabla} \times \vec{A}(\vec{r},t) . \tag{1.56}$$

The vector potential is used in Faraday's law $\vec{\nabla} \times \vec{E} + \partial_t \vec{B} = 0$ to yield

$$\vec{\nabla} \times \vec{E}(\vec{r},t) + \frac{\partial}{\partial t}\left[\vec{\nabla} \times \vec{A}(\vec{r},t)\right] = \vec{\nabla} \times \left[\vec{E}(\vec{r},t) + \frac{\partial}{\partial t}\vec{A}(\vec{r},t)\right] = \vec{0}. \tag{1.57}$$

Note that, on the right-hand side, we use the zero vector $\vec{0}$ instead of a simple zero in order to be fully consistent. It follows that the curl of the vector field $\vec{E}(\vec{r},t) + \partial_t \vec{A}(\vec{r},t)$ vanishes, which is why we can find a scalar potential $\Phi(\vec{r},t)$ such that

$$\vec{E}(\vec{r},t) + \frac{\partial}{\partial t}\vec{A}(\vec{r},t) = -\vec{\nabla}\Phi(\vec{r},t) \tag{1.58}$$

and

$$\vec{E}(\vec{r},t) = -\vec{\nabla}\Phi(\vec{r},t) - \frac{\partial}{\partial t}\vec{A}(\vec{r},t) , \tag{1.59}$$

which expresses the electric field in terms of the scalar and vector potentials. The fields constructed according to Eqs. (1.56) and (1.59) automatically fulfill the homogeneous Maxwell equations $\vec{\nabla} \cdot \vec{B} = 0$ and $\vec{\nabla} \times \vec{E} + \partial_t \vec{B} = \vec{0}$. Given a field configuration $\vec{B} = \vec{B}(\vec{r},t)$, we can obtain the vector potential $\vec{A}(\vec{r},t)$ by the integral (1.49b), setting $\vec{V} = \vec{B}$, and the scalar potential Φ by Eq. (1.49a), setting $\vec{V} = -\vec{E} - \partial_t \vec{A}$.

Gauss's law for the electric field and the Ampere–Maxwell law couple the vector and scalar potentials to their sources, which are the charge and current densities. First we consider Gauss' law and obtain

$$\vec{\nabla} \cdot \vec{E} = -\vec{\nabla}^2\Phi(\vec{r},t) - \frac{\partial}{\partial t}\vec{\nabla} \cdot \vec{A}(\vec{r},t) = \frac{1}{\epsilon_0}\rho(\vec{r},t) , \tag{1.60}$$

while the Ampere–Maxwell law

$$\vec{\nabla} \times \vec{B}(\vec{r},t) = \vec{\nabla} \times \left[\vec{\nabla} \times \vec{A}(\vec{r},t)\right] = \frac{1}{c^2}\frac{\partial}{\partial t}\vec{E}(\vec{r},t) + \mu_0 \vec{J}(\vec{r},t) \tag{1.61}$$

becomes

$$\vec{\nabla} \times \left[\vec{\nabla} \times \vec{A}\left(\vec{r},t\right) \right] + \frac{1}{c^2} \frac{\partial}{\partial t} \left[\frac{\partial}{\partial t} \vec{A}\left(\vec{r},t\right) + \vec{\nabla} \Phi\left(\vec{r},t\right) \right] = \mu_0 \, \vec{J}\left(\vec{r},t\right)$$

$$\Leftrightarrow \vec{\nabla} \left\{ \vec{\nabla} \cdot \vec{A}\left(\vec{r},t\right) + \frac{1}{c^2} \frac{\partial}{\partial t} \Phi\left(\vec{r},t\right) \right\} - \vec{\nabla}^2 \vec{A}\left(\vec{r},t\right) + \frac{1}{c^2} \frac{\partial^2}{\partial t^2} \vec{A}\left(\vec{r},t\right) = \mu_0 \, \vec{J}\left(\vec{r},t\right) . \quad (1.62)$$

From Eqs. (1.60) and (1.62), we can conclude that the scalar and vector potentials are coupled to the sources as follows, in a general gauge,

$$-\vec{\nabla}^2 \Phi\left(\vec{r},t\right) - \frac{\partial}{\partial t} \vec{\nabla} \cdot \vec{A}\left(\vec{r},t\right) = \frac{1}{\epsilon_0} \rho\left(\vec{r},t\right) , \quad (1.63a)$$

$$\vec{\nabla} \left\{ \vec{\nabla} \cdot \vec{A}\left(\vec{r},t\right) + \frac{1}{c^2} \frac{\partial}{\partial t} \Phi\left(\vec{r},t\right) \right\} - \vec{\nabla}^2 \vec{A}\left(\vec{r},t\right) + \frac{1}{c^2} \frac{\partial^2}{\partial t^2} \vec{A}\left(\vec{r},t\right) = \mu_0 \, \vec{J}\left(\vec{r},t\right) . \quad (1.63b)$$

These coupled second-order partial differential equations relate the potentials Φ and $\vec{A}$ to the sources ρ and $\vec{J}$.

1.2.3 *Lorenz Gauge*

We should precede the discussion of the Lorenz gauge with a brief remark on the difference in the naming conventions between the Lorentz force and Lorentz transformation on the one hand, and the Lorenz gauge on the other hand. Indeed, the Lorentz force and the Lorentz transformation are associated with LORENTZ[5] in contrast to the Lorenz gauge, which is due to LORENZ.[6] The two names are similar but not the same.

At the beginning of the preceding Sec. 1.2.2, we briefly discussed the paradigm of a gauge transformation, which is a local transformation of the vector potentials $\vec{A}(\vec{r},t) \to \vec{A}(\vec{r},t) + \vec{\nabla}\Lambda(\vec{r},t)$ (now, we assume a time-dependent field configuration). Another interpretation, which is somewhat global, for a gauge transformation is as follows: If one measures lengths or heights, one always has the freedom to choose the zero point of the measurement. If one is to hang a picture on a wall, a convenient zero point for the elevation measurement is the floor of the room, while, for measuring the altitude of a mountain, a convenient reference point is the sea level. A trivial gauge transformation which leaves the electric field invariant is simply $\Phi(\vec{r},t) \to \Phi(\vec{r},t) + \mathcal{C}$, where $\mathcal{C}$ is a constant potential, which of course does not affect the gradient ("global" gauge transformation). This amounts to choosing a zero for the measurement of the "altitude" of the potential. The local character of the gauge transformation implies that we are free to choose the zeros for the measurements of the potentials any way we want, and moreover, to choose these zero points differently at every space-time point, provided the fields $\vec{E}(\vec{r},t)$ and $\vec{B}(\vec{r},t)$ are left invariant. This would amount to measuring (in absolute terms) the elevation of the picture on the wall differently today and tomorrow, provided the difference of the elevation measurements is left invariant. Let us now explore this concept in detail.

---

[5] Hendrik Antoon Lorentz (1853–1928)
[6] Ludvig Valentin Lorenz (1829–1891)

First, we observe that the addition of a term $\vec{A}(\vec{r},t) \to \vec{A}(\vec{r},t) + \vec{\nabla}\Lambda(\vec{r},t)$ to the vector potential does not change the magnetic induction field, because

$$\vec{\nabla} \times \vec{\nabla}\Lambda(\vec{r},t) = \vec{0}. \tag{1.64}$$

In this case we obtain the gauge transformed vector potential $\vec{A}'(\vec{r},t)$,

$$\vec{A}'(\vec{r},t) = \vec{A}(\vec{r},t) + \vec{\nabla}\Lambda(\vec{r},t). \tag{1.65}$$

We now insert the equation $\vec{A}(\vec{r},t) = \vec{A}'(\vec{r},t) - \vec{\nabla}\Lambda(\vec{r},t)$ for the gauge transformed vector potential into the formula Eq. (1.58) for the electric field,

$$
\begin{aligned}
\vec{E}(\vec{r},t) &= -\frac{\partial}{\partial t}\vec{A}'(\vec{r},t) + \frac{\partial}{\partial t}\vec{\nabla}\Lambda(\vec{r},t) - \vec{\nabla}\Phi(\vec{r},t) \\
&= -\frac{\partial}{\partial t}\vec{A}'(\vec{r},t) - \vec{\nabla}\left\{\Phi(\vec{r},t) - \frac{\partial}{\partial t}\Lambda(\vec{r},t)\right\}.
\end{aligned}
\tag{1.66}
$$

Changing the vector potential will change the electric field unless the scalar potential is also changed to the term that appears in curly brackets, which amounts to gauge transform of the scalar potential,

$$\Phi'(\vec{r},t) = \Phi(\vec{r},t) - \frac{\partial}{\partial t}\Lambda(\vec{r},t). \tag{1.67}$$

The electric field remains unchanged under the gauge transformation,

$$\vec{E}(\vec{r},t) = -\frac{\partial}{\partial t}\vec{A}'(\vec{r},t) - \vec{\nabla}\Phi'(\vec{r},t) = -\frac{\partial}{\partial t}\vec{A}(\vec{r},t) - \vec{\nabla}\Phi(\vec{r},t). \tag{1.68}$$

Equations (1.65) and (1.67) can be summarized as the gauge transform of the vector and scalar potentials,

$$\vec{A}'(\vec{r},t) = \vec{A}(\vec{r},t) + \vec{\nabla}\Lambda(\vec{r},t), \qquad \Phi'(\vec{r},t) = \Phi(\vec{r},t) - \frac{\partial}{\partial t}\Lambda(\vec{r},t). \tag{1.69}$$

The electric and magnetic fields, which are invariant under the gauge transformations, are said to be gauge invariant. In fact, they *have* to be gauge invariant as they constitute physically observable field strengths.

We are now free to choose Λ so that the gauge transformed scalar and vector potentials fulfill certain conditions. One possible requirement on the new potentials is the Lorenz gauge condition,

$$\vec{\nabla} \cdot \vec{A}'(\vec{r},t) + \frac{1}{c^2}\frac{\partial}{\partial t}\Phi'(\vec{r},t) = 0. \tag{1.70}$$

One may ask how, given $\vec{A}$ and Φ, Λ can be constructed to make the gauge transformed vector and scalar potentials fulfill the gauge condition (1.70). The answer is as follows. Given potentials $\vec{A}$ and Φ which do not satisfy the Lorenz gauge condition, we have for the gauge-transformed potentials,

$$
\begin{aligned}
\vec{\nabla} \cdot \vec{A}'(\vec{r},t) + \frac{1}{c^2}\frac{\partial}{\partial t}\Phi'(\vec{r},t) &= \vec{\nabla} \cdot \left[\vec{A}(\vec{r},t) + \vec{\nabla}\Lambda(\vec{r},t)\right] \\
&+ \frac{1}{c^2}\frac{\partial}{\partial t}\left[\Phi(\vec{r},t) - \frac{\partial}{\partial t}\Lambda(\vec{r},t)\right] = 0,
\end{aligned}
\tag{1.71}
$$

or

$$\vec{\nabla} \cdot \vec{A} (\vec{r}, t) + \vec{\nabla}^2 \Lambda (\vec{r}, t) + \frac{1}{c^2} \frac{\partial}{\partial t} \Phi (\vec{r}, t) - \frac{1}{c^2} \frac{\partial^2}{\partial t^2} \Lambda (\vec{r}, t) = 0 . \tag{1.72}$$

Bringing now all terms with Λ to one side of the equation, we obtain a condition on the gauge transform function Λ which mediates the transition to the Lorenz gauge,

$$\left(\vec{\nabla}^2 - \frac{1}{c^2} \frac{\partial^2}{\partial t^2} \right) \Lambda (\vec{r}, t) = -\vec{\nabla} \cdot \vec{A} (\vec{r}, t) - \frac{1}{c^2} \frac{\partial}{\partial t} \Phi (\vec{r}, t) . \tag{1.73}$$

The gauge function Λ is required to satisfy the wave equation with the source determined by the original potentials, i.e., by Φ and $\vec{A}$ (the ones before the gauge transform). We shall see later that the differential operator $(\vec{\nabla}^2 - c^{-2} \partial_t^2)$ can be inverted in much the same right as the operator $\vec{\nabla}^2$. This implies that it is always possible to find an adequate solution Λ to Eq. (1.73), given Φ and $\vec{A}$. In other words, we do not exclude any important solutions to the Maxwell equations if we impose the Lorenz gauge condition.

A remark is in order: If the vector and scalar potentials $\vec{A}$ and Φ were to fulfill the Lorenz condition in the first place, then the right-hand side of Eq. (1.73) would vanish. The gauge transform function Λ has to fulfill an inhomogeneous wave equation in which the source term is just equivalent to the "left-hand side of the gauge condition Eq. (1.70)". That means that if, accidentally, $\vec{A}$ and Φ were to fulfill the Lorenz condition even before being gauge transformed, then we could choose for Λ any function that fulfills the homogeneous wave equation, and still retain fields that fulfill the Lorenz condition, after the gauge transformation. This illustrates that the gauge condition itself does not yet determine the potentials uniquely.

Let us now couple the gauge transformed potentials to their sources. We recall that according to Eqs. (1.63), Gauss's law and the Ampere–Maxwell law relate the sources to the potentials. In terms of the primed, gauge-transformed quantities, we have

$$-\vec{\nabla}^2 \Phi' (\vec{r}, t) - \frac{\partial}{\partial t} \vec{\nabla} \cdot \vec{A}' (\vec{r}, t) = \frac{1}{\epsilon_0} \rho (\vec{r}, t) \tag{1.74a}$$

for the coupling to the charge density, and

$$\vec{\nabla} \left\{ \underbrace{\vec{\nabla} \cdot \vec{A}' (\vec{r}, t) + \frac{1}{c^2} \frac{\partial}{\partial t} \Phi' (\vec{r}, t)}_{=0} \right\} - \vec{\nabla}^2 \vec{A}' (\vec{r}, t) + \frac{1}{c^2} \frac{\partial^2}{\partial t^2} \vec{A}'(\vec{r}, t) = \mu_0 \, \vec{J}(\vec{r}, t) \tag{1.74b}$$

for the coupling to the current density. If the primed potentials fulfill the Lorenz gauge condition (1.70),

$$\vec{\nabla} \cdot \vec{A}' (\vec{r}, t) + \frac{1}{c^2} \frac{\partial}{\partial t} \Phi' (\vec{r}, t) = 0 , \tag{1.75}$$

the equations are uncoupled. For the second equation, this is immediately obvious, and for the first equation, this becomes obvious as follows,

$$-\vec{\nabla}^2 \Phi' (\vec{r}, t) - \frac{\partial}{\partial t} \vec{\nabla} \cdot \vec{A}' (\vec{r}, t) = -\vec{\nabla}^2 \Phi' (\vec{r}, t) + \frac{1}{c^2} \frac{\partial}{\partial t} \left(\frac{\partial}{\partial t} \Phi \right) = \frac{1}{\epsilon_0} \rho (\vec{r}, t) . \tag{1.76}$$

The equations thus take the form of uncoupled, inhomogeneous wave equations, in Lorenz gauge,

$$\left(\frac{1}{c^2}\frac{\partial^2}{\partial t^2} - \vec{\nabla}^2\right)\Phi'(\vec{r},t) = \frac{1}{\epsilon_0}\rho(\vec{r},t)\,, \tag{1.77a}$$

$$\left(\frac{1}{c^2}\frac{\partial^2}{\partial t^2} - \vec{\nabla}^2\right)\vec{A}'(\vec{r},t) = \mu_0\vec{J}(\vec{r},t)\,. \tag{1.77b}$$

Once these are solved, we can compute the electric and magnetic fields according to Eqs. (1.59) and (1.56). In a source-free region of space-time, the scalar and vector potentials thus satisfy homogeneous wave equations. This finding suggests the existence of electromagnetic waves.

Let us add a remark on the relativistic formulation. It is customary to define the "quabla" operator as

$$\Box = \frac{1}{c^2}\frac{\partial^2}{\partial t^2} - \vec{\nabla}^2\,. \tag{1.78}$$

Furthermore, the 4-vector potential is $A^\mu = (\Phi, c\vec{A})$, which summarizes the scalar potential and the vector potential into a Lorentz-covariant 4-vector, denoted by a Greek superscript μ. This means that if we have a Lorentz transformation that transforms the space-time coordinates $x^\mu = (ct,\vec{r})$ according to

$$x'^\mu = L^\mu_\nu\, x^\nu \tag{1.79}$$

then the 4-vector potentials, as seen from the moving observer, need to be transformed in the same way,

$$A'^\mu = L^\mu_\nu\, A^\nu\,, \tag{1.80}$$

which clarifies the combined transformation properties of scalar and vector potential under Lorentz transformations. Note that here, the primed quantities refer to those seen in a different Lorentz frame (not the gauge transformed quantities). Then, in units with $\hbar = c = \epsilon_0 = 1$ (Heaviside–Lorentz units, or just "natural units"), the two equations (1.77a) and (1.77b) can be summarized into one equation, which takes a particularly simple form, namely

$$\Box A^\mu = \frac{1}{\epsilon_0}J^\mu\,, \tag{1.81}$$

where the components of the four-vector potential A^μ and the four-current density J^μ are

$$A^\mu = (\Phi, c\vec{A})\,, \qquad J^\mu = (\rho, c^{-1}\vec{J})\,. \tag{1.82}$$

Greek indices $\mu, \nu = 0,1,2,3$ describe space-time coordinates. Choosing potentials which satisfy the Lorentz condition not only uncouples the equations for the potentials but also yields equations which are invariant under the Lorentz/Poincaré transformations, as is evident from Eq. (1.81). Both sides of Eq. (1.81) carry only one Lorentz index.

1.2.4   *"Gauge Always Shoots Twice"*

In the preceding Sec. 1.2.3, we have already observed that the gauge condition (1.70), which we recall for convenience,

$$\vec{\nabla} \cdot \vec{A}'\,(\vec{r},t) + \frac{1}{c^2}\frac{\partial}{\partial t}\Phi'\,(\vec{r},t) = 0\,, \tag{1.83}$$

does not determine Φ and $\vec{A}$ uniquely. Indeed, the gauge transformed potentials, defined according to Eqs. (1.65) and (1.67),

$$\vec{A}'\,(\vec{r},t) = \vec{A}\,(\vec{r},t) + \vec{\nabla}\Lambda\,(\vec{r},t)\,, \qquad \Phi'\,(\vec{r},t) = \Phi\,(\vec{r},t) - \frac{\partial}{\partial t}\Lambda\,(\vec{r},t)\,, \tag{1.84}$$

fulfill the Lorenz gauge condition if the Λ function obeys Eq. (1.73), which constitutes an inhomogeneous wave equation. Once $\vec{A}'$ and Φ' fulfill the Lorentz condition, it is possible to do a second gauge transform, using a function Λ', and to still remain within the family of potentials that obey the Lorenz gauge condition.

The Lorenz gauge condition is a first-order partial differential equation imposed on the 4-vector potential. It is not an algebraic condition. We know that a partial differential equation allows for many solutions. Therefore, Eq. (1.70) does not determine the 4-vector potential uniquely, and even given the Lorentz condition, we still have a certain freedom to choose our 4-vector potential within the "Lorenz gauge family" without affecting the electric and magnetic fields. This is sometimes expressed by saying that "gauge always shoots twice."

Therefore, if $\vec{A}'$ and Φ' satisfy the Lorentz condition (1.70) and Λ' is yet another solution of the homogeneous wave equation,

$$\left(-\vec{\nabla}^2 + \frac{1}{c}\frac{\partial^2}{\partial t^2}\right)\Lambda'\,(\vec{r},t) = 0\,, \tag{1.85}$$

then the second gauge transform of vector and scalar potentials reads

$$\vec{A}''\,(\vec{r},t) = \vec{A}'\,(\vec{r},t) + \vec{\nabla}\Lambda'\,(\vec{r},t)\,, \tag{1.86a}$$

$$\Phi''\,(\vec{r},t) = \Phi'\,(\vec{r},t) - \frac{\partial}{\partial t}\Lambda'\,(\vec{r},t)\,. \tag{1.86b}$$

The second gauge transformation defines new vector and scalar potentials which still satisfy the Lorentz condition and give the same $\vec{E}$ and $\vec{B}$ fields. The class of all such combinations of vector and scalar potentials related by this restricted gauge transformation belong to the Lorenz gauge.

1.2.5   *Coulomb Gauge*

Another gauge which is often used in radiation theory is the Coulomb, transverse, or radiation gauge. The gauge condition is

$$\vec{\nabla} \cdot \vec{A}(\vec{r},t) = 0\,, \qquad \vec{A}\,(\vec{r},t) = \vec{A}_\perp\,(\vec{r},t)\,, \tag{1.87}$$

where by $\vec{A}_\perp$ we refer to the transverse component of the vector potential defined according to Eq. (1.48b). In general, the vector and scalar potential are coupled to the sources according to Eq. (1.63). In the radiation gauge, the scalar and vector potentials thus satisfy the equations

$$-\vec{\nabla}^2 \Phi\left(\vec{r},t\right) = \frac{1}{\epsilon_0} \rho\left(\vec{r},t\right), \tag{1.88a}$$

$$-\vec{\nabla}^2 \vec{A}\left(\vec{r},t\right) + \frac{1}{c^2}\frac{\partial^2}{\partial t^2}\vec{A}\left(\vec{r},t\right) = \mu_0 \vec{J}\left(\vec{r},t\right) - \frac{1}{c^2}\vec{\nabla}\left[\frac{\partial}{\partial t}\Phi\left(\vec{r},t\right)\right], \tag{1.88b}$$

respectively. Equation (1.88b) can be simplified further in terms of two simpler equations.

We first observe that the left-hand side of Eq. (1.88b) is equal to its own transverse part, because $\vec{A}\left(\vec{r},t\right) = \vec{A}_\perp\left(\vec{r},t\right)$ in Coulomb gauge [see Eq. (1.87)]; the transverse character of a vector field is not changed by taking the Laplacian, as is immediately obvious from Eq. (1.48a). Namely, for a general vector field $\vec{V}$, if $\vec{V}_\parallel$ vanishes then so does $\vec{\nabla}^2 \vec{V}_\parallel$, and because of the uniqueness of the separation of a vector field into longitudinal and transverse components, we then have $\vec{\nabla}^2 \vec{V} = \vec{\nabla}^2 \vec{V}_\perp$. Now we analyze the right-hand side of Eq. (1.88b) in terms of longitudinal and transverse components. First, we observe that the expression

$$\frac{1}{c^2}\vec{\nabla}\left[\frac{\partial}{\partial t}\Phi\left(\vec{r},t\right)\right] = \left(\frac{1}{c^2}\vec{\nabla}\left[\frac{\partial}{\partial t}\Phi\left(\vec{r},t\right)\right]\right)\Big|_\parallel \tag{1.89}$$

is a longitudinal vector field because the curl of a gradient vanishes,

$$\vec{\nabla} \times \left(\frac{1}{c^2}\vec{\nabla}\left[\frac{\partial}{\partial t}\Phi\left(\vec{r},t\right)\right]\right) = \vec{0}. \tag{1.90}$$

The longitudinal component of Eq. (1.88b) is thus equivalent to

$$\vec{J}_\parallel\left(\vec{r},t\right) = \epsilon_0 \vec{\nabla}\left[\frac{\partial}{\partial t}\Phi\left(\vec{r},t\right)\right], \tag{1.91}$$

while for the transverse component of Eq. (1.88b), only the transverse component of the current density $\vec{J}(\vec{r},t)$ contributes on the right-hand side. So, we can establish the three equations couple the scalar and vector potentials to their sources in the Coulomb or radiation gauge,

$$-\vec{\nabla}^2 \Phi\left(\vec{r},t\right) = \frac{1}{\epsilon_0}\rho\left(\vec{r},t\right), \tag{1.92a}$$

$$\left(\frac{1}{c^2}\frac{\partial^2}{\partial t^2} - \vec{\nabla}^2\right)\vec{A}_\perp\left(\vec{r},t\right) = \mu_0 \vec{J}_\perp\left(\vec{r},t\right), \tag{1.92b}$$

$$\epsilon_0 \vec{\nabla}\frac{\partial}{\partial t}\Phi\left(\vec{r},t\right) = \vec{J}_\parallel\left(\vec{r},t\right). \tag{1.92c}$$

Let us now consider, just like for the Lorenz gauge, a second gauge transformation within the family of the radiation gauge, writing Λ' as χ,

$$\vec{A}''\left(\vec{r},t\right) = \vec{A}'\left(\vec{r},t\right) + \vec{\nabla}\chi\left(\vec{r},t\right), \tag{1.93a}$$

$$\Phi''\left(\vec{r},t\right) = \Phi'\left(\vec{r},t\right) - \frac{\partial}{\partial t}\chi\left(\vec{r},t\right). \tag{1.93b}$$

The gauge condition imposed on the second transformed vector potential implies that

$$\vec{\nabla} \cdot \vec{A}''(\vec{r}, t) = 0, \qquad \vec{\nabla}^2 \chi(\vec{r}, t) = 0. \tag{1.93c}$$

Thus, any function $\chi = \chi(\vec{r}, t)$ that fulfills $\vec{\nabla}^2 \chi(\vec{r}, t) = 0$ describes a second gauge transform of the vector and scalar potentials within the radiation gauge.

The radiation gauge is well suited for describing the propagation of radiation in the absence of sources, i.e., in a region of space where there are no charges, and no currents. If sources are absent, then $\vec{\nabla}^2 \Phi(\vec{r}, t) = 0$, and therefore $\Phi(\vec{r}, t) = 0$, and so

$$\vec{E}(\vec{r}, t) = -\frac{\partial}{\partial t} \vec{A}(\vec{r}, t), \tag{1.94a}$$

$$\vec{B}(\vec{r}, t) = \vec{\nabla} \times \vec{A}(\vec{r}, t). \tag{1.94b}$$

Let us consider a typical vector potential describing a traveling electromagnetic wave, of the form

$$\vec{A}(\vec{r}, t) = \hat{e}_{\vec{k}\lambda} A_0 \cos(\vec{k} \cdot \vec{r} - \omega t), \tag{1.95}$$

where $\vec{k}$ is the wave vector, and $\omega = c|\vec{k}|$. The polarization vectors $\hat{e}_{\vec{k}\lambda}$ are two unit vectors perpendicular to $\vec{k}$ (in consequence, we have $\lambda = 1, 2$). Then,

$$\vec{\nabla} \cdot \vec{A}(\vec{r}, t) = -(\vec{k} \cdot \hat{e}_{\vec{k}\lambda}) A_0 \sin(\vec{k} \cdot \vec{r} - \omega t) = 0 \tag{1.96}$$

holds because the polarization vector is always perpendicular to the propagation vector, $\vec{k} \cdot \hat{e}_{\vec{k}\lambda}$.

However, the gauge is also called the Coulomb gauge. This is because the equation that couples the electrostatic potential to the source,

$$\vec{\nabla}^2 \Phi(\vec{r}, t) = -\frac{1}{\epsilon_0} \rho(\vec{r}, t), \tag{1.97}$$

has the instantaneous, action-at-a-distance solution

$$\Phi(\vec{r}, t) = \frac{1}{4\pi\epsilon_0} \int d^3 r' \frac{1}{|\vec{r} - \vec{r}'|} \rho(\vec{r}', t). \tag{1.98}$$

The similarity of this solution with the Coulomb law, i.e., the simple presence of the factor $\frac{1}{|\vec{r}-\vec{r}'|}$, implies the nomenclature of Coulomb gauge. Moreover, in a time-independent situation, the dependence on t cancels, and the Poisson equation from electrostatics is immediately recovered. However, for time-dependent problems, the action-at-a-distance solution poses important questions. If $\vec{r}'$ is in the Andromeda Galaxy and $\vec{r}$ is near Alpha Centauri, then we have an apparent problem with the causality principle because the field is generated at the same point in time as the source acts, namely, at time t. The task then is to show that the perceived action-at-a-distance character of the potential does not imply that the electric fields generated by the charges violate the causality principle (see Chap. 4.2).

1.3 Poynting Theorem and Maxwell Stress Tensor

1.3.1 *Electric and Magnetic Field Energies*

The introduction into basic properties of the electromagnetic field in the current chapter would not be complete without a discussion of the field energy, or, energy density stored in any nonvanishing electromagnetic field. NEWTON[7] invented the concept of a physical field, as an entity which exists at every point in space-time even if the existence of the field is not always detected or observed by a probe. In consequence, it is a natural question to ask if we can assign an "energy" to a field configuration, and if yes, how the corresponding "momentum flow" of the field could be described.

Imagine an electrically charged, heavy object in the middle of a coordinate system, and an oppositely charged test object far apart. The initial configuration has a nontrivial, nonvanishing electric field configuration. Then, the test object is accelerated toward the center of the heavy charged object. At the moment when they meet, the field vanishes because the system is effectively electrically neutral. (In practice, of course, there is still a small dipole moment.) The test object has gained some kinetic energy; the electric force has performed work on the test body. However, the potential energy of the initial configuration is gone. We can either ascribe the initial potential energy to the position of the charges in their potentials, or ascribe it to the field configuration as a whole. The latter aspect will be the center of discussion here.

We first derive a seemingly innocent vector identity,

$$\frac{1}{\mu_0} \vec{\nabla} \cdot \left[\vec{E}(\vec{r},t) \times \vec{B}(\vec{r},t) \right]$$

$$= -\frac{1}{\mu_0} \vec{E}(\vec{r},t) \cdot \left[\vec{\nabla} \times \vec{H}(\vec{r},t) \right] + \frac{1}{\mu_0} \vec{B}(\vec{r},t) \cdot \vec{\nabla} \times \left[\vec{E}(\vec{r},t) \right]$$

$$= -\vec{E}(\vec{r},t) \cdot \left[\vec{J}(\vec{r},t) + \epsilon_0 \frac{\partial}{\partial t} \vec{E}(\vec{r},t) \right] + \frac{1}{\mu_0} \left[\vec{B}(r,t) \cdot \left(-\frac{\partial}{\partial t} \vec{B}(\vec{r},t) \right) \right]$$

$$= -\vec{E}(\vec{r},t) \cdot \vec{J}(\vec{r},t) - \epsilon_0 \vec{E}(\vec{r},t) \cdot \frac{\partial}{\partial t} \vec{E}(\vec{r},t) - \frac{1}{\mu_0} \vec{B}(r,t) \cdot \frac{\partial}{\partial t} \vec{B}(\vec{r},t)$$

$$= -\vec{E}(\vec{r},t) \cdot \vec{J}(\vec{r},t) - \frac{\partial}{\partial t} \left[\frac{\epsilon_0}{2} \vec{E}(\vec{r},t) \cdot \vec{E}(\vec{r},t) + \frac{1}{2\mu_0} \vec{B}(\vec{r},t) \cdot \vec{B}(\vec{r},t) \right]. \quad (1.99)$$

In order to interpret the terms on the right-hand side of this we first recall that if the current density $\vec{J}(\vec{r},t)$ is to be nonvanishing, charges have to move. We can write

$$\vec{J}(\vec{r},t) = \frac{\Delta q}{\Delta t \, \Delta A} \, \hat{v}, \quad (1.100)$$

where Δq is the charge moving in the time interval Δt through the cross-sectional area ΔA, in a direction given by the unit velocity vector $\hat{v}$.

---

[7]Sir Isaac Newton (1642–1726)

The mechanical work done on the moving charges per unit time and unit volume, by the electric force acting on the charges due to the electric field, is

$$\vec{E}\left(\vec{r},t\right)\cdot\vec{J}\left(\vec{r},t\right)\approx\frac{\Delta q\,\vec{E}\left(\vec{r},t\right)\cdot\vec{v}}{\Delta t\,\Delta A}=\frac{\left(\Delta q\,\vec{E}\right)\cdot\Delta\vec{x}}{\Delta t\,\Delta V}$$

$$=\frac{\left(\Delta\vec{F}_{\text{elec}}\right)\cdot\Delta\vec{x}}{\Delta t\,\Delta V}=\frac{\Delta W}{\Delta t\,\Delta V}\,. \tag{1.101}$$

We therefore identify

$$\vec{E}\left(\vec{r},t\right)\cdot\vec{J}\left(\vec{r},t\right)=\frac{\partial}{\partial t}w(\vec{r},t) \tag{1.102}$$

as the time derivative of the work density $w(\vec{r},t)$, i.e., as the power density. Note that this is the work done by the electric field on the charges, i.e., the work done to accelerate the charges, not the work necessary to move the charges against the electric field.

We can naturally identify

$$u\left(\vec{r},t\right)=\frac{\epsilon_0}{2}\left[\vec{E}(\vec{r},t)\right]^2+\frac{1}{2\mu_0}\left[\vec{B}(\vec{r},t)\right]^2 \tag{1.103}$$

as the energy density of the electromagnetic field. With these identifications, Eq. (1.101) takes (almost) the form of a continuity equation,

$$\frac{1}{\mu_0}\vec{\nabla}\cdot\left[\vec{E}\left(\vec{r},t\right)\times\vec{B}\left(\vec{r},t\right)\right]+\frac{\partial u\left(\vec{r},t\right)}{\partial t}+\frac{\partial w\left(\vec{r},t\right)}{\partial t}=0\,. \tag{1.104}$$

We identify the Poynting vector $\vec{S}(\vec{r},t)$ as

$$\vec{S}\left(\vec{r},t\right)=\frac{1}{\mu_0}\,\vec{E}(\vec{r},t)\times\vec{B}(\vec{r},t)\,, \tag{1.105}$$

i.e., as the energy current density leaving a small test volume, where $\vec{S}$ has the physical dimension of energy per time per area. We can write

$$\vec{\nabla}\cdot\vec{S}(\vec{r},t)+\frac{\partial u\left(\vec{r},t\right)}{\partial t}+\frac{\partial w\left(\vec{r},t\right)}{\partial t}=0\,. \tag{1.106}$$

In view of the divergence theorem, this implies that

$$\int_{\partial V}\vec{S}(\vec{r},t)\cdot\mathrm{d}\vec{A}=-\int_V\vec{E}(\vec{r},t)\cdot\vec{J}(\vec{r},t)\,\mathrm{d}^3r-\int_V\frac{\partial u\left(\vec{r},t\right)}{\partial t}\mathrm{d}^3r\,. \tag{1.107}$$

The interpretation is clear: The rate at which energy leaves a region plus the rate at which field energy is converted to mechanical energy is compensated by a commensurate loss in field energy, i.e., equal to minus the rate at which the field energy changes. Note that $\vec{E}(\vec{r},t)\cdot\vec{J}(\vec{r},t)=\partial_t w(\vec{r},t)$ is positive when work is done on the charges by the field, i.e., when the work is done "in the direction" of the electromagnetic fields. This work leads to a loss in the field energy density, or otherwise to an increase in the kinetic energy of the charges inside the reference volume. Work done on the charges against the direction of the electromagnetic fields ("actual work") leads to a negative value for $\partial_t w(\vec{r},t)$, and to a positive value for the radiated energy $\int_{\partial V}\vec{S}(\vec{r},t)\cdot\mathrm{d}\vec{A}$ per time.

For vanishing power density $\partial_t w(\vec{r}, t) = 0$, let us write the energy dissipation equation (1.106) in a slightly different way,

$$\vec{\nabla} \cdot \vec{S}(\vec{r}, t) = -\frac{\partial u(\vec{r}, t)}{\partial t} \, , \qquad \frac{\partial w(\vec{r}, t)}{\partial t} = 0 \qquad (1.108)$$

and compare it to the continuity equation (1.29) (for the total current)

$$\vec{\nabla} \cdot \vec{J}(\vec{r}, t) = -\frac{\partial \rho(\vec{r}, t)}{\partial t} \, . \qquad (1.109)$$

Here, $\vec{J}$ is the current distribution, and ρ is the volume charge density. We can observe the analogy $\vec{S} \leftrightarrow \vec{J}$ and $u \leftrightarrow \rho$. The continuity equation states that a loss in the local charge distribution can only occur if charge leaves the area. The energy dissipation equation states that a loss in the local field energy distribution is compensated when field energy leaves the area (Poynting vector), or when performing work on the local charges.

1.3.2 Maxwell Stress Tensor

From the above discussion, it is clear that the Poynting vector describes the flow of electromagnetic field energy. Furthermore, the quantity $\vec{S}(\vec{r}, t)/c$ has the physical dimension of an energy density. Energy has the dimension of force multiplied by length, and so the quantity $(1/c^2)\partial \vec{S}(\vec{r}, t)/\partial t$ has the physical dimension of a force density, because ct has the dimension of length. One may thus ask if it is possible to formulate the force density acting on an ensemble of charges in terms of the electromagnetic fields, and conjecture that the resulting formula might contain a term proportional to $(1/c^2)\partial \vec{S}(\vec{r}, t)/\partial t$.

We start from the Lorentz force on a charged particle,

$$\vec{F} = q \left(\vec{E}(\vec{r}, t) + \vec{v} \times \vec{B}(\vec{r}, t) \right) . \qquad (1.110)$$

The force density, or force per unit volume, acting at point $\vec{r}$ and time t, is

$$\frac{\Delta \vec{F}}{\Delta V} = \frac{\Delta q}{\Delta V} \vec{E} + \frac{\Delta q \, \vec{v}}{\Delta V} \times \vec{B} . \qquad (1.111)$$

Defining $\vec{f}$ as the force density, and identifying $\Delta q/\Delta V \to \rho$ and $\Delta q \, \vec{v}/\Delta V \to \vec{J}$, in the limit of a small reference volume ΔV, we have

$$\vec{f}(\vec{r}, t) = \rho(\vec{r}, t) \, \vec{E}(\vec{r}, t) + \vec{J}(\vec{r}, t) \times \vec{B}(\vec{r}, t) . \qquad (1.112)$$

In the following, we suppress the space-time arguments $(\vec{r}, t)$ of the fields and sources. We now use Gauss's law and the Ampere–Maxwell law to express the charge density ρ and the current density $\vec{J}$ in terms of the fields,

$$\vec{f} = \left(\epsilon_0 \, \vec{\nabla} \cdot \vec{E} \right) \vec{E} + \left(\frac{1}{\mu_0} \vec{\nabla} \times \vec{B} - \epsilon_0 \frac{\partial \vec{E}}{\partial t} \right) \times \vec{B}$$

$$= \epsilon_0 \, \vec{E} \, \vec{\nabla} \cdot \vec{E} + \frac{1}{\mu_0} \left(\vec{\nabla} \times \vec{B} \right) \times \vec{B} - \epsilon_0 \frac{\partial \vec{E}}{\partial t} \times \vec{B} . \qquad (1.113)$$

The time derivative of the Poynting vector may be written as

$$\frac{\partial}{\partial t}\left(\vec{E}\times\vec{B}\right) = \frac{\partial\vec{E}}{\partial t}\times\vec{B} + \vec{E}\times\frac{\partial\vec{B}}{\partial t} = \frac{\partial\vec{E}}{\partial t}\times\vec{B} - \vec{E}\times\left(\vec{\nabla}\times\vec{E}\right), \tag{1.114}$$

where Faraday's law (1.2c) has been used. Solving this equation for $\partial_t\vec{E}\times\vec{B}$, one obtains

$$\frac{\partial\vec{E}}{\partial t}\times\vec{B} = \frac{\partial}{\partial t}\left(\vec{E}\times\vec{B}\right) + \vec{E}\times\left(\vec{\nabla}\times\vec{E}\right). \tag{1.115}$$

Using this relation in Eq. (1.113), we have

$$\vec{f} = \epsilon_0\,\vec{E}\left(\vec{\nabla}\cdot\vec{E}\right) + \frac{1}{\mu_0}\left(\vec{\nabla}\times\vec{B}\right)\times\vec{B} - \epsilon_0\,\frac{\partial}{\partial t}\left(\vec{E}\times\vec{B}\right) - \epsilon_0\,\vec{E}\times\left(\vec{\nabla}\times\vec{E}\right)$$

$$= \epsilon_0\left[\vec{E}\left(\vec{\nabla}\cdot\vec{E}\right) - \vec{E}\times\left(\vec{\nabla}\times\vec{E}\right)\right]$$

$$+ \frac{1}{\mu_0}\left[\vec{B}\left(\vec{\nabla}\cdot\vec{B}\right) - \vec{B}\times\left(\vec{\nabla}\times\vec{B}\right)\right] - \epsilon_0\,\frac{\partial}{\partial t}\left(\vec{E}\times\vec{B}\right), \tag{1.116}$$

where an "artificial zero" has been added in the form of the term $\vec{\nabla}\cdot\vec{B} = 0$ in order to render the (first part of the) resulting expression symmetric in $\vec{E}\leftrightarrow\vec{B}$.

For any vector field $\vec{V}$, one can write the identity

$$\vec{V}\times\left(\vec{\nabla}\times\vec{V}\right) = \hat{e}_i\,\epsilon_{ijk}V_j\epsilon_{k\ell m}\nabla_\ell V_m = \hat{e}_i\left(\delta_{i\ell}\,\delta_{jm} - \delta_{im}\,\delta_{j\ell}\right)V_j\nabla_\ell V_m$$

$$= V_j\,\hat{e}_i\,\nabla_i\,V_j - V_j\nabla_j\hat{e}_i\,V_i = \frac{1}{2}\vec{\nabla}\left(\vec{V}\cdot\vec{V}\right) - \left(\vec{V}\cdot\vec{\nabla}\right)\vec{V}, \tag{1.117}$$

where we carefully keep track of the fields on which the gradient operator acts (only those to the right of the operator). The force density can be written as follows,

$$\vec{f} = \epsilon_0\left[\vec{E}\left(\vec{\nabla}\cdot\vec{E}\right) + \left(\vec{E}\cdot\vec{\nabla}\right)\vec{E}\right] + \frac{1}{\mu_0}\left[\vec{B}\left(\vec{\nabla}\cdot\vec{B}\right) + \left(\vec{B}\cdot\vec{\nabla}\right)\vec{B}\right]$$

$$+ \vec{\nabla}\left(\frac{1}{2}\epsilon_0\,\vec{E}^2 + \frac{1}{2\mu_0}\vec{B}^2\right) - \epsilon_0\,\frac{\partial}{\partial t}\left(\vec{E}\times\vec{B}\right). \tag{1.118}$$

We are now in the position to define the stress tensor $\mathbb{T}$ of the electromagnetic field, with components $\mathbb{T}_{ij}$,

$$\mathbb{T}_{ij} = \epsilon_0\left(E_i\,E_j - \frac{1}{2}\delta_{ij}\,\vec{E}^2\right) + \frac{1}{\mu_0}\left(B_i\,B_j - \frac{1}{2}\delta_{ij}\,\vec{B}^2\right), \tag{1.119}$$

or in dyadic notation,

$$\mathbb{T} = \epsilon_0\left(\vec{E}\otimes\vec{E} - \frac{\vec{E}^2}{2}\,\mathbb{1}_{3\times3}\right) + \frac{1}{\mu_0}\left(\vec{B}\otimes\vec{B} - \frac{\vec{B}^2}{2}\,\mathbb{1}_{3\times3}\right) = \mathbb{T}_{ij}\,\hat{e}_i\otimes\hat{e}_j. \tag{1.120}$$

Then,

$$\vec{\nabla}\cdot\mathbb{T} = \nabla_i\mathbb{T}_{ij}\,\hat{e}_j = \epsilon_0\,\hat{e}_j\left(E_j\,\nabla_i E_i + E_i\,\nabla_i E_j\right) + \frac{1}{\mu_0}\left(B_j\,\nabla_i B_i + B_i\,\nabla_i B_j\right)$$

$$- \hat{e}_j\nabla_j\left(\frac{1}{2}\epsilon_0\,\vec{E}^2 + \frac{1}{2\mu_0}\vec{B}^2\right)$$

$$= \epsilon_0\left(\vec{E}\left(\vec{\nabla}\cdot\vec{E}\right) + \left(\vec{E}\cdot\vec{\nabla}\right)\vec{E}\right) + \frac{1}{\mu_0}\left(\vec{B}\left(\vec{\nabla}\cdot\vec{B}\right) + \left(\vec{B}\cdot\vec{\nabla}\right)\vec{B}\right)$$

$$- \vec{\nabla}\left(\frac{1}{2}\epsilon_0\,\vec{E}^2 + \frac{1}{2\mu_0}\vec{B}^2\right). \tag{1.121}$$

We can finally write Eq. (1.118) as follows,

$$\vec{f} = \vec{\nabla} \cdot \mathbb{T} - \epsilon_0 \mu_0 \frac{\partial}{\partial t} \frac{1}{\mu_0} \left(\vec{E} \times \vec{B} \right) = \vec{\nabla} \cdot \mathbb{T} - \frac{1}{c^2} \frac{\partial}{\partial t} \vec{S}. \tag{1.122}$$

Here, $\vec{S}$ is the Poynting vector, defined in Eq. (1.105). Equation (1.122) is dimensionally correct; all components of the stress tensor have a physical dimension of energy per volume, or force per area, which is the same as the physical dimension of pressure. The relativistic generalization of the stress tensor, the so-called stress–energy tensor, is discussed in Sec. 8.3.2.

1.4 Exercises

• **Exercise 1.1**: Repeat the derivation the equivalence of the integral and differential forms of Maxwell equations for the absence of magnetic monopoles, i.e., show that the following two equations are equivalent:

$$(i) \; \vec{\nabla} \cdot \vec{B}(\vec{r}, t) = 0, \qquad (ii) \int_{\partial V} \vec{B}(\vec{r}, t) \cdot \mathrm{d}\vec{A} = 0. \tag{1.123}$$

• **Exercise 1.2**: Show that the Kronecker symbol δ_{ij} is a tensor under rotations. Geometrically, this means that if two vectors $\vec{a}$ and $\vec{b}$ undergo a rotation $\vec{a} \rightarrow \vec{a}' = R \cdot \vec{a}$, $\vec{b} \rightarrow \vec{b}' = R \cdot \vec{b}$, then $\vec{a} \cdot \vec{b} = \vec{a}' \cdot \vec{b}'$ (the rotation matrix is denoted as R). Component-wise, it means that the components of the ϵ tensor are invariant under rotations, i.e., that

$$\delta_{ij}' = R_{ik} \, \delta_{k\ell} \, R_{\ell j}^{\mathrm{T}} = \delta_{ij}. \tag{1.124}$$

The sum over k and ℓ is implicitly assumed by the Einstein summation convention. Show the equivalence of both statements, refer to a general form of a rotation matrix, and relate Eq. (1.124) to a well-known property of rotation matrices.

• **Exercise 1.3**: Show that ϵ_{ijk} is a tensor. Geometrically, this means that if three vectors $\vec{a}$, $\vec{b}$ and $\vec{c}$ undergo a rotation, $\vec{a} \rightarrow \vec{a}' = R \cdot \vec{a}$, and $\vec{b} \rightarrow \vec{b}' = R \cdot \vec{b}$ as well as $\vec{c} \rightarrow \vec{c}' = R \cdot \vec{c}$, then $\vec{a} \times (\vec{b} \times \vec{c}) = \vec{a}' \cdot (\vec{b}' \times \vec{c}')$. Again, the rotation matrix is denoted as R. Component-wise, it means that the components of the δ tensor are invariant under rotations, i.e., that

$$\epsilon_{ijk}' = R_{i\ell} \, \epsilon_{\ell mn} \, R_{mj}^{\mathrm{T}} \, R_{nk}^{\mathrm{T}} = \epsilon_{ijk}. \tag{1.125}$$

The sum over ℓ, m and n is implicitly assumed by the Einstein summation convention. Show the equivalence of both statements, and the invariance of the components of the ϵ tensor under rotations.

• **Exercise 1.4**: Using your favorite graphical and numerical computer package (Mathematica, Maple, Matlab, Axiom, etc.) investigate the properties of the integral

$$I(\omega_0) = \int_0^{\pi/2} \mathrm{d}\omega \, \delta(\omega - \omega_0) \, f(\omega) \rightarrow I(\eta, \omega) = \int_0^{\pi/2} \mathrm{d}\omega \, \Delta(\eta, \omega - \omega_0) \, f(\omega) \tag{1.126}$$

$$\omega_0 = \frac{\pi}{2}, \qquad f(\omega) = \sin^2(\omega), \tag{1.127}$$

where Δ is defined as in Eq. (1.32). Observe how $I(\eta,\omega)$ converges to $I(\omega_0) = 1$ as $\eta \to 0$.

• **Exercise 1.5**: Show Eq. (1.34) by first convincing yourself that $\delta[a(x - x_0)] = \delta(x-x_0)/a$ for a positive constant a. Then, expand the argument $f(x)$ of $\delta(f(x))$ in terms of $\delta(f(x)) \approx \delta[f(x_0) + f'(x_0)(x - x_0)]$ and make an appropriate assumption (which one?) about x_0.

• **Exercise 1.6**: Show that, because the Fourier transform of the Dirac-δ distribution is just a constant, the function $g = g(\vec{r} - \vec{r}')$ defined in Eq. (1.37) can be interpreted as the inverse of the Laplacian operator. In particular, deliberate on how you could formulate the quantity $(1/\vec{\nabla}^2)G(\vec{r},\vec{r}')$ using the momentum-space representation? (Consult Chaps. 4 and 6 for a systematic introduction to Fourier transforms in space and time.)

• **Exercise 1.7**: Consider the scalar and vector potentials for a traveling wave [see Eq. (1.95)],

$$\Phi(\vec{r},t) = 0, \qquad \vec{A}(\vec{r},t) = \hat{e}_{\vec{k}\lambda} A_0\, e^{i(\vec{k}\cdot\vec{r}-\omega t)}, \tag{1.128}$$

where $\omega = c|\vec{k}|$. (We here anticipate the complex formalism for the description of fields; the real part of Eq. (1.128) is equivalent to Eq. (1.95).) *(i)* Show that the scalar and vector potentials given in Eq. (1.128) fulfill both the Lorenz and Coulomb gauge potentials. *(ii)* Consider the gauge function

$$\Lambda(\vec{r},t) = \chi_0\, e^{i(\vec{k}\cdot\vec{r}-\omega t)}. \tag{1.129}$$

Show that $\Lambda(\vec{r},t)$ fulfills the condition (1.73), and calculate the gauge transformed scalar and vector potentials Φ' and $\vec{A}'$. Show that these give rise to the same fields $\vec{E}$ and $\vec{B}$. **Hint:** You might obtain something like

$$\Phi'(\vec{r},t) = i\omega\, \chi_0\, e^{i(\vec{k}\cdot\vec{r}-\omega t)}, \tag{1.130}$$

and

$$\vec{A}'(\vec{r},t) = i\vec{k}\, \chi_0\, e^{i(\vec{k}\cdot\vec{r}-\omega t)} + \vec{A}(\vec{r},t). \tag{1.131}$$

• **Exercise 1.8**: Consider the scalar and vector potentials,

$$\Phi(\vec{r},t) = -\vec{E}(t)\cdot\vec{r}, \qquad \vec{A}(\vec{r},t) = \vec{0}. \tag{1.132}$$

Show that under the gauge transformation

$$\Lambda(\vec{r},t) = \vec{A}(t)\cdot\vec{r}, \tag{1.133}$$

the following gauge-transformed potentials are obtained,

$$\Phi'(\vec{r},t) = 0, \qquad \vec{A}'(\vec{r},t) = \vec{A}(t). \tag{1.134}$$

You have transformed from a situation with a vanishing vector potential, to a gauge with a vanishing scalar potential. Equation (1.132) is referred to as the length gauge, and Eq. (1.134) is called the velocity gauge. A decisive feature is that $\vec{E}(t)$ and $\vec{A}(t)$ are time-dependent, but independent of the spatial coordinates.

• **Exercise 1.9**: Under which conditions can a gauge transformation be found which implements the *temporal gauge* $\Phi'(\vec{r},t) = 0$? *This is a difficult exercise.*

Chapter 2

Green Functions of Electrostatics

2.1 Overview

In the current chapter, we shall encounter basic techniques involved in the calculation of so-called Green functions. These functions are central to theoretical physics.
They have many applications; in particular, Green functions connect a physical signal to its source, in space or time. We shall start by discussing the Green function
of the harmonic oscillator (see Sec. 2.2), which connects the position of a classical
particle bound in a harmonic potential, to a "source" term, which for mechanical
problems, is simply a force acting on the particle. The Green function is calculated
by Fourier transformation with respect to time, or alternatively by matching solutions of the homogeneous equation near the cusp which defines the Green function.
Both techniques for calculating the Green functions will be used in the following
chapters of this book.

The "matching near the cusp" technique will be used in Secs. 2.3 and 2.4, for
the calculation of the electrostatic Green function in spherical and cylindrical coordinates. Here, the connection of the "signal" (the electrostatic potential) to the
source (the charge density) is mediated in three-dimensional space as opposed to the
time domain (as for the harmonic oscillator). The calculation of the Green function
in spherical coordinates (see Sec. 2.3) directly leads to the multipole expansion of
electrostatics. Finally, in Sec. 2.5, we shall encounter a third, alternative method
for the calculation of a Green function, namely, a so-called eigenfunction expansion,
which is inspired by ideas originally developed within quantum mechanics.

2.2 Green Function of the Harmonic Oscillator

One of the most basic applications of the Green function formalism concerns the
harmonic oscillator, which is introduced here as an example for the basic paradigm
of the formalism. Indeed, from our elementary education on electric circuits, we
know that a series resonant circuit consisting of a coil with inductance L, a resistor
with resistance R, and a capacitor with capacitance C fulfills

$$L \frac{\mathrm{d}I}{\mathrm{d}t} + I R + \frac{Q}{C} = V(t), \tag{2.1}$$

where $V(t)$ is the applied, possibly time-dependent voltage. If one uses the fact that the current I through the coil is equal to the time derivative of the charge Q, then the equation reads

$$L \frac{d^2 Q}{dt^2} + R \frac{dQ}{dt} + \frac{Q}{C} = V(t) . \tag{2.2}$$

The mechanical analogue (damped harmonic oscillator) reads as follows,

$$m \frac{d^2 z}{dt^2} + r \frac{dz}{dt} + D z = 0 . \tag{2.3}$$

Here, D is the spring constant, r is the coefficient of the damping term (linear in the velocity), m is the inertia term (kinetic energy, corresponds to the inductance L for the electric circuit), and z is the time-dependent elevation (z coordinate) of the test object attached to the spring.

Both Eqs. (2.2) and (2.3) can be brought into the following form by a trivial scaling and reidentification of the parameters,

$$\ddot{x}(t) + \gamma \dot{x}(t) + \omega_0^2 x(t) = f(t) , \tag{2.4}$$

where the dot means differentiation with respect to the time. The Green function $g = g(t - t')$ for this equation fulfills the equation

$$\ddot{g}(t - t') + \gamma \dot{g}(t - t') + \omega_0^2 g(t - t') = \delta(t - t') , \tag{2.5}$$

where δ is the Dirac-δ distribution and the dot again denotes the differentiation with respect to t (not t'). Introducing the variable $s \equiv t - t'$, we could write the defining equation as

$$\frac{d^2}{ds^2} g(s) + \gamma \frac{d}{ds} g(s) + \omega_0^2 g(s) = \delta(s) . \tag{2.6}$$

The primary use of the Green function is as follows: It describes the response of the system to a point-like (in time) perturbation, as described by a Dirac δ function. The response of the system to an arbitrary perturbation can be obtained by summing (integrating) the response of the system to the perturbation $f(t')$ weighted with the Green function $g(t - t')$,

$$x(t) = \int g(t - t') f(t') dt' . \tag{2.7}$$

In view of Eq. (2.5), $x(t)$ defined according to Eq. (2.7) fulfills Eq. (2.4). The relation (2.7) illustrates why the "natural" argument of the Green function is the time difference $s = t - t'$. We note that Eq. (2.5) does not define the Green function uniquely. Given a particular solution g, we can always add a solution of the homogeneous equation

$$\frac{d^2}{ds^2} h(s) + \gamma \frac{d}{ds} h(s) + \omega_0^2 h(s) = 0 , \tag{2.8}$$

and $g + h$ will also be a Green function. This finding illustrates in very basic terms that the Green function is only defined uniquely if it is supplied with suitable boundary conditions. For our problem, the general solution to the homogeneous equation reads ($s = t - t'$)

$$h(s) = A \exp\left(-\frac{1}{2}\gamma s\right) \sin\left(\frac{1}{2}\sqrt{4\omega_0^2 - \gamma^2}\, s\right)$$
$$+ B \exp\left(-\frac{1}{2}\gamma s\right) \cos\left(\frac{1}{2}\sqrt{4\omega_0^2 - \gamma^2}\, s\right), \tag{2.9}$$

where A and B are integration constants that can be adjusted to fulfill specific boundary conditions.

Let us now consider the Fourier representation of the Green function. Indeed, writing the Green function as a Fourier transform

$$g(t - t') = \int_{-\infty}^{\infty} \frac{d\omega}{2\pi} e^{-i\omega(t-t')} g(\omega), \tag{2.10}$$

we note that $g(t - t')$ and $g(\omega)$ are different functions and have different physical dimension; one of them is obtained from the other by integration over $d\omega$. One might ask if this can be mathematically correct. The answer is that it can be correct and unique if we interpret the symbols g to be "overloaded" in the sense of an overloaded operator in "Fortran 90" or "C++" notation: The physical dimension of the argument (time or frequency) defines the functional form to be used for g. It is easy to show that

$$g(\omega) = -\frac{1}{\omega^2 + i\gamma\omega - \omega_0^2} = -\frac{1}{(\omega - \omega_+)(\omega - \omega_-)}, \tag{2.11}$$

where the poles are given by

$$\omega_\pm = \frac{1}{2}\left(\pm\sqrt{4\omega_0^2 - \gamma^2} - i\gamma\right). \tag{2.12}$$

For light damping ($4\omega_0^2 > \gamma^2$), both of these poles $\omega_\pm$ have a negative imaginary part. In order to carry out the integration in Eq. (2.10), we now need to calculate the residues

$$\operatorname*{Res}_{\omega=\omega_\pm} g(\omega) = \operatorname*{Res}_{\omega=\omega_\pm} \left(-\frac{1}{\omega^2 + i\gamma\omega - \omega_0^2}\right) = \mp\frac{1}{\sqrt{4\omega_0^2 - \gamma^2}}. \tag{2.13}$$

For $t - t' > 0$, we have to close the contour in Eq. (2.10) in the lower half of the complex plane because the expression $\exp[-i\omega(t - t')]$ is exponentially damped for $\mathrm{Im}(\omega) < 0$. In that case, we pick up nonvanishing residues at $\omega = \omega_\pm$ and obtain

$$g(t - t') = 2\,\Theta(t - t')\, e^{-\gamma(t-t')/2}\, \frac{\sin\left(\frac{1}{2}\sqrt{4\omega_0^2 - \gamma^2}\,(t - t')\right)}{\sqrt{4\omega_0^2 - \gamma^2}}, \tag{2.14}$$

where Θ is the Heaviside step function, which is unity for nonnegative argument, and zero otherwise. A nonvanishing result is obtained only for $t > t'$. For $t < t'$, the contour needs to be closed in the upper half of the complex plane and no pole contributions are picked up. The Green function (2.14) is the retarded Green function; it vanishes when $t < t'$. Furthermore, it vanishes in the limit $t - t' \to \infty$.

An alternative way of constructing the Green function is as follows. We first observe that the homogeneous and inhomogeneous Eqs. (2.8) and (2.6), which we recall for convenience,

$$\frac{d^2}{ds^2} g(s) + \gamma \frac{d}{ds} g(s) + \omega_0^2 \, g(s) = \delta(s) \,, \tag{2.15a}$$

$$\frac{d^2}{ds^2} h(s) + \gamma \frac{d}{ds} h(s) + \omega_0^2 \, h(s) = 0 \,, \tag{2.15b}$$

are actually equivalent to each other except for the immediate vicinity of $s = 0$, because $\delta(s \neq 0) = 0$. Now, the general solution of the homogeneous equation is given in Eq. (2.9). Hence, we use an *ansatz* of the form

$$\begin{aligned}
g(s) = \Theta(s) &\left[A_+ \, e^{-\gamma s/2} \sin\left(\frac{1}{2} \sqrt{4\omega_0^2 - \gamma^2} \, s \right) \right. \\
&\left. + B_+ \, e^{-\gamma s/2} \cos\left(\frac{1}{2} \sqrt{4\omega_0^2 - \gamma^2} \, s \right) \right] \\
+ \Theta(-s) &\left[A_- \, e^{-\gamma s/2} \sin\left(\frac{1}{2} \sqrt{4\omega_0^2 - \gamma^2} \, s \right) \right. \\
&\left. + B_- \, e^{-\gamma s/2} \cos\left(\frac{1}{2} \sqrt{4\omega_0^2 - \gamma^2} \, s \right) \right] .
\end{aligned} \tag{2.16}$$

The integration constants $A_\pm$ and $B_\pm$ can be determined by *(i)* boundary conditions and *(ii)* integrating Eq. (2.6) in an infinitesimal interval about $s = 0$. For the retarded Green function, we have to require that $A_- = B_- = 0$. This also follows from the regularity requirement at $s = -\infty$; the numerical value of the Green function would otherwise diverge in that limit. Furthermore, as the Green function needs to be continuous at $s = 0$, we have to impose the condition $B_+ = 0$; the cosine would otherwise induce a kink. After setting $A_- = B_- = B_+ = 0$, the remaining parameter A_+ can be determined by integrating Eq. (2.15a) in an infinitesimal interval around $s = 0$,

$$\begin{aligned}
1 &= \int_{-\epsilon}^{\epsilon} \delta(s) \, ds = \int_{-\epsilon}^{\epsilon} \left[\frac{d^2}{ds^2} g(s) + \gamma \frac{d}{ds} g(s) + \omega_0^2 \, g(s) \right] ds \\
&= \frac{d}{ds} g(s) \Big|_{s=\epsilon} - \frac{d}{ds} g(s) \Big|_{s=-\epsilon} + \gamma \left[g(\epsilon) - g(-\epsilon) \right] + \omega_0^2 \, g(0) \, \epsilon + \mathcal{O}(\epsilon^2) \\
&= \frac{d}{ds} g(s) \Big|_{s=\epsilon} - \frac{d}{ds} g(s) \Big|_{s=-\epsilon} + \mathcal{O}(\epsilon) = \left(\frac{1}{2} A_+ \sqrt{4\omega_0^2 - \gamma^2} \right) - (0) + \mathcal{O}(\epsilon) . \tag{2.17}
\end{aligned}$$

We have assumed that the Green function itself is continuous while its derivative may have a kink. The result thus reads

$$A_+ = \frac{2}{\sqrt{4\omega_0^2 - \gamma^2}} \,. \tag{2.18}$$

Inserting this result into Eq. (2.16), we recover (2.14),

$$g(s) = 2 \, \Theta(s) \, e^{-\gamma s/2} \, \frac{\sin\left(\frac{1}{2} \sqrt{4\omega_0^2 - \gamma^2} \, s \right)}{\sqrt{4\omega_0^2 - \gamma^2}} \,. \tag{2.19}$$

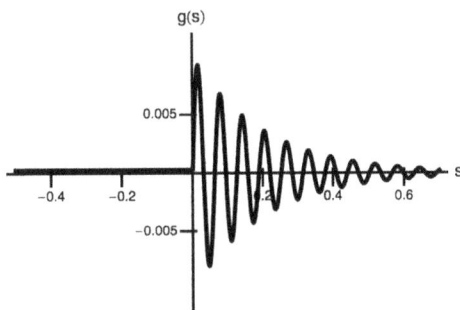

Fig. 2.1 Plot of the Green function $g(s)$ given in Eqs. (2.14) and (2.19) for $\gamma = \omega_0/10$ and $\omega_0 = 100$.

A plot of the Green function for typical parameters is given in Fig. 2.1. We shall use the "concatenation approach" for the construction of the Green function in future derivations, related to electrodynamics.

2.3 Electrostatic Green Function and Spherical Coordinates

2.3.1 *Poisson and Laplace Equations in Electrostatics*

In generalizing the result of the previous section on damped harmonic oscillators to electrostatics, we encounter the first example of a Green function relevant to electrodynamics. Below, we formulate the Poisson equation of electrostatics as the central inhomogeneous equation which gives rise to a Green function. In spherical coordinates, the equations of electrostatics naturally split into angular and radial parts. We shall find that the angular part of the equations is easily solved in terms of so-called spherical harmonics, which constitute some of the most important special functions of theoretical physics. For the radial equation, we shall first find the solutions to the homogeneous equation. Near the turning point, the solutions to the homogeneous equation will be matched such as to fulfill the defining inhomogeneous equation of the Green function, in an approach similar to that used in Eqs. (2.16)–(2.19). The calculation of the Green function for electrostatics is more difficult than for the damped harmonic oscillator because it is manifestly three-dimensional.

In electrostatics (no time dependence), the electric field $\vec{E}(\vec{r})$ is obtained from the potential $\Phi(\vec{r})$ as $\vec{E}(\vec{r}) = -\vec{\nabla}\Phi(\vec{r})$. If the charge distribution vanishes, $\rho(\vec{r}) = 0$, then the potential fulfills the Laplace equation

$$\vec{\nabla}^2\,\Phi(\vec{r}) = 0 \tag{2.20}$$

by virtue of Gauss's law (1.2a). For nonvanishing $\rho(\vec{r})$, one obtains the Poisson equation

$$\vec{\nabla}^2\,\Phi(\vec{r}) = -\frac{1}{\epsilon_0}\,\rho(\vec{r})\,. \tag{2.21}$$

The Green function G of the Poisson equation therefore has to fulfill the defining equation

$$\vec{\nabla}^2 G(\vec{r},\vec{r}') = \vec{\nabla}^2 G(\vec{r}-\vec{r}') = -\frac{1}{\epsilon_0} \delta^{(3)}(\vec{r}-\vec{r}'), \tag{2.22}$$

where $\vec{\nabla}^2$ is the Laplace operator, which acts on $\vec{r}$ (not $\vec{r}'$) unless stated otherwise. The three-dimensional Dirac-δ function has been defined in Eq. (1.35). In view of Eqs. (1.37), (1.38) and (1.42), the desired solution can immediately be written down as

$$G(\vec{r},\vec{r}') = G(\vec{r}-\vec{r}') = \frac{1}{4\pi\epsilon_0 |\vec{r}-\vec{r}'|}. \tag{2.23}$$

The Green function provides a convenient way to evaluate integrals of the form

$$\Phi(\vec{r}) = \int \mathrm{d}^3 r' \, G(\vec{r},\vec{r}') \rho(\vec{r}'), \qquad \vec{\nabla}^2 \Phi(\vec{r}) = -\frac{1}{\epsilon_0} \rho(\vec{r}), \tag{2.24}$$

for extended charge distributions $\rho(\vec{r}')$, which are solutions to the Poisson equation (2.21).

We here anticipate the central result of our discussion of the electrostatic Green function in spherical coordinates, which is the multipole decomposition

$$G(\vec{r},\vec{r}') = \frac{1}{4\pi\epsilon_0 |\vec{r}-\vec{r}'|} = \frac{1}{\epsilon_0} \sum_{\ell=0}^{\infty} \sum_{m=-\ell}^{\ell} \frac{1}{2\ell+1} \frac{r_<^\ell}{r_>^{\ell+1}} Y_{\ell m}(\theta,\varphi) \, Y_{\ell m}^*(\theta',\varphi'). \tag{2.25}$$

Here, $r_< = \min(r,r')$, $r_> = \max(r,r')$, while θ and φ are the polar and azimuth angles of the vector $\vec{r}$, and correspondingly for θ' and φ'. The $Y_{\ell m}$ functions are spherical harmonics, as defined in Eq. (2.53). We shall now work toward deriving Eq. (2.25), using spherical coordinates. To this end, we first have to analyze the Laplace equation in spherical coordinates, before "concatenating" the expression (2.25), which is a solution of the Poisson equation, from the solutions of the Laplace equation, at the cusp $r_< = r_>$, in a way similar to the derivation outlined in Eqs. (2.16)–(2.19).

2.3.2 *Laplace Equation in Spherical Coordinates*

The Laplacian $\vec{\nabla}^2 G(\vec{r},\vec{r}')$ of the Green function G vanishes almost everywhere. Indeed, it vanishes for all $\vec{r}$ except at the Dirac-δ source $\delta^{(3)}(\vec{r}-\vec{r}')$. We thus have to investigate the Laplacian operator $\vec{\nabla}^2$ in spherical coordinates, which are defined as

$$x = r \sin\theta \cos\varphi, \qquad y = r \sin\theta \sin\varphi, \qquad z = r \cos\theta, \tag{2.26a}$$

with

$$r = \sqrt{x^2+y^2+z^2}, \qquad \theta = \arcsin\left(\frac{z}{\sqrt{x^2+y^2}}\right), \qquad \varphi = \arccos\left(\frac{x}{\sqrt{x^2+y^2}}\right). \tag{2.26b}$$

One writes, by the chain rule,

$$\frac{\partial}{\partial x} = \frac{\partial r}{\partial x}\frac{\partial}{\partial r} + \frac{\partial\theta}{\partial x}\frac{\partial}{\partial\theta} + \frac{\partial\varphi}{\partial x}\frac{\partial}{\partial\varphi}, \tag{2.27}$$

thereby expressing the partial derivative operator $\partial/\partial x$ as a linear combination of $\partial/\partial r$, $\partial/\partial\theta$, and $\partial/\partial\varphi$. By a tedious, but straightforward calculation, one then obtains the conversion of the Laplacian operator in spherical coordinates,

$$
\begin{aligned}
\vec{\nabla}^2 &= \frac{\partial^2}{\partial x^2} + \frac{\partial^2}{\partial y^2} + \frac{\partial^2}{\partial z^2} \\
&= \frac{\partial^2}{\partial r^2} + \frac{2}{r}\frac{\partial}{\partial r} + \frac{1}{r^2\sin\theta}\frac{\partial}{\partial\theta}\sin\theta\frac{\partial}{\partial\theta} + \frac{1}{r^2\sin^2\theta}\frac{\partial^2}{\partial\varphi^2}.
\end{aligned}
\tag{2.28}
$$

(In Chap. 8, we shall explore Christoffel symbols which provide a more systematic view on the curvilinear coordinate systems.)

In quantum mechanics, one often introduces the angular momentum operator $\vec{L} = \vec{r}\times\vec{p} = -i\hbar\vec{r}\times\vec{\nabla}$, where $\vec{p} = -i\hbar\vec{\nabla}$ is the momentum operator. This book is not about quantum mechanics; however, the designation of the first subscript of the spherical harmonic $Y_{\ell m}$ actually is inspired by the angular momentum quantum number. In the context of classical electrodynamics, it is thus helpful to introduce a dimensionless version $\vec{L}$ of the angular momentum operator as the vector product

$$
\vec{L} = -i\vec{r}\times\vec{\nabla},
\tag{2.29}
$$

so that the Cartesian components of $\vec{L}$ can be expressed in terms of the spherical coordinates as

$$
L_x = -i\left(y\frac{\partial}{\partial z} - z\frac{\partial}{\partial y}\right) = i\left(\sin\varphi\frac{\partial}{\partial\theta} + \cos\varphi\cot\theta\frac{\partial}{\partial\varphi}\right),
\tag{2.30a}
$$

$$
L_y = -i\left(z\frac{\partial}{\partial x} - x\frac{\partial}{\partial z}\right) = i\left(-\cos\varphi\frac{\partial}{\partial\theta} + \sin\varphi\cot\theta\frac{\partial}{\partial\varphi}\right),
\tag{2.30b}
$$

$$
L_z = -i\left(x\frac{\partial}{\partial y} - y\frac{\partial}{\partial x}\right) = -i\frac{\partial}{\partial\varphi}.
\tag{2.30c}
$$

The square of the $\vec{L}$ operator reads as

$$
\vec{L}^2 = \vec{L}\cdot\vec{L} = -(\vec{r}\times\vec{\nabla})^2 = -\frac{1}{\sin\theta}\frac{\partial}{\partial\theta}\sin\theta\frac{\partial}{\partial\theta} - \frac{1}{\sin^2\theta}\frac{\partial^2}{\partial\varphi^2}.
\tag{2.31}
$$

We can thus write the Laplacian operator as follows,

$$
\vec{\nabla}^2 = \frac{\partial^2}{\partial r^2} + \frac{2}{r}\frac{\partial}{\partial r} - \frac{1}{r^2}\vec{L}^2.
\tag{2.32}
$$

This identity separates radial and angular variables in spherical coordinates. The homogeneous (Laplace) equation in spherical coordinates becomes

$$
\vec{\nabla}^2\Phi(r,\theta,\varphi) = 0, \qquad \Phi(r,\theta,\varphi) = R(r)\,Y(\theta,\varphi),
\tag{2.33}
$$

where the latter equation constitutes an *ansatz* for the solution, obtained by separation of variables. Inserting this *ansatz* into Eq. (2.32), one obtains

$$
\frac{1}{R(r)}\left\{\frac{\partial}{\partial r}\left[r^2\frac{\partial}{\partial r}R(r)\right]\right\} = \frac{1}{Y(\theta,\varphi)}\vec{L}^2\,Y(\theta,\varphi).
\tag{2.34}
$$

This equation must be satisfied for all (r, θ, φ), which are independent variables, Equation (2.34) can only be satisfied if both sides are equal to a constant which we write as $\ell(\ell+1)$. Then,

$$\vec{L}^2 Y(\theta, \varphi) = \ell(\ell+1) Y(\theta, \varphi), \tag{2.35}$$

and the radial part of the function fulfills

$$\frac{d}{dr}\left[r^2 \frac{d}{dr} R(r)\right] - \ell(\ell+1) R(r) = 0, \qquad R(r) = a_\ell r^\ell + b_\ell r^{-\ell-1}, \tag{2.36}$$

where we indicate the power law solutions. Equations (2.35) and (2.36) imply that if we can find eigenfunctions Y of the dimensionless angular momentum operator $\vec{L}^2$ satisfying (2.35), then we obtain solutions for the radial equation (2.36) as simple power laws.

It is useful to observe that the angular equation (2.35) can be further separated into polar and azimuthal parts by setting $Y(\theta, \varphi) = V(\theta) W(\varphi)$,

$$\frac{1}{V(\theta)} \sin\theta \frac{\partial}{\partial\theta}\left(\sin\theta \frac{\partial}{\partial\theta} V(\theta)\right) + \frac{1}{V(\theta)} \sin^2\theta \, \ell(\ell+1) \, V(\theta) = -\frac{1}{W(\varphi)} \frac{\partial^2}{\partial\varphi^2} W(\varphi). \tag{2.37}$$

Again, since θ and φ are independent variables, both sides must be equal to a constant (the second separation constant) which we take to be $-m^2$ and write

$$\frac{\partial^2}{\partial\varphi^2} W(\varphi) = -m^2 W(\varphi), \qquad W_m(\varphi) = A_m e^{im\varphi} + B_m e^{-im\varphi}. \tag{2.38}$$

Here, A_m and B_m are arbitrary integration constants, and m needs to be an integer in order for the function $W(\varphi)$ to be uniquely defined. So, the remaining equation in the polar angle θ is given as

$$\sin\theta \frac{\partial}{\partial\theta}\left(\sin\theta \frac{\partial}{\partial\theta} V(\theta)\right) + \left(\sin^2\theta \, \ell(\ell+1) - m^2\right) V(\theta) = 0. \tag{2.39}$$

Substituting $u = \cos\theta$ and defining $F(u) = V(\theta)$, we obtain the associated Legendre equation,

$$\frac{d}{du}\left[(1-u^2)\frac{d}{du} F(u)\right] + \left(\ell(\ell+1) - \frac{m^2}{1-u^2}\right) F(u) = 0. \tag{2.40}$$

It has two linearly independent solutions called the associated Legendre functions of the first, $P_\ell^{|m|}(u)$, and second kind, $Q_\ell^{|m|}(u)$. The general solution to the polar equation reads as

$$F_{\ell,m}(u) = A_{\ell,m} P_\ell^m(u) + B_{\ell,m} Q_\ell^m(u), \qquad -\ell < m \le \ell. \tag{2.41}$$

For $-\ell < m \le \ell$, with $\ell = 0, 1, 2, \ldots$, the $P_\ell^m(\cos\theta)$ are finite in the interval $0 \le \theta \le \pi$. By contrast, the $Q_\ell^m(\cos\theta)$ diverge at $u = \cos\theta = \pm 1$ or $\theta = 0, \pi$ (i.e., on the "pole caps" of the unit sphere). While the $Q_\ell^m(\cos\theta)$ are possible solutions when we can manifestly restrict the range of allowed polar angles to $|\cos\theta| < 1$, we must exclude these solutions if we seek functions that converge to finite values on the entire unit sphere. The functions $P_\ell^m(u)$ and $P_\ell^{-m}(u)$ are not linearly independent, but rather, directly proportional to each other,

$$P_\ell^{-|m|}(u) = (-1)^{|m|} \frac{(\ell - |m|)!}{(\ell + |m|)!} P_\ell^{|m|}(u). \tag{2.42}$$

Given a solution $F_{\ell,m}(u)$, the integration constants $A_{\ell,m}$ and $B_{\ell,m}$ from Eq. (2.41) are thus not uniquely determined, because $P_\ell^{|m|}(u)$ and $P_\ell^{-|m|}(u)$ are not linearly independent. However, the form of the solution given in Eq. (2.41) is still general. One can make the expansion unique by choosing between $P_\ell^{-|m|}(u)$ and $P_\ell^{|m|}(u)$ depending on the sign of m in the $e^{im\varphi}$ phase term, i.e., $e^{i|m|\varphi}$ is multiplied by $P_\ell^{|m|}(u)$, while $e^{-i|m|\varphi}$ is multiplied by $P_\ell^{-|m|}(u)$. This freedom of choice is incorporated into the definition of the spherical harmonics to be discussed in Eq. (2.53).

We shall assume that the solution $\Phi = \Phi(r, \theta, \varphi)$ of the Laplace equation (2.33) is well defined on the entire unit sphere and so, we will write the general solution to the homogeneous equation as

$$\Phi(r, \theta, \varphi) = \sum_{\ell=0}^{\infty} \sum_{m=-\ell}^{\ell} \left(a_{\ell,m} \, r^\ell + b_{\ell,m} \, r^{-\ell-1} \right) P_\ell^m(\cos\theta) \, e^{im\varphi}, \tag{2.43}$$

with arbitrary constants $a_{\ell,m}$ and $b_{\ell,m}$. We have absorbed the integration constants $A_{\ell,m}$ from Eq. (2.41) into the $a_{\ell,m}$ and $b_{\ell,m}$ that multiply the radial factors. From Eq. (2.38), we might otherwise assume that the allowed range for m contains the entire set of the positive and negative integers, and zero; however, the m in Eq. (2.38) has to be the same as the one in Eq. (2.41), and thus have to fulfill the requirement that $-\ell < m \leq \ell$. For $|m| > \ell$, the associated Legendre polynomials vanish because the $(|m| + 1)$th derivative annihilates $P_\ell(\cos\theta)$ [see Eq. (2.47a)].

Based on Eq. (2.43), we may proceed to solve the inhomogeneous equation defining the Green function (2.22) by matching the solution regular at the origin (r^ℓ) with the solution regular at infinity $(r^{-\ell-1})$ near the turning point, where the Dirac-δ function is nonvanishing, in analogy to the derivation outlined in Eqs. (2.16)–(2.19). However, before we proceed to this endeavor, we first dwell a little on the Legendre functions, and spherical harmonics, which constitute examples of the "special functions" of mathematical physics.

2.3.3 *Legendre Functions and Spherical Harmonics*

Let us first discuss the Legendre functions of the first kind. In our problem, we expect the functions to be well behaved for all $|x| = |\cos\theta| \leq 1$. A special case is found when $|m| = 0$. These solutions,

$$P_\ell^0(x) = P_\ell(x), \qquad P_\ell(x) = \frac{1}{2^\ell \, \ell!} \frac{\mathrm{d}^\ell}{\mathrm{d}x^\ell} [(x^2 - 1)]^\ell, \tag{2.44}$$

are called Legendre polynomials. They fulfill the recursion relation

$$(\ell + 1) P_{\ell+1}(x) = (2\ell + 1) \, x \, P_\ell(x) - \ell \, P_{\ell-1}(x). \tag{2.45}$$

For non-integer ℓ, the Legendre polynomials are generalized to Legendre functions, which in turn are expressible in terms of so-called hypergeometric functions. It is useful to indicate the first few $P_\ell(x)$ which read

$$P_0(x) = 1, \qquad P_1(x) = x, \qquad P_2(x) = \tfrac{3}{2}x^2 - \tfrac{1}{2}, \qquad (2.46a)$$

$$P_3(x) = \tfrac{5}{2}x^3 - \tfrac{3}{2}x, \qquad P_4(x) = \tfrac{35}{8}x^4 - \tfrac{15}{4}x^2 + \tfrac{3}{8}. \qquad (2.46b)$$

The associated Legendre polynomials $P_\ell^m(x)$ of degree ℓ and order m are related to the Legendre polynomials by the relations,

$$P_\ell^{|m|}(x) = \left(1 - x^2\right)^{|m|/2} \frac{\mathrm{d}^{|m|} P_\ell(x)}{\mathrm{d}x^{|m|}}, \qquad (2.47a)$$

$$P_\ell^{-|m|}(x) = (-1)^{|m|} \frac{(\ell - |m|)!}{(\ell + |m|)!} P_\ell^{|m|}(x), \qquad (2.47b)$$

where the first relation applies to $P_\ell^m(x)$ with $m \geq 0$, and the second one, which we have already encountered in Eq. (2.42), is to be used for $m < 0$. Because $P_\ell(x)$ is a polynomial of order ℓ, the associated Legendre polynomials are only non-zero when $\ell \geq |m|$. There are orthogonality relations,

$$\int_{-1}^{1} \mathrm{d}x \, P_{\ell'}^m(x) P_\ell^m(x) = \frac{2}{2\ell + 1} \frac{(\ell + m)!}{(\ell - m)!} \delta_{\ell\ell'}. \qquad (2.48)$$

Finally, we note that $P_\ell^0(1) = 1$, $P_\ell^0(-1) = (-1)^\ell$, and if $m \neq 0$, $P_\ell^m(\pm 1) = 0$. The general symmetry relation of these functions for $x \to -x$ is

$$P_\ell^m(-x) = (-1)^{\ell+m} P_\ell^m(x). \qquad (2.49)$$

The irregular solutions are the Legendre functions of the second kind, which are the functions $Q_\ell(x)$ that have singularities at $x = \pm 1$. They are expressed as follows ($|x| \leq 1$),

$$Q_\ell(x) = \frac{P_\ell(x)}{2} \ln\left(\frac{1+x}{1-x}\right) - W_{\ell-1}(x), \qquad (2.50a)$$

$$W_{\ell-1}(x) = \sum_{s=0}^{\ell-1} \frac{(n+s)! \left[\psi(n+1) - \psi(s+1)\right]}{2^s (n-s)! (s!)^2} (x-1)^s, \qquad (2.50b)$$

$$W_{-1}(x) = 0, \qquad W_0(x) = 1, \qquad W_1(x) = \tfrac{3}{2}x, \qquad W_2(x) = \tfrac{5}{2}x^2 - \tfrac{2}{3}. \qquad (2.50c)$$

Here, $\psi(x) = \mathrm{d}\ln[\Gamma(x)]/\mathrm{d}x$ is the logarithmic derivative of the Gamma function. We note the logarithmic divergences of $Q_\ell(x)$ for $x \to \pm 1$. They fulfill a recursion relation analogous to Eq. (2.45),

$$(\ell + 1) Q_{\ell+1}(x) = (2\ell + 1) x \, Q_\ell(x) - \ell \, Q_{\ell-1}(x), \qquad (2.51)$$

and the associated Legendre functions $Q_\ell^m(x)$ of degree ℓ and order m are obtained by relations analogous to (2.47),

$$Q_\ell^{|m|}(x) = \left(1 - x^2\right)^{|m|/2} \frac{\mathrm{d}^{|m|} Q_\ell(x)}{\mathrm{d}x^{|m|}}, \qquad (2.52a)$$

$$Q_\ell^{-|m|}(x) = (-1)^{|m|} \frac{(\ell - |m|)!}{(\ell + |m|)!} Q_\ell^{|m|}(x). \qquad (2.52b)$$

The logarithmic singularity of the Legendre Q functions near the poles $x = \cos\theta = \pm 1$ implies that we can ignore them in the physically relevant solutions of the Laplace equation.

In Eq. (2.43), we have already seen that the solutions to the angular equation (2.35) can be expressed as being proportional to $P_\ell^m(\cos\theta)\, e^{im\varphi}$. The normalization is canonically chosen to define the spherical harmonics $Y_{\ell m}(\theta, \varphi)$ as

$$Y_{\ell m}(\theta, \varphi) = (-1)^m \sqrt{\frac{2\ell + 1}{4\pi} \frac{(\ell - m)!}{(\ell + m)!}}\; P_\ell^m(\cos\theta)\, e^{im\varphi}. \tag{2.53}$$

We note the occurrence of the factor $(\ell-m)!/(\ell+m)!$ under the square root. Together with the prefactor in Eq. (2.47b), it ensures that the case of positive and negative m are consistently defined. The spherical harmonics $Y_{\ell m}(\theta, \varphi)$ are solutions to Eq. (2.35), which we recall for convenience,

$$\vec{L}^2\, Y_{\ell m}(\theta, \varphi) = \ell\,(\ell + 1)\, Y_{\ell m}(\theta, \varphi), \tag{2.54}$$

and they have the following property,

$$\vec{L}_z\, Y_{\ell m}(\theta, \varphi) = -i\frac{\partial}{\partial\varphi} Y_{\ell m}(\theta, \varphi) = m\, Y_{\ell m}(\theta, \varphi), \tag{2.55}$$

where $L_z = -i\partial/\partial\varphi$ is the z component of the operator defined in Eq. (2.29) [see also Eq. (2.30)]. Other useful properties are given by

$$Y_{\ell\,-m}(\theta, \varphi) = (-1)^m\, Y_{\ell m}^*(\theta, \varphi), \tag{2.56a}$$

$$Y_{\ell\, m}(\pi - \theta, 2\pi - \varphi) = (-1)^\ell\, Y_{\ell\,-m}(\theta, \varphi), \tag{2.56b}$$

$$Y_{\ell\, m}(\pi - \theta, \pi + \varphi) = (-1)^\ell\, Y_{\ell\, m}(\theta, \varphi). \tag{2.56c}$$

Equation (2.56c) is the parity transformation, which is mediated by the replacement $\theta \to \pi - \theta$ and $\varphi \to \varphi + \pi$. Based on the convention (2.53), we can establish the orthonormality conditions

$$\int_0^{2\pi} \int_0^\pi Y_{\ell'\, m'}^*(\theta, \varphi)\, Y_{\ell m}(\theta, \varphi)\, \sin\theta\, d\theta\, d\varphi = \delta_{\ell\ell'}\, \delta_{mm'}, \tag{2.57}$$

where the integral over the solid angle $d\Omega = \sin\theta\, d\theta\, d\varphi$ defines a scalar product. We appeal to Eq. (2.48) for the derivation. The scalar product may also also be written as $\langle Y_{\ell'm'}|Y_{\ell m}\rangle$, if we use Dirac's bra-ket notation, and we can interpret the value of $Y_{\ell m}(\theta, \varphi)$ as the "matrix element"

$$Y_{\ell m}(\theta, \varphi) = \langle\theta, \varphi|Y_{\ell m}\rangle, \tag{2.58}$$

i.e., as the scalar product of the eigenvector $|\theta, \varphi\rangle$ of the position on the unit sphere, and the eigenket $|Y_{\ell m}\rangle$ of the modulus square of the total angular momentum $\vec{L}^2$, and the z component L_z. Furthermore, the integration over the solid angle in Eq. (2.57) can be understood as a summation over the quantum numbers of the eigenkets $|\theta, \varphi\rangle$ of the position operator. With this interpretation, Eq. (2.57) can be written as

$$\oint_{\theta,\varphi} \langle Y_{\ell'm'}|\theta, \varphi\rangle \langle\theta, \varphi|Y_{\ell m}\rangle = \delta_{\ell\ell'}\, \delta_{mm'}, \tag{2.59}$$

The scalar product of two eigenkets $|\theta, \varphi\rangle$ and $|\theta', \varphi'\rangle$ simply is the Dirac-δ function on the unit sphere,

$$\langle \theta', \varphi' | \theta, \varphi \rangle = \frac{1}{\sin \theta} \delta (\theta - \theta') \delta (\varphi - \varphi') . \tag{2.60}$$

These eigenkets also form a complete basis for functions defined on the unit sphere, and we can thus write a completeness relation analogous to Eq. (2.59),

$$\sum_{\ell,m} \langle \theta, \varphi | Y_{\ell m} \rangle \langle Y_{\ell m} | \theta', \varphi' \rangle = \langle \theta', \varphi' | \theta, \varphi \rangle . \tag{2.61}$$

The aim of the detour into Dirac's bra-ket notation, comprising Eqs. (2.58)–(2.61), has been to motivate the completeness condition

$$\sum_{\ell=0}^{\infty} \sum_{m=-\ell}^{\ell} Y_{\ell m} (\theta, \varphi) Y_{\ell m}^* (\theta', \varphi') = \frac{1}{\sin \theta} \delta (\theta - \theta') \delta (\varphi - \varphi')$$

$$= \delta (\varphi - \varphi') \delta (\cos \theta - \cos \theta') , \tag{2.62}$$

which is equivalent to Eq. (2.61). We have used Eq. (1.34). For $\ell = 0$, the spherical harmonic with $m = 0$ is only a constant, $Y_{00} (\theta, \varphi) = (4\pi)^{-1/2}$. For $\ell = 1$, the spherical harmonics read as follows,

$$Y_{10} (\theta, \varphi) = \sqrt{\frac{3}{4\pi}} \cos \theta , \tag{2.63a}$$

$$Y_{11} (\theta, \varphi) = -\sqrt{\frac{3}{8\pi}} \sin \theta \, e^{i\varphi} , \qquad Y_{1-1} (\theta, \varphi) = \sqrt{\frac{3}{8\pi}} \sin \theta \, e^{-i\varphi} . \tag{2.63b}$$

The spherical harmonics with $\ell = 2$ are of second order in the trigonometric functions $\sin \theta$ and $\cos \theta$, and Y_{20} is proportional to $P_2(\cos \theta) = \frac{3}{2} \cos^2 \theta - \frac{1}{2}$,

$$Y_{20} (\theta, \varphi) = \sqrt{\frac{5}{4\pi}} \left(\frac{3}{2} \cos^2 \theta - \frac{1}{2} \right) , \tag{2.64a}$$

$$Y_{21} (\theta, \varphi) = -\sqrt{\frac{15}{8\pi}} \sin \theta \, \cos \theta \, e^{i\varphi} , \qquad Y_{2-1} (\theta, \varphi) = \sqrt{\frac{15}{8\pi}} \sin \theta \, \cos \theta \, e^{-i\varphi} , \tag{2.64b}$$

$$Y_{22} (\theta, \varphi) = \sqrt{\frac{15}{32\pi}} \sin^2 \theta \, e^{2i\varphi} , \qquad Y_{2-2} (\theta, \varphi) = \sqrt{\frac{15}{32\pi}} \sin^2 \theta \, e^{-2i\varphi} . \tag{2.64c}$$

These functions are illustrated in Fig. 2.2.

2.3.4 Expansion of the Green Function in Spherical Coordinates

We have found the general solution to the Laplace equation in spherical coordinates, given in Eq. (2.43),

$$\Phi(r, \theta, \varphi) = \sum_{\ell=0}^{\infty} \sum_{m=-\ell}^{\ell} \left(c_{\ell,m} \, r^\ell + d_{\ell,m} \, r^{-\ell-1} \right) Y_{\ell m}(\theta, \varphi) , \tag{2.65}$$

$\ell = 0$:

$\ell = 1$:

$\ell = 2$:

$\ell = 3$:

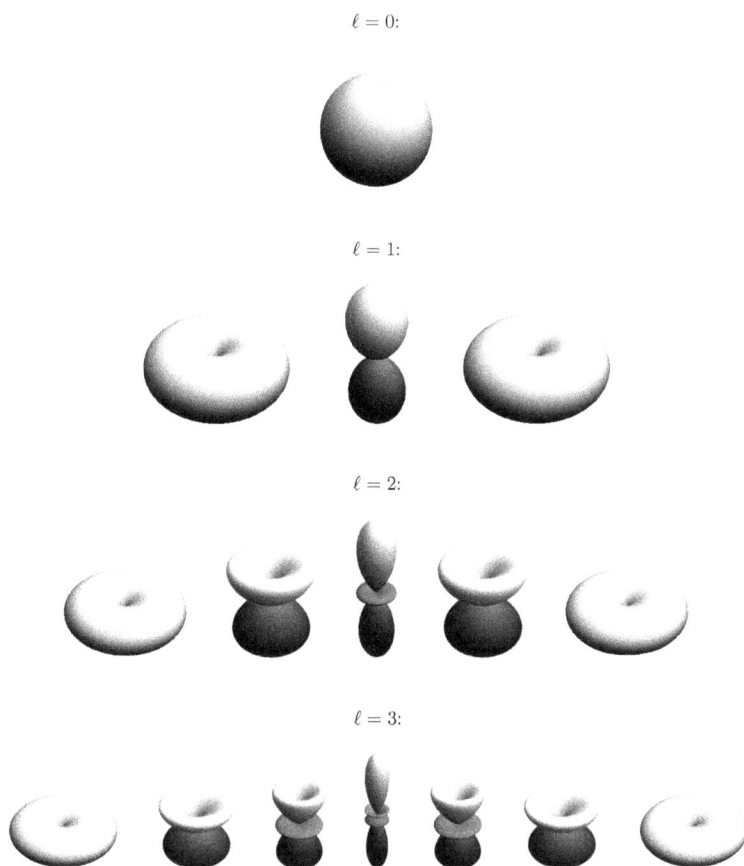

Fig. 2.2 We illustrate the shape of the spherical harmonic functions $Y_{\ell m}(\theta, \varphi)$ for $\ell = 0, 1, 2, 3$ by plotting $f(\theta, \varphi) = |Y_{\ell m}(\theta, \varphi)|^2$ as a function of the polar angle θ and of the azimuth angle φ. The sequence of magnetic components is from $m = -\ell$ to $m = \ell$.

with arbitrary constants $c_{\ell,m}$ and $d_{\ell,m}$, which are related to the $a_{\ell,m}$ and $b_{\ell,m}$ used in Eq. (2.43) by the prefactor given in Eq. (2.53). For convenience, we recall the Poisson equation (2.22) fulfilled by the Green function, which reads as

$$\vec{\nabla}^2 G(\vec{r}, \vec{r}') = -\frac{1}{\epsilon_0} \delta^{(3)}(\vec{r} - \vec{r}').\tag{2.66}$$

In order to solve this equation, we recall that the Dirac-δ function on the right-hand side of Eq. (2.66) vanishes everywhere except for the immediate vicinity of the region where $\vec{r} - \vec{r}'$. Hence, we can hope to construct a solution of Eq. (2.66) by concatenating solutions of the homogeneous equation $\vec{\nabla}^2 \Phi = 0$, given in Eq. (2.43), using a method previously outlined in Eqs. (2.16)–(2.19).

However, before we engage in this endeavor, we should first clarify the explicit form of the three-dimensional Dirac-δ function used in Eq. (2.66), in spherical coordinates. In passing, we shall obtain an *a posteriori* justification for the form of the Dirac-δ function on the unit sphere, given in Eq. (2.62). In order to express the three-dimensional Dirac-δ function in spherical coordinates, we use the *ansatz*

$$\delta^{(3)}(\vec{r} - \vec{r}') = F(r,\theta,\varphi)\,\delta(r - r')\,\delta(\theta - \theta')\,\delta(\varphi - \varphi'),\tag{2.67}$$

which is justified because the coordinates r, θ, and φ describe the position uniquely, just like x, y and z. We fix the function F by the condition

$$1 = \int \mathrm{d}^3r\,\delta^{(3)}(\vec{r} - \vec{r}') = \int \mathrm{d}r\,r^2\,\mathrm{d}\theta\,\sin\theta\,\mathrm{d}\varphi\,F(r,\theta,\varphi)\,\delta(r - r')\,\delta(\theta - \theta')\,\delta(\varphi - \varphi').\tag{2.68}$$

If this relation is to hold for all r', θ', and φ', then we must have

$$F(r,\theta,\varphi) = \frac{1}{r^2\,\sin\theta}.\tag{2.69}$$

Finally, the three-dimensional Dirac-δ function is expressed as

$$\begin{aligned}
\delta^{(3)}(\vec{r} - \vec{r}') &= \frac{1}{r^2\,\sin\theta}\,\delta(r - r')\,\delta(\theta - \theta')\,\delta(\varphi - \varphi') \\
&= \frac{1}{r^2}\,\delta(r - r')\,\delta(\cos\theta - \cos\theta')\,\delta(\varphi - \varphi'),
\end{aligned}\tag{2.70}$$

where again, use is made of Eq. (1.34). We now re-examine the defining equation for the Green function (2.22), which is equivalent to the Poisson equation (2.21) with a "charge distribution" (in incorrect physical dimension) $\rho(\vec{r}) = \epsilon_0\,\delta^{(3)}(\vec{r} - \vec{r}')$ located at the point $\vec{r}'$. We recall that the Laplace and Poisson equations are defined in Eqs. (2.21) and (2.20).

In spherical coordinates and with the convention

$$G_{\vec{r}'}(r,\theta,\varphi) \equiv G(\vec{r} - \vec{r}') = G(\vec{r},\vec{r}'),\tag{2.71}$$

we have

$$\vec{\nabla}^2\,G(\vec{r} - \vec{r}') = \vec{\nabla}^2\,G_{\vec{r}'}(r,\theta,\varphi) = -\frac{1}{\epsilon_0\,r^2\,\sin\theta}\,\delta(r - r')\,\delta(\theta - \theta')\,\delta(\varphi - \varphi').\tag{2.72}$$

The subscript $\vec{r}'$ on $G_{\vec{r}'}(r,\theta,\varphi)$ is meant to indicate that the Green function G is proportional to the electrostatic potential generated by a point charge located at $\vec{r}'$, measured at the point $\vec{r}$ whose spherical coordinates are given by r, θ, and φ. Because the angular component of the Laplace operator takes a particularly simple form when acting on a spherical harmonic, we use the following *ansatz* for the Green function,

$$G_{\vec{r}'}(r,\theta,\varphi) = \sum_{\ell=0}^{\infty}\sum_{m=-\ell}^{\ell} f_{\ell m}(r;r',\theta',\varphi')\,Y_{\ell m}(\theta,\varphi).\tag{2.73}$$

Inserting this into the defining equation (2.22), using Eqs. (2.3.2) and (2.35), we obtain

$$\begin{aligned}
\sum_{\ell m}\left(\frac{\partial^2}{\partial r^2} + \frac{2}{r}\frac{\partial}{\partial r} - \frac{\ell(\ell+1)}{r^2}\right) &f_{\ell m}(r;r',\theta',\varphi')Y_{\ell m}(\theta,\varphi) \\
&= -\frac{1}{\epsilon_0\,r^2}\,\delta(r - r')\,\delta(\cos\theta - \cos\theta')\,\delta(\varphi - \varphi').
\end{aligned}\tag{2.74}$$

One may wonder how it is possible that the right-hand side of this equation includes Dirac-δ functions in the angular variables while the left-hand side does not. In fact, a further *ansatz* for the angular dependence of $f_{\ell m}(r; r', \theta', \varphi')$, given in Eq. (2.79), implies that the solution for $f_{\ell m}(r; r', \theta', \varphi')$ includes a spherical harmonic $Y^*_{\ell m}(\theta', \varphi')$. The completeness relation (2.62) implies that the summation over ℓ and m generates a term which is equivalent to a Dirac-δ function in the angular variables, on the left-hand side of Eq. (2.74), corresponding to the equivalent term on the right-hand side. We thus multiply both sides of Eq. (2.74) by $Y^*_{\ell' m'}(\theta, \varphi)$, and integrate over the solid angle

$$\int d\Omega = \int_{-1}^{1} d(\cos\theta) \int_{0}^{2\pi} d\varphi = \int_{0}^{\pi} d\theta \sin\theta \int_{0}^{2\pi} d\varphi. \tag{2.75}$$

Using the orthonormality of the spherical harmonics, we find that

$$\frac{1}{r^2} \sum_{\ell=0}^{\infty} \sum_{m=-\ell}^{\ell} \left[\left(\frac{\partial}{\partial r} r^2 \frac{\partial}{\partial r} - \ell(\ell+1) \right) f_{\ell m}(r; r', \theta', \varphi') \right] \delta_{\ell \ell'} \delta_{m m'}$$
$$= -\frac{q}{\epsilon_0 \, r'^2} \delta(r - r') \int_{-1}^{1} d(\cos\theta) \int_{0}^{\pi} d\varphi \, \delta(\varphi - \varphi') \, \delta(\cos\theta - \cos\theta') \, Y^*_{\ell' m'}(\theta, \varphi), \tag{2.76}$$

where we use the operator identity

$$\frac{\partial^2}{\partial r^2} + \frac{2}{r} \frac{\partial}{\partial r} = \frac{1}{r^2} \frac{\partial}{\partial r} r^2 \frac{\partial}{\partial r}. \tag{2.77}$$

After the θ and φ integrations and the use of the Kronecker symbols, we obtain

$$\frac{1}{r^2} \left[\left(\frac{\partial}{\partial r} r^2 \frac{\partial}{\partial r} - \ell'(\ell'+1) \right) f_{\ell' m'}(r; r', \theta', \varphi') \right] = -\frac{1}{\epsilon_0 \, r'^2} \delta(r - r') \, Y^*_{\ell' m'}(\theta', \varphi'). \tag{2.78}$$

It is suggestive to assume that the angular dependence of $f_{\ell m}(r; r', \theta', \varphi')$ is given by $Y^*_{\ell m}(\theta', \varphi')$. We now redefine $\ell' \to \ell$ and $m' \to m$ and write

$$f_{\ell m}(r; r', \theta', \varphi') = G_\ell(r, r') \, Y^*_{\ell m}(\theta', \varphi'). \tag{2.79}$$

Our *ansatz* (2.73) now looks like

$$G(\vec{r}, \vec{r}') = G_{\vec{r}'}(r, \theta, \varphi) = \sum_{\ell=0}^{\infty} \sum_{m=-\ell}^{\ell} G_\ell(r, r') \, Y_{\ell m}(\theta, \varphi) \, Y^*_{\ell m}(\theta', \varphi'), \tag{2.80}$$

where the radial equation that must be satisfied by the G_ℓ thus reads

$$\frac{1}{r^2} \left(\frac{\partial}{\partial r} r^2 \frac{\partial}{\partial r} - \ell(\ell+1) \right) G_\ell(r, r') = -\frac{1}{\epsilon_0 \, r^2} \delta(r - r'). \tag{2.81}$$

We now solve Eq. (2.81) by the method outlined previously in Eqs. (2.16)–(2.19), obtaining solutions to the homogeneous equations for $r < r'$ and $r > r'$. Finally, we render $G_\ell(r, r')$ continuous at $r = r'$, and satisfy the condition placed on $G_\ell(r, r')$ obtained by integrating Eq. (2.81) from $r = r' - \varepsilon$ to $r = r' + \varepsilon$. In Sec. 2.3.2, we have already seen that the solution to the homogeneous equation in r has the general form

$$G_\ell(r, r') = \alpha_\ell \, r^\ell + \beta_\ell \, r^{-\ell-1}, \tag{2.82}$$

where α_ℓ and β_ℓ may depend on r'. Assuming regularity of the Green function for large and small arguments (boundary conditions), we choose the solutions as follows: For $r < r'$, we choose $G_\ell(r, r') = \alpha_\ell r^\ell$, and for $r > r'$, we have $G_\ell(r, r') = \beta_\ell r^{-\ell-1}$. Continuity at $r = r'$ requires that

$$G_\ell(r, r') = a_\ell \frac{r_<^\ell}{r_>^{\ell+1}} = \begin{cases} a_\ell \dfrac{r^\ell}{r'^{\ell+1}}, & r < r' \\[2mm] a_\ell \dfrac{r'^\ell}{r^{\ell+1}}, & r > r' \end{cases} \tag{2.83}$$

where $r_< = \min(r, r')$, and $r_> = \max(r, r')$. The constant a_ℓ needs to be determined. Integrating Eq. (2.81) from $r = r' - \varepsilon$ to $r = r' + \varepsilon$, after multiplying both sides by r^2, we obtain

$$\int_{r'-\varepsilon}^{r'+\varepsilon} \left[\frac{\partial}{\partial r} \left(r^2 \frac{\partial}{\partial r} G_\ell(r, r') \right) - \ell(\ell+1) G_\ell(r, r') \right] dr = -\frac{1}{\epsilon_0} \int_{r'-\varepsilon}^{r'+\varepsilon} \delta(r - r') \, dr. \tag{2.84}$$

The second term on the left is continuous at $r = r'$ and drops out of the equation in the limit $\epsilon \to 0$, and so

$$r^2 \frac{\partial}{\partial r} G_\ell(r, r') \Big|_{r=r'+\varepsilon} - r^2 \frac{\partial}{\partial r} G_\ell(r, r') \Big|_{r=r'-\varepsilon} = -\frac{1}{\epsilon_0}, \tag{2.85}$$

where we ignore higher-order terms in ε. The radial Green function $G_\ell(r, r')$ has to be continuous at the cusp $r = r'$, but the derivative may have a discontinuity. We can thus approximate $r^2 \approx r'^2$ in the prefactors in Eq. (2.85) and write

$$r'^2 \left(\frac{\partial}{\partial r} G_\ell(r, r') \Big|_{r=r'+\varepsilon} - \frac{\partial}{\partial r} G_\ell(r, r') \Big|_{r=r'-\varepsilon} \right) = -\frac{1}{\epsilon_0}, \tag{2.86}$$

where of course higher-order terms in ε have been neglected. For $r = r' + \varepsilon$, r is a little greater than r', so $r = r_>$, and we have to use the functional form $r^{-\ell-1}$ for G_ℓ, whereas for $r = r' - \varepsilon$, we have $r = r_<$, and we need to use the functional form r^ℓ for G_ℓ. Thus, the condition (2.86) translates into the equation

$$r'^2 a_\ell \left(\frac{\partial}{\partial r} \frac{r'^\ell}{r^{\ell+1}} \Big|_{r=r'} - \frac{\partial}{\partial r} \frac{r^\ell}{r'^{\ell+1}} \Big|_{r=r'} \right) = -\frac{1}{\epsilon_0}, \tag{2.87}$$

and therefore

$$r'^2 a_\ell \left[-(\ell+1) \frac{r'^\ell}{r'^{\ell+2}} - \ell \frac{r'^{\ell-1}}{r'^{\ell+1}} \right] = a_\ell \left(-(\ell+1) - \ell \right) = -a_\ell (2\ell+1) = -\frac{1}{\epsilon_0}. \tag{2.88}$$

We finally obtain the prefactor in Eq. (2.83) as

$$a_\ell = \frac{1}{\epsilon_0 (2\ell+1)}, \tag{2.89}$$

and find that

$$G_\ell(r, r') = \frac{1}{\epsilon_0 (2\ell+1)} \frac{r_<^\ell}{r_>^{\ell+1}}, \qquad r_< = \min(r, r'), \qquad r_> = \max(r, r'). \tag{2.90}$$

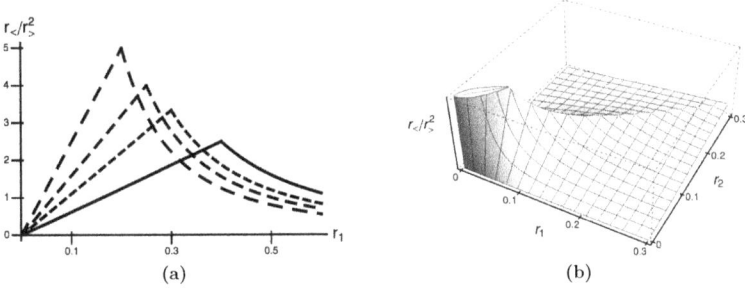

Fig. 2.3 Panel (a) has a plot of the function $f(r_1, r_2) = r_</r_>^2$ as a function of r_1 for $r_2 = 0.2$ (long dashes), $r_2 = 0.25$ (medium dashes), $r_2 = 0.3$ (short dashes) and $r_2 = 0.4$ (solid line). The maximum is reached at $r_1 = r_2$, with $f(r_1, r_1) = 1/r_1$. This implies that for $r_2 = 0.2$, the maximum (kink) is at $f = 5$. In panel (b), we have a plot of the function $f(r_1, r_2) = r_</r_>^2$ in the range $0 < r_1 < 0.3$ and $0 < r_2 < 0.3$.

Our *ansatz* (2.73) with the intermediate result (2.80) and the result (2.90) finally leads to the rather compact formula

$$G(\vec{r}, \vec{r}') = G_{\vec{r}'}(r, \theta, \varphi) = \frac{1}{\epsilon_0} \sum_{\ell=0}^{\infty} \sum_{m=-\ell}^{\ell} \frac{1}{2\ell+1} \frac{r_<^\ell}{r_>^{\ell+1}} Y_{\ell m}(\theta, \varphi) Y_{\ell m}^*(\theta', \varphi'), \quad (2.91)$$

confirming Eq. (2.25). On the other hand, we know that

$$G(\vec{r} - \vec{r}') = G_{\vec{r}'}(r, \theta, \varphi) = \frac{1}{4\pi\epsilon_0 |\vec{r} - \vec{r}'|}. \quad (2.92)$$

Combining Eqs. (2.91) and (2.92), we can thus write the formula

$$\frac{1}{|\vec{r} - \vec{r}'|} = \sum_{\ell=0}^{\infty} \sum_{m=-\ell}^{\ell} \frac{4\pi}{2\ell+1} \frac{r_<^\ell}{r_>^{\ell+1}} Y_{\ell m}(\theta, \varphi) Y_{\ell m}^*(\theta', \varphi'), \quad (2.93)$$

which is the basis for the so-called multipole expansion in electrostatics. This is a good point to include a specific remark. Namely, the terms in ascending order of ℓ are canonically referred to as the monopole ($\ell = 0$), dipole ($\ell = 1$), quadrupole ($\ell = 2$), octupole ($\ell = 3$), and hexadecupole ($\ell = 4$) terms. The designation is inspired by the numerals for the 0, 1, 4, 8 and 16-pole contributions. One may easily convince oneself that in order to generate a 2^ℓ pole term, one has to form a symmetric arrangement of positive and negative charges $\pm q$, a linear distance d apart, and a finite 2^ℓ pole is obtained upon letting $d \to 0$, keeping the quantity qd^ℓ constant.

The case-by-case differentiation according to $r_<$ and $r_>$ for $\ell = 1$ is illustrated in Fig. 2.3. The kink at the turning point $r = r'$ is a characteristic property of radial Green Functions.

2.3.5 *Multipole Expansion of Charge Distributions*

From the multipole expansion of the Green function [see Eq. (2.91)],

$$G(\vec{r}, \vec{r}') = \frac{1}{4\pi\epsilon_0 |\vec{r} - \vec{r}'|} = \frac{1}{\epsilon_0} \sum_{\ell=0}^{\infty} \sum_{m=-\ell}^{\ell} \frac{1}{2\ell+1} \frac{r_<^\ell}{r_>^{\ell+1}} Y_{\ell m}(\theta, \varphi) Y_{\ell m}^*(\theta', \varphi'), \quad (2.94)$$

and the formula

$$\Phi(\vec{r}) = \int d^3r' \, G(\vec{r},\vec{r}') \, \rho(\vec{r}'), \qquad (2.95)$$

one can derive the multipole expansion of an electrostatic potential. If the charge distribution is localized and we calculate the potential outside of the localized distribution ($r > r'$), then the relations $r_< = r'$ and $r_> = r$ are simultaneously fulfilled, and we may express $\Phi(\vec{r})$ as

$$\Phi(\vec{r}) = \frac{1}{\epsilon_0} \sum_{\ell=0}^{\infty} \sum_{m=-\ell}^{\ell} \frac{1}{2\ell+1} \frac{Y_{\ell m}(\theta,\varphi)}{r^{l+1}} \left(\int d^3r' \, \rho(\vec{r}') \, r'^{\ell} \, Y_{\ell m}^*(\theta',\varphi') \right). \qquad (2.96)$$

The integral in round brackets in this expression defines the multipole moments $q_{\ell m}$ of the charge distribution,

$$q_{\ell m} = \int d^3r \, \rho(\vec{r}) \, r^{\ell} \, Y_{\ell m}^*(\theta,\varphi). \qquad (2.97)$$

Furthermore, because of the relationship $Y_{\ell-m}(\theta,\varphi) = (-1)^m \, Y_{\ell m}^*(\theta,\varphi)$, it follows that

$$q_{\ell-m} = (-1)^m \, q_{\ell m}^*. \qquad (2.98)$$

The potential is given as a sum over the multipole moments of the localized charge distribution, as follows,

$$\Phi(\vec{r}) = \frac{1}{\epsilon_0} \sum_{\ell=0}^{\infty} \sum_{m=-\ell}^{\ell} \frac{q_{\ell m}}{2\ell+1} \frac{Y_{\ell m}(\theta,\varphi)}{r^{l+1}}. \qquad (2.99)$$

The set of elements $\{q_{\ell m}; \; m = -\ell, \, -\ell+1, \, \ldots, \, \ell-1, \, \ell\}$ form a spherical tensor of rank ℓ. There are two common representations of tensors: the familiar rectilinear representation (in Cartesian coordinates) and the spherical representation.

A discussion of two example cases for the multipole decomposition of charge distributions in spherical coordinates is now in order. The main use of the multipole decomposition (2.99) lies in the calculation of potentials outside the generating charge distributions which are centered around the origin; i.e., one assumes that the coordinate r' always is the lesser radial coordinate as compared to the observation point r. In consequence, results obtained using Eq. (2.99) are valid only for $r > R$, where R is a radial variable that characterizes the extension of the charge distribution.

As a first example (see Fig. 2.4), let us consider a charge distribution

$$\rho(\vec{r}') = \frac{Q}{a^2} \cos\theta' \, \delta(r'-a), \qquad \cos\theta' = \sqrt{\frac{4\pi}{3}} \, Y_{10}(\theta',\varphi'), \qquad (2.100)$$

where we have identified the cosine as a spherical harmonic. This charge distribution describes a dipole-shaped charge distribution centered on a sphere. Because of the orthogonality of the spherical harmonics, the only nonvanishing multipole component is

$$q_{10} = \int d^3r' \, \rho(\vec{r}') \, r' \, Y_{10}^*(\theta',\varphi') = \sqrt{\frac{4\pi}{3}} \, Q \, a, \qquad (2.101)$$

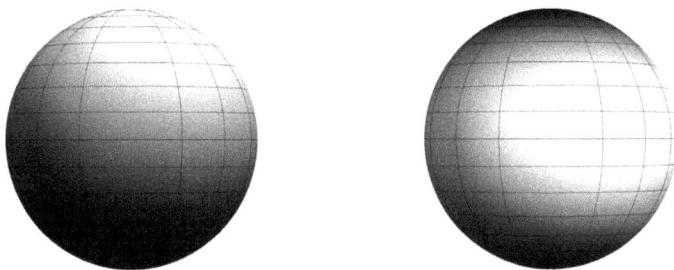

Fig. 2.4 The left plot shows a sphere with a charge distribution $\rho(\vec{r}') = \dfrac{Q}{a^2} \cos\theta'\, \delta(r'-a)$, proportional to $\cos\theta'$. Positively charged areas are lighter than negatively charged areas. The right plot displays a sphere with a charge distribution $\rho(\vec{r}') = \dfrac{Q}{a^2} \sin^2\theta'\, \delta(r'-a)$.

which has the correct physical dimension (charge × length) of a dipole moment. In terms of the multipole moments of the charge distribution, the potential for $r > a$ is found to be

$$\Phi(\vec{r}) = \frac{1}{\epsilon_0} \sum_{\ell=0}^{\infty} \sum_{m=-\ell}^{\ell} \frac{q_{\ell m}}{2\ell+1} \frac{Y_{\ell m}(\theta,\varphi)}{r^{\ell+1}} = \frac{1}{\epsilon_0} \frac{q_{10}}{3} \frac{Y_{10}(\theta,\varphi)}{r^2} = \frac{q_{10}}{3\epsilon_0 r^2} \sqrt{\frac{3}{4\pi}} \cos\theta,$$

$$(2.102)$$

and thus, in view of Eq. (2.101),

$$\Phi(\vec{r}) = \frac{Q\,a}{3\epsilon_0 r^2} \cos\theta. \qquad (2.103)$$

As a second example (see also Fig. 2.4), we consider the charge distribution

$$\rho(\vec{r}) = \frac{Q}{a^2} \sin^2\theta\, \delta(r-a). \qquad (2.104)$$

In this case, the potential involves a monopole and a quadrupole term. Calculating the moments q_{00} and q_{20}, we may convince ourselves that the dipole moments vanish, and we obtain the result as a sum of a monopole and a quadrupole term,

$$\Phi(\vec{r}) = \frac{2Q}{3\,\epsilon_0\, r} - \frac{Q\,a^2}{15\,\epsilon_0\, r^3} \left(3\cos^2\theta - 1\right). \qquad (2.105)$$

2.3.6 *Addition Theorem for Spherical Harmonics*

Suppose that in the expansion (2.93), which we recall for convenience,

$$\frac{1}{|\vec{r}-\vec{r}'|} = \sum_{\ell=0}^{\infty} \sum_{m=-\ell}^{\ell} \frac{4\pi}{2\ell+1} \frac{r_<^\ell}{r_>^{\ell+1}} Y_{\ell m}(\theta,\varphi)\, Y_{\ell m}^*(\theta',\varphi'), \qquad (2.106)$$

we let $\vec{r}' = r'\hat{e}_z$, which is tantamount to letting $\theta' = 0$. Only the $m = 0$ terms will appear in the sum because $Y_{\ell m}(0,\varphi')$ vanishes for $m \neq 0$. This holds because the associated Legendre polynomials $P_\ell^m(\cos\theta = 1)$ vanish except for the case $m = 0$, where $P_\ell^0(\cos\theta) = P_\ell(\cos\theta)$. In this case, we use the expression for $Y_{\ell m}(\theta, \varphi)$. In view of the equality $P_\ell(1) = P_\ell^0(1) = 1$, we have

$$\frac{1}{|\vec{r} - r'\,\hat{e}_z|} = \sum_{\ell=0}^{\infty} \frac{4\pi}{2\ell+1} \frac{r_<^\ell}{r_>^{\ell+1}} \, Y_{\ell 0}(\theta, \varphi) \, Y_{\ell 0}^*(0, \varphi')$$

$$= \sum_{\ell=0}^{\infty} \frac{4\pi}{2\ell+1} \frac{r_<^\ell}{r_>^{\ell+1}} \sqrt{\frac{2\ell+1}{4\pi}} \, P_\ell(\cos\theta) \sqrt{\frac{2\ell+1}{4\pi}} \, P_\ell(1) = \sum_{\ell=0}^{\infty} \frac{r_<^\ell}{r_>^{\ell+1}} \, P_\ell(\cos\theta) . \tag{2.107}$$

However, we also have the relation

$$\frac{1}{|\vec{r} - \vec{r}'|} = \frac{1}{\sqrt{r^2 + r'^2 - 2\,r\,r'\,\cos\theta}} , \tag{2.108}$$

where θ is the polar angle of $\vec{r}$, which is equal to the angle $\angle(\vec{r}, \vec{r}')$ between the unit vector $\hat{r}$ and $\hat{r}'$ because $\vec{r}' = r'\hat{e}_z$. We can now use the derived relations in two ways. First, we observe that we have chosen $\vec{r}'$ to lie along the z axis somewhat arbitrarily. We can thus conclude that, more generally,

$$\frac{1}{|\vec{r} - \vec{r}'|} = \frac{1}{\sqrt{r^2 + r'^2 - 2\,r\,r'\,\cos\angle(\vec{r}, \vec{r}')}} = \sum_{\ell=0}^{\infty} \frac{r_<^\ell}{r_>^{\ell+1}} \, P_\ell(\cos\angle(\vec{r}, \vec{r}')) , \tag{2.109}$$

where $\cos\angle(\vec{r}, \vec{r}') = \hat{r} \cdot \hat{r}'$. Here, $\hat{r}$ is the unit vector pointing in the direction of $\vec{r}$. The angle γ connects the position vector locating the charge, $\vec{r}'$, and the vector locating the observation point, $\vec{r}$. Second, the original multipole decomposition must still hold for the expression $1/|\vec{r} - \vec{r}'|$,

$$\frac{1}{|\vec{r} - \vec{r}'|} = \sum_{\ell=0}^{\infty} \sum_{m=-\ell}^{\ell} \frac{4\pi}{2\ell+1} \frac{r_<^\ell}{r_>^{\ell+1}} \, Y_{\ell m}(\theta, \varphi) \, Y_{\ell m}^*(\theta', \varphi') . \tag{2.110}$$

The equality of (2.109) and (2.110) has to hold for each ℓ component separately, and we finally obtain the addition theorem for spherical harmonics, which reads

$$P_\ell(\cos\angle(\hat{r}, \hat{r}')) = \frac{4\pi}{2\ell+1} \sum_{m=-\ell}^{\ell} Y_{\ell m}(\theta, \varphi) \, Y_{\ell m}^*(\theta', \varphi') . \tag{2.111}$$

By expressing all direction cosines in spherical coordinates, we also obtain

$$\cos\angle(\hat{r}, \hat{r}') = \hat{r} \cdot \hat{r}' = \sin\theta \, \sin\theta' \, \cos(\varphi - \varphi') + \cos\theta \, \cos\theta' . \tag{2.112}$$

2.3.7 Multipole Expansion in Cartesian Coordinates

The multipole decomposition, in Cartesian coordinates, is based on the expansion

$$\frac{1}{|\vec{r} - \vec{r}'|} = \frac{1}{r} + \frac{\vec{r} \cdot \vec{r}'}{r^3} + \frac{3(\vec{r} \cdot \vec{r}')^2 - r^2\,r'^2}{2r^5} + \mathcal{O}(r^{-4}) . \tag{2.113}$$

This leads to the following expansion for the potential,

$$\Phi(\vec{r}) = \frac{1}{4\pi\epsilon_0} \int d^3r' \frac{1}{|\vec{r} - \vec{r}'|} \rho(\vec{r}')$$

$$\approx \frac{1}{4\pi\epsilon_0} \int d^3r' \left(\frac{1}{r} + \frac{\vec{r} \cdot \vec{r}'}{r^3} + \frac{3(\vec{r} \cdot \vec{r}')^2 - r^2 r'^2}{2r^5} + \cdots \right) \rho(\vec{r}')$$

$$\approx \frac{Q}{4\pi\epsilon_0\, r} + \frac{\vec{r} \cdot \vec{p}}{4\pi\epsilon_0\, r^3} + \frac{1}{4\pi\epsilon_0} \frac{Q_{ij}\left(3r_i r_j - \delta_{ij}\, r^2\right)}{2r^5} + \cdots, \tag{2.114}$$

where the total charge Q, the dipole moment vector $\vec{p}$, and the components of the quadrupole tensor Q_{ij} are given as

$$Q = \int d^3r' \rho(\vec{r}'), \tag{2.115a}$$

$$\vec{p} = \int d^3r' \vec{r}' \rho(\vec{r}'), \tag{2.115b}$$

$$Q_{ij} = \int d^3r' \left(3r'_i r'_j - \delta_{ij} \vec{r}'^2\right) \rho(\vec{r}'). \tag{2.115c}$$

Let us focus on the dipole term, which is given by the term with $\ell = 1$ in Eq. (2.99),

$$\Phi(\vec{r})|_{\ell=1} \sim \frac{1}{\epsilon_0} \sum_{m=-1}^{1} \frac{1}{3} \frac{Y_{\ell m}(\theta, \varphi)}{r^2} \left(\int d^3r'\, r'\, \rho(\vec{r}')\, Y_{\ell m}^*(\theta', \varphi') \right), \tag{2.116}$$

where we have already inserted the definition of the multipole moment according to Eq. (2.97). The addition theorem (2.111) implies that for $\ell = 1$,

$$P_1\left(\cos \angle(\vec{r}, \vec{r}')\right) = \hat{r} \cdot \hat{r}' = \frac{4\pi}{3} \sum_{m=-1}^{1} Y_{1m}(\theta, \varphi)\, Y_{1m}^*(\theta', \varphi'). \tag{2.117}$$

Inserting this relationship into Eq. (2.116), one obtains

$$\Phi(\vec{r})|_{\ell=1} \sim \frac{1}{3\,\epsilon_0\, r^2} \left(\int d^3r' \frac{3}{4\pi} \hat{r} \cdot \hat{r}'\, r'\, \rho(\vec{r}') \right)$$

$$\sim \frac{1}{4\pi\epsilon_0} \frac{1}{r^3} \left(\int d^3r' \vec{r} \cdot \vec{r}'\, \rho(\vec{r}') \right) = \frac{\vec{r} \cdot \vec{p}}{4\pi\epsilon_0\, r^3}. \tag{2.118}$$

Equations (2.116) and (2.118) express the conversion of the dipole term from spherical to Cartesian coordinates.

It remains to clarify the relation of the components of the dipole moment vector, in the Cartesian versus the spherical representation. Indeed, in the spherical representation, we find that

$$q_{1\pm 1} = \mp\sqrt{\frac{3}{8\pi}}\, (p_1 \mp i p_2), \qquad q_{10} = \sqrt{\frac{3}{4\pi}}\, p_3. \tag{2.119}$$

We here appeal to the fact that the multipole moments are defined with the complex conjugate of the spherical harmonic entering the integrand, according to Eq. (2.97).

The quadrupole term, by contrast, is given by the term with $\ell = 2$ in Eq. (2.99),

$$\Phi\left(\vec{r}\right)\big|_{\ell=2} \sim \frac{1}{\epsilon_0} \sum_{m=-2}^{2} \frac{1}{5} \frac{Y_{2m}\left(\theta,\varphi\right)}{r^3} \left(\int d^3r'\, r'^2\, \rho\left(\vec{r}'\right) Y_{\ell m}^*\left(\theta',\varphi'\right)\right). \qquad (2.120)$$

The addition theorem implies that

$$P_2\left(\cos\angle\left(\vec{r},\vec{r}'\right)\right) = \frac{1}{2}\left(3(\hat{r}\cdot\hat{r}')^2 - 1\right) = \frac{4\pi}{5}\sum_{m=-2}^{2} Y_{2m}\left(\theta,\varphi\right) Y_{2m}^*\left(\theta',\varphi'\right). \qquad (2.121)$$

Inserting this relationship into Eq. (2.120), one obtains

$$\Phi\left(\vec{r}\right)\big|_{\ell=2} \sim \frac{1}{\epsilon_0}\frac{1}{5}\left(\int d^3r'\, r'^2\, \rho\left(\vec{r}'\right)\frac{5}{4\pi}\frac{1}{2}\left[3(\hat{r}\cdot\hat{r}')^2 - 1\right]\right)$$

$$= \frac{1}{4\pi\epsilon_0}\frac{1}{2r^5}\left(\int d^3r'\, r'^2\, \rho\left(\vec{r}'\right)\left(3(\vec{r}\cdot\vec{r}')^2 - \vec{r}^2\, \vec{r}'^2\right)\right)$$

$$= \frac{1}{4\pi\epsilon_0}\frac{Q_{ij}\left(3r_i\, r_j - \delta_{ij}\,\vec{r}^2\right)}{2r^5}. \qquad (2.122)$$

Equations (2.120) and (2.122) variously express the quadrupole term from the charge distribution, in spherical versus Cartesian coordinates.

Let us count. We have five components of the quadrupole tensor q_{2m} with $m = -2,\ldots,2$. The quadrupole tensor in Cartesian coordinates has the form

$$(Q_{ij}) = \begin{pmatrix} Q_{11} & Q_{12} & Q_{13} \\ Q_{21} & Q_{22} & Q_{32} \\ Q_{31} & Q_{32} & Q_{33} \end{pmatrix} = \begin{pmatrix} Q_{11} & Q_{12} & Q_{13} \\ Q_{12} & Q_{22} & Q_{32} \\ Q_{13} & Q_{2l} & -Q_{11}-Q_{22} \end{pmatrix}, \qquad (2.123)$$

where we have used the symmetry $Q_{ij} = Q_{ji}$ and the tracelessness of the quadrupole tensor $\sum_{i=1}^{3} Q_{ii} = 0$ which holds in view of $\sum_{i=1}^{3}\delta_{ii} = 3$. Thus, because the quadrupole moment of the charge distribution is a symmetric, traceless, second-rank tensor, it is therefore specified by five parameters, as the second form in Eq. (2.123) suggests.

The simplest relationship occurs for q_{20} and Q_{33}. This simplicity is obtained because the spherical tensors we are using have taken the z axis as a preferred axis,

$$Q_{33} = \int\left(3\,z^2 - r^2\right)\rho\left(\vec{r}\right)d^3r = \int\left(3\cos^2\theta - 1\right)r^2\,\rho\left(\vec{r}\right)d^3r$$

$$= \sqrt{\frac{16\pi}{5}}\int\sqrt{\frac{5}{16\pi}}\left(3\cos^2\theta - 1\right)r^2\,\rho\left(\vec{r}\right)d^3r$$

$$= 4\sqrt{\frac{\pi}{5}}\int Y_{20}(\theta,\varphi)\,r^2\,\rho\left(\vec{r}\right)d^3r = 4\sqrt{\frac{\pi}{5}}\,q_{20}. \qquad (2.124)$$

2.3.8   *Multipole Expansion in an External Field*

Up to now, we have considered the multipole decomposition in spherical coordinates, and we have considered the original charge distribution to be centered about the point $\vec{r}' \approx 0$, with a distant observation point $\vec{r}$. An expansion of $1/|\vec{r} - \vec{r}'|$ into multipoles (spherical coordinates) then leads to the multipole decomposition. The total charge of the charge distribution that generates the potential at $\vec{r}$ is the monopole.

In order to switch the formalism of the multipole decomposition to Cartesian coordinates, we consider an opposite situation, in some sense. A charge distribution is centered about the point $\vec{r} = \vec{0}$, with the potential being generated by a distant charge distribution that is centered about the point $\vec{r}' \neq \vec{0}$. The distant charge distribution generates a potential $\Phi(\vec{r})$ which interacts with the charge distribution $\rho(\vec{r})$. The potential energy of the configuration is given by the overlap integral

$$W = \int \rho(\vec{r})\, \Phi_{\text{ext}}(\vec{r})\, d^3r \,, \tag{2.125}$$

$$\Phi_{\text{ext}}(\vec{r}) = \frac{1}{4\pi\epsilon_0} \int d^3r' \frac{1}{|\vec{r} - \vec{r}'|} \rho_{\text{ext}}(\vec{r}') \,, \tag{2.126}$$

where $\rho_{\text{ext}}(\vec{r}')$ is the "external" charge distribution, which is located at a large distance from the origin, while the "probe" charge distribution $\rho(\vec{r})$ is located near $\vec{r} \approx \vec{0}$. Hence, in particular,

$$\vec{\nabla}^2 \Phi_{\text{ext}}(\vec{r})\big|_{\vec{r}=0} = \sum_{i=1}^{3}\sum_{j=1}^{3} \delta_{ij} \frac{\partial^2 \Phi_{\text{ext}}(\vec{r})}{\partial r_i \partial r_j}\bigg|_{\vec{r}=0} = 0 \,, \tag{2.127}$$

because the potential $\Phi_{\text{ext}}(\vec{r})$ is supposed to be due to charges external to the volume in which $\rho(\vec{r})$ is nonzero.

Because the sources of the potential are external to the volume containing the probe charge, the electric potential can be expanded in a Taylor series,

$$\Phi_{\text{ext}}(\vec{r}) = \Phi_{\text{ext}}(0) + \sum_{i=1}^{3} r_i \frac{\partial \Phi_{\text{ext}}(\vec{r}')}{\partial r_i'}\bigg|_{\vec{r}'=\vec{0}} + \frac{1}{2}\sum_{i=1}^{3}\sum_{j=1}^{3} r_i r_j \frac{\partial^2 \Phi_{\text{ext}}(\vec{r}')}{\partial r_i' \partial r_j'}\bigg|_{\vec{r}'=\vec{0}} + \dots . \tag{2.128}$$

Expressing the gradient of the external potential in terms of the electric field,

$$\vec{E}(\vec{0}) = - \frac{\partial \Phi(\vec{r}')}{\partial r_i'}\bigg|_{\vec{r}'=\vec{0}} \,, \tag{2.129}$$

we obtain the expansion of the potential energy in Cartesian coordinates,

$$\begin{aligned} W &= \Phi(\vec{0}) \int d^3r\, \rho(\vec{r}) - \vec{E}(\vec{0}) \cdot \int d^3r\, \vec{r}\, \rho(\vec{r}) \\ &\quad + \frac{1}{2}\sum_{i=1}^{3}\sum_{j=1}^{3} \frac{\partial^2 \Phi(\vec{r}')}{\partial r_i' \partial r_j'}\bigg|_{\vec{r}'=\vec{0}} \int d^3r\, r_i r_j\, \rho(\vec{r}) + \dots \\ &= \Phi(\vec{0})\, Q - \vec{E}(\vec{0}) \cdot \vec{p} + \frac{1}{6}\sum_{i=1}^{3}\sum_{j=1}^{3} \frac{\partial^2 \Phi(\vec{r})}{\partial r_i \partial r_j}\bigg|_{\vec{r}=\vec{0}} Q_{ij} + \dots . \end{aligned} \tag{2.130}$$

The first term is a kind of average potential energy, the second is the interaction of the electric dipole distribution with the applied electric field, and the third term gives the interaction of the quadrupole moment of the charge distribution with the gradient of the electric field. The formula for the total charge Q, the dipole moment vector $\vec{p}$, and the components of the quadrupole tensor Q_{ij}, have been given in Eq. (2.115). Equation (2.127) has been used in order to eliminate the term with $i = j$ in Eq. (2.115c).

2.4 Electrostatic Green Function and Cylindrical Coordinates

2.4.1 *Laplace Equation in Cylindrical Coordinates*

Up to now, we have analyzed the Green function in spherical coordinates, and obtained a result which involves a case-by-case differentiation in terms of the mapping $(r, r') \to (r_< = \min(r, r'), r_> = \min(r, r'))$. It is very interesting to generalize the formalism to cylindrical coordinates. In analogy with Eq. (2.26), let us first define cylindrical coordinates as follows,

$$x = \rho \cos\varphi, \qquad y = \rho \sin\varphi, \qquad z = z, \tag{2.131}$$

with the backtransformation [cf. Eq. (2.26b)] into Cartesian coordinates,

$$\rho = \sqrt{x^2 + y^2}, \qquad \varphi = \arccos\left(\frac{x}{\sqrt{x^2 + y^2}}\right), \qquad z = z. \tag{2.132}$$

The Laplacian reads as

$$\vec{\nabla}^2 = \frac{1}{\rho}\frac{\partial}{\partial\rho}\rho\frac{\partial}{\partial\rho} + \frac{1}{\rho^2}\frac{\partial^2}{\partial\varphi^2} + \frac{\partial^2}{\partial z^2}. \tag{2.133}$$

The homogeneous equation [cf. (2.33)] is

$$\vec{\nabla}^2\Phi(\rho, \varphi, z) = 0. \tag{2.134}$$

As in the case of spherical coordinates, this equation is solved by a separation of variables,

$$\Phi(\rho, \varphi, z) = U(\rho)\, V(\varphi)\, W(z). \tag{2.135}$$

Substituting this *ansatz* into Eq. (2.133), we can separate all variable dependence:

$$\frac{1}{U}\frac{1}{\rho}\frac{d}{d\rho}\rho\frac{d}{d\rho}U + \frac{1}{V}\frac{1}{\rho^2}\frac{d^2}{d\varphi^2}V + \frac{1}{W}\frac{d^2}{dz^2}W = 0. \tag{2.136}$$

The integration constants may be defined as

$$\frac{1}{V}\frac{d^2}{d\varphi^2}V = -m^2, \qquad \frac{1}{W}\frac{d^2}{dz^2}W = k^2. \tag{2.137}$$

Thus, $W(z)$ can be written as a linear combination of two solutions,

$$W(z) = A_k\, e^{kz} + B_k\, e^{-kz}, \tag{2.138}$$

which are regular at $z = -\infty$ and at $z = +\infty$, respectively (provided k is positive). However, if the z domain is confined to a finite interval, then k may be complex. The dependence on the azimuthal angle is given by

$$V(\varphi) = C_m\, e^{im\varphi} + D_m\, e^{-im\varphi}. \tag{2.139}$$

Again, m has to be an integer, because otherwise the function $V(\varphi)$ is not uniquely defined. Using Eq. (2.137) in Eq. (2.136), one may show that the equation fulfilled by U is [see Eq. (3.176)]

$$\left(\rho^2 \frac{\mathrm{d}^2}{\mathrm{d}\rho^2} + \rho \frac{\mathrm{d}}{\mathrm{d}\rho} + k^2 \rho^2 - m^2\right) U(\rho) = 0. \qquad (2.140)$$

This equation is solved by

$$U(\rho) = a_m J_m(k\rho) + b_m Y_m(k\rho), \qquad (2.141)$$

where the J_m and Y_m are called Bessel functions of the first and second kind, respectively.

The regular solution (at $\rho = 0$) is given by the Bessel J functions, while the Bessel Y function (Neumann function) diverges for $\rho \to 0$. The general solution to Eq. (2.134) is thus given as

$$\Phi(\rho, \varphi, z) = \sum_k \sum_{m=-\infty}^{\infty} [a_{k,m} J_m(k\rho) + b_{k,m} Y_m(k\rho)] \, \mathrm{e}^{kz} \, \mathrm{e}^{im\varphi}, \qquad (2.142)$$

where the expansion coefficients are $a_{k,m}$ and $b_{k,m}$. The sum over k entails all permissible values (including negative ones, as determined by the boundary conditions). It would thus be redundant to indicate both terms $\exp(kz)$ and $\exp(-kz)$ in the general solution; they are contained in the sum over k. Typically, the system fills the region from $\varphi = 0$ to $\varphi = 2\pi$. As already stated, the continuity of $V(\varphi)$ requires that m be an integer. The values of k are determined by boundary conditions. If we require the solutions to be periodic in z, then k must be purely imaginary. If the solutions are exponential in z, then k is purely real. Also, the sum over k may be discrete, or continuous, in which case one has to integrate over k.

If k is continuous and the allowed k values are the real numbers, and the solution is required to be regular at the origin, then it can be written as

$$\Phi(\rho, \varphi, z) = \sum_{m=-\infty}^{\infty} \int_0^{\infty} \mathrm{d}k \, \left(a(k, m) J_m(k\rho) \, \mathrm{e}^{kz} \, \mathrm{e}^{im\varphi}\right.$$
$$\left. + b(k, m) J_m(k\rho) \, \mathrm{e}^{-kz} \, \mathrm{e}^{im\varphi}\right), \qquad (2.143)$$

where the discrete subscripts become continuous functions of their arguments and we have denoted the expansion coefficient for positive versus negative k by a and b, respectively. This form will be used later in Eq. (2.184). However, before we get to this point in our discussion, we shall first discuss a number of properties of the Bessel functions, which belong to the most important special functions of mathematical physics.

2.4.2 Cylindrical Coordinates and Bessel Functions

In the case of the representation of the Green function in the spherical basis, the general solution to the homogeneous equation involves Legendre polynomials and spherical harmonics [see Eq. (2.65)]. We recall that we explored properties of the

Legendre polynomials and spherical harmonics in Sec. 2.3.3, before discussing the Green function in spherical coordinates, in Sec. 2.3.4. Let us attempt the same procedure in the cylindrical basis, where the general solution to the homogeneous equation involves BESSEL[1] and NEUMANN[2] functions [see Eq. (2.142)].

Indeed, Bessel's differential equation, which is fulfilled by the Bessel and Neumann functions, reads [see Eq. (10.2.1) of Ref. [5]],

$$\left(z^2 \frac{\partial^2}{\partial z^2} + z \frac{\partial}{\partial z} + z^2 - \nu^2 \right) J_\nu(z) = 0 . \tag{2.144}$$

Here, ν is not necessarily integer-valued (in contrast to m), and the argument z may be complex. The Bessel J function may be defined via the series expansion [see Eq. (10.2.2) of Ref. [5]],

$$J_\nu(z) = \sum_{j=0}^{\infty} \frac{(-1)^j}{j! \, \Gamma(\nu + j + 1)} \left(\frac{z}{2} \right)^{2j+\nu} , \qquad z \in \mathbb{C} . \tag{2.145}$$

which converges for all $z \in \mathbb{C}$. The Gamma function is defined as [see Eq. (5.2.1) of Ref. [5]],

$$\Gamma(x) = \int_0^\infty dt \, t^{x-1} \, e^{-t} , \qquad x > 0 . \tag{2.146}$$

For complex argument, the Gamma function has to be defined by a complex contour integral (see Ref. [6] for a comprehensive discussion). For integer argument, the Gamma function has the property [see Eq. (5.4.1) of Ref. [5]],

$$\Gamma(m + 1) = m! , \qquad m \in \mathbb{N} . \tag{2.147}$$

The complementary solution Y_ν is defined as

$$Y_\nu(z) = \lim_{\alpha \to \nu} \left[\frac{1}{\sin(\alpha\pi)} \left(J_\alpha(z) \cos(\alpha\pi) - J_{-\alpha}(z) \right) \right] , \tag{2.148}$$

where the limit is nontrivial if ν is an integer. The Neumann function Y_ν (Bessel function of the second kind), which is sometimes denoted as N_ν, is fundamentally different from the spherical harmonic $Y_{\ell m}$. The notation Y_ν has found its way into the literature and is currently adopted universally (including many current computer algebra systems), and the symbol N_ν is used only very rarely.

The recurrence relations for the J functions (analogous relations are fulfilled by the Y's) read as follows [see Eq. (10.6.1) of Ref. [5]],

$$J_{\nu-1}(z) + J_{\nu+1}(z) = \frac{2\nu}{z} J_\nu(z) , \tag{2.149a}$$

$$J_{\nu-1}(z) - J_{\nu+1}(z) = 2 J_\nu'(z) . \tag{2.149b}$$

---

[1] Friedrich Wilhelm Bessel (1784–1846)
[2] Carl Gottfried Neumann (1832–1925)

The latter equation expresses the derivative of the Bessel function of order ν in terms of Bessel functions of order $\nu - 1$ and $\nu + 1$. Alternatively, one has

$$J_\nu'(z) = J_{\nu-1}(z) - \frac{\nu}{z} J_\nu(z) = -J_{\nu+1}(z) + \frac{\nu}{z} J_\nu(z). \tag{2.150}$$

For non-integer order ν and complex argument z with $\mathrm{Re}(z) > 0$, an integral representation of the Bessel J function is [see Eq. (10.9.6) of Ref. [5]]

$$J_\nu(z) = \int_0^\pi \frac{\mathrm{d}\theta}{\pi} \cos\left[z\sin(\theta) - \nu\theta\right] - \sin(\nu\pi) \int_0^\infty \frac{\mathrm{d}\theta}{\pi} e^{-z\sinh(\theta) - \nu\theta}, \tag{2.151}$$

while under the same conditions, the Bessel Y function has the integral representation [see Eq. (10.9.7) of Ref. [5]]

$$Y_\nu(z) = \int_0^\pi \frac{\mathrm{d}\theta}{\pi} \sin\left[z\sin(\theta) - \nu\theta\right] - \int_0^\infty \frac{\mathrm{d}\theta}{\pi} e^{-z\sinh(t)} \left(e^{\nu t} + e^{-\nu t}\cos(\nu\pi)\right). \tag{2.152}$$

The asymptotic behavior for small argument of the Bessel J and Y functions is given as

$$J_\nu(z) \sim \frac{1}{\Gamma(\nu+1)} \left(\frac{z}{2}\right)^\nu, \qquad Y_\nu(z) \sim -\frac{\Gamma(\nu)}{\pi} \left(\frac{2}{z}\right)^\nu, \qquad |z| \to 0. \tag{2.153}$$

The asymptotic formula is for $\nu > 0$ for the Y function. For $\nu < 0$, one has to consider a second term,

$$Y_\nu(z) \sim -\frac{\cos(\pi\nu)\Gamma(-\nu)}{\pi} \left(\frac{z}{2}\right)^\nu - \frac{\Gamma(\nu)}{\pi} \left(\frac{2}{z}\right)^\nu, \qquad |z| \to 0. \tag{2.154}$$

In the limit $\nu \to 0$, the asymptotic formula reduces to

$$Y_0(z) \sim \frac{2}{\pi} \ln\left(\frac{z}{2}\right) + \frac{2}{\pi} \gamma_E + \mathcal{O}\left(z^2 \ln(z)\right), \qquad |z| \to 0. \tag{2.155}$$

Here, $\gamma_E = 0.577\,215\,66\ldots$ is the Euler–Mascheroni constant. For $\nu = -n$ with positive integer n, one has

$$Y_{-n}(z) \sim -(-1)^n \frac{(n-1)!}{\pi} \left(\frac{2}{z}\right)^n, \qquad |z| \to 0. \tag{2.156}$$

For large arguments $x > 0$, the asymptotic behavior of the Bessel functions is given by a damped harmonic oscillation, with a nontrivial phase,

$$J_\nu(x) \sim \sqrt{\frac{2}{\pi x}} \cos\left(x - \frac{\nu\pi}{2} - \frac{\pi}{4}\right), \qquad x \to \infty. \tag{2.157a}$$

$$Y_\nu(x) \sim \sqrt{\frac{2}{\pi x}} \sin\left(x - \frac{\nu\pi}{2} - \frac{\pi}{4}\right), \qquad x \to \infty. \tag{2.157b}$$

For intermediate values of x, the situation is more complicated. This is illustrated in Fig. 2.5 for $J_0(x)$ and $J_1(x)$. Otherwise, it is helpful to know that the literature on Bessel functions is abundant, with basic formulas being summarized in Refs. [7, 5] and more intricate treatments in Refs. [8, 9].

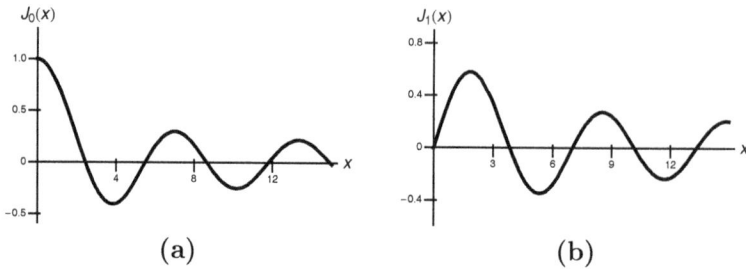

Fig. 2.5   Plot of the first oscillations of the Bessel functions $J_0(x)$ and $J_1(x)$ [panels (a) and (b), respectively]. For $x \to 0$, one discerns that $\lim_{x\to 0} J_1(x) = 0$ and the linear asymptotics for $J_1(x)$ in the regime of small x.

2.4.3   *Orthogonality Properties of Bessel Functions*

While this is a textbook on physics, not on the theory of special functions, a certain familiarity with concepts related to the most important special functions of mathematical physics is of universal importance for any theoretical physicist, and the slight detour we are taking now will prove useful for a number of applications. Indeed, of particular importance are the zeros of the Bessel and Neumann functions. Let us denote the zeros of the Bessel and Neumann functions of integer order as x_{mn} and y_{mn}, respectively. These fulfill the relations

$$J_m(x_{mn}) = 0 \tag{2.158}$$

and $Y_m(y_{mn}) = 0$, where x_{mn} is the nth zero of the mth Bessel J function, and y_{mn} is the nth zero of the mth Neumann Y function. In particular, because $J_m(x) \propto x^m$ for small x, a trivial zero of $J_m(x)$ for $m \geq 1$ is $x = 0$. Furthermore, in view of Eq. (2.157), there exists an infinite number of zeros of the Bessel functions on the real axis. However, it is rather nontrivial to calculate these numerically, for intermediate ranges of the argument. Selected numerical values for x_{mn} are given in Table 2.1. We also indicate numerical values for the zeros of derivative of the Bessel functions, defined according to

$$J'_m(\bar{x}_{mn}) = 0 \tag{2.159}$$

In computer algebra systems (e.g., Ref. [10]), one sometimes finds built-in algorithms like `BesselJZero` or `BesselYZero` for the computation of zeros of Bessel and Neumann functions; these can be used directly without recourse to any additional asymptotic formulas. Still, it is useful to know which asymptotic properties determine the behavior of the roots of the Bessel functions in large order. A modern treatise on the intricate aspects of the calculation of Bessel functions, with numerical examples in extreme parameter ranges, is given in [11].

Let now investigate whether the basis functions in our *ansatz* (2.142) for a general solution of the Laplace equation in cylindrical coordinates can be interpreted as orthogonal with respect to an appropriate integration measure. For a potential

Table 2.1 Numerical values for the nth zeros x_{mn} of the Bessel functions of order m, with $J_m(x = x_{mn}) = 0$, for $m = 0, \ldots, 5$ and $n = 1, \ldots, 5$.

n	$m = 0$	$m = 1$	$m = 2$	$m = 3$	$m = 4$	$m = 5$
1	2.405	3.832	5.135	6.379	7.588	8.771
2	5.520	7.016	8.417	9.761	11.065	12.339
3	8.654	10.173	11.620	13.015	14.373	15.700
4	11.792	13.323	14.796	16.224	17.616	18.980
5	14.931	16.470	17.960	19.409	20.827	22.218

Table 2.2 Numerical values for the nth zeros $\bar{x}_{mn}$ of the derivatives of the Bessel functions of order m, with $J'_m(x = \bar{x}_{mn}) = 0$, for $m = 0, \ldots, 5$ and $n = 1, \ldots, 5$.

n	$m = 0$	$m = 1$	$m = 2$	$m = 3$	$m = 4$	$m = 5$
1	3.832	1.841	3.054	4.201	5.317	6.415
2	7.016	5.331	6.706	8.015	9.282	10.520
3	10.174	8.536	9.969	11.346	12.682	13.987
4	13.324	11.706	13.170	14.585	13.540	17.313
5	16.471	14.864	16.348	17.789	14.970	20.576

regular on the cylinder axis, with $\rho = 0$, we must concentrate on the terms with J functions. Furthermore, the zeros of the Bessel functions play a special role in the orthogonality properties. Based on Eq. (2.144), for an arbitrary scale parameter λ, it is straightforward to see that the function $J_n(\lambda x)$ fulfills

$$x \frac{\mathrm{d}}{\mathrm{d}x} \left[x \frac{\mathrm{d}}{\mathrm{d}x} J_n(\lambda x) \right] + \left(\lambda^2 x^2 - n^2 \right) J_n(\lambda x) = 0. \tag{2.160}$$

Now, we assume that λ and μ are two distinct zeros of the Bessel function $J_n(x)$,

$$J_n(\lambda) = J_n(\mu) = 0, \qquad \lambda \neq \mu, \tag{2.161}$$

so that $J_n(\lambda x) = J_n(\mu x) = 0$ for $x = 1$. Multiplying Eq. (2.160) by $J_n(\mu x)/x$, we obtain

$$J_n(\mu x) \frac{\mathrm{d}}{\mathrm{d}x} \left[x \frac{\mathrm{d}}{\mathrm{d}x} J_n(\lambda x) \right] + \frac{\lambda^2 x^2 - n^2}{x} J_n(\mu x) J_n(\lambda x) = 0, \tag{2.162}$$

and a trivial generalization is obtained by the replacement $\mu \leftrightarrow \lambda$,

$$J_n(\lambda x) \frac{\mathrm{d}}{\mathrm{d}x} \left[x \frac{\mathrm{d}}{\mathrm{d}x} J_n(\mu x) \right] + \frac{\mu^2 x^2 - n^2}{x} J_n(\mu x) J_n(\lambda x) = 0. \tag{2.163}$$

Furthermore, manipulating Eqs. (2.162) and (2.163), one can show that a few terms cancel in the difference,

$$J_n(\mu x) \frac{\mathrm{d}}{\mathrm{d}x} \left[x \frac{\mathrm{d}}{\mathrm{d}x} J_n(\lambda x) \right] - J_n(\lambda x) \frac{\mathrm{d}}{\mathrm{d}x} \left[x \frac{\mathrm{d}}{\mathrm{d}x} J_n(\mu x) \right]$$
$$+ \left(\lambda^2 - \mu^2 \right) x J_n(\mu x) J_n(\lambda x) = 0, \tag{2.164}$$

which can be rewritten as

$$\frac{\mathrm{d}}{\mathrm{d}x} \left[J_n(\mu x) x \frac{\mathrm{d}}{\mathrm{d}x} J_n(\lambda x) \right] - \frac{\mathrm{d}}{\mathrm{d}x} \left[J_n(\lambda x) x \frac{\mathrm{d}}{\mathrm{d}x} J_n(\mu x) \right]$$
$$+ \left(\lambda^2 - \mu^2 \right) x J_n(\mu x) J_n(\lambda x) = 0. \tag{2.165}$$

Finally, one integrates over $x \in (0,1)$ to show the orthogonality

$$\int_0^1 \mathrm{d}x\, x\, J_n(\mu x) J_n(\lambda x) = 0, \qquad \lambda \neq \mu, \qquad J_n(\lambda) = J_n(\mu) = 0, \tag{2.166}$$

which concerns integrations up to specific zeros λ and μ, with $\lambda \neq \mu$.

The case $\lambda = \mu$ in Eq. (2.166) must be treated separately. First, we sidestep the question and derive a helpful identity. From the recursion relations of the Bessel function and its derivative, given in Eq. (2.150), one can conclude that, at a zero $x = \lambda$ of the Bessel function,

$$\left. \frac{\mathrm{d}}{\mathrm{d}x} J_n(x) \right|_{x=\lambda} = -J_{n+1}(\lambda). \tag{2.167}$$

In Eq. (2.165), one now sets $\mu = \lambda + \epsilon$, i.e., one slightly displaces one parameter from the zero, integrates over $x \in (0,1)$ and obtains

$$[J_n((\lambda + \epsilon))\lambda J_n'(\lambda)] + \left(\lambda^2 - (\lambda + \epsilon)^2 \right) \int_0^1 \mathrm{d}x\, x\, J_n((\lambda + \epsilon) x) J_n(\lambda x) = 0. \tag{2.168}$$

One then expands to first order in ϵ to obtain

$$\epsilon \lambda \left[J_n'(\lambda) \right]^2 - 2 \lambda \epsilon \int_0^1 \mathrm{d}x\, x\, \left[J_n(\lambda x) \right]^2 = 0, \tag{2.169}$$

and thus shows that

$$\int_0^1 \mathrm{d}x\, x\, \left[J_n(\lambda x) \right]^2 = \frac{1}{2} \left[J_{n+1}(\lambda) \right]^2. \tag{2.170}$$

Equations (2.166) and (2.170) allow us to write

$$\int_0^1 \mathrm{d}x\, x\, J_n(\lambda x) J_n(\mu x) = \frac{1}{2} \delta_{\lambda\mu} \left[J_{n+1}(\lambda) \right]^2, \tag{2.171}$$

where the Kronecker symbol is unity if the two (in this case, real rather than integer) arguments are equal to each other and zero otherwise. Let now $x_{mn} = k_{mn} a$ where a is the radius of a cylinder. Then, a simple scaling of the integration variable in Eq. (2.171) leads to the result (see Fig. 2.6)

$$\int_0^a \mathrm{d}\rho\, \rho\, J_m(k_{mn'}\rho) J_m(k_{mn}\rho) = \frac{1}{2} \delta_{n'n} \left[a\, J_{m+1}(k_{mn}\, a) \right]^2. \tag{2.172}$$

(a)   **(b)**

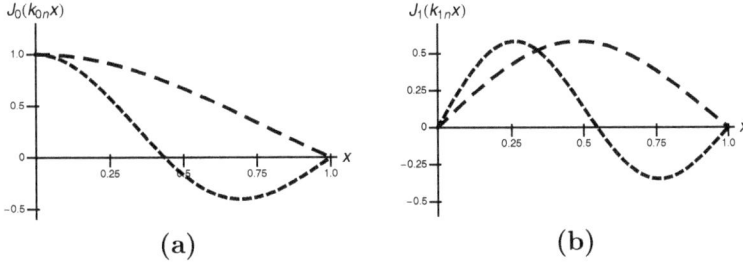

Fig. 2.6   Illustration of the property (2.172) for the Bessel functions $f_n(\rho) = J_0(k_{0n}\rho)$ and $f_n(\rho) = J_1(k_{1n}\rho)$, with $n = 1, 2$ (long dashes are for $n = 1$, short dashes denote the curves for $n = 2$). We set $a = 1$, so that $k_{0n} = x_{0n}$ and $k_{1n} = x_{1n}$. As n is increased, the number of oscillations increases, and the Bessel functions scaled with different x_{0n} and x_{1n} are orthogonal to each other upon integration with respect to the measure given in Eq. (2.172).

Now, the orthogonality and normalization has should be generalized for an integrations over the real axis, $x \in (0, \infty)$. One starts once more from Eq. (2.165), but with the replacements $\lambda \to k$, and $\mu \to k'$,

$$\frac{\mathrm{d}}{\mathrm{d}x}\left[J_n(k'x)\, x\, \frac{\mathrm{d}}{\mathrm{d}x} J_n(kx) \right] - \frac{\mathrm{d}}{\mathrm{d}x}\left[J_n(kx)\, x\, \frac{\mathrm{d}}{\mathrm{d}x} J_n(k'x) \right]$$
$$+ \left(k^2 - k'^2 \right) x J_n(kx)\, J_n(k'x) = 0. \tag{2.173}$$

The well-known asymptotics of the Bessel function given in Eq. (2.157) imply that $\rho = \infty$ is a zero of the Bessel function. Thus, integrating Eq. (2.173) within the interval $x \in (0, \infty)$, one easily shows that

$$\int_0^\infty \mathrm{d}x\, x\, J_n(kx)\, J_n(k'x) = 0, \qquad k \neq k'. \tag{2.174}$$

The limit $k \to k'$ remains to be evaluated. The only region which can sizeably contribute to the integral in this limit is the one for very large x; otherwise an infinitesimal displacement $k = k' + \epsilon$ will lead to a vanishing integral due to orthogonality. One writes the asymptotics (2.157) as an exponential,

$$J_n(kx) \sim \frac{1}{2}\sqrt{\frac{2}{\pi k x}}\left\{ \exp\left[\mathrm{i}\left(kx - \frac{(n + \frac{1}{2})\pi}{2} \right) \right] \right.$$
$$\left. + \exp\left[-\mathrm{i}\left(kx - \frac{(n + \frac{1}{2})\pi}{2} \right) \right] \right\}, \qquad x \to \infty, \tag{2.175}$$

and analogously for k replaced by k'. The only relevant terms in the integrand $x J_n(kx)\, J_n(k'x)$ in Eq. (2.174) is given by the following replacement,

$$x J_n(kx)\, J_n(k'x) \to \frac{1}{2\pi\sqrt{k k'}}\left\{ \exp\left(\mathrm{i}(k - k')\, x \right) + \exp\left(-\mathrm{i}(k - k')\, x \right) \right\} + \dots, \tag{2.176}$$

where the ellipsis denotes terms proportional to $\exp(i(k+k')x)$ and $\exp(-i(k+k')x)$. When integrated over the interval $x \in (0,\infty)$, with the help of a convergent factor $\exp(-\eta x)$, these terms give only finite corrections to the Dirac-δ function term proportional to $\delta(k-k')$ which dominates in the limit $k \to k'$. As a last step, one symmetrize the the expression given in Eq. (2.176) for the interval $x \in (-\infty, \infty)$ to show that

$$\int_0^\infty dx\, x\, J_n(k\,x)\, J_n(k'\,x) \overset{k \to k'}{=} \frac{1}{2} \int_{-\infty}^\infty dx\, \frac{1}{2\pi\sqrt{k\,k'}} \left(e^{i(k-k')\,x} + e^{-i(k-k')\,x} \right)$$

$$= \frac{1}{2\pi\sqrt{k\,k'}}\, 2\pi\delta(k-k') = \frac{1}{k}\,\delta(k-k'), \qquad (2.177)$$

where we have used the fact that $k = k'$ at the peak of the Dirac-δ function. Equations (2.174) and (2.177) conclude the proof of the orthogonality relation

$$\int_0^\infty d\rho\, \rho\, J_n(k\,\rho)\, J_n(k'\,\rho) = \frac{1}{k}\,\delta(k-k'). \qquad (2.178)$$

For spherical Bessel functions, which are defined via the relation

$$j_\ell(x) = \sqrt{\frac{\pi}{2\,x}}\, J_{\ell+1/2}(x), \qquad (2.179)$$

the orthogonality relations are somewhat different. Identifying $n = \ell + 1/2$ in Eq. (2.178), one writes the relation

$$\int_0^\infty dx\, x^2 \left(\sqrt{\frac{\pi}{2\,k\,x}}\, J_{\ell+1/2}(k\,x)\right) \left(\sqrt{\frac{\pi}{2\,k'\,x}}\, J_{\ell+1/2}(k'\,x)\right) = \left(\sqrt{\frac{\pi}{2}}\right)^2 \frac{1}{k\,k'}\,\delta(k-k'), \qquad (2.180)$$

which implies that

$$\int_0^\infty dx\, x^2\, j_\ell(k\,x)\, j_\ell(k'\,x) = \frac{\pi}{2\,k\,k'}\,\delta(k-k'). \qquad (2.181)$$

The above derivation has been lengthy but it serves to alert us to the beauty of the theory of special functions, whose power we shall bring to bear on the calculation of Green functions.

2.4.4   *Expansion of the Green Function in Cylindrical Coordinates*

We now repeat the analysis of Sec. 2.3.4 regarding the construction of a Green function, but this time in cylindrical coordinates. The integration measure reads

$$d^3r = \rho\, d\rho\, d\varphi\, dz, \qquad (2.182)$$

and the Green function satisfies

$$\vec{\nabla}^2 G(\vec{r}, \vec{r}') = -\frac{1}{\epsilon_0}\delta^{(3)}(\vec{r} - \vec{r}') = -\frac{1}{\epsilon_0\,\rho}\,\delta(\rho-\rho')\,\delta(z-z')\,\delta(\varphi-\varphi'). \qquad (2.183)$$

The right-hand side of Eq. (2.183) vanishes almost everywhere, except for the point $\vec{r} = \vec{r}'$. We refer to Eq. (2.143) and write the following general form for

$G(\vec{r}, \vec{r}')$, leaving one set of variables (in this case z and z') in terms of an unknown function, which we refer to as $g_m(k; z, z')$. Then,

$$G(\vec{r}, \vec{r}') = \frac{1}{2\pi} \sum_{m=-\infty}^{\infty} e^{im(\varphi-\varphi')} \int_0^\infty dk \, k \, g_m(k; z, z') \, J_m(k\rho) \, J_m(k\rho') \,. \quad (2.184)$$

We now use the Laplace operator in cylindrical coordinates, given in Eq. (2.133),

$$\vec{\nabla}^2 G(\vec{r}, \vec{r}') = \left(\frac{1}{\rho} \frac{\partial}{\partial\rho} \rho \frac{\partial}{\partial\rho} + \frac{1}{\rho^2} \frac{\partial^2}{\partial\varphi^2} + \frac{\partial^2}{\partial z^2} \right) \left(\frac{1}{2\pi} \sum_{m=-\infty}^{\infty} e^{im(\varphi-\varphi')} \right.$$

$$\left. \times \int_0^\infty dk \, k \, g_m(k; z, z') \, J_m(k\rho) \, J_m(k\rho') \, dk \right) \,. \quad (2.185)$$

In view of the equation fulfilled by the Bessel function, we can replace

$$\frac{1}{\rho} \frac{\partial}{\partial\rho} \rho \frac{\partial}{\partial\rho} + \frac{1}{\rho^2} \frac{\partial^2}{\partial\varphi^2} \to \left(\frac{m^2}{\rho^2} - k^2 - \frac{m^2}{\rho^2} \right) \to -k^2 \,. \quad (2.186)$$

Then,

$$\vec{\nabla}^2 G(\vec{r}, \vec{r}') = \frac{1}{2\pi} \sum_{m=-\infty}^{\infty} e^{im(\varphi-\varphi')} \int_0^\infty dk \, k \, J_m(k\rho) \, J_m(k\rho')$$

$$\times \left(-k^2 + \frac{\partial^2}{\partial z^2} \right) g_m(k; z, z') \stackrel{!}{=} -\frac{1}{\epsilon_0 \rho} \delta(\rho - \rho') \, \delta(z - z') \, \delta(\varphi - \varphi') \,. \quad (2.187)$$

We now apply the integral operator

$$\int_0^{2\pi} d\varphi \, e^{-im'\varphi} \int_0^\infty d\rho\rho \, J_{m'}(k'\rho) \,, \quad (2.188)$$

and integrate both sides of Eq. (2.187) over $d\varphi$ and $d\rho$. The left-hand side (LHS) of Eq. (2.187) becomes

$$\text{LHS} = \frac{1}{2\pi} \int_0^{2\pi} d\varphi \sum_{m=-\infty}^{\infty} e^{-i(m'-m)\varphi} e^{-im\varphi'} \int_0^\infty dk \, k \, J_m(k\rho')$$

$$\times \left(\int_0^\infty d\rho \, \rho \, J_m(k\rho) \, J_{m'}(k'\rho) \right) \left(-k^2 + \frac{\partial^2}{\partial z^2} \right) g_m(k; z, z') \,. \quad (2.189)$$

We now use the orthogonality condition

$$\frac{1}{2\pi} \int_0^{2\pi} e^{-i(m'-m)\varphi} \, d\varphi = \delta_{mm'} \quad (2.190)$$

to reduce the sum over m to a single term with $m = m'$, and the integral (2.178),

$$\int_0^\infty d\rho \, \rho \, J_m(k\rho) \, J_m(k'\rho) = \frac{1}{k} \delta(k - k') \quad (2.191)$$

to carry out the integration over k. Then, the left-hand side of Eq. (2.187) becomes

$$\text{LHS} = e^{-im'\varphi'} \int_0^\infty dk\, k\, J_{m'}(k\rho') \frac{1}{k}\, \delta(k-k') \left(-k^2 + \frac{\partial^2}{\partial z^2}\right) g_{m'}(k;z,z')$$

$$= e^{-im'\varphi'} J_{m'}(k'\rho') \left(-k^2 + \frac{\partial^2}{\partial z^2}\right) g_{m'}(k;z,z') . \tag{2.192}$$

The right-hand side of Eq. (2.187) becomes

$$\text{RHS} = -\frac{1}{\epsilon_0} \int_0^{2\pi} d\varphi \int_0^\infty d\rho\, \delta(\rho-\rho')\, \delta(z-z')\, \delta(\varphi-\varphi')\, e^{-im'\varphi} J_{m'}(k'\rho)$$

$$= -\frac{1}{\epsilon_0} e^{-im'\varphi'} J_{m'}(k'\rho')\, \delta(z-z') . \tag{2.193}$$

Equating LHS = RHS, we finally extract the differential equation for $g_{m'}(k;z,z')$ and substitute $m' \to m$,

$$\left(\frac{\partial^2}{\partial z^2} - k^2\right) g_m(k;z,z') = -\frac{1}{\epsilon_0}\, \delta(z-z') . \tag{2.194}$$

This equation is easy to solve by integrating over the kink at $r = r'$. While the function $g_m(k;z,z')$ is continuous, its derivative with respect to z should have a discontinuity at $z = z'$.

As in Eq. (2.84), we now integrate both sides over the interval $z \in (z'-\varepsilon, z'+\varepsilon)$ and obtain

$$\int_{z'-\varepsilon}^{z'+\varepsilon} dz \left(\frac{\partial^2}{\partial z^2} - k^2\right) g_m(k;z,z') = -\frac{1}{\epsilon_0} \int_{z'-\varepsilon}^{z'+\varepsilon} \delta(z-z')\, dz = -\frac{1}{\epsilon_0} . \tag{2.195}$$

The left-hand side can be reformulated as

$$\int_{z'-\varepsilon}^{z'+\varepsilon} \frac{\partial}{\partial z} \left(\frac{\partial}{\partial z} g_m(k;z,z')\right) dz = \frac{\partial}{\partial z} g_m(k;z,z') \Big|_{z=z'-\varepsilon}^{z=z'+\varepsilon} . \tag{2.196}$$

The following *ansatz*

$$g_m(k;z,z') = C_{m,k} \exp(k z_< - k z_>) , \tag{2.197}$$

with $z_< = \min(z,z')$, $z_> = \max(z,z')$, is inspired by the general solution [Eq. (2.143)] to the homogeneous equation. Differentiation leads to

$$\frac{\partial}{\partial z} g_m(z;z') \Big|_{z'-\varepsilon}^{z'+\varepsilon} = C_{m,k} \left[e^{k z'} \left\{ \frac{\partial}{\partial z'} e^{-k(z'+\varepsilon)} \right\} - \left\{ \frac{\partial}{\partial z'} e^{k(z'-\varepsilon)} \right\} e^{-k z'} \right] = -\frac{1}{\epsilon_0} , \tag{2.198}$$

and thus

$$C_{m,k} \left(e^{k z'} (-k) e^{-k(z'+\varepsilon)} - k e^{k(z'-\varepsilon)} e^{-k z'} \right) = 1 . \tag{2.199}$$

As $\varepsilon \to 0$,

$$C_{m,k} \left(-2k\, e^{k z'-k z'}\right) = -\frac{1}{\epsilon_0} , \qquad C_{m,k} = \frac{1}{2\epsilon_0 k} . \tag{2.200}$$

We now insert the solution

$$g_m(k; z, z') = \frac{1}{2\epsilon_0 k} \exp\left[-k\left(z_> - z'_<\right)\right] \tag{2.201}$$

into the expansion (2.184) for the Green function and obtain

$$G(\vec{r}, \vec{r}') = \frac{1}{2\pi} \sum_{m=-\infty}^{\infty} e^{im(\varphi - \varphi')} \int_0^\infty dk\, k \left(\frac{1}{2\epsilon_0 k}\right) e^{-k(z_> - z_<)} J_m(k\rho)\, J_m(k\rho') \,. \tag{2.202}$$

The latter result can be simplified slightly,

$$G(\vec{r}, \vec{r}') = \frac{1}{4\pi\,\epsilon_0} \sum_{m=-\infty}^{\infty} e^{im(\varphi - \varphi')} \int_0^\infty dk\, e^{-k(z_> - z_<)} J_m(k\rho)\, J_m(k\rho')$$

$$= \frac{1}{4\pi\epsilon_0\, |\vec{r} - \vec{r}'|} \,. \tag{2.203}$$

Multiplication by $4\pi\epsilon_0$ leads to the identity

$$\frac{1}{|\vec{r} - \vec{r}'|} = \sum_{m=-\infty}^{\infty} e^{im(\varphi - \varphi')} \int_0^\infty dk\, e^{-k(z_> - z_<)} J_m(k\rho)\, J_m(k\rho') \,. \tag{2.204}$$

A comparison of Eqs. (2.203) and (2.204) with the corresponding expansions (2.91) and (2.93) in spherical coordinates implies that in the spherical coordinate system, we have to integrate over k in contrast to a summation over discrete values of ℓ and m. The case differentiation is not carried out in the radial variables $r_<$ and $r_>$, but in the $z_>$ and $z_<$ coordinates. The dependence on z is such that the argument of the exponential leads to exponential damping for both $z, z' \to \pm\infty$, in agreement with the boundary conditions imposed on the Green function.

2.5 Electrostatic Green Function and Eigenfunction Expansions

2.5.1 *Eigenfunction Expansions for the Green Function*

The calculation of Green functions may be mapped onto the inversion of linear operators. This becomes clear if we take into account that the Dirac-δ function, which appears in the defining equation for a typical Green function, is equal to a constant in Fourier space. We consider the inversion of the linear operator Ξ,

$$\Xi = L - \lambda \,, \tag{2.205}$$

where λ is a constant and L is a linear differential operator. For the Green function of electrostatics, we would have $L = -\epsilon_0\, \vec{\nabla}^2$. We wish to solve the defining equation (2.22) for the Green function, which can be mapped onto the special case $\lambda = 0$ of the generalized equation

$$(L - \lambda)\, G(\vec{r}, \vec{r}') = \delta(\vec{r} - \vec{r}') \,. \tag{2.206}$$

Let us suppose that we have a set $\{\psi_n(\vec{r})\}_{n=0}^{\infty}$ of eigenfunctions that satisfy

$$L\, \psi_n(\vec{r}) = \lambda_n\, \psi_n(\vec{r}) \,, \tag{2.207}$$

with eigenvalues λ_n and with appropriate boundary conditions on the $\psi_n(\vec{r})$ so that the eigenvalue equation for λ_n is properly defined.

A small detour for those readers who have already heard a lecture on quantum mechanics is in order. If L is equal to the Hamiltonian for a quantum system, then $L = H$ has energy eigenvalues $\lambda_n = E_n$. The discreteness of the eigenvalues for, e.g., the hydrogen atom, follows from the boundary condition for the bound wave functions at infinity, i.e., from the condition that a bound state should be normalizable. Otherwise, it would be possible to solve the Schrödinger equation for any energy "eigenvalue". The discreteness of the bound spectrum follows *exclusively* from the normalization condition, which therefore is of preeminent importance.

We assume that the $\{\psi_n(\vec{r})\}_{n=0}^{\infty}$ provide a complete set of eigenfunctions, in the sense of Eq. (2.207), so that

$$\int_V \mathrm{d}^3 r \, \psi_m^*(\vec{r}) \, \psi_n(\vec{r}) = \delta_{mn} . \tag{2.208}$$

Completeness implies that

$$\sum_n \psi_n(\vec{r}) \, \psi_n^*(\vec{r}') = \delta^{(3)}(\vec{r} - \vec{r}') , \tag{2.209}$$

for $\vec{r}$, $\vec{r}'$ in V. The solution is

$$G(\lambda, \vec{r}, \vec{r}') = \sum_n \frac{\psi_n(\vec{r}) \, \psi_n^*(\vec{r}')}{\lambda_n - \lambda} , \tag{2.210}$$

because

$$(L - \lambda) \, G(\lambda, \vec{r}, \vec{r}') = \sum_n (\lambda_n - \lambda) \frac{\psi_n(\vec{r}) \, \psi_n^*(\vec{r}')}{\lambda_n - \lambda} = \sum_n \psi_n(\vec{r}) \, \psi_n^*(\vec{r}') = \delta^{(3)}(\vec{r} - \vec{r}') . \tag{2.211}$$

In the case $\lambda = 0$, we have

$$G(\lambda, \vec{r}, \vec{r}') = \sum_n \frac{\psi_n(\vec{r}) \, \psi_n^*(\vec{r}')}{\lambda_n} . \tag{2.212}$$

This is a relation common to the Green function, because

$$L \, G(\lambda, \vec{r}, \vec{r}') = \sum_n \psi_n(\vec{r}) \, \psi_n^*(\vec{r}') = \delta(\vec{r} - \vec{r}') . \tag{2.213}$$

In practical cases, one may construct a Green function like so: One takes a differential operator and formulates a discrete approximation to its spectrum and eigenfunctions within a particular basis set of functions. Using the approximate eigenstates, one can use Eq. (2.210) to calculate the Green function immediately.

One decisive feature of the Green function expansion in terms of eigenfunctions is as follows. Let us reconsider Eqs. (2.91) and (2.202) for the Green function of electrostatics (in spherical and cylindrical coordinates, respectively). In both cases, we have had to implement the "cusp" of the Green function at equal arguments by an explicit dependence on $r_>$ and $r_<'$, or $z_>$ and $z_<'$, respectively. In an eigenfunction decomposition, this cusp is naturally implemented in the sum over eigenstates $\psi_n(\vec{r})$, leading to some sort of simplification. However, the sum over states leads to an additional level of complexity in the calculation which we shall study in the following.

2.5.2 *Application to Electrostatics in Spherical Coordinates*

We discuss the calculation of the Green function of electrostatics in terms of an eigenfunction decomposition of solutions of a corresponding eigenvalue problem. In order to apply the formalism developed in the previous Sec. 2.5.1, we bring the defining Eq. (2.22) of the Green function into the standard form (2.206),

$$\vec{\nabla}^2 G(\vec{r},\vec{r}') = -\frac{1}{\epsilon_0}\delta^{(3)}(\vec{r}-\vec{r}')\,, \qquad \vec{\nabla}^2\left[-\epsilon_0\,G(\vec{r},\vec{r}')\right] = \delta^{(3)}(\vec{r}-\vec{r}')\,, \qquad (2.214)$$

so that

$$(L-\lambda)\,G(\vec{r},\vec{r}') = \delta^{(3)}(\vec{r}-\vec{r}')\,, \qquad L = -\epsilon_0\vec{\nabla}^2\,, \qquad \lambda = 0\,. \qquad (2.215)$$

One parameterizes the eigenvalue as

$$\lambda = \lambda_k = \epsilon_0\,k^2\,. \qquad (2.216)$$

The corresponding eigenvalue problem is

$$(L-\lambda_k)\,\psi_{k\ell m}(\vec{r}) = \left(-\epsilon_0\vec{\nabla}^2 - \epsilon_0 k^2\right)\psi_{k\ell m}(\vec{r}) = 0\,. \qquad (2.217)$$

We will seek to find solutions of the form

$$\left(\vec{\nabla}^2 + k^2\right)\psi_{k\ell m}(\vec{r}) = 0\,, \qquad \psi_{k\ell m}(\vec{r}) = R_{k\ell}(r)\,Y_{\ell m}(\theta,\varphi)\,. \qquad (2.218)$$

The radial equation for $R_{k\ell}(r)$ reads

$$\frac{\partial^2}{\partial r^2}R_{k\ell}(r) + \frac{2}{r}\frac{\partial}{\partial r}R_{k\ell}(r) - \frac{\ell(\ell+1)}{r^2}R_{k\ell}(r) + k^2\,R_{k\ell}(r) = 0. \qquad (2.219)$$

Solutions regular at the origin are given by the spherical Bessel functions $j_\ell(k\,r)$; these vary as r^ℓ for small r. These are defined according to Eq. (2.179),

$$R_{k\ell}(r) = j_\ell(k\,r)\,, \qquad j_\ell(k\,r) = \sqrt{\frac{\pi}{2\,k\,r}}\,J_{\ell+\frac{1}{2}}(k\,r)\,. \qquad (2.220)$$

A more thorough discussion on spherical Bessel functions follows in Sec. 5.2.2. According to Eq. (2.181), the eigenfunctions satisfy the following orthogonality condition

$$\langle\psi_{k\ell m}|\psi_{k'\ell'm'}\rangle = \int_0^\infty dr\,r^2\,j_\ell(k\,r)\,j_{\ell'}(k'\,r)\int d\Omega\,Y_{\ell m}^*(\theta,\varphi)\,Y_{\ell'm'}(\theta,\varphi)$$

$$= \frac{\pi}{2\,k\,k'}\,\delta(k-k')\,\delta_{\ell\ell'}\,\delta_{mm'}\,. \qquad (2.221)$$

Of course, we could have written replaced $k\,k' \to k^2$ in the prefactor, but the above expression is explicitly symmetric. For a discrete set of eigenfunctions, Eq. (2.221) becomes analogous to Eq. (2.208).

So, alternatively, the completeness relationship is given by

$$\sum_{\ell=0}^\infty \sum_{m=-\ell}^\ell \int_0^\infty dk\,\frac{2\,k^2}{\pi}\,j_\ell(k\,r)\,j_\ell(k\,r')\,Y_{\ell m}(\theta,\varphi)\,Y_{\ell m}^*(\theta',\varphi') = \delta^{(3)}(\vec{r}-\vec{r}')\,.$$

$$(2.222)$$

For a discrete set of eigenfunctions, Eq. (2.222) is the analogue of Eq. (2.209). The Green function thus has the form

$$G(\vec{r},\vec{r}') = \sum_{\ell=0}^{\infty} \sum_{m=-\ell}^{\ell} \int_0^{\infty} dk \, \frac{1}{\lambda_k} \, \frac{2\,k^2}{\pi} \, j_\ell(k\,r) \, j_\ell(k\,r') \, Y_{\ell m}(\theta,\varphi) \, Y_{\ell m}^*(\theta',\varphi') \,,$$
(2.223)

where we have simply inserted the reciprocal of the eigenvalue according to Eq. (2.212). It follows that

$$G(\vec{r},\vec{r}') = \frac{2}{\pi\,\epsilon_0} \sum_{\ell=0}^{\infty} \sum_{m=-\ell}^{\ell} \int_0^{\infty} dk \, j_\ell(k\,r) \, j_\ell(k\,r') \, Y_{\ell m}(\theta,\varphi) \, Y_{\ell m}^*(\theta',\varphi') \,.$$
(2.224)

In this formula, the cusp is formulated in terms of the eigenfunction expansion, as anticipated near the end of Sec. 2.5.1. The radial component of the Green function is formulated in terms of r and r', not $r_<$ and $r_>$. However, we now have to sum over the eigenfunctions, i.e., integrate over k. Comparing with the familiar formula

$$G(\vec{r},\vec{r}') = \frac{1}{4\pi\epsilon_0\,|\vec{r}-\vec{r}'|} = \frac{1}{\epsilon_0} \sum_{\ell=0}^{\infty} \sum_{m=-\ell}^{\ell} \frac{1}{2\ell+1} \, \frac{r_<^\ell}{r_>^{\ell+1}} \, Y_{\ell m}(\theta,\varphi) \, Y_{\ell m}^*(\theta',\varphi') \,,$$
(2.225)

we find that the cusp at equal argument is implemented in the formula

$$\int_0^{\infty} dk \, j_\ell(kr) \, j_\ell(kr') = \frac{\pi}{2\,(2\ell+1)} \, \frac{r_<^\ell}{r_>^{\ell+1}} \,.$$
(2.226)

We may perform a consistency check on the result (2.224), by considering a point charge q_0 located on the z axis at $z = a$. In order to describe a charge at the origin, we shall consider the limit $a \to 0$ near the end of the calculation. The charge density for a point charge located at the point $\vec{r}_0 = a\,\hat{e}_z$ is given as

$$\rho(\vec{r}) = q_0\,\delta^{(3)}(\vec{r}-\vec{r}_0) = \frac{q_0}{a^2\,\sin\theta} \, \delta(r-r_0) \, \delta(\theta-\theta_0) \, \delta(\varphi-\varphi_0)$$

$$= \frac{q_0}{a^2} \, \delta(r-a) \, \delta(\cos\theta-1) \, \delta(\varphi-\varphi_0) \,.$$
(2.227)

We have used the fact that $\theta_0 = 0$ for a charge located at $\vec{r}_0 = a\,\hat{e}_z$, with $a > 0$ (hence, also, $r_0 = |\vec{r}_0| = a$). In terms of the Green function, expanded in the spherical Bessel functions and spherical harmonics, the potential is

$$\Phi(\vec{r}) = \frac{1}{4\pi\epsilon_0} \int d^3r' \, \frac{\rho(\vec{r}')}{|\vec{r}-\vec{r}'|} = \int_0^{\infty} dr'\,r'^2 \int_{-1}^{1} d(\cos\theta') \int_0^{2\pi} d\varphi' \, G(\vec{r},\vec{r}') \, \rho(\vec{r}')$$

$$= \frac{2q_0}{\pi\epsilon_0} \int_0^{\infty} dr' \int_{-1}^{1} d(\cos\theta') \int_0^{2\pi} d\varphi' \, \delta(r'-a) \, \delta(\cos\theta'-1) \, \delta(\varphi'-\varphi_0) \left(\frac{r'}{a}\right)^2$$

$$\times \sum_{\ell=0}^{\infty} \sum_{m=-\ell}^{\ell} \int_0^{\infty} dk \, j_\ell(k\,r) \, j_\ell(k\,r') \, Y_{\ell m}(\theta,\varphi) \, Y_{\ell m}^*(\theta',\varphi')$$

$$= \frac{2q_0}{\pi\epsilon_0} \sum_{\ell=0}^{\infty} \sum_{m=-\ell}^{\ell} \int_0^{\infty} dk \, j_\ell(k\,r) \, j_\ell(k\,a) \, Y_{\ell m}(\theta,\varphi) \, Y_{\ell m}^*(\theta_0,\varphi_0)$$

$$= \frac{2q_0}{\pi\epsilon_0} \sum_{\ell=0}^{\infty} \frac{2\ell+1}{4\pi} \, P_\ell(\cos(\angle(\vec{r},\vec{r}_0))) \int_0^{\infty} dk \, j_\ell(k\,r) \, j_\ell(k\,a) \,.$$
(2.228)

The integration over φ' leads to the replacement $\varphi' \to \varphi_0$ in the argument of the spherical harmonic. We have used the addition theorem (2.111). We now consider the limit $a \to 0$. Since $j_\ell(0) = 0$ for $\ell \neq 0$ and only the spherical Bessel function of order zero survives, with unit value $j_0(0) = 1$, the only surviving term in the sum over ℓ reads as

$$\Phi(\vec{r}) \overset{a\to 0}{=} \frac{2q_0}{\pi \epsilon_0} \frac{1}{4\pi} P_0\left(\cos\left(\angle(\vec{r},\vec{r}_0)\right)\right) \int_0^\infty dk\, j_\ell(k\,r) = \frac{q_0}{2\pi^2\epsilon_0} \int_0^\infty dk\, j_\ell(k\,r),$$
(2.229)

where we have used the fact that $P_0(x) = 1$. We thus have for $a \to 0$,

$$\Phi(\vec{r}) \overset{a\to 0}{=} \frac{q_0}{2\pi^2\epsilon_0} \int_0^\infty dk\, j_0(k\,r) = \frac{q_0}{2\pi^2\epsilon_0 r} \int_0^\infty dk\, \frac{\sin(k\,r)}{k}$$

$$= \frac{q_0}{2\pi^2\epsilon_0 r}\left(\int_0^\infty \frac{d\rho}{\rho} \sin(\rho)\right) = \frac{q_0}{2\pi^2\epsilon_0\, r}\left(\frac{\pi}{2}\right) = \frac{q_0}{4\pi\epsilon_0\, r}.$$
(2.230)

This is equal to the familiar result for a point charge located at the origin.

2.6 Summary: Green Function of Electrostatics

The Green function of electrostatics, first encountered in Eq. (2.23),

$$\vec{\nabla}^2 G(\vec{r},\vec{r}') = -\frac{1}{\epsilon_0}\delta^{(3)}(\vec{r}-\vec{r}'), \qquad G(\vec{r},\vec{r}') = \frac{1}{4\pi\epsilon_0\,|\vec{r}-\vec{r}'|},$$
(2.231)

has been the subject of our discussions. Let us summarize the three different representations we have derived for the Green function of electrostatics. First, we have the familiar decomposition in spherical coordinates, as derived in Sec. 2.3.4,

$$G(\vec{r},\vec{r}') = \frac{1}{\epsilon_0}\sum_{\ell=0}^\infty \sum_{m=-\ell}^\ell \frac{1}{2\ell+1}\frac{r_<^\ell}{r_>^{\ell+1}} Y_{\ell m}(\theta,\varphi)\, Y_{\ell m}^*(\theta',\varphi').$$
(2.232)

It relies on a matching of the regular solutions at the origin ($r = 0$) and at infinity ($r = \infty$). The corresponding representation in cylindrical coordinates is derived in Sec. 2.4.4 and reads

$$G(\vec{r};\vec{r}') = \frac{1}{4\pi\,\epsilon_0}\sum_{m=-\infty}^\infty e^{im(\varphi-\varphi')} \int_0^\infty dk\, e^{-k(z_>-z_<)} J_m(k\rho)\, J_m(k\rho').$$
(2.233)

It is based on a matching of the regular solutions at $z = -\infty$ and $z = +\infty$. The decomposition into eigenfunctions of the Laplacian operator is discussed in Sec. 2.5.2 and leads to Eq. (2.224),

$$G(\vec{r},\vec{r}') = \frac{2}{\pi\epsilon_0}\sum_{\ell=0}^\infty \sum_{m=-\ell}^\ell \int_0^\infty dk\, j_\ell(k\,r)\, j_\ell(k\,r')\, Y_{\ell m}(\theta,\varphi)\, Y_{\ell m}^*(\theta',\varphi').$$
(2.234)

The results (2.232) and (2.224) are matched using the result (2.226), which we recall for convenience,

$$\int_0^\infty dk\, j_\ell(k\,r)\, j_\ell(k\,r') = \frac{\pi}{2\,(2\ell+1)}\frac{r_<^\ell}{r_>^{\ell+1}}.$$
(2.235)

This result implements the cusp of the Green function at equal argument.

2.7   Exercises

• **Exercise 2.1**: By directly inserting the result given in Eq. (2.14) into the left-hand side of Eq. (2.5), verify the validity of the result given for the harmonic oscillator Green function.

• **Exercise 2.2**: We investigate the harmonic oscillator using Green function techniques and apply the Green function formalism to the harmonic oscillator. The "displacement" $x = x(t)$ of a damped, harmonic oscillator with unit mass $m = 1$ satisfies the equation

$$\ddot{x} + \gamma\,\dot{x} + \omega_0^2\,x = f(t) \,. \tag{2.236}$$

(i) Obtain the Green function $g = g(t - t')$ for this equation. Note that it has to fulfill the equation

$$\ddot{g}(t - t') + \gamma\,\dot{g}(t - t') + \omega_0^2\,g(t - t') = \delta(t - t') \,. \tag{2.237}$$

(ii) Why is this Green function naturally obtained as the retarded Green function? Why do you have to change the sign of the damping term in order to obtain the advanced Green function? Consider the effect of time reversal on the equation of motion, and provide an illustrative discussion.

(iii) In the case that the "force" is given by $f(t) = f_0\,[\Theta(t) + \Theta(-t)\exp(t/\tau)]$, with vanishing boundary conditions in the infinite past, $x(-\infty) = 0$, and $\dot{x}(-\infty) = 0$, evaluate $x(t)$.

(iv) Evaluate the work done by the driving force on the oscillator (per unit mass) in the time interval from $-\infty$ to $+\infty$ as a function of ω_0, γ, and τ.

(v) Now let $\gamma = \frac{1}{10}\omega_0$. Plot the work done by the driving force divided by the final (potential) energy stored in the oscillator as a function of $L = \ln(\omega_0\tau)$ from $L = -4$ to $L = 4$.

• **Exercise 2.3**: Starting from Eq. (2.30), derive Eq. (2.31) by applying the operators given in Eq. (2.30) to a test function.

• **Exercise 2.4**: Show Eq. (2.57) by using the explicit definitions of the spherical harmonics in terms of Legendre polynomials given in Eqs. (2.44)–(2.47), as well as Eq. (2.53), and the orthogonality properties of the Legendre polynomials [see Eq. (2.48)].

• **Exercise 2.5**: Derive Eq. (2.77) by acting with it on a test function.

• **Exercise 2.6**: For the charge distribution given in Eq. (2.100), calculate the potential in Eq. (2.103), i.e., fill the missing steps in the derivation.

• **Exercise 2.7**: Use the expansion (2.93) in order to generalize the potential (2.103) for $r < a$, still assuming the charge distribution (2.100).

• **Exercise 2.8**: Consider the charge distribution

$$\rho(\vec{r}) = \frac{Q}{a^2}\,\sin\theta\,\delta(r - a) \,. \tag{2.238}$$

Evaluate all multipoles of the given charge distribution with angular momenta $\ell =$ $0, 1, 2$ and write down the final potential. **Hint:** The potential should finally be obtained as follows,

$$\Phi(\vec{r}) = \frac{\pi Q}{4 \epsilon_0 r} - \frac{\pi Q a^2}{64 \epsilon_0 r^3} \left(3 \cos^2 \theta - 1\right). \tag{2.239}$$

- **Exercise 2.9**: For the charge distribution given in Eq. (2.104), carry out the multipole decomposition explicitly and calculate the potential in Eq. (2.105).
- **Exercise 2.10**: Show Eq. (2.119) based on the explicit form of the spherical harmonics for $\ell = 1$, given in Eq. (2.63).
- **Exercise 2.11**: Show Eq. (2.147) based on Eq. (2.146).
- **Exercise 2.12**: Show how to obtain Eq. (2.150) from Eq. (2.149).
- **Exercise 2.13**: Write a computer program in the language of your choice, implementing the Newton–Raphson method for finding the zeros of Bessel functions. Choosing the starting values of the Newton–Raphson method as integer-valued, verify the entries in Table 2.1.
- **Exercise 2.14**: In establishing Eq. (2.166), we have ignored the fact that we must actually treat the case $n = 0$ separately; it requires special attention at the lower limit of integration because $J_n(0) = \delta_{n0}$ for $n \in \mathbb{N}_0$. Fill this gap in our derivation. Hint: Use the known fact that $J_n'(0) = 0$.
- **Exercise 2.15**: Treat the special case $n = 0$ in Eq. (2.174), observing that the slope of $J_0(x)$ vanishes at $x = 0$.
- **Exercise 2.16**: Show the relation

$$\delta^{(3)}(\vec{r} - \vec{r}') = \frac{1}{\rho} \, \delta(\rho - \rho') \, \delta(z - z') \, \delta(\varphi - \varphi') \tag{2.240}$$

using similar arguments to those in the derivation of Eq. (2.70).
- **Exercise 2.17**: Use the expansions (2.203) and conceivably (2.204) in order to calculate the potential (in the upper half space $z > 0$) generated by a charge distribution in a "ring of Saturn" geometry located in the xy plane,

$$\rho(\vec{r}) = \frac{Q}{a^2} \, \mathbb{1}_{a < \rho < b} \, \delta(z). \tag{2.241}$$

Here, $\mathbb{1}_{a < \rho < b}$ is the characteristic function of the interval $\rho \in (a, b)$, i.e.,

$$\mathbb{1}_{a < \rho < b} = \Theta(\rho - a) - \Theta(\rho - b). \tag{2.242}$$

Generalize your result for the lower half space, i.e., $z < 0$.
- **Exercise 2.18**: Generalize the $\psi_{k\ell m}$ to eigenfunctions of the free Schrödinger Hamilton operator $H = -\vec{\nabla}^2/(2\mu)$ and establish their orthogonality properties. **Hint:** A remark is in order. The function

$$\psi_{k\ell m}(r, \theta, \varphi) = j_\ell(k r) \, Y_{\ell m}(\theta, \varphi) \tag{2.243}$$

fulfills the Schrödinger equation

$$H_0 \, \psi_{k\ell m}(r, \theta, \varphi) = \left(-\frac{\hbar^2 \vec{\nabla}^2}{2\mu}\right) \psi_{k\ell m}(r, \theta, \varphi) = \frac{p^2}{2\mu} \, \psi_{k\ell m}(r, \theta, \varphi), \qquad p = \hbar k. \tag{2.244}$$

In view of Eq. (2.178), the eigenfunctions satisfy the following orthogonality condition,

$$\langle \psi_{k\ell m} | \psi_{k\ell'm'} \rangle = \int_0^\infty dr\, r^2\, j_\ell(kr)\, j_{\ell'}(k'r) \int d\Omega\, Y_{\ell m}(\theta, \varphi)^* \, Y_{\ell'm'}(\theta, \varphi)$$

$$= \frac{\pi}{2\, k\, k'}\, \delta(k - k')\, \delta_{\ell\ell'}\, \delta_{mm'}\,. \tag{2.245}$$

This is immediately obvious if we take the prefactor $\sqrt{\pi/(2x)}$ in the definition of the spherical Bessel function into account.

• **Exercise 2.19**: Using Eq. (2.224), calculate the potential generated by a charge distribution

$$\rho(\vec{r}') = \frac{q_0}{a^2}\, \delta(r' - a) \tag{2.246}$$

where a is the radius of the sphere.

• **Exercise 2.20**: Using Eq. (2.224), reduce the potential generated by a charge distribution without compact support, given in spherical coordinates,

$$\rho(\vec{r}') = \frac{q_0}{a^3}\, \exp(-\lambda r)\, \cos\theta'\,, \tag{2.247}$$

to a one-dimensional integral (a is a parameter). Choose convenient numerical values for the parameters q_0, a and λ, and evaluate the remaining integral numerically for selected values of r and θ. Calculate the dipole moment of the charge distribution, and show analytically that it is equal to the long-range limit of the potential just calculated, even if part of the charge distribution is always in the outer volume $r' > r$. **Hint:** Observe that $\cos\theta'$ is proportional to a spherical harmonic in the primed coordinates, and use the result

$$\int_0^\infty dx\, J_{\ell+1/2}(x)\, r^n\, e^{-\lambda r} = \frac{\left(\ell + \frac{3}{2}\right)_n}{2^{\ell+1/2}\, \lambda^{n+\ell+3/2}}$$

$$\times\, {}_2F_1\left(\tfrac{1}{4}(2n + 2\ell + 3), \tfrac{1}{4}(2n + 2\ell + 5), \ell + \tfrac{3}{2}, -\frac{1}{\lambda^2}\right),$$

$$\tag{2.248}$$

where $(a)_n = \Gamma(a + n)/\Gamma(a)$ is the Pochhammer symbol. The complete (Gaussian) hypergeometric function is denoted as ${}_2F_1$. Part of the exercise is to look up the definition of the hypergeometric function, and to become familiar with its numerical properties.

Chapter 3

Paradigmatic Calculations in Electrostatics

3.1 Overview

3.1.1 *Differential Equations of Electrostatics*

The previous chapter dealt with the Green functions of electrostatics, which fulfill the differential equation (2.22), which we recall for convenience,

$$\vec{\nabla}^2 G(\vec{r} - \vec{r}') = -\frac{1}{\epsilon_0} \delta^{(3)}(\vec{r} - \vec{r}').\tag{3.1}$$

Given a formula for G, one can calculate the electrostatic potential due to charge distribution $\rho(\vec{r}')$ with the help of the formula (2.24), which we write as

$$\Phi(\vec{r}) = \int \mathrm{d}^3 r'\, G(\vec{r}, \vec{r}')\, \rho(\vec{r}') = \frac{1}{4\pi\epsilon_0} \int \mathrm{d}^3 r'\, \frac{\rho(\vec{r}')}{|\vec{r} - \vec{r}'|}.\tag{3.2}$$

Given Φ, one can calculate the electric field according to Eq. (1.59),

$$\vec{E}(\vec{r}) = -\vec{\nabla}\Phi(\vec{r}),\tag{3.3}$$

for the time-independent (static) case where the time derivative of the vector potential $\vec{A}$ vanishes.

In principle, one might assume that formula (2.24) solves all physically relevant problems in electrostatics. However, that is not the case. Let us look at the Poisson equation

$$\vec{\nabla}^2 \Phi(\vec{r}) = -\frac{1}{\epsilon_0} \rho(\vec{r}),\tag{3.4}$$

and analyze its properties, being a differential equation. We know that the solutions of differential equations are never uniquely defined without the specification of boundary conditions. Let us assume that the charge distribution $\rho(\vec{r}')$ in Eq. (3.2) has no significant long-range tails, and that it is concentrated in the region around the origin, $\vec{r}' \approx 0$. In this case, $\Phi(\vec{r})$ vanishes for $r = |\vec{r}| \to \infty$, as a potential proportional to $1/r$, commensurate with the leading (monopole) term in the multipole expansion (2.99). The boundary condition implemented in Eq. (3.2) therefore reads as

$$\Phi(\vec{r})|_{r\to\infty} = 0,\tag{3.5}$$

and Φ is defined for all $\vec{r} \in \mathbb{R}$.

Yet, this cannot be the whole story. E.g., one may ask how potentials should be calculated in a volume V of space whose surface manifold ∂V, consisting of a perfect conductor, is held at a constant potential $\Phi(\vec{r})|_{\vec{r} \in \partial V} = \Phi_0$. In this case, if there are no charges present inside the volume V, the simple solution $\Phi(\vec{r}) = \Phi_0$ will solve the differential equation (3.4) with $\rho = 0$. However, if the charge distribution inside V is nonvanishing, then the boundary-value problem cannot be solved using Eq. (3.2) because the boundary condition $\Phi(\vec{r})|_{\vec{r} \in \partial V} = \Phi_0$ is not fulfilled.

Let us also ask about the solution to a boundary-value problem given by a cube, held at zero potential on five of the six side surfaces, with the remaining sixth side surface held at potential $\Phi_0 \neq 0$. In this case, even if there are no charges present inside the cube, then the solution to the Poisson equation $\vec{\nabla}^2 \Phi(\vec{r}) = 0$ cannot be given by the simple *ansatz* $\Phi(\vec{r}) = 0$, because it does not fulfill the boundary condition on the sixth side surface. Even more generally, one may ask how to calculate the solution to a boundary-value problem

$$\vec{\nabla}^2 \Phi(\vec{r}) = -\frac{1}{\epsilon_0} \rho(\vec{r}), \qquad \Phi(\vec{r})|_{\vec{r} \in \partial V} = f(\vec{r}), \tag{3.6}$$

with a nontrivial function $f(\vec{r})$ describing the potential on the surface manifold.

In principle, all of the above questions can be traced to the availability of non-trivial solutions to the Laplace equation (2.20)

$$\vec{\nabla}^2 \Phi(\vec{r}) = 0, \tag{3.7}$$

which can be added to the solutions of the Poisson equation (3.4), to fulfill the boundary conditions. However, there are more sophisticated techniques available in comparison to a simple guessing of the correct nontrivial solution of the Laplace equation which will lead to a solution of a given boundary-value problem. We start by temporarily falling back to a simpler one-dimensional model problem.

3.1.2 *Boundary-Value Problems: One-Dimensional Analogy*

We investigate a one-dimensional analogue of the Laplace equation $\vec{\nabla}^2 \Phi(\vec{r}) = 0$,

$$f''(x) = \frac{\partial^2}{\partial x^2} f(x) = 0. \tag{3.8}$$

It is known that the solution of a differential equation in one variable can be made unique on the basis of two boundary conditions. As an example, Eq. (3.8) is solved by any function of the form

$$f(x) = \mathcal{C} + \mathcal{D} x, \tag{3.9}$$

with constant coefficients $\mathcal{C}$ and $\mathcal{D}$. A differential equation for $f = f(x)$ is an equation describing a function on a one-dimensional interval $x \in [a, b]$. The boundary conditions are related to a zero-dimensional manifold of points, i.e., to the specific points $x = a$ and $x = b$. Let us consider a slightly more complex "boundary-value problem",

$$f''(x) = s(x), \qquad f(a) = s_a, \qquad f(b) = s_b, \tag{3.10}$$

where $s(x)$ is a source term. We shall later see that this situation corresponds to a so-called Dirichlet boundary-value problem. We are free to choose functions $g(x)$ with $g''(x) = 0$ and add them to a particular solution $f(x)$,

$$f(x) \to f(x) + \mathcal{C} + \mathcal{D}\,x. \tag{3.11}$$

The boundary conditions then read as follows,

$$f(a) = \mathcal{C} + \mathcal{D}\,a = s_a, \qquad f(b) = \mathcal{C} + \mathcal{D}\,b = s_b, \tag{3.12}$$

and they fix the integration constants $\mathcal{C}$ and $\mathcal{D}$ uniquely, even if we have not used any boundary conditions that involve the derivative $f'(x)$. For a differential equation in a single variable, the boundary of the integration interval is the set of points $[a, b]$ that define the interval $a \le x \le b$. Boundary conditions that fix the derivative $f'(x)$ of the function at the end points $x = a$ and $x = b$ determine $f(x)$ up to an additive constant. This situation corresponds to a so-called Neumann boundary-value problem.

The integration interval in this case is one-dimensional, whereas the boundary conditions are defined on a zero-dimensional "surface", namely, at the end points of the interval (a, b). We can now figure an analogy: Namely, for the three-dimensional case, the integration interval is three-dimensional, and the boundary conditions must be defined on a two-dimensional surface. Indeed, the time-independent Poisson equation $\vec{\nabla}^2 \Phi(\vec{r}) = -\rho(\vec{r})/\epsilon_0$ is a second-order differential equation whose solution is defined for $\vec{r} \in V$. Consequently, its boundary conditions are relevant for the boundary ∂V of a three-dimensional volume V.

3.1.3 Green's Theorem: Dirichlet and Neumann Green Functions

We have already stressed that the solution of the Poisson equation (3.4) is not unique; it may be modified by the addition of a solution of the homogeneous (Laplace) equation. This property can be taken advantage of in the analysis of the Green function. Let us recall the defining differential Eq. (2.22),

$$\vec{\nabla}^2 G(\vec{r}, \vec{r}') = -\frac{1}{\epsilon_0}\,\delta^{(3)}(\vec{r} - \vec{r}'), \qquad G(\vec{r}, \vec{r}') = \frac{1}{4\pi\epsilon_0\,|\vec{r} - \vec{r}'|}, \tag{3.13}$$

as a suitable electrostatic Green function. Indeed, we may always add a solution of the homogeneous equation, $\vec{\nabla}^2 G(\vec{r}, \vec{r}') = 0$. Green functions can be unique by the specification of boundary conditions, much in the same way as potentials are uniquely specified by differential equations and boundary conditions.

This statement needs to be illustrated. We have already encountered, in Eq. (3.6), what we now identify as a Dirichlet boundary-value problem,

$$\vec{\nabla}^2 \Phi(\vec{r}) = -\frac{1}{\epsilon_0}\,\rho(\vec{r}), \qquad \Phi(\vec{r})\big|_{\vec{r} \in \partial V} = f(\vec{r}). \tag{3.14}$$

One specifies the charge distribution $\rho(\vec{r})$ and the value $f(\vec{r})$ on the boundary ∂V of the volume V. A Neumann boundary-value problem is specified as

$$\vec{\nabla}^2 \Phi(\vec{r}) = -\frac{1}{\epsilon_0}\,\rho(\vec{r}), \qquad \vec{\nabla}\Phi(\vec{r})\big|_{\vec{r} \in \partial V} = \vec{g}(\vec{r}), \tag{3.15}$$

with a vector-valued function $\vec{g}(\vec{r})$. This corresponds to a boundary condition for the electric field instead of the potential on the surface ∂V. Actually, we could have stated the Neumann boundary-value problem as

$$\vec{\nabla}^2 \Phi(\vec{r})\Big|_{\vec{r} \in V} = -\frac{1}{\epsilon_0}\, \rho(\vec{r}) \,, \qquad \hat{n}(\vec{r}) \cdot \vec{\nabla}\Phi(\vec{r})\Big|_{\vec{r} \in \partial V} = h(\vec{r}) \,, \qquad (3.16)$$

where $\hat{n}(\vec{r})$ is the unit normal pointing away from the surface.

We shall now attempt to relate the boundary conditions fulfilled by the potential, to the boundary conditions fulfilled by the Green function. It is useful to consider Green's theorem in that context, which will enable us to accomplish the goal. On our way to Green's theorem, we first observe that the divergence theorem (Gauss's theorem) states that for any $\vec{r}\,' \in V$,

$$\int_V \mathrm{d}^3 r\, \vec{\nabla} \cdot \left[\Phi(\vec{r})\vec{\nabla}\, G(\vec{r},\vec{r}\,') - G(\vec{r},\vec{r}\,')\, \vec{\nabla}\Phi(\vec{r})\right]$$

$$= \int_{\partial V} \mathrm{d}\vec{A} \cdot \left[\Phi(\vec{r})\vec{\nabla} G(\vec{r},\vec{r}\,') - G(\vec{r},\vec{r}\,')\, \vec{\nabla}\Phi(\vec{r})\right] . \qquad (3.17)$$

By carrying out the differentation on the left-hand side explicitly, we see that

$$\int_V \mathrm{d}^3 r\, \vec{\nabla} \cdot \left[\Phi(\vec{r})\vec{\nabla}\, G(\vec{r},\vec{r}\,') - G(\vec{r},\vec{r}\,')\, \vec{\nabla}\Phi(\vec{r})\right]$$

$$= \int_V \mathrm{d}^3 r\, \left[\Phi(\vec{r})\, \vec{\nabla}^2 G(\vec{r},\vec{r}\,') - G(\vec{r},\vec{r}\,')\, \vec{\nabla}^2\Phi(\vec{r})\right]$$

$$= -\frac{1}{\epsilon_0} \int_V \mathrm{d}^3 r\, \delta^{(3)}(\vec{r}-\vec{r}\,')\, \Phi(\vec{r}) - \int_V \mathrm{d}^3 r\, G(\vec{r},\vec{r}\,')\left[-\frac{1}{\epsilon_0}\rho(\vec{r})\right]$$

$$= -\frac{1}{\epsilon_0}\Phi(\vec{r}\,') + \frac{1}{\epsilon_0} \int_V \mathrm{d}^3 r\, G(\vec{r},\vec{r}\,')\, \rho(\vec{r}) \qquad (3.18)$$

where we assume that $\vec{r}\,' \in V$. We solve Eq. (3.18) for $\Phi(\vec{r}\,')$,

$$\Phi(\vec{r}\,') = \int_V \mathrm{d}^3 r\, G(\vec{r},\vec{r}\,')\, \rho(\vec{r}) - \epsilon_0 \int_{\partial V} \mathrm{d}\vec{A} \cdot \left[\Phi(\vec{r})\, \vec{\nabla} G(\vec{r},\vec{r}\,') - G(\vec{r},\vec{r}\,')\, \vec{\nabla}\Phi(\vec{r})\right] .$$
$$(3.19)$$

This formulation involves a volume term (convolution term with the charge distribution) and a surface term (integral over ∂V). It is known as Green's theorem.

Equation (3.19) still does not solve the boundary-value problem. Moreover, the right-hand side of Eq. (3.19) actually contains two terms, the evaluation of which requires knowledge of both the values of the potential and of its gradient on the surface ∂V. However, one can show that knowledge of either $\Phi(\vec{r})$ or of $\vec{\nabla}\Phi(\vec{r})$ is sufficient in order to specify a unique boundary-value problem (in the latter case, up to an additive constant).

At this stage, we recall that the Green function, (or just "Green's function" but without the "the") is named after George Green, who was mentioned in this book previously, and does not carry the Irish national color (which happens to be green). Likewise, the PLANCK constant[1] is denoted as $\hbar$, and sometimes referred

---

[1] Max Karl Ernst Ludwig Planck (1858–1947)

to as "Planck's constant" (but in the latter case, without the "the" in front of the word).

In order to evaluate appropriate solutions to a boundary-value problem of the form Eq. (3.14), (3.15), or (3.16), it is useful to employ a Green function that fulfills additional boundary conditions. The goal is to discard one of the terms on the right-hand side of Eq. (3.19). The defining equation for the so-called Dirichlet Green function G_D is

$$\vec{\nabla}^2 G_D\left(\vec{r},\vec{r}'\right)\Big|_{\vec{r},\,\vec{r}'\in V} = -\frac{1}{\epsilon_0}\delta^{(3)}\left(\vec{r}-\vec{r}'\right), \qquad G_D\left(\vec{r},\vec{r}'\right)\Big|_{\vec{r}\in V,\vec{r}'\in\partial V} = 0. \qquad (3.20)$$

In this case, the solution to the boundary-value problem (3.14) becomes

$$\Phi_D(\vec{r}') = \int_V d^3r\, G_D\left(\vec{r},\vec{r}'\right)\rho(\vec{r}) - \epsilon_0 \int_{\partial V} d\vec{A}\cdot\vec{\nabla}G_D\left(\vec{r},\vec{r}'\right)\Phi(\vec{r})$$

$$= \int_V d^3r\, G_D\left(\vec{r},\vec{r}'\right)\rho(\vec{r}) - \epsilon_0 \int_{\partial V} d\vec{A}\cdot\vec{\nabla}G_D\left(\vec{r},\vec{r}'\right) f(\vec{r}). \qquad (3.21)$$

The second integral requires the values of the potential on the boundary. We have fulfilled our paradigm: The function G_D depends only on the shape of the reference volume V and can be used to integrate any given Dirichlet boundary-value problem in the given volume. The solution is given by the general formula (3.21).

We now turn our attention to a Neumann boundary-value problem. In this case, the gradient of the potential is given on the surface ∂V, as indicated in Eq. (3.15). Let us write Green's theorem (3.19) in a slightly different form,

$$\Phi(\vec{r}') = \int_V d^3r\, G\left(\vec{r},\vec{r}'\right)\rho(\vec{r}) - \epsilon_0 \int_{\partial V} d\vec{A}\cdot\vec{\nabla}G\left(\vec{r},\vec{r}'\right)\Phi(\vec{r})$$

$$+ \epsilon_0 \int_{\partial V} d\vec{A}\cdot\vec{\nabla}\Phi(\vec{r})\, G\left(\vec{r},\vec{r}'\right). \qquad (3.22)$$

Under Neumann boundary conditions, we do not know the potential $\Phi(\vec{r})$ itself on the surface. So, we would like the second term on the right-hand side of Eq. (3.22) to become irrelevant if the Green function G fulfills special properties, summarized in the Neumann Green function G_N. This is in analogy to the Dirichlet boundary-value problem, where we had arranged G_D so that the third term on the right-hand side of (3.22) becomes irrelevant.

Now, in a typical case, we cannot simply assume that $\vec{\nabla}G\left(\vec{r},\vec{r}'\right)$ vanishes on the surface ∂V; this would imply that all its three Cartesian components vanish, or in other words, that the Green function $G\left(\vec{r},\vec{r}'\right)$, as a function of $\vec{r}$, for an arbitrary but fixed $\vec{r}'$, is an extremum on the surface ∂V in all three spatial directions. There is a way out of this dilemma. It is instructive to observe that the second term on the right-hand side of Eq. (3.22) involves the expression

$$d\vec{A}\cdot\vec{\nabla}G\left(\vec{r},\vec{r}'\right) = dA\left(\hat{n}(\vec{r})\cdot\vec{\nabla}G\left(\vec{r},\vec{r}'\right)\right). \qquad (3.23)$$

For reasons to be explained in the following, we shall assume that we can assign a constant value to the projection of the gradient of the Neumann Green function

$G_N(\vec{r}, \vec{r}')$ onto the surface normal $\hat{n}(\vec{r})$,

$$\vec{\nabla}^2 G_N(\vec{r}, \vec{r}')\big|_{\vec{r}, \vec{r}' \in V} = -\frac{1}{\epsilon_0} \delta^{(3)}(\vec{r} - \vec{r}'), \qquad (3.24a)$$

$$\vec{\nabla} G_N(\vec{r}, \vec{r}') \cdot \hat{n}(\vec{r})\big|_{\vec{r} \in V, \vec{r}' \in \partial V} = -\frac{1}{\epsilon_0 A_{\text{total}}} = \left(-\epsilon_0 \int_{\partial V} dA\right)^{-1}, \qquad (3.24b)$$

where A_{total} is the total surface area of ∂V. We first convince ourselves that the second term on the right-hand side of Eq. (3.22) assumes the form of an irrelevant constant term, independent of the position on the surface,

$$\int_{\partial V} \Phi(\vec{r}) \, d\vec{A} \cdot \vec{\nabla} G_N(\vec{r}, \vec{r}') = \int_{\partial V} dA \, \hat{n}(\vec{r}) \cdot \vec{\nabla} G_N(\vec{r}, \vec{r}') \, \Phi(\vec{r})$$

$$= -\frac{\int_{\partial V} dA \, \Phi(\vec{r})}{\epsilon_0 \int_{\partial V} dA} \equiv -\frac{\langle \Phi \rangle}{\epsilon_0}. \qquad (3.25)$$

This corresponds to the average value $\langle \Phi \rangle$ of the potential Φ, taken on the surface ∂V. With these assumptions on the Neumann Green function G_N, Eq. (3.19) becomes

$$\Phi_N(\vec{r}') = \int_V d^3 r \, G_N(\vec{r}, \vec{r}') \, \rho(\vec{r}) + \langle \Phi \rangle + \epsilon_0 \int_{\partial V} d\vec{A} \cdot \vec{\nabla} \Phi(\vec{r}) \, G_N(\vec{r}, \vec{r}')$$

$$= \int_V d^3 r \, G_N(\vec{r}, \vec{r}') \, \rho(\vec{r}) + \langle \Phi \rangle + \epsilon_0 \int_{\partial V} dA \, \hat{n}(\vec{r}) \cdot \vec{g}(\vec{r}) \, G_N(\vec{r}, \vec{r}')$$

$$= \int_V d^3 r \, G_N(\vec{r}, \vec{r}') \, \rho(\vec{r}) + \langle \Phi \rangle + \epsilon_0 \int_{\partial V} dA \, h(\vec{r}) \, G_N(\vec{r}, \vec{r}'), \qquad (3.26)$$

where we have assumed Neumann boundary conditions as given in Eq. (3.15) and (3.16). It becomes clear that in order write down a Neumann boundary-value problem, it is sufficient to specify the component of the gradient of the potential along the surface normal. Furthermore, it becomes clear that the solution to the Neumann problem (3.26) involves a physically irrelevant constant term $\langle \Phi \rangle$, which can be chosen to be equal to the average value of the potential on the surface ∂V if the condition (3.24) is fulfilled by the Green function.

3.1.4 *Boundary-Value Problems and Laplace Equation*

A special status must be attributed to electrostatic boundary-value problems where the charge distribution vanishes, $\rho(\vec{r}) = 0$. In this case, the potential fulfills the Laplace equation $\vec{\nabla}^2 \Phi(\vec{r}) = 0$ for $\vec{r} \in V$. As already mentioned, this equation still allows for nontrivial solutions specified by boundary conditions. The first way of solving this kind of problem refers to Green's theorem as discussed in the previous section. In this case, the solution is formally given by setting ρ to zero in Eqs. (3.21) and (3.26). For a Dirichlet problem, we have

$$\Phi_D(\vec{r}') = -\epsilon_0 \int_{\partial V} d\vec{A} \cdot \vec{\nabla} G_D(\vec{r}, \vec{r}') \, \Phi(\vec{r}), \qquad \rho(\vec{r}) = 0, \qquad (3.27)$$

and for a Neumann problem, the solution is given as

$$\Phi_N(\vec{r}') = \epsilon_0 \int_{\partial V} \mathrm{d}\vec{A} \cdot \vec{\nabla}\Phi(\vec{r})\, G_N(\vec{r}, \vec{r}') \,, \qquad \rho(\vec{r}) = 0 \,. \tag{3.28}$$

However, the explicit calculation of the Dirichlet or Neumann Green functions can be difficult for nontrivial geometries, and it is noteworthy that two alternative methods exist which can be used in order to tackle boundary-value problems with a vanishing charge distribution. These methods are based on two very fundamental ideas which constitute recurrent themes throughout theoretical physics and therefore deserve special mention; the fundamental ideas are *(i)* the variational principle and *(ii)* series expansions.

The electrostatic field energy E, in the absence of charges, is given by

$$E = \frac{\epsilon_0}{2} \int_V \mathrm{d}^3 r \left(\vec{\nabla}\Phi(\vec{r}) \right)^2 \,. \tag{3.29}$$

A slight change ("variation") of the integrand $[\vec{\nabla}\Phi(\vec{r})]^2$ will lead to a concomitant change in the field energy. As we shall see, variational calculus shows that the physically relevant, correct potential actually minimizes the integral E. The situation bears a certain analogy to the shape of a soap film clamped within an irregularly shaped closed wire loop: The wire acts as a boundary condition, and the soap film minimizes its total surface area, which is tantamount to saying that its Gaussian curvature vanishes locally [12]. The latter condition of vanishing local curvature, translated into the variation of the electrostatic field energy, reads as

$$\vec{\nabla}^2 \Phi(\vec{r}) = 0 \,, \tag{3.30}$$

which is, in fact, identical to the Laplace equation. Boundary conditions can be implemented by the use of so-called Lagrange multipliers. In typical cases, a solution to the boundary-value problem can be obtained through an explicit variation of $\Phi(\vec{r})$. For example, one may assume for Φ, a particular functional form with free parameters, and then find the particular set of expansion parameters that minimizes the electrostatic field energy.

The other method is based on series expansions. Let us assume that we can expand $\Phi(\vec{r})$ in terms of a series of mutually orthogonal functions, say, in terms of a Fourier series,

$$\Phi(\vec{r}) = \sum_n a_n \, \exp(\mathrm{i}\vec{k}_n \cdot \vec{r}) \,, \tag{3.31}$$

where the $\vec{k}_n$ are drawn from an "allowed" set of Fourier expansion coefficients. The $\vec{k}_n$ are complex, so that

$$\vec{\nabla}^2 \exp(\mathrm{i}\vec{k}_n \cdot \vec{r}) = 0 \,, \qquad \vec{k}_n^2 = 0 \,. \tag{3.32}$$

This condition does not imply that all components of $\vec{k}_n$ need to vanish; in fact, the components of $\vec{k}_n$ can be complex-valued. However, it simply means that, say, oscillatory solutions in the x and y directions must be accompanied by exponential behavior in the z direction. Provided the surface ∂V is such that the Fourier components remain orthogonal upon integration over the surface area, we can typically project out the expansion coefficient a_n with the help of an integral proportional to

$$a_n \propto \int_{\partial V} \mathrm{d}S \, \Phi(\vec{r}) \, \mathrm{e}^{-\mathrm{i}\vec{k}_n \cdot \vec{r}} \,. \tag{3.33}$$

Using this approach, given appropriate boundary conditions on ∂V, we can uniquely determine the expansion coefficient a_n. The basic idea for the series expansion approach to the solution of boundary-value problems in the context of the Laplace equation can thus be given as follows: One writes a suitable expansion for the potential $\Phi(\vec{r})$, hopefully in terms of mutually orthogonal functions (with respect to a suitably defined scalar product, to be discussed in more detail in the following). This renders the determination of the expansion coefficients unique. Then, one projects out the expansion coefficients by evaluating overlap integrals like the one in Eq. (3.33). Finally, all expansion coefficients are determined, and one can gradually obtain better approximations to $\Phi(\vec{r})$ by summing more terms of the (typically infinite) series that describes the potential.

The main difficulties of the variational method and the series expansion method are different; one difficulty lies in the implementation of structurally complex boundary condition. In the case of the series expansion, the conditions imposed on the expansion coefficients are defined uniquely only if the functions in the set are orthogonal with respect to the integration measure (i.e., one must use a symmetry-adapted basis). In the variational method, the form of the boundary conditions, expressed in terms of the variational parameters, is easily tractable only if, again, a symmetry-adapted basis is used.

In summary, we have three possibilities for solving Dirichlet problems in the context of the Laplace equation. *(i)* The Dirichlet Green function method is very convenient if we know the Dirichlet Green function. This function is uniquely defined for a given reference volume, and incorporates the boundary conditions naturally. *(ii)* The variational method is guaranteed to converge provided we use a sufficiently large basis for our trial functions. *(iii)* The series expansion method also is guaranteed to converge provided we use a complete set of functions, which hopefully is orthogonal with respect to an integral measure. The latter two methods will be illustrated by example calculations in the following Secs. 3.2 and 3.3, while a nontrivial example for the conceptually difficult first method (Dirichlet Green function) is relegated to Sec. 3.5.

3.2   Laplace Equation and Variational Calculations

3.2.1   *Definition of the Functional Derivative*

Let us first recall the formalism of multivariate functions, i.e., functions of many variables. A function S of N arguments, denoted here as f_i with $i = 1, \ldots, N$, can be written as

$$S = S(f_1, \ldots, f_N). \tag{3.34}$$

If S just singles out a particular argument (say, of subscript j), then, for example,

$$S(f_1, \ldots, f_N) = f_j, \qquad \frac{\partial S}{\partial f_i} = \delta_{ij}, \tag{3.35}$$

where δ_{ij} is the Kronecker delta. For a function defined as the sum of the squares of its (many) arguments, one has

$$S = S(f_1, \ldots, f_N) = \sum_{i=1}^{N} F(f_i) = \sum_{i=1}^{N} (f_i)^2, \qquad \frac{\partial S}{\partial f_i} = F'(f_i) = 2 f_i. \tag{3.36}$$

Let us now interpret the sum, multiplied by a "standard step size" Δx, as an approximation to an integral,

$$\sum_{i=1}^{N} (f_i)^2 \, \Delta x \approx \int [f(x)]^2 \, \mathrm{d}x = S[f]. \tag{3.37}$$

Here, F is reinterpreted (with a slightly ambiguous notation) as a "function of the function f." A function of a function is commonly referred to as a "functional."

Any function is specified by its values on an arbitrarily dense set of points, which can be explained by the following correspondence (in a hopefully somewhat self-explanatory notation),

$$\{f_1, \ldots, f_N\} \triangleq \{f(x_1), \ldots, f(x_N)\} \triangleq f = f(x). \tag{3.38}$$

The generalization of Eq. (3.35) to the case of a continuous argument reads respectively,

$$S[f] = \int f(x) \delta(x - x') \, \mathrm{d}x' = f(x'), \qquad \frac{\delta S}{\delta f(x)} = \delta(x - x'), \tag{3.39}$$

and for Eq. (3.36), the generalization reads as

$$S[f] = \int \mathrm{d}x \, F(f(x))^2 \mathrm{d}x = \int \mathrm{d}x \, [f(x)]^2, \qquad \frac{\delta S}{\delta f(x)} = F'(f(x)) = 2 f(x). \tag{3.40}$$

This defines the functional derivative $\delta S / \delta f(x)$ for a functional of the form

$$S[f] = \int F(f(x)) \, \mathrm{d}x, \qquad \frac{\delta S}{\delta f(x)} = \frac{\partial F}{\partial f}\bigg|_{f=f(x)} = F'(f(x)). \tag{3.41}$$

An intuitive understanding is gained if we consider a discretized approximation to the integral and consider a "standard step size" of $\Delta x = 1$ for the integral. Alternatively, we may investigate the variation

$$\delta S = S[f + \delta f] - S[f] = \int [F(f(x) + \delta f(x)) - F(f(x))] \, \mathrm{d}x \tag{3.42}$$

$$\approx \int F'(f(x)) \, \delta f(x)) \, \mathrm{d}x = \langle F'(f(x)), \delta f(x) \rangle, \tag{3.43}$$

and define the scalar product of functions, as the scalar product in the sense of Hilbert space vectors, i.e. $\langle f, g \rangle \equiv \int \mathrm{d}x\, f(x)\, g(x)$. Then, we can define the functional derivative as the expression multiplying $\delta f(x)$ under the integral sign, i.e.

$$\frac{\delta S}{\delta f(x)} \equiv \left. \frac{\partial F}{\partial f} \right|_{f = f(x)} = F'(f(x)). \tag{3.44}$$

A more systematic way of defining the functional derivative is as follows. Because $\partial S/\partial f(x)$ probes the function $f(x)$ near x, we can formulate it as

$$\frac{\delta S}{\delta f(x)} = \lim_{\epsilon \to 0} \int_V \mathrm{d}x' \, \frac{F\left(f(x') + \epsilon \delta(x' - x)\right) - F\left(f(x')\right)}{\epsilon}. \tag{3.45}$$

Of course, if $F = F(f(x), f'(x))$ is a function which depends not only on f itself, but also, on the derivative f', then we can calculate the variation as follows,

$$\frac{\delta S}{\delta f(x)} = \lim_{\epsilon \to 0} \int_V \frac{F\left(f(x') + \epsilon \delta(x' - x),\, f'(x') + \epsilon \delta'(x' - x)\right) - F\left(f(x'),\, f'(x')\right)}{\epsilon} \, \mathrm{d}x', \tag{3.46}$$

i.e., if the variation of $f(x')$ is given by the Dirac-δ term $\epsilon \delta(x'-x)$, then the variation of $f'(x')$ is given by the term $\epsilon \delta'(x' - x)$. Below we shall carry out the variation of a functional $S[\Phi]$, which is proportional to the electrostatic field energy, and we shall find that the fulfillment of the Laplace equation is equivalent to the vanishing of the variation. Here, $\Phi = \Phi(\vec{r})$ is the electrostatic potential.

3.2.2 *Variational Principle and Euler–Lagrange Equations*

In order to understand basic aspects of the variational technique, it is instructive to consider the functional derivative of a functional $S = S[\psi]$, where $\psi = \psi(\vec{r})$ replaces f as defined in Sec. 3.2.1 as the argument of the functional. So, ψ is to be understood as a function of a vector-valued argument $\vec{r}$. More formally, we consider the variation of a functional S,

$$S[\psi] = \int_V F(\psi(\vec{r}), \vec{\nabla}\psi(\vec{r})) \, \mathrm{d}^3 r. \tag{3.47}$$

The variation $\psi \to \psi + \delta\psi$ implies the variation $\vec{\nabla}\psi \to \vec{\nabla}\psi + \vec{\nabla}\delta\psi$, so that $\delta\vec{\nabla}\psi = \vec{\nabla}[\delta\psi]$ is the variation of $\vec{\nabla}\psi$. Thus, the variation δS of S is found to be

$$\delta S = S\left[\psi + \delta\psi, \vec{\nabla}\psi + \delta\vec{\nabla}\psi\right] - S\left[\psi, \vec{\nabla}\psi\right], \tag{3.48}$$

for small $\delta\psi$. To first order in $\delta\psi$, the variation δS reads as

$$\delta S = \int_V \left[\frac{\partial F}{\partial \psi} \, \delta\psi(\vec{r}) + \vec{\nabla}\delta\psi \, \frac{\partial F}{\partial \vec{\nabla}\psi} \right] \mathrm{d}^3 r, \tag{3.49}$$

where in Cartesian coordinates,

$$\vec{\nabla}[\delta\psi] \cdot \frac{\partial F}{\partial \vec{\nabla}\psi} \equiv \sum_{i=1}^{3} \left(\frac{\partial}{\partial x_i} \delta\psi \right) \frac{\partial F}{\partial(\partial\psi/x_i)}. \tag{3.50}$$

Using an integration by parts, one obtains

$$\delta S = \int_V d^3 r \left(\delta\psi(\vec{r}) \frac{\partial F}{\partial \psi} + \vec{\nabla} \cdot \left(\delta\psi \frac{\partial F}{\partial \vec{\nabla}\psi} \right) - \delta\psi(\vec{r}) \, \vec{\nabla} \cdot \frac{\partial F}{\partial \vec{\nabla}\psi} \right). \tag{3.51}$$

The divergence theorem implies that

$$\delta S = \int_V d^3 \left[\frac{\partial F}{\partial \psi} - \vec{\nabla} \cdot \frac{\partial F}{\partial \vec{\nabla}\psi} \right] \delta\psi(\vec{r}) + \int_{\partial V} dS \, \delta\psi(\vec{r}) \left(\frac{\partial F}{\partial \vec{\nabla}\psi} \cdot \hat{n} \right). \tag{3.52}$$

Generally, the values of the function $\psi(\vec{r})$ are uniquely specified on the boundary ∂V of the volume. Only those variations $\delta\psi(\vec{r})$ that vanish on the boundary are permitted. If we could even vary the boundary conditions, then the variational problem would be trivial: we would just choose the boundary conditions to be such that F vanishes on the boundary (and perhaps everywhere), or even shrink the volume V to zero. The functional S immediately vanishes and becomes optimal.

Under the assumption that $\delta\psi(\vec{r}) = 0$ for $\vec{r} \in \partial V$, the functional S has an extremum when

$$\delta S = \int_V d^3 r \, \delta\psi(\vec{r}) \left[\frac{\partial F}{\partial \psi} - \sum_{i=1}^{3} \frac{\partial}{\partial x_i} \frac{\partial F}{\partial (\partial\psi/\partial x_i)} \right] = 0 \tag{3.53}$$

for all "permissible" variations $\delta\psi(\vec{r})$.

So, the requirement that the first-order variation vanishes implies that $\psi(\vec{r})$ must be a solution of the familiar Euler-Lagrange equation,

$$\frac{\partial F}{\partial \psi} = \sum_{i=1}^{3} \frac{\partial}{\partial x_i} \frac{\partial F}{\partial (\partial\psi/\partial x_i)} = \vec{\nabla} \frac{\partial F}{\partial \vec{\nabla}\psi(\vec{r})}. \tag{3.54}$$

In order to determine whether the extremum is a minimum, maximum, or stationary point, the second order term in $\delta\psi$ must be calculated.

The Euler–Lagrange equation (3.54) corresponds to a situation where the function ψ is a function of $\vec{r}$, that is, of three arguments $x = x_1$, $y = x_2$ and $z = x_3$. E.g., for a functional of the form (3.29),

$$S[\Phi] = \frac{2E}{\epsilon_0} = \int_V d^3 r \left(\vec{\nabla}\Phi(\vec{r}) \right)^2, \tag{3.55}$$

the Euler–Lagrange equations can be verified to read

$$\vec{\nabla}^2 \Phi(\vec{r}) = 0, \tag{3.56}$$

which is simply equal to the Laplace equation.

Other cases also are of practical interest. For example, let us consider the variation of $S[\vec{r}(t)] = \int L(\vec{r}(t), \dot{\vec{r}}(t)) \, dt$. Here, the argument function $\vec{r} = \vec{r}(t)$ has three components but only one argument. This situation yields

$$\frac{\partial L}{\partial \vec{r}(t)} = \sum_{i=1}^{3} \frac{\partial}{\partial t} \frac{\partial L}{\partial \dot{\vec{r}}(t)}. \tag{3.57}$$

For the classical action of a single particle moving under the influence of a potential $V = V(\vec{r})$, the Lagrange function reads as

$$L(\vec{r}(t), \dot{\vec{r}}(t)) = \frac{m}{2}\dot{\vec{r}}^2(t) - V(\vec{r}(t)). \tag{3.58}$$

Under variation, the Euler–Lagrange equations then become the Newtonian equations of classical mechanics,

$$m\ddot{\vec{r}}(t) = -\frac{\partial}{\partial\vec{r}}V(\vec{r}) = -\vec{\nabla}V(\vec{r}(t)). \tag{3.59}$$

3.2.3   Variations with Constraints

In an extension of the variational calculations, the function $\psi(\vec{r})$ is required to satisfy constraints in addition to the boundary conditions. Generally, these have an integral form,

$$C_\alpha[\psi] = \int_V K_\alpha(\psi(\vec{r}), \vec{\nabla}\psi(\vec{r}))\,d^3r = D_\alpha, \tag{3.60}$$

where $\{D_\alpha, \alpha = 1, ..., N\}$ is a set of N constants. This places restrictions on the $\delta\psi(\vec{r})$ used in the variation of $\psi(\vec{r})$ in the previous formalism. Instead of

$$S[\psi] = \int_V F(\psi(\vec{r}), \vec{\nabla}\psi(\vec{r}))\,d^3r, \tag{3.61}$$

we now use the functional

$$S[\{\lambda_\alpha\}_{\alpha=1,...,N};\psi] = \int_V F(\psi(\vec{r}), \vec{\nabla}\psi(\vec{r}))\,d^3r + \sum_{\alpha=1}^N \lambda_\alpha(C_\alpha[\psi] - D_\alpha). \tag{3.62}$$

Varying this functional with respect to the λ_α, which are normal arguments, we obtain the N constraints back,

$$\frac{\partial}{\partial\lambda_\alpha}S[\{\lambda_\alpha\}_{\alpha=1,...,N};\psi] = C_\alpha[\psi] - D_\alpha = 0. \tag{3.63}$$

Proceeding as before to set the $\delta C_\alpha = 0$, we have $\alpha = 1, 2, 3, ..., N$ equations. Variation with respect to $\psi(\vec{r})$ leads to the modified variational equation,

$$\frac{\partial F}{\partial\psi} - \sum_{i=1}^3 \frac{\partial}{\partial x_i}\frac{\partial F}{\partial(\partial\psi/\partial x_i)} + \sum_\alpha \lambda_\alpha u_\alpha(\vec{r}) = 0, \tag{3.64}$$

where ψ is the function to be determined, and

$$u_\alpha(\vec{r}) = \frac{\partial K_\alpha}{\partial\psi} - \sum_{i=1}^3 \frac{\partial}{\partial x_i}\frac{\partial K_\alpha}{\partial(\partial\psi/\partial x_i)} \tag{3.65}$$

is a set of "vectors" in a function space. Let us mention a subtle point. The variation of the functional (3.60) with respect to the parameters λ_α gives us back the constraints, given in Eq. (3.63). If the variation is carried out to respect the constraints, then we must demand that $\delta\psi(\vec{r})$ lie in the subspace orthogonal to the $u_\alpha(\vec{r})$, in the sense that $C_\alpha[\psi+\delta\psi] = C_\alpha[\psi]$, which implies that $d^3r\,u_\alpha(\vec{r})\,\delta\psi(\vec{r}) = 0$. The "expansion coefficients" λ_α are constants, called Lagrange multipliers, which are determined such that the constraint equations are satisfied.

3.2.4 *Second Functional Derivative and Hilbert Space*

Let's consider once more the variation of the functional

$$S[\Phi] = \int_V d^3r \left(\vec{\nabla} \Phi(\vec{r}) \right)^2 , \tag{3.66}$$

this time using an explicit calculation, including the second-order variation. We calculate

$$
\begin{aligned}
S[\Phi + \delta\Phi] &= \int_V d^3r \left(\vec{\nabla}\Phi(\vec{r}) + \vec{\nabla}\delta\Phi(\vec{r}) \right)^2 \\
&= \int_V d^3r \left(\left(\vec{\nabla}\Phi(\vec{r}) \right)^2 + 2\,\vec{\nabla}\Phi(\vec{r})\,\vec{\nabla}\delta\Phi(\vec{r}) + \left(\vec{\nabla}\delta\Phi(\vec{r}) \right)^2 \right) \\
&= \int_V d^3r \left(\left(\vec{\nabla}\Phi(\vec{r}) \right)^2 - 2\vec{\nabla}^2\Phi(\vec{r})\,\delta\Phi(\vec{r}) + \delta\Phi(\vec{r}) \left(-\vec{\nabla}^2 \right) \delta\Phi(\vec{r}) \right) . \tag{3.67}
\end{aligned}
$$

So, the following result actually is exact (no higher-order corrections),

$$S[\Phi + \delta\Phi] - S[\Phi] = -2 \int d^3r \left(\vec{\nabla}^2\Phi(\vec{r}) \right) \delta\Phi(\vec{r}) + \int d^3r\, \delta\Phi(\vec{r}) \left(-\vec{\nabla}^2 \right) \delta\Phi(\vec{r}) , \tag{3.68}$$

where we ignore boundary terms due to the obvious restrictions on the variation $\delta\Phi(\vec{r})$, which is assumed to vanish on the boundary. The first- and second-order variational derivatives read as follows,

$$\frac{\delta S}{\partial \Phi(\vec{r})} = -2\vec{\nabla}^2\Phi(\vec{r}) , \qquad \frac{\delta^2 S}{\partial \Phi(\vec{r})\,\partial \Phi(\vec{r}')} = -2\delta^{(3)}(\vec{r} - \vec{r}')\,\vec{\nabla}^2 . \tag{3.69}$$

The first functional derivative should vanish at the extremum. This implies that

$$\vec{\nabla}^2\Phi(\vec{r}) = 0 , \tag{3.70}$$

at the minimum of the functional J, which is just the Laplace equation. However, we may ask, as well, how we intend to show that the second-order variation, i.e., the second term on the right hand side of Eq. (3.68), is positive. The second functional derivative is distribution-valued; it involves the Dirac-δ function (distribution). First we need to define what the positivity of the operator $-\vec{\nabla}^2$ could mean. The operator $-\vec{\nabla}^2$ acts on a space of functions which, according to Chaps. 6 and 7 of Ref. [13], is infinite-dimensional and constitutes a so-called Hilbert space. The determinant of an operator acting on an infinite-dimensional space is a so-called Fredholm determinant; however, even the positivity of the Fredholm determinant does not imply that the operator $-\vec{\nabla}^2$ is positive. Namely, we can understand the operators $-\vec{\nabla}^2$ in terms of its matrix representation, as it acts on a suitable set of basis functions that span the Hilbert space. If we can show that all the eigenvalues of $-\vec{\nabla}^2$ in that basis are positive, then the matrix representation of $-\vec{\nabla}^2$ corresponds to a positive definite matrix, and in that sense, we can understand the positivity of the operator $-\vec{\nabla}^2$ that occurs in the second functional derivative. It should be positive for the solution of the Laplace equation given in Eq. (3.70) to be a minimum of the functional $S[\Phi]$ given in Eq. (3.66).

Let us consider a one-dimensional analogue in order to heuristically show that $-\vec{\nabla}^2$ is a positive definite operator. On any finite interval $\{-L, L\}$, any function can uniquely be decomposed into

$$f(x) = \sum_{n=-\infty}^{\infty} a_n \exp\left(\frac{in}{L} x\right), \qquad (3.71)$$

in the sense of a Fourier decomposition. The eigenvectors of $-\vec{\nabla}^2$ are the functions $\exp(in x/L)$, because they fulfill

$$-\vec{\nabla}^2 \exp\left(\frac{in}{L} x\right) = \left(\frac{n}{L}\right)^2 \exp\left(\frac{in}{L} x\right). \qquad (3.72)$$

The eigenvalues are

$$\left(\frac{n}{L}\right)^2, \qquad n = -\infty, \ldots, \infty. \qquad (3.73)$$

If an operator only has positive eigenvalues, then it is positive definite. Generalizing from the one-dimensional example case to three dimensions, and letting $L \to \infty$, we can thus conclude that all eigenvalues of the operator $-\vec{\nabla}^2$ are positive, which again confirms that any solution of the Laplace equation indeed minimizes the functional S given in Eq. (3.66).

3.2.5 *Variational Calculation of the Capacitance of Plates*

We have learned that variational calculus corresponds to the search for a solution of an optimization problem in the infinite-dimensional Hilbert space of functions. One may require the functions to fulfill additional conditions, such as being square-integrable, or impose additional boundary-conditions, or other restrictions. Temporarily, we simplify the problem a bit, and limit the function space to a finite number of parameters. In the simplest case, we let our potential be $\Phi(\vec{r}) = w(\beta, \vec{r})$, where β is a single free parameter, and the class of functions $\Phi(\vec{r}) = w(\beta, \vec{r})$ is chosen to satisfy the boundary conditions on $\Phi(\vec{r})$ but is otherwise arbitrary.

The functional equation now is a function of only a vector of variables $\vec{\beta}$, and we therefore replace square brackets by round brackets,

$$S(\vec{\beta}) = \int_V d^3r \, F(\vec{r}, w(\vec{\beta}, \vec{r}), \vec{\nabla}w(\vec{\beta}, \vec{r})). \qquad (3.74)$$

If $w(\vec{\beta}, \vec{r}) \to \Phi(\vec{r})$ is just a "trial function" for the electrostatic potential, then

$$E(\vec{\beta}) = S(\vec{\beta}) = \frac{\epsilon_0}{2} \int d^3r \, \left[\vec{\nabla}w(\vec{\beta}, \vec{r})\right]^2. \qquad (3.75)$$

We now aim to find the extremum of $S(\vec{\beta})$. Let us suppose that there are boundary conditions, which relate to the surfaces $\partial V = \partial V_1 \cup \partial V_2$ of V,

$$w(\vec{\beta}, \vec{r}_1)\big|_{\vec{r}_1 \in \partial V_1} = 0, \qquad w(\vec{\beta}, \vec{r}_2)\big|_{\vec{r}_2 \in \partial V_2} = V_0, \qquad (3.76)$$

which can be implemented into our trial function $w = w(\vec{\beta}, \vec{r})$ via the use of Lagrange multipliers. In the presence of boundary conditions, some of the variational parameters (components of $\vec{\beta}$) constitute Lagrange multipliers. We then find the extremum value of $\vec{\beta}$, say $\vec{\beta}_{\min}$, for which

$$\left.\frac{\partial E}{\partial \beta_i}\right|_{\vec{\beta}=\vec{\beta}_{\min}} = 0, \qquad i = 1, 2, \dots, n. \tag{3.77}$$

If the matrix of second derivatives with entries

$$M_{ij} = \left.\frac{\partial^2 E}{\partial \beta_i \, \partial \beta_j}\right|_{\vec{\beta}=\vec{\beta}_{\min}} \tag{3.78}$$

is positive definite (all eigenvalues of M are positive), then the type of the extremum is as required, and $w(\vec{\beta}, \vec{r})$ is an approximation to $\Phi(\vec{r})$.

The variational program should now be illustrated by way of an example. We consider the potential between capacitor plates at $z = 0$ and $z = d$. The (large) plates are held at zero potential and at V_0, respectively. We start from the following *ansatz* for the potential inside the plates,

$$w(A, B, C, \vec{r}) = w(z) = A + B z + C z^2. \tag{3.79}$$

The energy E stored in the field assumes the role of a functional

$$E = E(A, B, C) = \frac{\epsilon_0}{2} \int_V |\vec{\nabla} w(A, B, C; \vec{r})|^2 \, d^3 r = \frac{\epsilon_0}{2} S \int_{z=0}^{z=d} \left(\frac{\partial w}{\partial z}\right)^2 dz$$

$$= \frac{\epsilon_0 S}{6} d \left(3B^2 + 6dBC + 4d^2 C^2\right). \tag{3.80}$$

Here, S is the surface area of the capacitor plates. The boundary conditions are

$$w(z = 0) = A = 0, \qquad w(z = d) = V_0 = A + B d + C d^2. \tag{3.81}$$

Including Lagrange multipliers, the functional thus is

$$E = E(A, B, C, \lambda_1, \lambda_2) = \frac{\epsilon_0 S}{6} d \left(3B^2 + 6 d B C + 4d^2 C^2\right)$$

$$+ \lambda_1 A + \lambda_2 (A + B d + C d^2 - V_0). \tag{3.82}$$

We now need to perform variations with respect to the five components of the vector

$$\vec{\beta} = (A, B, C, \lambda_1, \lambda_2). \tag{3.83}$$

The two equations

$$\frac{\partial}{\partial \lambda_1} E = 0, \qquad \frac{\partial}{\partial \lambda_2} E = 0, \tag{3.84}$$

just give us the boundary conditions back. The other derivatives read

$$\frac{\partial}{\partial A} E = \lambda_1 + \lambda_2,$$

$$\frac{\partial}{\partial B} E = d \left[S \epsilon_0 (B + C d) + \lambda_2\right],$$

$$\frac{\partial}{\partial C} E = \frac{d^2}{3} \left(S \epsilon_0 (3B + 4C d) + 3\lambda_2\right). \tag{3.85}$$

The solution of the system of equations

$$\frac{\partial}{\partial A} E = \frac{\partial}{\partial B} E = \frac{\partial}{\partial C} E = \frac{\partial}{\partial \lambda_1} E = \frac{\partial}{\partial \lambda_2} E = 0 \tag{3.86}$$

reads

$$A = C = 0, \qquad B = \frac{V_0}{d}, \qquad \lambda_1 = -\lambda_2 = \frac{\epsilon_0 S V_0}{d}. \tag{3.87}$$

Our trial function turned solution function thus becomes

$$w(z) = w_{\min}(z) = \frac{V_0 z}{d}. \tag{3.88}$$

The value of the energy at the extremum (minimum) becomes

$$E_{\min} = \frac{\epsilon_0}{2} S \int_{z=0}^{z=d} \left(\frac{\partial w_{\min}}{\partial z} \right)^2 dz = \frac{1}{2} \frac{\epsilon_0 S}{d} V_0^2. \tag{3.89}$$

Furthermore, from the equation

$$E_{\min} = \frac{1}{2} C V_0^2, \tag{3.90}$$

we read off the capacitance

$$C = \frac{\epsilon_0 S}{d}, \tag{3.91}$$

which is of course a well-known result. In our example calculation, the exact result is contained in the set of trial functions and is recovered by the variational calculus.

We still need to check that the extremum really is a minimum (not a maximum or a saddle point) of the energy functional. In order to write the second-order variation of the energy functional in compact form, we write

$$\beta_1 = A, \qquad \beta_2 = B, \qquad \beta_3 = C, \qquad \beta_4 = \lambda_1, \qquad \beta_5 = \lambda_2. \tag{3.92}$$

Furthermore,

$$\delta E = \sum_{i=1}^{5} \frac{\partial E}{\partial \beta_i} \delta \beta_i + \frac{1}{2} \sum_{i=1}^{5} \sum_{j=1}^{5} \frac{\partial^2 E}{\partial \beta_i \partial \beta_j} \delta \beta_i \delta \beta_j \tag{3.93}$$

up to the second order in the $\delta \beta_i$. At the extremum values for the parameters given in Eq. (3.87), we have

$$\frac{\partial E}{\partial \beta_i}\bigg|_{\vec{\beta}=\vec{\beta}_{\min}} = 0, \qquad \delta E = \frac{1}{2} \sum_{i=1}^{5} \sum_{j=1}^{5} \frac{\partial^2 E}{\partial \beta_i \partial \beta_j}\bigg|_{\vec{\beta}=\vec{\beta}_{\min}} \delta \beta_i \, \delta \beta_j. \tag{3.94}$$

Let us define the matrix M with components

$$M_{ij} = \frac{\partial^2 E}{\partial \beta_i \partial \beta_j}\bigg|_{\vec{\beta}=\vec{\beta}_{\min}} = \begin{pmatrix} 0 & 0 & 0 & 1 & 1 \\ 0 & \epsilon_0 S d & \epsilon_0 S d^2 & 0 & d \\ 0 & \epsilon_0 S d^2 & \frac{4}{3} \epsilon_0 S d^3 & 0 & d^2 \\ 1 & 0 & 0 & 0 & 0 \\ 1 & d & d^2 & 0 & 0 \end{pmatrix}. \tag{3.95}$$

According to a theorem from linear algebra, in order to establish that a matrix is positive definite, it is sufficient to show that all its leading principal minors are positive (Sec. 6.5 of Ref. [14], and Chap. 7 of Ref. [15]). The leading principal minors are given as the determinants of the leading $(k \times k)$ submatrices of a given matrix M, i.e., as

$$\det(M_1) = \det(0) = 0 ,$$

$$\det(M_2) = \begin{pmatrix} 0 & 0 \\ 0 & \epsilon_0 Sd \end{pmatrix} = 0 ,$$

$$\det(M_3) = \det \begin{pmatrix} 0 & 0 & 0 \\ 0 & \epsilon_0 Sd & \epsilon_0 Sd^2 \\ 0 & \epsilon_0 Sd^2 & \frac{4}{3}\epsilon_0 Sd^3 \end{pmatrix} = 0 ,$$

$$\det(M_4) = \det \begin{pmatrix} 0 & 0 & 0 & 1 \\ 0 & \epsilon_0 Sd & \epsilon_0 Sd^2 & 0 \\ 0 & \epsilon_0 Sd^2 & \frac{4}{3}\epsilon_0 Sd^3 & 0 \\ 1 & 0 & 0 & 0 \end{pmatrix} = -\frac{1}{3}\epsilon_0^2 S^2 d^4 ,$$

$$\det(M_5) = \det \begin{pmatrix} 0 & 0 & 0 & 1 & 1 \\ 0 & \epsilon_0 Sd & \epsilon_0 Sd^2 & 0 & d \\ 0 & \epsilon_0 Sd^2 & \frac{4}{3}\epsilon_0 Sd^3 & 0 & d^2 \\ 1 & 0 & 0 & 0 & 0 \\ 1 & d & d^2 & 0 & 0 \end{pmatrix} = \frac{1}{3}\epsilon_0 S d^5 . \tag{3.96}$$

The fourth principal minor determinant is negative, indicating that M cannot be positive definite: Namely, the product of the eigenvalues of M_4 must be negative, indicating the presence of at least one negative eigenvalue (of M_4). A further indication can be found by considering the case (setting aside physical units for the moment) $d = 1$, $S = 1$, and considering the limit $\epsilon_0 \to 0$. Then,

$$M \to D = \begin{pmatrix} 0 & 0 & 0 & 1 & 1 \\ 0 & 0 & 0 & 0 & 1 \\ 0 & 0 & 0 & 0 & 1 \\ 1 & 0 & 0 & 0 & 0 \\ 1 & 1 & 1 & 0 & 0 \end{pmatrix} , \qquad D\vec{w}_i = d_i \vec{w}_i 1 w , \qquad i = 1, \ldots, 5 . \tag{3.97}$$

(We do not consider the explicit form of the vectors $\vec{w}_i$.) The eigenvalues are

$$d_i = \pm\sqrt{2 \pm \sqrt{2}} , \qquad i = 1, \ldots, 4 , \qquad d_5 = 0 . \tag{3.98}$$

Two of these are negative, namely,

$$d_2 = -\sqrt{2 + \sqrt{2}} , \qquad d_4 = -\sqrt{2 - \sqrt{2}} , \tag{3.99}$$

indicating again that D cannot be positive definite.

What happened here? We started from a minimization problem which, as we showed in Sec. 3.2.4, concerns an operator which is manifestly positive definite, and now we see that some eigenvalues of the matrix of second partial derivatives are negative. We notice that the first three minors $\det(M_i)$ with $i = 1, 2, 3$ are nonnegative. These concern the variational parameters A, B, and C, which enter the *ansatz* for the potential (3.79). The introduction of the Lagrange multipliers changes the nature of the problem; in fact, it is easy to see that the term $\lambda_1 A$ in Eq. (3.82) does not attain any global (or even local) minimum near the values attained by the parameters in the solution (3.87). Or, in other words, one observes that the function $f(\lambda_1, A) = \lambda_1 A$ has a hyperbolic structure as a function of its two arguments λ_1 and A and therefore cannot obtain a minimum. The Lagrange multipliers are not "physical" terms that need to be minimized; they are just included in order to recover the boundary conditions.

If we use the fact that $A = 0$ by virtue of the boundary conditions, then the "physical" portion of M can be identified as the submatrix of M consisting of the second and third rows and columns, namely,

$$M_{\text{phys}} = \begin{pmatrix} \epsilon_0 S d & \epsilon_0 S d^2 \\ \epsilon_0 S d^2 & \frac{4}{3} \epsilon_0 S d^3 \end{pmatrix}, \qquad \det(M_{\text{phys}}) = \frac{1}{3} (\epsilon_0 S d^2)^2 > 0. \tag{3.100}$$

All the leading principal minors of the matrix M_{phys} are positive; so it is indeed positive definite.

3.2.6 *Variational Calculation for Coaxial Cylinders*

Our solution for the energy stored in a plane-plate capacitor was exact, because the exact solution was contained in the space of the trial functions. By contrast, the problem of two coaxial, long cylinders is not so trivial. For a very cylindrical capacitor, one may calculate the potential using Gauss's law. Here, we would like to use this exactly soluble problem in order to demonstrate the utility of the variational method for a case where the exact solution is not contained in the space of trial functions.

Consider two coaxial cylinders, aligned along the z axis, with radii b and c, extending from $z = -L/2$ to $z = L/2$. We use cylindrical coordinates and set

$$x = \rho \cos\varphi, \qquad y = \rho \sin\varphi, \qquad z = z. \tag{3.101}$$

In a numerical example, we will later use the inner surface of the cylinder to lie at $b = 0.2$ cm and the outer surface to be at $c = 4b = 0.8$ cm. Let $x^2 + y^2 = \rho^2$ and let us assume that the z axis defines the symmetry axis of the cylinders. Because of the symmetry of the problem, the potential should only depend on the radial variable ρ. Our trial potential therefore has the form

$$w(A, B, C, \rho) = A + B\rho + C\rho^2. \tag{3.102}$$

Our task is to estimate the potential between the plates, and eventually the capacitance per unit length. The inner surface is grounded, and the outer surface is held at a potential V_0,

$$w(A, B, C, b) = A + Bb + Cb^2 = 0, \qquad w(A, B, C, c) = A + Bc + Cc^2 = V_0. \tag{3.103}$$

We work in cylindrical coordinates, as defined in Eq. (3.101). In order to calculate the energy functional, we need the gradient operator in cylindrical coordinates. It is given by

$$\vec{\nabla} = \hat{e}_\rho \, \frac{\partial \Phi}{\partial \rho} + \hat{e}_\varphi \, \frac{1}{\rho} \frac{\partial \Phi}{\partial \varphi} + \hat{e}_z \, \frac{\partial \Phi}{\partial z} \,. \tag{3.104}$$

The energy stored in the space in between the coaxial cylinders of radii b and c is

$$E = \frac{\epsilon_0}{2} \int_{b \le \rho \le c} \left[\vec{\nabla}\Phi(\vec{r})\right]^2 \rho \, d\rho \, d\varphi \, dz \,, \tag{3.105}$$

where $\rho \, d\rho \, d\varphi \, dz$ is the volume element in cylindrical coordinates. We pull out the constant prefactor $\epsilon_0/2$ and a further factor $2\pi L$ from E; these do not affect the minimization process. The result is the functional S,

$$E = \frac{\epsilon_0}{2} \, 2\pi L S = \epsilon_0 \pi L \, S \,, \tag{3.106}$$

which can be expressed as

$$S = \frac{1}{2\pi L} \int_0^{2\pi} d\varphi \int_{-L/2}^{L/2} dz \int_b^c d\rho \, \rho \left[\hat{e}_\rho \, \frac{\partial w}{\partial \rho} + \hat{e}_\varphi \, \frac{1}{\rho} \frac{\partial w}{\partial \varphi} + \hat{e}_z \, \frac{\partial w}{\partial z} \right]$$

$$\times \left[\hat{e}_\rho \, \frac{\partial w}{\partial \rho} + \hat{e}_\varphi \, \frac{1}{\rho} \frac{\partial w}{\partial \varphi} + \hat{e}_z \, \frac{\partial w}{\partial z} \right]$$

$$= \frac{1}{2\pi L} \int_0^{2\pi} d\varphi \int_{-L/2}^{L/2} dz \int_b^c d\rho \left(\frac{\partial w}{\partial \rho}\right)^2 \rho$$

$$= \frac{1}{2}\mathcal{B}^2 \left(c^2 - b^2\right) + \frac{4}{3}\,\mathcal{B}\mathcal{C} \left(c^3 - b^3\right) + \mathcal{C}^2 \left(c^4 - b^4\right) \,. \tag{3.107}$$

The two boundary conditions at $\rho = b$ and at $\rho = c$ are added with the help of Lagrange multipliers,

$$S(\mathcal{A},\mathcal{B},\mathcal{C},\lambda_1,\lambda_2) = \frac{1}{2}\mathcal{B}^2 \left(c^2 - b^2\right) + \frac{4}{3}\,\mathcal{B}\mathcal{C} \left(c^3 - b^3\right) + \mathcal{C}^2 \left(c^4 - b^4\right)$$

$$+ \lambda_1 \left(\mathcal{A} + \mathcal{B} b + \mathcal{C} b^2\right) + \lambda_2 \left(\mathcal{A} + \mathcal{B} c + \mathcal{C} c^2 - V_0\right) \,. \tag{3.108}$$

The vector β of variational parameters and the extremum condition read as follows,

$$\vec{\beta} = (\mathcal{A},\mathcal{B},\mathcal{C},\lambda_1,\lambda_2) \,, \qquad \frac{\partial S}{\partial \beta_i} = 0 \,, \qquad i = 1,\ldots,5 \,. \tag{3.109}$$

The system of equations

$$\frac{\partial}{\partial \mathcal{A}} S = \lambda_1 + \lambda_2 = 0 \,,$$

$$\frac{\partial}{\partial \mathcal{B}} S = \mathcal{B} \left(c^2 - b^2\right) + \frac{4}{3} \mathcal{C} \left(c^3 - b^3\right) + \lambda_1 \, b + \lambda_2 \, c = 0 \,,$$

$$\frac{\partial}{\partial \mathcal{C}} S = \frac{4}{3} \mathcal{B}(c^3 - b^3) + 2\mathcal{C} \left(c^4 - b^4\right) + \lambda_1 \, b^2 + \lambda_2 \, c^2 = 0 \,,$$

$$\frac{\partial}{\partial \lambda_1} S = \mathcal{A} + b \left(\mathcal{B} + b\mathcal{C}\right) = 0 \,,$$

$$\frac{\partial}{\partial \lambda_2} S = \mathcal{A} + c \left(\mathcal{B} + c\mathcal{C}\right) - V_0 = 0 \,, \tag{3.110}$$

has the solution

$$A = \frac{b\,(b+2\mathcal{C})}{b^2 - c^2}\,V_0\,, \qquad B = \frac{1}{c-b}\,V_0\,,$$

$$C = \frac{1}{b^2 - c^2}\,V_0\,, \qquad \lambda_1 = -\lambda_2 = -\frac{2}{3}\,\frac{b^2 + 4\,b\,c + c^2}{b^2 - c^2}\,V_0\,. \tag{3.111}$$

The solution $w = w_{\min}$ therefore reads

$$w_{\min}(\rho) = \frac{(b-\rho)\,(b+2\,c-\rho)}{b^2 - c^2}\,V_0\,. \tag{3.112}$$

The value of S at the minimum is

$$S_{\min} = \frac{b^2 + 4\,b\,c + c^2}{c^2 - b^2}\,V_0^2\,. \tag{3.113}$$

Restoring the prefactors, this translates into a field energy of

$$E_{\min} = \frac{\epsilon_0}{2}\,2\pi L\,S_{\min} = \epsilon_0\,\pi L\,\frac{b^2 + 4\,b\,c + c^2}{c^2 - b^2}\,V_0^2\,. \tag{3.114}$$

Solving the equation

$$E_{\min} = \frac{1}{2}\,C\,V_0^2 \tag{3.115}$$

for the capacitance C, we obtain

$$C_{\min} = \frac{2\pi L\,\epsilon_0}{3}\,\frac{b^2 + 4\,b\,c + c^2}{c^2 - b^2}\,. \tag{3.116}$$

It remains to show that the optimum solution for the potential within the space of our trial functions, $w_{\min}(\rho)$, approximates the exact solution

$$w_{\min}(\rho) = \frac{(b-\rho)\,(b+2\,c-\rho)}{b^2 - c^2}\,V_0 \approx w_{\text{exact}}(\rho) = \frac{\ln(\rho/b)}{\ln(c/b)}\,V_0\,. \tag{3.117}$$

Furthermore, the approximate and exact formulas for the capacitance read

$$C_{\min} = \frac{2\pi L\,\epsilon_0}{3}\,\frac{b^2 + 4bc + c^2}{c^2 - b^2} \approx C_{\text{exact}} = \frac{2\pi L\,\epsilon_0}{\ln(c/b)}\,. \tag{3.118}$$

For $b = 0.2\,\text{cm}$ and $c = 0.8\,\text{cm}$, the approximate and exact formulas for the capacitance read

$$C_{\min} = 4.608\,L\,\epsilon_0\,, \qquad C_{\text{exact}} = 4.532\,L\,\epsilon_0\,, \tag{3.119}$$

with a 1.6 % deviation, which is quite good for a rational approximation with three parameters (see also Fig. 3.1).

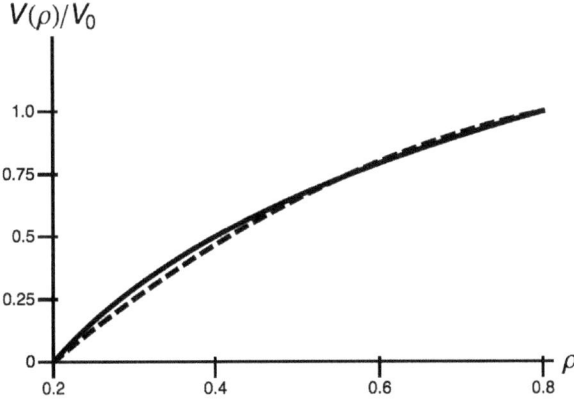

Fig. 3.1   The variational solution $w_{\min}$ for the potential is shown for $b = 0.2$ cm and $c = 0.8$ cm. The optimal approximation $w_{\min}(\rho)$ given in Eq. (3.117) (dashed curve) approximates the exact solution quite well. The latter is given by the solid curve, with $V(\rho) = w_{\text{exact}}(\rho)$.

3.2.7   Exact Integration of the Capacitance of Coaxial Cylinders

The variational calculation of the potential in between the capacitor plates led to the polynomial approximation $w_{\min}(\rho)$ [see Eq. (3.117)] to the exact result

$$\Phi(\rho) = V_0 \frac{\ln(\rho/b)}{\ln(c/b)}, \tag{3.120}$$

which is valid in the limit of long cylinders, $L \to \infty$, where again, L is the cylinder length measured parallel to the symmetry axis. When L is finite, the exact result for the coaxial cylinders is non-trivial. It is extremely instructive to perform an explicit and exact integration over the charge distributions. The two cylindrical surfaces are assumed to be located at distances $\rho = b$ and $\rho = c$ from the cylinder axis. We assume a constant charge density of $-\sigma_0$ on the inner cylinder at $\rho = b$ and a constant charge density of $+\sigma_0 b/c$ on the outer cylinder at $\rho = c$, in order to make the arrangement electrostatically neutral. Let us briefly comment on the relation of the constant surface charge and the equipotential surfaces used in our variational problem, which was based on the additional assumption of a long cylinder $L \to \infty$. Namely, the constant surface charge density does not necessarily imply that the cylindrical surfaces are equipotential surfaces. It is only in the limit of large L that the equipotential surfaces will be surfaces of constant surface charge density. For finite L, we here assume the constant surface charge density.

The solution can be obtained as follows. We have formulated the problem in terms of the surface charge distributions on the two capacitor plates, not in terms of the boundary conditions of the potential. Therefore, we may use the Poisson equation (2.21) with the Green function $G(\vec{r}, \vec{r}') = 1/(4\pi\epsilon_0|\vec{r} - \vec{r}'|)$ [see Eq. (2.23)] in order to integrate the potential,

$$\Phi(\vec{r}) = \int d^3 r' \, G(\vec{r}, \vec{r}') \, \rho(\vec{r}'), \qquad \vec{\nabla}^2 \Phi(\vec{r}) = -\frac{1}{\epsilon_0} \rho(\vec{r}). \qquad (3.121)$$

Here, $\rho(\vec{r}')$ is the volume charge density corresponding to the uniform surface charge density on the coaxial cylinder plates. A major difficulty is encountered in how to express the volume charge density $\rho(\vec{r})$ in terms of the surface charge density $\sigma(\vec{r}) = \pm\sigma_0$ (here, we assume it to be uniform). This problem can be formulated more generally. Let $\sigma(\vec{r})$ be a surface charge density on the boundary $\vec{r} \in \partial V$ defined by an implicit equation $F(\vec{r}) = 0$. In our case, we would simply have $F(\vec{r}) = \rho - b$ and $F(\vec{r}) = \rho - c$ for the inner and outer cylindrical surfaces. How is $\rho(\vec{r})$ related to $\sigma(\vec{r})$ and $F(\vec{r})$? This problem is of general interest and needs to be discussed.

The answer is as follows. On the surface defined by $F(\vec{r}) = 0$, we have $dF(\vec{r}) = 0$ when the argument $\vec{r}$ is varied within the surface according to $\vec{r} \to \vec{r} + d\vec{s}$. Thus,

$$dF(\vec{r}) = F(\vec{r} + d\vec{s}) - F(\vec{r}) = d\vec{s} \cdot \vec{\nabla} F(\vec{r}) = 0, \qquad (3.122)$$

where $d\vec{s}$ is an infinitesimal displacement within the surface, and $\vec{r}$ is a point on the surface. Because $d\vec{s}$ is arbitrary, $\vec{\nabla} F(\vec{r})$ must be normal to the surface. Now, let $\xi \hat{n}$ be the displacement of a point from the surface at $\vec{r}$, where ξ measures the distance from the surface. Then,

$$d^3 r = d\vec{A} \cdot (d\xi \, \hat{n}) = [dS(\vec{r}) \, \hat{n}] \cdot (d\xi \, \hat{n}). \qquad (3.123)$$

Here, $\hat{n}$ is the surface normal. If $\vec{r}$ is a point on the surface, then, using definition of the $\vec{\nabla}$ operator, one writes

$$\hat{u} \cdot \vec{\nabla} F(\vec{r}) = \left. \frac{\partial F(\vec{r}' + \xi \, \hat{u})}{\partial \xi} \right|_{\xi=0}, \qquad (3.124)$$

where $\hat{u}$ is an arbitrary unit vector. If $\hat{u} = \hat{t}$ were a tangential vector to the surface, then we would have $dF(\vec{r}) = 0$ in the sense of Eq. (3.122), with $d\vec{s} = \hat{t} d\xi$. We conclude that the modulus of $dF(\vec{r}) = 0$ becomes maximal (as a function of the angle between $\vec{\nabla} F(\vec{r})$ and $\hat{n}$) when $\hat{n}$ is the surface normal. We now choose $\hat{u} = \hat{n}$ to be the unit normal, parallel (not antiparallel) to $\vec{\nabla} F(\vec{r})$, and write

$$\hat{n} \cdot \vec{\nabla} F(\vec{r}) = \left. \frac{\partial F(\vec{r} + \xi \, \hat{n})}{\partial \xi} \right|_{\xi=0} = \left. \frac{\partial F(\xi)}{\partial \xi} \right|_{\xi=0}, \qquad (3.125)$$

with an appropriately defined function $F = F(\xi) = F(\vec{r} + \xi \, \hat{n})$. So,

$$\left. \frac{\partial F(\xi)}{\partial \xi} \right|_{\xi=0} = \hat{n} \cdot \vec{\nabla} F(\vec{r}) = \frac{(\vec{\nabla} F(\vec{r}))}{|\vec{\nabla} F(\vec{r})|} \cdot \vec{\nabla} F(\vec{r}) = |\vec{\nabla} F(\vec{r})|. \qquad (3.126)$$

We have thus reduced our problem to a one-dimensional problem, replacing $\vec{r} \to \xi$. Application of Eq. (1.34) then shows that

$$\delta(\xi) = \delta(F(\xi)) \left| \frac{\partial F(\xi)}{\partial \xi} \right| = \delta(F(\xi)) \, |\vec{\nabla} F(\xi)| = \delta(F(\vec{r})) \, |\vec{\nabla} F(\vec{r})|. \qquad (3.127)$$

Because ξ measures the distance from the surface, we finally obtain a suitable expression for the volume charge density,

$$\rho(\vec{r}) = \sigma(\vec{r})\,\delta(\xi) = \sigma(\vec{r})\,\delta(F(\vec{r}))\,|\vec{\nabla}F(\vec{r})|\,. \tag{3.128}$$

In our case, the two surfaces are simply given by

$$F(\vec{r}) = \rho - b = 0\,, \qquad F(\vec{r}) = \rho - c = 0\,. \tag{3.129}$$

For the inner cylinder surface, we thus have

$$\rho(\vec{r}) = -\sigma_0\,\delta(\rho - b)\,|\vec{\nabla}(\rho - b)|\,. \tag{3.130}$$

In Cartesian coordinates, the gradient is expressed as

$$\rho = \sqrt{x^2 + y^2}\,, \qquad \vec{\nabla}F(\vec{r}) = \vec{\nabla}(\rho - b) = \hat{e}_x \frac{x}{\sqrt{x^2 + y^2}} + \hat{e}_y \frac{x}{\sqrt{x^2 + y^2}}\,, \tag{3.131}$$

and thus

$$|\vec{\nabla}(\rho - b)| = 1\,, \qquad \rho(\vec{r}) = \sigma(\vec{r})\,\delta(\rho - b) = -\sigma_0\,\delta(\rho - b)\,. \tag{3.132}$$

We denote the potential as $\Phi(\vec{r}) = \Phi(\rho, \varphi, z)$ in cylindrical coordinates. The source point $\vec{r}'$ is expressed as

$$\vec{r}' = \hat{e}_x\,\rho'\,\cos\varphi' + \hat{e}_y\,\rho'\,\sin\varphi' + \hat{e}_z\,z'\,, \tag{3.133}$$

Of special interest is the point $\vec{r} = \hat{e}_x\,\rho$ which is at a distance ρ from the z axis. Then,

$$\vec{r} - \vec{r}' = \hat{e}_x\,(\rho - \rho'\,\cos\varphi') + \hat{e}_y\,(-\rho'\,\sin\varphi') - \hat{e}_z\,z'\,. \tag{3.134}$$

This implies that

$$|\vec{r} - \vec{r}'| = \sqrt{\rho^2 + \rho'^2 - 2\rho\,\rho'\,\cos\varphi' + z'^2}\,. \tag{3.135}$$

We thus need to calculate the following contribution to the potential from the inner cylindrical surface at $\rho = b$, taking into account that the observation point $\vec{r} = \hat{e}_x\,\rho$ has cylindrical coordinates $\varphi = 0$ and $z = 0$,

$$
\begin{aligned}
\Phi_{\text{inner}}(\rho, 0, 0) &= \frac{1}{4\pi\epsilon_0} \int\limits_0^\infty d\rho'\rho' \int\limits_0^{2\pi} d\varphi' \int\limits_{-L/2}^{L/2} dz'\,(-\sigma_0)\,\delta(\rho' - b)\,\frac{1}{|\vec{r} - \vec{r}'|} \\[2mm]
&= -\frac{\sigma_0}{4\pi\epsilon_0} \int\limits_0^{2\pi} d\varphi' \int\limits_{-L/2}^{L/2} dz'\,\frac{b}{\sqrt{\rho^2 + b^2 - 2b\rho\cos\varphi' + z'^2}} \\[2mm]
&= -\frac{\sigma_0}{2\pi\epsilon_0} \int\limits_0^{2\pi} d\varphi' \int\limits_0^{L/2} dz'\,\frac{b}{\sqrt{\rho^2 + b^2 - 2b\rho\cos\varphi' + z'^2}} \\[2mm]
&= -\frac{\sigma_0 b}{2\pi\epsilon_0} \int\limits_0^{2\pi} d\varphi' \left[\ln\left(2\left(z' + \sqrt{\rho^2 + b^2 - 2b\rho\cos\varphi' + z'^2}\right)\right)\right]_{z'=0}^{z'=L/2}\,.
\end{aligned}
\tag{3.136}
$$

In principle, this expression already expresses the exact result for finite L in terms of an integral, which has to be evaluated numerically. The full result $\Phi(\rho,0,0)$ is obtained after adding the contributions from the inner and outer surfaces, $\Phi(\rho,0,0) = \Phi_{\text{inner}}(\rho,0,0) + \Phi_{\text{outer}}(\rho,0,0)$, where $\Phi_{\text{outer}}(\rho,0,0)$ is obtained from $\Phi_{\text{inner}}(\rho,0,0)$ by the replacement $b \to c$ and a sign change in σ_0. Then,

$$
\Phi(\rho,0,0) = \frac{\sigma_0 c}{2\pi\epsilon_0} \int_0^{2\pi} d\varphi' \left[\ln\left(2\left(z' + \sqrt{\rho^2 + c^2 - 2c\rho\cos\varphi' + z'^2} \right) \right) \right]_{z'=0}^{z'=L/2}
$$

$$
- \frac{\sigma_0 b}{2\pi\epsilon_0} \int_0^{2\pi} d\varphi' \left[\ln\left(2\left(z' + \sqrt{\rho^2 + b^2 - 2b\rho\cos\varphi' + z'^2} \right) \right) \right]_{z'=0}^{z'=L/2} . \quad (3.137)
$$

It is difficult to express these integrals in closed analytic form; a numerical evaluation of Eq. (3.137) for given parameters b, c and L is to be preferred.

Let us now consider the limit $L \to \infty$ in Eq. (3.136), which corresponds to the situation considered in our variational calculation described in Sec. 3.2.6. At the upper limit, the logarithm in the integrand of Eq. (3.136) assumes the form

$$
\ln\left(2\left(z' + \sqrt{\rho^2 + b^2 - 2b\rho\cos\varphi' + z'^2} \right) \right) \stackrel{z'=\frac{L}{2}}{=} \ln(2L) + \frac{b^2 + \rho^2 - 2b\rho\cos\varphi'}{L^2} + O\left(\frac{1}{L^3}\right).
$$
$$(3.138)$$

The constant $\ln(2L)$ is independent of ρ and can be absorbed into an additive constant potential Φ_0 which can always be added to any electrostatic potential without changing the physics. The term of order $1/L^2$ vanishes in the limit $L \to \infty$, which we are interested in here. We can thus replace, in the integrand of Eq. (3.136), the logarithm by its value at the lower limit $z' = 0$,

$$
\left[\ln\left(2\left(z' + \sqrt{\rho^2 + b^2 - 2b\rho\cos\varphi' + z'^2} \right) \right) \right]_{z'=0}^{z'=L/2} \to -\ln\left(2\sqrt{\rho^2 + b^2 - 2b\rho\cos\varphi'} \right).
$$
$$(3.139)$$

The remaining L-independent integral

$$
\Phi_{\text{inner}}(\rho,0,0) = \frac{\sigma_0 b}{2\pi\epsilon_0} \int_0^{2\pi} d\varphi' \ln\left(2\sqrt{\rho^2 + b^2 - 2b\rho\cos\varphi'} \right) \quad (3.140)
$$

is still difficult; in particular, a direct evaluation leads to an elliptic function.

The integral (3.140) becomes easier if we first differentiate with respect to ρ, then integrate over φ', and then, integrate back again with respect to ρ. This only adds a physically irrelevant constant to the potential. In particular, we have

$$
\frac{\partial}{\partial\rho} \Phi(\rho,0,0) = \frac{\sigma_0}{4\pi\epsilon_0} \int_0^{2\pi} d\varphi' \frac{b\,(2\rho - 2b\cos\varphi')}{b^2 + \rho^2 - 2b\rho\cos\varphi'}. \quad (3.141)
$$

A partial-fraction decomposition of the integrand leads to

$$
\frac{b\,(2\rho - 2b\cos\varphi')}{b^2 + \rho^2 - 2b\rho\cos\varphi'} = \frac{b}{\rho} - \frac{b\,(b^2 - \rho^2)}{\rho\,(b^2 + \rho^2 - 2b\rho\cos\varphi')}. \quad (3.142)
$$

The integral of this expression with respect to φ' is

$$\frac{b\varphi'}{\rho} - \frac{2b}{\rho} \arctan\left(\frac{b+\rho}{b-\rho} \tan\left(\frac{\varphi'}{2}\right)\right). \tag{3.143}$$

At face value, we have

$$\left[\frac{b\varphi'}{\rho} - \frac{2b}{\rho} \arctan\left(\frac{b+\rho}{b-\rho} \tan\left(\frac{\varphi'}{2}\right)\right)\right]_{\varphi'=0}^{\varphi'=2\pi} \overset{?}{=} \frac{2\pi b}{\rho} - \frac{2b}{\rho}\left[\arctan\left(0\right) - \arctan\left(0\right)\right], \tag{3.144}$$

where we simply insert the upper and lower integration limits; this would lead to a value of $2\pi b/\rho$ for the integral. However, this result is incorrect. At $\varphi' = \pi$, the second term with the arctan function receives a jump by

$$\left[-\arctan\left(\frac{b+\rho}{b-\rho} \tan\left(\frac{\varphi'}{2}\right)\right)\right]_{\varphi'=\pi-\epsilon}^{\varphi'=\pi+\epsilon} = -\frac{\pi}{2} - \left(+\frac{\pi}{2}\right) = -\pi, \qquad b < \rho, \tag{3.145}$$

(downward) because of the singularity of the tangent function. In order to define the integral so that it becomes smooth at $\varphi' = \pi$, we have to add a compensating term in the form of a Heaviside Θ step function,

$$\int d\varphi' \left(\frac{b}{\rho} - \frac{b\left(b^2 - \rho^2\right)}{\rho\left(b^2 + \rho^2 - 2b\rho \cos\varphi'\right)}\right)$$

$$= \frac{b\varphi'}{\rho} + \frac{2b}{\rho}\left[-\arctan\left(\frac{b+\rho}{b-\rho} \tan\left(\frac{\varphi'}{2}\right)\right) + \pi\,\Theta(\varphi' - \pi)\right]. \tag{3.146}$$

Then,

$$\int_0^{2\pi} d\varphi' \left(\frac{b}{\rho} - \frac{b\left(b^2 - \rho^2\right)}{\rho\left(b^2 + \rho^2 - 2b\rho \cos\varphi'\right)}\right) = \frac{4\pi b}{\rho}. \tag{3.147}$$

The final result for the expression given in Eq. (3.141) therefore is

$$\frac{\partial}{\partial\rho}\Phi_{\text{inner}}(\rho,0,0) = \frac{\sigma_0}{4\pi\epsilon_0}\frac{4\pi b}{\rho}, \tag{3.148}$$

and so

$$\Phi_{\text{inner}}(\rho,0,0) = \frac{\sigma_0 b}{\epsilon_0} \ln\left(\frac{\rho}{b}\right). \tag{3.149}$$

The argument of the logarithm has to be dimensionless for physical reasons; in principle, one can ascertain that $\ln(\rho/C)$ is a valid integral for any constant length C; our choice implements the condition $\Phi(\rho = b, 0, 0) = 0$.

We recall that in view of Eq. (3.145), the result (3.149) is valid for $\rho > b$. On the outer layer, we have a charge density $+\sigma_0\, b/c$ in order to make the arrangement neutral, and calculate its contribution to the potential.

At radial distance ρ, we have a corresponding integral which replaces Eq. (3.145) by

$$\left[-\arctan\left(\frac{c+\rho}{c-\rho} \tan\left(\frac{\varphi'}{2}\right)\right)\right]_{\varphi'=\pi-\epsilon}^{\varphi'=\pi+\epsilon} = \frac{\pi}{2} - \left(-\frac{\pi}{2}\right) = \pi, \qquad \rho < c, \tag{3.150}$$

[the arctan($\cdot$) function jumps in the other direction here], so that

$$\int_0^{2\pi} d\varphi' \left(\frac{c}{\rho} - \frac{c\,(c^2 - \rho^2)}{\rho\,(c^2 + \rho^2 - 2c\rho\cos\varphi')} \right) = 0, \qquad \rho < c. \tag{3.151}$$

Hence,

$$\Phi_{\text{outer}}(\rho, 0, 0) = \begin{cases} 0 & (\rho < c) \\ -\dfrac{\sigma_0 b}{\epsilon_0} \ln\left(\dfrac{\rho}{b}\right) & (\rho > c) \end{cases} . \tag{3.152}$$

We thus verify that

$$\Phi(\rho, 0, 0) = \Phi_{\text{inner}}(\rho, 0, 0) + \Phi_{\text{outer}}(\rho, 0, 0) = \begin{cases} \dfrac{\sigma_0 b}{\epsilon_0} \ln\left(\dfrac{\rho}{b}\right) & (b < \rho < c) \\ 0 & (\rho > c) \end{cases} , \tag{3.153}$$

a result which could have been obtained by the application of the integral form of Gauss's law, given in Eq. (1.5a), to appropriate cylindrical surfaces inserted in between the plates. In this case, only the inner cylinder matters as the "charge inside" the Gaussian surface.

Using the result from Eq. (3.149) and solving the equation

$$V_0 = \Phi(c, 0, 0) - \Phi(b, 0, 0) = \frac{\sigma_0 b}{\epsilon_0} \ln\left(\frac{c}{b}\right) = \frac{Q}{2\pi L \epsilon_0} \ln\left(\frac{c}{b}\right) = \frac{Q}{C}, \tag{3.154}$$

we obtain the capacitance as

$$C = \frac{2\pi L\,\epsilon_0}{\ln\,(c/b)}, \tag{3.155}$$

as already noted and used in Eq. (3.118). We recall that the area of the inner cylinder is $2\pi b\,L$, so that $\sigma_0 = Q/(2\pi b\,L)$. We thus verify Eqs. (3.117) and (3.118) by an alternative calculation.

We have already stated that the approach based on Eq. (3.137) is easily generalizable to cases where L is finite, e.g., within a numerical approach. One just has to keep L finite at the places where we have taken the limit $L \to \infty$ within the z integration. Let us illustrate this procedure by calculating the first nonvanishing correction term to the potential for large, but finite L. To this end, we keep the term of relative order $1/L^2$ in Eq. (3.155) and carry out the integration over φ' as indicated. The same procedure has to be applied to the outer layer, i.e., we have to replace $b \to c$ and $(-\sigma_0) \to (+\sigma_0)b/c$ in the first line in Eq. (3.136). One finds that the correction term of order $1/L^2$ to the potential (3.149) vanishes. Taking into account the next order $(1/L^4)$, one obtains after some algebra

$$\Phi(\rho, 0, 0) = \frac{\sigma_0 b}{\epsilon_0} \ln\left(\frac{\rho}{b}\right) + \frac{6\,\sigma_0\,b\,(b^2 - c^2)\,\rho^2}{L^4\,\epsilon_0} = \frac{Q}{2\pi L\,\epsilon_0} \ln\left(\frac{\rho}{b}\right) + \frac{3\,Q\,(b^2 - c^2)\,\rho^2}{\pi\,L^5\,\epsilon_0} . \tag{3.156}$$

This result leads to the following correction term for the capacitance,

$$C = \frac{2\pi L\,\epsilon_0}{\ln\,(c/b) - 6\,(b^2 - c^2)^2/L^4} . \tag{3.157}$$

The denominator of the latter expression involves the difference of the potential $\Phi(\rho, 0, 0)$ given in Eq. (3.156) for $\rho = c$ and $\rho = b$.

3.3 Laplace Equation and Series Expansions

3.3.1 *Coordinate Systems and Special Functions*

After discussing the variational approach to the solution of the Laplace equation $\vec{\nabla}^2\Phi(\vec{r}) = 0$, we now turn our attention to the series expansion method. The basic paradigm is that one writes down mutually orthogonal functions (with respect to a particular integration measure); all of these fulfill the Laplace equation individually. The expansion coefficients are fixed by the boundary conditions. The relevant equations look fundamentally different in the Cartesian coordinate system (x, y, z), in the spherical system (r, θ, φ), and in the cylindrical coordinates (ρ, φ, z).

Suppose we are given the problem of solving the Laplace equation in a system with a given symmetry. In general, functions may be expanded in terms of a complete set of other functions, suitably chosen so that the chosen "basis set" reflects the defining properties of the function. Let us assume that we know that $f(x)$ has a period length L. Then, the (discrete) Fourier decomposition of $f(x)$ is given by

$$f(x) = \sum_{n=-\infty}^{\infty} F(n) \exp\left(\mathrm{i}\frac{2\pi}{L}nx\right), \tag{3.158}$$

where

$$F(n) = \frac{1}{2\pi} \int_0^{2\pi} \mathrm{d}x\, f(x) \exp\left(-\mathrm{i}\frac{2\pi}{L}nx\right). \tag{3.159}$$

The basis functions in this case are the exponentials

$$\exp\left(\mathrm{i}\frac{2\pi}{L}nx\right) \tag{3.160}$$

for integer n, all of which are periodic functions with period length L. Here, we will encounter different sets of functions in which the solutions of the Laplace equation are expanded.

We will also encounter the $\vec{\nabla}^2$ operator in different coordinate systems. Let us suppose that we transform the gradient operator $\vec{\nabla}$ from the Cartesian coordinate system with coordinates x, y, and z, to a different coordinate system with coordinate ξ, θ, and φ (here, ξ, θ, and φ can be any coordinates in any system, say, spherical, in which case we would replace $\xi \to r$, while in cylindrical coordinates, we would replace $\xi \to \rho$, $\theta \to \varphi$, and $\varphi \to z$). Then, we can rewrite the differentials as follows,

$$\frac{\partial}{\partial x} = \frac{\partial\xi}{\partial x}\frac{\partial}{\partial\xi} + \frac{\partial\theta}{\partial x}\frac{\partial}{\partial\theta} + \frac{\partial\varphi}{\partial x}\frac{\partial}{\partial\varphi}, \tag{3.161}$$

and likewise for $\partial/\partial y$ and $\partial/\partial z$. Also, the unit vector in the ξ direction becomes

$$\hat{e}_\xi = \left(\left|\frac{\partial\vec{r}}{\partial\xi}\right|\right)^{-1}\frac{\partial\vec{r}}{\partial\xi} = \left(\left|\frac{\partial\vec{r}}{\partial\xi}\right|\right)^{-1}\left(\frac{\partial x}{\partial\xi}\hat{e}_x + \frac{\partial y}{\partial\xi}\hat{e}_y + \frac{\partial z}{\partial\xi}\hat{e}_z\right). \tag{3.162}$$

Inverting these relations, we obtain $\hat{e}_x$, $\hat{e}_y$, and $\hat{e}_z$ as functions of $\hat{e}_\xi$, $\hat{e}_\theta$ and $\hat{e}_\varphi$. So, we can rewrite the entire gradient operator $\vec{\nabla}$ in terms of the "new" unit vectors and "new" differential operators as

$$\vec{\nabla} = \hat{e}_x \frac{\partial}{\partial x} + \hat{e}_y \frac{\partial}{\partial y} + \hat{e}_z \frac{\partial}{\partial z} = \hat{e}_\xi \, f_1 \left(\xi, \theta, \varphi; \frac{\partial}{\partial \xi}, \frac{\partial}{\partial \theta}, \frac{\partial}{\partial \varphi} \right)$$

$$+ \hat{e}_\theta \, f_2 \left(\xi, \theta, \varphi; \frac{\partial}{\partial \xi}, \frac{\partial}{\partial \theta}, \frac{\partial}{\partial \varphi} \right) + \hat{e}_\varphi \, f_3 \left(\xi, \theta, \varphi; \frac{\partial}{\partial \xi}, \frac{\partial}{\partial \theta}, \frac{\partial}{\partial \varphi} \right), \tag{3.163}$$

where f_1, f_2 and f_3 are functions that may contain derivative operators. This is a general algorithm for reformulating the gradient operator for a general non-Cartesian coordinate system.

The calculation of $\vec{\nabla}^2$ then proceeds in a straightforward manner, but we have to pay attention: the differentiations may incur additional terms because of the intertwined base vectors $\hat{e}_\xi$, $\hat{e}_\theta$ and $\hat{e}_\varphi$. For example, the partial derivative $\partial \hat{e}_\xi / \partial \theta$ does not necessarily vanish, nor does $\partial \hat{e}_\theta / \partial \xi$. Additional insight into the problem can be gained through the definition of Christoffel symbols, see Ref. [16]. The entire program has been carried through a long time ago. While respective formulas are contained in reference pages of other textbooks on related subjects (e.g., Refs. [17–19]), it is useful to summarize them, for the gradient operator (in Cartesian, cylindrical and spherical coordinates)

$$\vec{\nabla} f = \hat{e}_x \frac{\partial f}{\partial x} + \hat{e}_y \frac{\partial f}{\partial y} + \hat{e}_z \frac{\partial f}{\partial z}$$

$$= \hat{e}_\rho \frac{\partial f}{\partial \rho} + \hat{e}_\varphi \frac{1}{\rho} \frac{\partial f}{\partial \varphi} + \hat{e}_z \frac{\partial f}{\partial z}$$

$$= \hat{e}_r \frac{\partial f}{\partial r} + \hat{e}_\theta \frac{1}{r} \frac{\partial f}{\partial \theta} + \hat{e}_\varphi \frac{1}{r \sin \theta} \frac{\partial f}{\partial \varphi}, \tag{3.164a}$$

for the Laplacian operator,

$$\vec{\nabla}^2 = \frac{\partial^2 f}{\partial x^2} + \frac{\partial^2 f}{\partial y^2} + \frac{\partial^2 f}{\partial z^2}$$

$$= \frac{\partial^2 f}{\partial \rho^2} + \frac{1}{\rho} \frac{\partial f}{\partial \rho} + \frac{1}{\rho^2} \frac{\partial^2 f}{\partial \varphi^2} + \frac{\partial^2 f}{\partial z^2}$$

$$= \frac{\partial^2 f}{\partial r^2} + \frac{2}{r} \frac{\partial f}{\partial r} + \frac{1}{r^2 \sin \theta} \frac{\partial}{\partial \theta} \sin \theta \frac{\partial f}{\partial \theta} + \frac{1}{r^2 \sin^2 \theta} \frac{\partial^2 f}{\partial \varphi^2}, \tag{3.164b}$$

for the divergence,

$$\vec{\nabla} \cdot \vec{A} = \frac{\partial A_x}{\partial x} + \frac{\partial A_y}{\partial y} + \frac{\partial A_z}{\partial z}$$

$$= \frac{\partial A_\rho}{\partial \rho} + \frac{A_\rho}{\rho} + \frac{1}{\rho} \frac{\partial A_\varphi}{\partial \varphi} + \frac{\partial A_z}{\partial z}$$

$$= \frac{\partial A_r}{\partial r} + \frac{2 A_r}{r} + \frac{1}{r} \frac{\partial A_\theta}{\partial \theta} + \frac{\cot \theta \, A_\theta}{r} + \frac{1}{r \sin \theta} \frac{\partial A_\varphi}{\partial \varphi}, \tag{3.164c}$$

and the curl,

$$\vec{\nabla} \times \vec{A} = \hat{e}_x \left(\frac{\partial A_z}{\partial y} - \frac{\partial A_y}{\partial z} \right) + \hat{e}_y \left(\frac{\partial A_x}{\partial z} - \frac{\partial A_z}{\partial x} \right) + \hat{e}_z \left(\frac{\partial A_y}{\partial x} - \frac{\partial A_z}{\partial x} \right)$$

$$= \hat{e}_\rho \left(\frac{1}{\rho} \frac{\partial A_z}{\partial \varphi} - \frac{\partial A_\varphi}{\partial z} \right) + \hat{e}_\varphi \left(\frac{\partial A_\rho}{\partial z} - \frac{\partial A_z}{\partial \rho} \right) + \hat{e}_z \left(\frac{\partial A_\varphi}{\partial \rho} + A_\varphi - \frac{\partial A_\rho}{\partial \varphi} \right)$$

$$= \hat{e}_r \left(\frac{1}{r} \frac{\partial A_\varphi}{\partial \theta} + \frac{\cot \theta \, A_\varphi}{r} - \frac{1}{r \sin \theta} \frac{\partial A_\theta}{\partial \varphi} \right) + \hat{e}_\theta \left(\frac{1}{r \sin \theta} \frac{\partial A_r}{\partial \varphi} - \frac{\partial A_\varphi}{\partial r} - \frac{A_\varphi}{r} \right)$$

$$+ \hat{e}_\varphi \left(\frac{\partial A_\theta}{\partial r} + \frac{A_\theta}{r} - \frac{1}{r} \frac{\partial A_r}{\partial \theta} \right). \tag{3.164d}$$

In our investigations on the Laplace equation, let us start with Cartesian coordinates, where

$$\vec{\nabla}^2 \Phi(x, y, z) = \left(\frac{\partial^2}{\partial x^2} + \frac{\partial^2}{\partial y^2} + \frac{\partial^2}{\partial z^2} \right) \Phi(x, y, z) = 0. \tag{3.165}$$

The separation constant is the wave vector $\vec{k} = k_x \hat{e}_x + k_y \hat{e}_y + k_z \hat{e}_z$. The general solution reads

$$\Phi(x, y, z) = \sum_{\vec{k} \cdot \vec{k} = 0} \left[c(k) e^{\vec{k} \cdot \vec{r}} + d(k) e^{\vec{k} \cdot \vec{r}} \left\{ (ax + b) \delta_{k_x, 0} + (cy + d) \delta_{k_y, 0} + (ez + f) \delta_{k_z, 0} \right\} \right]$$

$$+ (a'x + b')(c'y + d')(e'z + f'). \tag{3.166}$$

The $\vec{k}$ vector may have complex rather than real components, and the summation is over all permissible vectors $\vec{k}$ (as defined by the geometry). Note that if the components of $\vec{k}$ are complex, then the condition $\vec{k} \cdot \vec{k} = 0$ does not necessarily imply that $\vec{k} = \vec{0}$. One possibility for $\vec{k} = k_x \hat{e}_x + k_y \hat{e}_y + k_z \hat{e}_z$ is to be an exponential $e^{\vec{k} \cdot \vec{r}}$ with $k_{x,y} = \pm i |k_{x,y}|$ being purely imaginary and k_z being real. This leads to an exponential factor

$$e^{\vec{k} \cdot \vec{r}} = e^{i(|k_x| x + |k_y| y)} e^{\pm k_z z}. \tag{3.167}$$

We thus have a sinusoidal dependence in x and y and an exponential in z. The special terms in Eq. (3.166) are generated as polynomial solutions to the equation $f''(x) = 0$, in the directions where specific components of the $\vec{k}$ vector vanish.

The condition $\vec{k} \cdot \vec{k} = 0$ implies that if k_x and k_y are given, then k_z can be determined (up to a sign). So, the summation over the possible values of $\vec{k}$ (a discrete set of values, in a typical case) will in general be a double summation, not a triple summation. Another possibility is $k_z = \pm i |k_z|$, with k_x and k_y being real. Then,

$$\exp(\vec{k} \cdot \vec{r}) = \exp(k_z x + k_y y) \exp(\pm i |k_z| z), \tag{3.168}$$

with a sinusoidal dependence in x and y, and an exponential in z.

Let us now switch to spherical coordinates. From Sec. 2.3.1, we recall that the Laplace equation reads

$$\vec{\nabla}^2 \Phi(r,\theta,\varphi) = \left(\frac{1}{r} \frac{\partial^2}{\partial r^2} r + \frac{1}{r^2 \sin\theta} \frac{\partial}{\partial \theta} \sin\theta \frac{\partial}{\partial \theta} + \frac{1}{r^2 \sin\theta} \frac{\partial^2}{\partial \varphi^2} \right) \Phi(r,\theta,\varphi)$$

$$= \left(\frac{1}{r} \frac{\partial^2}{\partial r^2} r - \frac{\vec{L}^2}{r^2} \right) \Phi(r,\theta,\varphi) = \left(\frac{1}{r} \frac{\partial^2}{\partial r^2} r - \frac{\ell(\ell+1)}{r^2} \right) \Phi(r,\theta,\varphi) = 0,$$

$$(3.169)$$

with $|m| \le \ell$, and $\ell \in \mathbb{N}_0$. The operator $\vec{L} = -i\vec{r} \times \vec{\nabla}$ has already been discussed in Sec. 2.3.1. The general solution

$$\Phi(r,\theta,\varphi) = \sum_{\ell=0}^{\infty} \sum_{m=-\ell}^{\ell} \left(a_{\ell m} r^\ell + b_{\ell m} r^{-\ell-1} \right) Y_{\ell,m}(\theta,\varphi) \tag{3.170}$$

involves the spherical harmonics,

$$\vec{L}^2 Y_{\ell m}(\theta,\varphi) = \ell(\ell+1) Y_{\ell,m}(\theta,\varphi), \qquad \vec{L}^2 = -\frac{1}{\sin\theta} \frac{\partial}{\partial \theta} \sin\theta \frac{\partial}{\partial \theta} - \frac{1}{\sin\theta} \frac{\partial^2}{\partial \varphi^2}, \tag{3.171}$$

whose explicit form we recall from Eq. (2.53),

$$Y_{\ell m}(\theta,\varphi) = (-1)^m \left(\frac{2\ell+1}{4\pi} \frac{(\ell-m)!}{(\ell+m)!} \right)^{1/2} P_\ell^{|m|}(\cos\theta)\, e^{im\varphi}. \tag{3.172}$$

The associated Legendre polynomials $P_\ell^{|m|}(\cos\theta)$ enter this expression. Using the fact that

$$\left(\frac{1}{r} \frac{\partial^2}{\partial r^2} r \right) r^\ell = \ell(\ell+1) r^\ell, \qquad \left(\frac{1}{r} \frac{\partial^2}{\partial r^2} r \right) r^{-\ell-1} = -\ell(-\ell-1) r^{-\ell-1} = \ell(\ell+1) r^{-\ell-1},$$

$$(3.173)$$

it becomes clear that all basis functions in the expansion (3.170) fulfill the Laplace equation separately. In cylindrical coordinates, the Laplace equation reads

$$\vec{\nabla}^2 \Phi(\rho,\varphi,z) = \left(\frac{1}{\rho} \frac{\partial}{\partial \rho} \rho \frac{\partial}{\partial \rho} + \frac{1}{\rho^2} \frac{\partial^2}{\partial \varphi^2} + \frac{\partial^2}{\partial z^2} \right) \Phi(\rho,\theta,z) = 0. \tag{3.174}$$

With $\alpha, m = 0, \pm 1, \pm 2, \ldots$, the solution reads

$$\Phi(\rho,\varphi,z) = \sum_{\alpha,m} (a_{\alpha m} J_m(\alpha\rho) + b_{\alpha m} Y_m(\alpha\rho))\, e^{im\varphi}\, e^{\pm\alpha z}. \tag{3.175}$$

The α parameters may be complex. The basis functions of the expansion (3.175) fulfill the Laplace equation provided

$$\left(\frac{\partial^2}{\partial \rho^2} + \frac{1}{\rho} \frac{\partial}{\partial \rho} - \frac{m^2}{\rho^2} + \alpha^2 \right) J_m(\alpha\rho) = 0. \tag{3.176}$$

Bessel's differential equation [Eq. (2.144)] and other important properties of the Bessel functions have been summarized in Sec. 2.4.2. The equivalence of Eqs. (3.176) and (2.144) can be shown by explicit differentiation with respect to ρ. Bessel functions have the orthogonality properties (2.171) and (2.178), which we recall for convenience,

$$\int_0^1 \mathrm{d}x \, x \, J_n(\lambda x) \, J_n(\mu x) = \frac{1}{2} \delta_{\lambda\mu} \left[J_{n+1}(\lambda) \right]^2 , \tag{3.177a}$$

$$\int_0^\infty \mathrm{d}\rho \, \rho \, J_n(k \, \rho) \, J_n(k' \, \rho) = \frac{1}{k} \, \delta(k - k') , \tag{3.177b}$$

where $J_n(\lambda) = J_n(\mu) = 0$. This property ensures the orthogonality of the Bessel functions with respect to the volume integral in cylindrical coordinates. Indeed, the general idea is to expand the boundary condition to the solution of a Laplace equation in cylindrical coordinates, into the basis functions of the solution later used for the entire space, and then, to use the orthogonality properties of the Bessel functions in order to project out specific components.

3.3.2 *Laplace Equation in a Rectangular Parallelepiped*

We have discussed solutions of the Laplace equation in Cartesian, spherical and cylindrical coordinates. It is illustrative to consider a nontrivial example, and we shall choose the case of a rectangular box, which corresponds to Cartesian coordinates. The potential is specified on each of the six rectangular surfaces of the box. A simplified version of this problem is given by the solutions to Laplace's equation in a rectangle, which is tantamount to assuming uniformity of the potential in the z direction. The reference volume V is chosen as follows with the x and y directions being restricted to finite intervals, and the z direction extending from $z = 0$ to $z = \infty$,

$$0 \le x \le a, \quad 0 \le y \le b, \quad z \ge 0. \tag{3.178}$$

We assume that the charge density inside the volume V vanishes, so that the electrostatic potential fulfills the Laplace equation $\vec{\nabla}^2 \Phi(\vec{r}) = 0$. Furthermore, in the sense of Dirichlet boundary conditions, the potential is held constant near the boundaries in the x and y directions,

$$\Phi(0, y, z) = \Phi(a, y, z) = \Phi(x, 0, z) = \Phi(x, b, z) = \Phi_0 . \tag{3.179}$$

The potential tends to a constant value at infinity,

$$\Phi(x, y, z \to \infty) = \Phi_0, \tag{3.180}$$

and in the xy plane ($z = 0$), we have a given, possibly non-constant, distribution for the potential

$$\Phi(x, y, 0) = \Phi_0 f(x, y), \tag{3.181}$$

where $f(x, y)$ is a given function. This defines our problem. In the first step of the calculation, one identifies the expansion (3.166) as the appropriate expansion for Cartesian coordinates. So,

$$\Phi(x, y, z) = \sum_{\vec{k}^2=0} \left[c(k) e^{\vec{k}\cdot\vec{r}} + d(k) e^{\vec{k}\cdot\vec{r}} \left\{ (ax + b)\, \delta_{k_1,0} + (cy + d)\, \delta_{k_2,0} + (ez + f)\, \delta_{k_3,0} \right\} \right]$$
$$+ (ax + b)(cy + d)(ez + f). \tag{3.182}$$

First, we note that any terms linear in x, y, or z in our approach given by Eq. (3.182) do not satisfy the boundary conditions (3.179). If we eliminate the linear terms, then the general solution can be written as a sum over the exponential terms alone, and a physically irrelevant constant C_0,

$$\Phi(x, y, z) = \sum_{\vec{k}^2=0} c(\vec{k})\, e^{\vec{k}\cdot\vec{r}} + C_0, \qquad \text{where} \qquad \vec{k}^2 = k_x^2 + k_y^2 + k_z^2 = 0. \tag{3.183}$$

The overall constant term C_0 is not eliminated (yet).

In the second step of the calculation, we observe that the deviation of the solution from the constant C_0 must vanish in the limit $z \to \infty$. So, we take a solution which is periodic in x and y and exponentially damped for large, positive z. This implies, in particular, that

$$\vec{k} = i\kappa_x \,\hat{e}_x + i\kappa_y \,\hat{e}_y + k_3 \,\hat{e}_z, \qquad k_3 = \sqrt{\kappa_x^2 + \kappa_y^2}. \tag{3.184}$$

Then, the solution becomes

$$\Phi(x, y, z) = \sum_{\vec{k}} c(\vec{k})\, e^{i(\kappa_x\, x + \kappa_y\, y)}\, e^{-\sqrt{\kappa_x^2 + \kappa_y^2}\,z} + C_0. \tag{3.185}$$

In general, if (k_1, k_2) is an admissible combination of wave vectors, then $(-k_1, -k_2)$ will also be admissible. We here restrict our approach, at first somewhat arbitrarily, to the combinations

$$\Phi(x, y, z) = \sum_{\vec{k},\pm} c_\pm(\vec{k}) \left(e^{i\kappa_x x} \pm e^{-i\kappa_x x} \right) \left(e^{i\kappa_y y} \pm e^{-i\kappa_y y} \right) e^{-\sqrt{\kappa_x^2 + \kappa_y^2}\,z} + C_0. \tag{3.186}$$

In a third step of the calculation, we let $(k_1, k_2) = (u, v)$ where u and v are real positive numbers. Then, our *ansatz* for the potential becomes

$$\Phi(x, y, z) = \sum_{u,v,\pm} c_\pm(u, v) \left[e^{iux} \pm e^{-iux} \right] \left[e^{ivy} \pm e^{-ivy} \right] \exp\left(-[u^2 + v^2]^{1/2} z \right) + C_0. \tag{3.187}$$

Finally, according to our assumption (3.179), we have

$$\Phi(0, y, z) = \Phi(a, y, z) = \Phi(x, 0, z) = \Phi(x, b, z) = \Phi_0. \tag{3.188}$$

We now force the periodic terms to vanish on the x and y boundaries and identify $C_0 = \Phi_0$. We are now in the position to specify the admissible wave vectors even more accurately, because they have to fulfill the condition that all terms except for

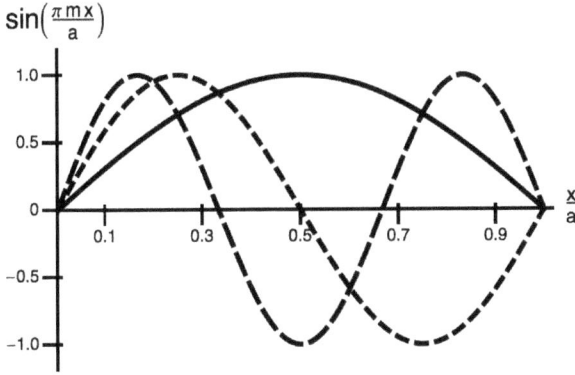

$$\sin\left(\frac{\pi m x}{a}\right)$$

Fig. 3.2   Plot of the functions $\sin(\pi m x/a)$ for $a = 1$ in the interval $x \in (0, 1)$ with parameters/indices $m = 1, 2, 3$. The basis functions $\sin(\pi m x/a)$ vanish at the boundaries and are mutually orthogonal when integrated over $0 \le x \le a$.

$C_0 = \Phi_0$ vanish on the boundary. Let $a u = m\pi$ and $b v = n\pi$ where $m, n \in \mathbb{N}$. This singles out the following solutions as admissible,

$$\Phi(x, y, z) = \sum_{m=1}^{\infty} \sum_{n=1}^{\infty} d(m, n) \frac{1}{2i} \left[\exp\left(i\frac{\pi m}{a} x\right) - \exp\left(-i\frac{\pi m}{a} x\right) \right]$$

$$\times \frac{1}{2i} \left[\exp\left(i\frac{\pi n}{b} y\right) - \exp\left(-i\frac{\pi n}{b} y\right) \right] + \Phi_0 \,, \tag{3.189}$$

where the $d(m, n)$ represent expansion coefficients which will be determined later. Then,

$$\Phi(x, y, z) = \sum_{m=1}^{\infty} \sum_{n=1}^{\infty} d(m, n) \sin\left(\frac{\pi m}{a} x\right) \sin\left(\frac{\pi n}{b} y\right) \exp\left(-\pi z \sqrt{\left(\frac{m}{a}\right)^2 + \left(\frac{n}{b}\right)^2}\right) + \Phi_0 \,.$$
$$\tag{3.190}$$

A plot of the first few basis functions $\sin(\pi m x/a)$ for $a = 1$ is given in Fig. 3.2. In the last step, we recall that at $z = 0$, we have to satisfy the boundary condition $\Phi(x, y, 0) = \Phi_0 f(x, y)$. To this end, we note that at $z = 0$, the exponential suppression factor in the z direction is just unity,

$$\left. \exp\left(-\pi z \sqrt{\left(\frac{m}{a}\right)^2 + \left(\frac{n}{b}\right)^2}\right) \right|_{z=0} = 1 \,. \tag{3.191}$$

The boundary condition $\Phi(x, y, 0) = \Phi_0 f(x, y)$ then reduces to

$$\sum_{m=1}^{\infty} \sum_{n=1}^{\infty} d(m, n) \sin\left(\frac{\pi m}{a} x\right) \sin\left(\frac{\pi n}{b} y\right) = \Phi_0 \left\{ f(x, y) - 1 \right\} \,. \tag{3.192}$$

In order to determine the coefficients $d(m,n)$, we now multiply both sides by $\sin(\pi m'x/a)\,\sin(\pi n'y/b)$ and integrate over x and y,

$$\Phi_0 \int_0^a dx \int_0^b dy\, [f(x,y)-1]\, \sin\left(\frac{\pi m'}{a}x\right) \sin\left(\frac{\pi n'}{n}y\right)$$

$$= \sum_{m=1}^\infty \sum_{n=1}^\infty d(m,n) \left(\int_0^a dx \sin\left(\frac{\pi m}{a}x\right) \sin\left(\frac{\pi m'}{a}x\right)\right)\left(\int_0^b dy \sin\left(\frac{\pi n}{b}y\right) \sin\left(\frac{\pi n'}{b}y\right)\right)$$

$$= \sum_{m=1}^\infty \sum_{n=1}^\infty d(m,n) \left(\frac{a}{\pi}\int_0^\pi d\xi\, \sin(m\xi)\sin(m'\xi)\right)\left(\frac{b}{\pi}\int_0^\pi d\chi\, \sin(n\chi)\sin(n'\chi)\right),$$

$$(3.193)$$

where $\xi \equiv \dfrac{\pi}{a}x$ and $\chi \equiv \dfrac{\pi}{b}y$. Using the orthogonality of the sine and cosine functions,

$$\int_0^\pi \sin(m\xi)\,\sin(m'\xi)\,d\xi = \frac{\pi}{2}\delta_{mm'}, \qquad \int_0^\pi \cos(n\chi)\,\cos(n'\chi)\,d\chi = \frac{\pi}{2}\delta_{nn'},$$

$$(3.194)$$

the right-hand side of Eq. (3.193) thus becomes

$$\sum_{m=1}^\infty \sum_{n=1}^\infty d(m,n)\,\frac{a}{\pi}\left(\frac{\pi}{2}\delta_{mm'}\right)\frac{b}{\pi}\left(\frac{\pi}{2}\delta_{nn'}\right) = d(m',n')\,\frac{ab}{4}. \qquad (3.195)$$

Renaming $m' \to m$, $n' \to n$, we can thus write a general expression for all expansion coefficients $d(m,n)$,

$$d(m,n) = \frac{4}{ab}\Phi_0 \int_0^a dx \int_0^b dy\, [f(x,y)-1]\, \sin\left(\frac{\pi m}{a}x\right) \sin\left(\frac{\pi n}{b}y\right). \qquad (3.196)$$

We recall that the general solution for the potential $\Phi(x,y,z)$, in terms of the $d(m,n)$, is then given as

$$\Phi(x,y,z) = \sum_{m=1}^\infty \sum_{n=1}^\infty d(m,n)\,\sin\left(\frac{\pi m}{a}x\right)\sin\left(\frac{\pi n}{b}y\right)\exp\left(-\pi z\sqrt{\left(\frac{m}{a}\right)^2 + \left(\frac{n}{b}\right)^2}\right) + \Phi_0.$$

$$(3.197)$$

Our boundary value problem involves the boundary conditions $\Phi(0,y,z) = \Phi(a,y,z) = \Phi(x,0,z) = \Phi(x,b,z) = \Phi_0$, with $\Phi(x,y,z \to \infty) = 0$ and $\Phi(x,y,0) = \Phi_0 f(x,y)$. We consider it for the choice

$$f(x,y) = 1 + \left[\frac{1}{a^2}\left(x - \frac{a}{2}\right)^2 - \frac{1}{4}\right] = 1 - \frac{x}{a} + \left(\frac{x}{a}\right)^2, \qquad 0 < x < a, \qquad 0 < y < b. \quad (3.198)$$

This boundary condition depends only on x [see Fig. 3.3(a)]. The expansion coefficients $d(m,n)$ in this case are given by

$$d(m,n) = -\frac{8}{m^3 n \pi^4}\,[1-(-1)^m]\,[1-(-1)^n] = -\frac{32}{(2\,m+1)^3\,(2\,n+1)\,\pi^4}, \qquad (3.199)$$

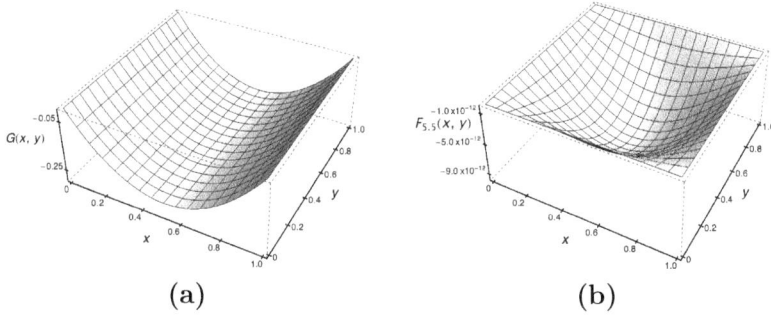

(a)   (b)

Fig. 3.3   Panel (a) illustrates the boundary condition (3.198) for $a = b = 1$. The function $G(x,y)$ is defined in Eq. (3.201). The right panel (b) displays the function $F_z(x,y) = \Phi(x,y,z) - \Phi_0$ for $z = 5.5$, for $a = b = 1$. Note the smaller scale of the ordinate axis. Indeed, the departure from the boundary value assumed near $x = 0, a$ and $y = 0, b$ becomes numerically smaller as we go up the z axis; observe the scale of the ordinate axis in panel (b).

where $m = 2\overline{m} + 1$ and $n = 2\overline{n} + 1$. The full solution is then given as

$$\Phi(x,y,z) = -\sum_{\overline{m}=0}^{\infty}\sum_{\overline{n}=0}^{\infty} \frac{32\,\Phi_0}{(2\,\overline{m}+1)^3\,(2\,\overline{n}+1)\,\pi^4}$$

$$\times \sin\left(\frac{\pi(2\,\overline{m}+1)}{a}\,x\right)\sin\left(\frac{\pi(2\,\overline{n}+1)}{b}\,y\right)$$

$$\times \exp\left(-\pi z\sqrt{\left(\frac{2\,\overline{m}+1}{a}\right)^2 + \left(\frac{2\,\overline{n}+1}{b}\right)^2}\right) + \Phi_0. \qquad (3.200)$$

For $z = 0$, we finally have to convince ourselves that the boundary condition is fulfilled. To this end we investigate the function

$$G(x,y) = \Phi(x,y,0) - \Phi_0$$

$$= -\sum_{\overline{m}=0}^{\infty}\sum_{\overline{n}=0}^{\infty} 32\,\Phi_0\, \frac{\sin\left(\frac{\pi(2\,\overline{m}+1)}{a}\,x\right)\sin\left(\frac{\pi(2\,\overline{n}+1)}{b}\,y\right)}{(2\,\overline{m}+1)^3\,(2\,\overline{n}+1)\,\pi^4}$$

$$= \Phi_0\left(-\frac{x}{a} + \left(\frac{x}{a}\right)^2\right) \qquad 0 \le x \le a, \qquad 0 \le y \le b, \qquad (3.201)$$

but still $\Phi(0,y,0) = \Phi(a,y,0) = \Phi(x,0,0) = \Phi(x,b,0) = \Phi_0$. We define $G_n(x,y)$ to be the function obtained from the defining formula (3.201) by truncating the summations over $\overline{m}$ and $\overline{n}$ at n,

$$G(x,y) = \lim_{n\to\infty} G_n(x,y), \qquad (3.202\text{a})$$

$$G_n(x,y) = \sum_{\overline{m}=0}^{n}\sum_{\overline{n}=0}^{n} 32\,\Phi_0\, \frac{\sin\left(\frac{\pi(2\,\overline{m}+1)}{a}\,x\right)\sin\left(\frac{\pi(2\,\overline{n}+1)}{b}\,y\right)}{(2\,\overline{m}+1)^3\,(2\,\overline{n}+1)\,\pi^4}. \qquad (3.202\text{b})$$

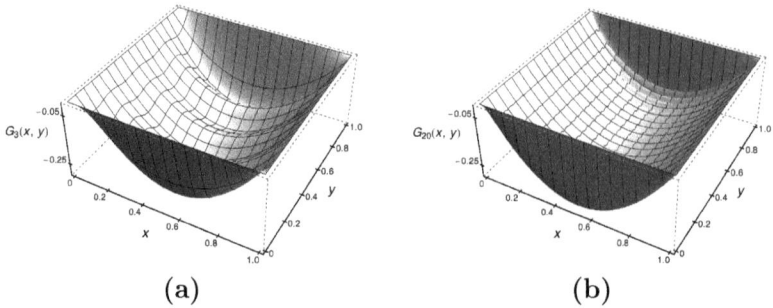

Fig. 3.4   Panel (a) illustrates how the partial sums over $\overline{m}$ and $\overline{n}$ approximate the solution (3.200), by plotting the function $G_n(x, y)$ with $n = 3$, given by Eq. (3.202). The sum is terminated at $n = \overline{m} = \overline{n} = 3$. Panel (b) displays a better approximation with the sum being terminated at $n = \overline{m} = \overline{n} = 20$.

As evident from the left panel in Fig. 3.4, with both summations over $\overline{m}$ and $\overline{n}$ truncated at $n = 3$, the boundary condition on the edges of the rectangle are fulfilled by our solution (3.200), at $x = 0$, $x = a$, $y = 0$, $y = b$. However, as soon as we depart from the lines with $y = 0$ and $y = b$, there is a discontinuity because the term $-\frac{x}{a} + \left(\frac{x}{a}\right)^2$ does not vanish for $0 \leq x \leq a$ and $0 \leq x \leq b$. The approximation of this discontinuity implies oscillations in the y direction for the truncated approximation, as visible in Fig. 3.4(a) and Fig. 3.4(b).

It is interesting to consider what happens as we depart from the $z = 0$ plane and go up to z axis, to arrive, say, at $z = 5.5$. This is illustrated in Fig. 3.3(b). The value of the expression $F(x, y, z) = \Phi(x, y, z) - \Phi_0$ becomes much smaller (of order 10^{-12}) at $z = 5.5$ than at $z = 0$, where it amounts to $G(x, y) = F(x, y, 0)$ and is of order unity. Because of the smallness of the function $G(x, y)$, the discontinuity near the edges of the rectangle $x = 0$, $x = a$, $y = 0$, and $y = b$ gradually disappears as we go up the z axis. This is compatible with the boundary condition $\Phi(x, y, z \to \infty) = \Phi_0$.

3.3.3   Laplace Equation in a Two-Dimensional Rectangle

A simplified version of the zero-charge-density problem is given by the solutions to Laplace's equation in a rectangle, which is tantamount to assuming uniformity of the potential in the z direction. In this case, the Laplace equation $\vec{\nabla}^2 \Phi(\vec{r}) = 0$ reduces to an equation with differential operators that only act in the x and y directions. The technique used for the two-dimensional problem (for a rectangle) thus is a specialization of the previously discussed problem for the rectangular parallelepiped. Let us consider boundary conditions given on a two-dimensional rectangle, with a potential $\Phi = \Phi(x, y)$ that fulfills

$$\left(\frac{\partial^2}{\partial x^2} + \frac{\partial^2}{\partial y^2}\right) \Phi(x, y) = 0, \qquad 0 < x < a, \qquad 0 < y < b. \qquad (3.203)$$

We take our boundary conditions to be

$$\Phi\left(0,y\right) = \Phi\left(a,y\right) = \Phi\left(x,0\right) = 0\,, \qquad \Phi\left(x,b\right) = f\left(x\right)\,, \tag{3.204}$$

where $f(x)$ is a given function. According to the above sections, we should search for a series solution involving exponentials,

$$\Phi\left(x,y\right) = \sum_{\vec{k}} c(\vec{k})\, e^{\pm i|k_1|x}\, e^{\pm k_2 y} + C_0\,, \tag{3.205}$$

where each term in the series satisfies the Laplace equation in two dimensions and so $|k_1| = |k_2|$. In order to select the proper solution to these equations we refer to the boundary conditions. Since $\Phi\left(x,b\right)$ is specified in the boundary we use a (complete) set of orthogonal functions for the x dependence in the interval $x \in (0,a)$. Since $\Phi\left(0,y\right) = \Phi\left(a,y\right) = 0$, we use sine functions which are equal to zero at $x = 0$ and $x = a$. The y dependence has to be adjusted so that the functions vanish at $y = 0$, which singles out the hyperbolic sine (as opposed to the cosine) functions. It follows that

$$\Phi\left(x,y\right) = \sum_{n=1}^{\infty} a_n \sin\left(\frac{n\pi}{a}x\right) \frac{\exp\left(\frac{\pi n}{a}y\right) - \exp\left(-\frac{\pi n}{a}y\right)}{\exp\left(\frac{\pi n}{a}b\right) - \exp\left(-\frac{\pi n}{a}b\right)}$$

$$= \sum_{n=1}^{\infty} a_n \sin\left(\frac{n\pi x}{a}\right) \frac{\sinh\left(\frac{n\pi y}{a}\right)}{\sinh\left(\frac{n\pi b}{a}\right)}\,. \tag{3.206}$$

The denominator term $\sinh\left(n\pi b/a\right)$ is a normalization. This *ansatz* automatically fulfills

$$\Phi\left(0,y\right) = \Phi\left(a,y\right) = \Phi\left(x,0\right) = 0\,. \tag{3.207}$$

In order to also fulfill

$$\Phi\left(x,b\right) = f\left(x\right)\,, \tag{3.208}$$

we have to demand that

$$a_n = \frac{2}{a} \int_0^a f\left(x\right) \sin\left(\frac{n\pi x}{a}\right) dx\,. \tag{3.209}$$

If we now specialize further to $f\left(x\right) = \Phi_0$, then

$$a_n = \frac{2}{a} \int_0^a \Phi_0 \sin\left(\frac{n\pi x}{a}\right) dx = \frac{2\Phi_0}{n\pi} \left\{1 - (-1)^n\right\}\,. \tag{3.210}$$

Finally, the potential is found to be

$$\Phi\left(x,y\right) = \frac{4\Phi_0}{\pi} \sum_{n=0}^{\infty} \frac{\sin\left[(2n+1)\pi x/a\right]}{2n+1} \frac{\sinh\left[(2n+1)\pi y/a\right]}{\sinh\left[(2n+1)\pi b/a\right]}\,, \tag{3.211}$$

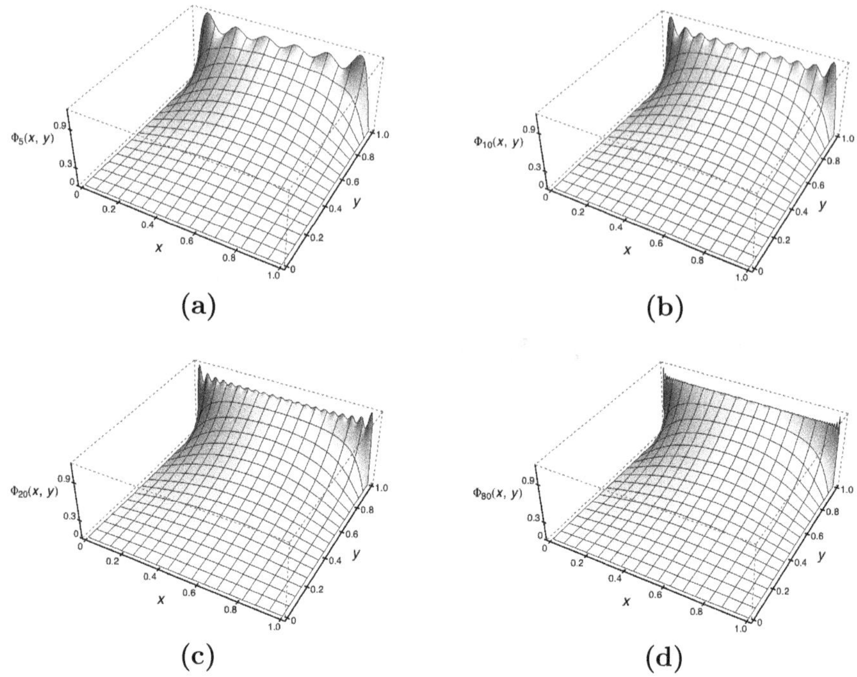

Fig. 3.5   Plot of the function $\Phi(x,y)$ as defined in Eq. (3.211) for $\Phi_0 = 1$ and $a = b = 1$. We define $\Phi_n(x,y)$ to be equal to the series expansion given in Eq. (3.211), where the sum is truncated at n. Panels (a), (b), (c) , and (d) show the truncated functions at $n = 5$, $n = 10$, $n = 20$ and $n = 80$.

where the sum is over the odd integers $2n + 1$, with $n = 0, 1, 2, \ldots$. For $0 \le y < a$, this series and its first derivatives converge absolutely. A plot of the sum of the terms up to $n = 20$ is shown in Fig. 3.5. Quite naturally, there are oscillations in the (slow) convergence pattern near the discontinuities of $\Phi(x,y)$ which occur near the points $(0, b)$ and (a, b).

3.3.4   Boundary Conditions on the Finite Part of a Long Strip

We also consider a two-dimensional boundary problem, but this time, one of the regions covered by the potential has an infinite extent, in contrast to the example discussed in Sec. 3.3.3. The problem is given as

$$\left(\frac{\partial^2}{\partial x^2} + \frac{\partial^2}{\partial y^2} \right) \Phi(x,y) = 0 , \qquad 0 < x < a \to \infty , \qquad 0 < y < b . \tag{3.212}$$

We define the boundary conditions to be

$$\Phi(x,0) = \Phi(x,b) = 0 , \qquad \Phi(0,y) = g(y) . \tag{3.213}$$

For these boundary conditions, the appropriate choice is

$$\Phi(x,y) = \sum_{n=1}^{\infty} \alpha_n \sin\left(\frac{n\pi y}{b}\right) \exp\left(-\frac{n\pi x}{b}\right). \tag{3.214}$$

The boundary condition $\Phi(x,0) = \Phi(x,b) = 0$ is automatically implemented via the sine function, and the boundary condition $\Phi(0,y) = g(y)$ implies that

$$g(y) = \sum_{n=1}^{\infty} \alpha_n \sin\left(\frac{n\pi y}{b}\right). \tag{3.215}$$

In view of the orthogonality of the sine functions, we have

$$\alpha_n = \frac{2}{b} \int_0^b dy\, g(y) \sin\left(\frac{n\pi y}{b}\right). \tag{3.216}$$

The coefficients b_n depend on the particular choice of the boundary condition. If we let

$$g(y) = \Phi_0 \frac{y}{b}, \tag{3.217}$$

then the expansion coefficients are given by

$$\alpha_n = \frac{2\Phi_0}{b^2} \int_0^b dy\, y \sin\left(\frac{n\pi y}{b}\right) \tag{3.218}$$

$$= \frac{2\Phi_0}{b^2} \frac{b}{n\pi} \int_0^b dy\, y \left(-\frac{d}{dy}\left\{\cos\left(\frac{n\pi y}{b}\right)\right\}\right) \tag{3.219}$$

$$= \frac{2\,\Phi_0}{b\,n\,\pi} \left[-y\left\{\cos\left(\frac{n\pi y}{b}\right)\right\}\Big|_0^b - \int_0^b dy\left(-\cos\left(\frac{n\pi y}{b}\right)\right)\right] \tag{3.220}$$

$$= -\frac{2\Phi_0}{an\pi} b \cos(n\pi) = -2\,(-1)^n \frac{\Phi_0}{n\pi}. \tag{3.221}$$

The potential thus is given as

$$\Phi(x,y) = -\frac{2\Phi_0}{\pi} \sum_{n=1}^{\infty} \frac{(-1)^n}{n} \sin\left(\frac{n\pi y}{b}\right) \exp\left(-\frac{n\pi x}{b}\right). \tag{3.222}$$

The sum over n can be performed by writing the entire expression in terms of exponentials,

$$\Phi(x,y) = -\frac{2\Phi_0}{\pi} \sum_{n=1}^{\infty} \frac{(-1)^n}{n} \left[\left\{e^{i\pi y/b}\right\}^n - \left\{e^{i\pi y/b}\right\}^{-n}\right] \frac{1}{2i} \left\{\exp\left(-\frac{\pi x}{b}\right)\right\}^n$$

$$= -\frac{2\Phi_0}{2i\pi} \sum_{n=1}^{\infty} \frac{(-1)^n}{n} \left[(uv)^n - \left(\frac{v}{u}\right)^n\right],$$

$$u = e^{i\pi y/b}, \qquad v = e^{-\pi x/b}. \tag{3.223}$$

We now use the formula

$$\sum_{n=1}^{\infty} \frac{(-1)^n}{n} x^n = -\ln(1+x), \tag{3.224}$$

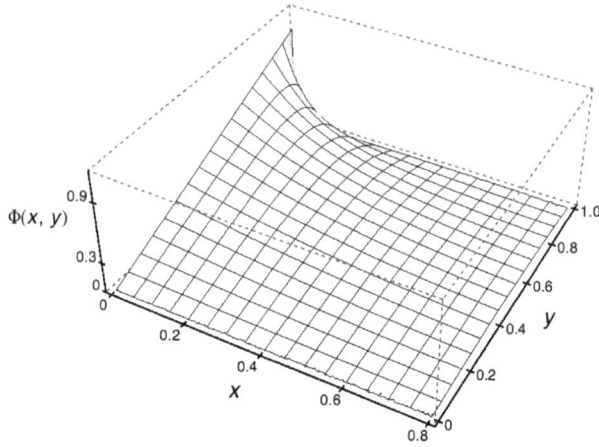

Fig. 3.6   Plot of the function $\Phi(x, y)$ given in Eq. (3.225), for $\Phi_0 = 1$ and $b = 1$.

to obtain

$$
\Phi\left(x, y\right) = -\frac{\Phi_0}{i\pi} \left\{ \ln\left(1 + \frac{v}{u}\right) - \ln\left(1 + u\,v\right) \right\} = -\frac{\Phi_0}{i\pi} \ln\left(\frac{u+v}{u\left(1+u\,v\right)}\right)
$$

$$
= -\frac{\Phi_0}{i\pi} \ln\left(\frac{e^{i\pi y/b} + e^{-\pi x/b}}{e^{i\pi y/b}\left(1 + e^{i\pi y/b - \pi x/b}\right)}\right) = -\frac{\Phi_0}{i\pi} \ln\left(\frac{1 + e^{-\pi x/b - i\pi y/b}}{1 + e^{i\pi y/b - \pi x/b}}\right)
$$

$$
= -\frac{\Phi_0}{i\pi} \ln\left(\frac{e^{-i\pi y/b} + e^{\pi x/b}}{e^{i\pi y/b} + e^{\pi x/b}}\right) = \frac{\Phi_0}{i\pi} \ln\left(\frac{e^{i\pi y/b} + e^{\pi x/b}}{e^{-i\pi y/b} + e^{\pi x/b}}\right). \tag{3.225}
$$

Somewhat counter-intuitively, this result is purely real, not complex. This can be seen as follows. If we consider the complex conjugate of the expression, the imaginary unit in the prefactor changes sign, but the numerator and the denominator in the argument of the logarithm are also exchanged. Inverting the argument of the logarithm, one restores the original expression. A plot of the solution is found in Fig. 3.6.

3.3.5 *Cauchy's Residue Theorem: A Small Digression*

Helpful mathematical techniques are sometimes compared to the "toolbox" of a theoretical physicist; within this analogy, one might identify the Cauchy residue theorem with a "leatherman". It has many physical applications; e.g., it becomes important to the understanding of time ordering in electrodynamics. Another very important application concerns the evaluation of integrals over an infinite domain of integration, closing the integration contour along an appropriate curve in the complex plane. This latter technique is important in the evaluation of certain integrals occurring in Fourier transformations, which naturally arise in boundary-value problems in electrostatics. Preparations for a graduate course in electrodynamics should in principle include deep familiarity with the Cauchy theorem; yet experience shows that this preparation may be lacking. A detour on this subject therefore is in order.

Let us start innocently, recalling that the Taylor expansions of a function $f = f(x)$ about the origin reads as

$$f(x) = f(0) + f'(0)\,x + \frac{1}{2!}f''(0)\,x^2 + \cdots = \sum_{n=0}^{\infty} \frac{f^{(n)}(0)}{n!}\,x^n = \sum_{n=0}^{\infty} a_n\,x^2. \qquad (3.226)$$

Here, the a_n with $n = 0, 1, 2, \ldots$ are constant coefficients, the coefficients of the Taylor expansion. However, functions like

$$g(z) = \frac{1}{z^4 \sin(z)} \qquad (3.227)$$

obviously cannot be expanded in a Taylor series about $z = 0$, because the first (leading) term is not a constant; the function diverges for $z \to 0$. At best, we can use the well-known expansion, $\sin(z) = z - z^3/3! + z^5/5! + \ldots$ and write

$$g(z) = \frac{1}{z^4 \sin(z)} = \frac{1}{z^4 \left(z - \dfrac{z^3}{3!} + \dfrac{z^5}{5!} + \ldots \right)} = \frac{1}{z^5 \left(1 - \dfrac{z^2}{3!} + \dfrac{z^4}{5!} + \ldots \right)}$$

$$= \frac{1}{z^5} + \frac{1}{6\,z^3} + \frac{7}{360\,z} + \frac{31\,z}{15120} + \mathcal{O}(z^3). \qquad (3.228)$$

This expansion obviously does not start from the term of order zero, but from a term proportional to z^{-5}. It is called a Laurent expansion. The general formula for the Laurent expansion reads

$$g(z) = \sum_{n=-\infty}^{\infty} a_n\,z^n. \qquad (3.229)$$

In contrast to the Taylor expansion (3.226), the Laurent expansion starts at $n = -\infty$ instead of $n = 0$. We can formulate a Laurent expansion about a point z_0, which reads

$$g(z) = \sum_{n=-\infty}^{\infty} b_n\,(z - z_0)^n. \qquad (3.230)$$

If all coefficients b_n with $n < 0$ vanish, we say that $g(z)$ is analytic about the point $z = z_0$. If all a_n's with $n < -M$ vanish, then we say that g has an Mth order pole at the origin.

Of preeminent importance is the integral

$$I = \oint g(z)\,\mathrm{d}z = \int_C g(z)\,\mathrm{d}z = \int_C \left(\sum_{n=-\infty}^{\infty} a_n\,z^n \right) \mathrm{d}z, \qquad (3.231)$$

where the contour C is an anticlockwise circle (i.e., in the mathematically positive sense) of radius R around the origin. We shall see that only a very limited number of a_n coefficients (in fact, just one of them) contributes to the integral. Let us therefore consider the integral

$$J = \oint_C dz \, \frac{1}{z^m} \, , \qquad z = R \exp(i\theta) \, , \qquad \theta \in (0, 2\pi) \, , \qquad (3.232)$$

where $z = z(\theta)$ thus describes an anticlockwise circle about the origin. Furthermore, m is assumed to be integer. Now, for $m \neq 1$, we have

$$J = \int_0^{2\pi} d\theta \, (iR) \exp(i\theta) \, [R \exp(i\theta)]^{-m} = i R^{1-m} \int_0^{2\pi} d\theta \, \exp[-i(m-1)\theta]$$

$$= i R^{1-m} \left[\frac{\exp[-i(m-1)\theta]}{-(m-1)i} \right]_{\theta=0}^{\theta=2\pi} = -R^{1-m} \left[\frac{\exp[-i(m-1)\theta]}{(m-1)} \right]_{\theta=0}^{\theta=2\pi} = 0 \, . \quad (3.233)$$

However, for $m = 1$, the correct result is obtained by first replacing $m \to 1 + \epsilon$ and doing a Taylor expansion,

$$\lim_{m \to 1} J = - R^{-\epsilon} \left[\lim_{\epsilon \to 0} \frac{1}{\epsilon} \exp(-i\epsilon\theta) \right]_{\theta=0}^{\theta=2\pi} = - \left[\frac{1 - i\epsilon\theta}{\epsilon} \right]_{\theta=0}^{\theta=2\pi}$$

$$= - \left[\frac{1}{\epsilon} - i\theta \right]_{\theta=0}^{\theta=2\pi} = [i\theta]_{\theta=0}^{\theta=2\pi} = 2\pi i \, . \qquad (3.234)$$

Of course, this result could have been obtained by a direct calculation, without letting $m \to 1$, but the above derivation illustrates that the special case $m = 1$ actually follows from the more general case by a limiting process. For a function $g(z)$ that has a Laurent expansion of the form $g(z) = \sum_{n=-\infty}^{\infty} a_n z^n$ about the point $z_0 = 0$, we therefore have

$$I = \oint_C dz \, f(z) = \oint_C dz \left(\sum_{m=-\infty}^{\infty} a_m z^m \right) = 2\pi i \, a_{-1} \equiv 2\pi \operatorname*{Res}_{z=0} f(z) \, , \qquad (3.235)$$

where the last term constitutes a definition of the residue at the point $z = z_0 = 0$. Only the coefficient a_{-1} contributes. We write the residue as

$$\operatorname*{Res}_{z=0} f(z) \equiv a_{-1} \, , \qquad \oint_C dz \, f(z) = 2\pi i \operatorname*{Res}_{z=0} f(z) \, . \qquad (3.236)$$

This result can be generalized easily, as follows. Let us assume that we have an expansion about an arbitrary point z_0,

$$f(z) = \sum_{m=-\infty}^{\infty} b_m \, (z - z_0)^m \, . \qquad (3.237)$$

In view of the above considerations, a contour integral that encircles z_0 in the mathematically positive sense can be written as

$$\oint_{C_0} dz \, f(z) = 2\pi i \, b_{-1} = 2\pi i \operatorname*{Res}_{z=z_0} f(z) \, . \qquad (3.238)$$

Only the coefficient b_{-1} contributes. Closed contour integrals in regions where the integrand is analytic, simply vanish. This enables us to generalize the result as follows. For a contour C that encircles n residues located at the points $z = z_i$ with

$i \in \{1, \ldots, n\}$, the result is

$$\oint_C dz \, f(z) = 2\pi \sum_{i=1}^{n} \operatorname*{Res}_{z=z_i} f(z). \tag{3.239}$$

All the z_i lie in the interior of the contour C. This is Cauchy's residue theorem.

The calculation of residues and the application of this theorem should be illustrated on a few example cases. For functions that have a single pole, like

$$g(z) = \frac{1}{\sin z} \approx \frac{1}{z - \frac{1}{3}z^3 + \ldots}, \qquad \operatorname*{Res}_{z=0} g(z) = 1, \tag{3.240}$$

the calculation of the residue is obvious. For functions that have poles of higher order, it becomes more complicated. In Eq. (3.228), we have already derived the result

$$\operatorname*{Res}_{z=0} \left(\frac{1}{z^4 \sin(z)} \right) = \frac{7}{360}. \tag{3.241}$$

Let us also consider

$$g(z) = \frac{\cos(z)}{z^4 \sin(z)} = \left(\frac{1}{z^5} \right) \left(\frac{z \cos(z)}{\sin(z)} \right). \tag{3.242}$$

We have split the function into a prefactor $1/z^5$ and term which is unity for $z \to 0$. A simple Taylor expansion of numerator and denominator shows that

$$\frac{z \cos(z)}{\sin(z)} = 1 - \frac{z^2}{3} - \frac{z^4}{45} + \ldots \tag{3.243}$$

and so, for $z \to 0$,

$$g(z) \approx \frac{1}{z^5} \left(1 - \frac{z^2}{3} - \frac{z^4}{45} + \ldots \right), \qquad \operatorname*{Res}_{z=0} g(z) = -\frac{1}{45}, \tag{3.244}$$

where the latter result is simply obtained by an expansion of the former, and reading off the term that multiplies z^{-1}. The residue always is the factor multiplying $1/(z - z_0)$ in the Laurent expansion of $F(z)$, independent of how many other terms of the form $(z - z_0)^n$, with positive or negative integer n, occur in the Laurent expansion of $g(z)$. One just has to expand $g(z)$ to the required order and read off the term of order $(z - z_0)^{-1}$.

Let us now investigate a generalization of this result, namely, the case where $f(z)$ has an Mth order pole. For the example in Eq. (3.244), we have $M = 5$. We are interested in finding out about the term of order $(z - z_0)^{-1}$ in $f(z)$. To this end, one can follow the *ad hoc* procedure outlined above: We just expand numerator and denominator to the required order and then read off the term of order $(z - z_0)^{-1}$. However, it is also possible to follow a more systematic approach. If one multiplies the terms $\{a_{-M}(z-z_0)^{-M}, a_{-M+1}(z-z_0)^{-M+1}, \ldots, a_{-1}(z-z_0)^{-1}, \ldots\}$ by a factor $(z-z_0)^M$, then they transform into the set $\{a_{-M}, a_{-M+1}(z-z_0), \ldots, a_{-1}(z-z_0)^{M-1}, \ldots\}$. Expressed differently,

$$f(z) = \frac{a_{-M}}{(z - z_0)^M} + \frac{a_{-M+1}}{(z - z_0)^{M-1}} + \cdots + \frac{a_{-1}}{(z - z_0)} + \cdots + a_n(z - z_0)^n + \ldots, \tag{3.245}$$

$$(z - z_0)^M f(z) = a_{-M} + (z - z_0) a_{-M+1} + \cdots + (z - z_0)^{M-1} a_{-1} + \ldots. \qquad (3.246)$$

In consequence, if we differentiate the expression $(z - z_0)^M f(z)$ a total of $M-1$ times with respect to z, and set $z = z_0$ after the differentiation, then the only contributing term is the one proportional to a_{-1}. Terms of lower order are explicitly annihilated by the differential operator, and terms of higher order give rise to positive powers of $(z - z_0)$; these are annihilated if we set $z = z_0$ after the differentiation. This leads to the general formula, valid for an Mth order pole,

$$\operatorname*{Res}_{z=z_0} f(z) = \frac{1}{(M-1)!} \left[\frac{d^{M-1} \left\{ (z - z_0)^M f(z) \right\}}{dz^{M-1}} \right]_{z=z_0}. \qquad (3.247)$$

A paradigmatic application of the above consideration is as follows. For an integral such as

$$\int_{-\infty}^{\infty} d\omega \, \frac{\eta}{\omega^2 + \eta^2} = \int_{-\infty}^{\infty} d\omega \, \frac{\eta}{(\omega + i\eta)(\omega - i\eta)}, \qquad (3.248)$$

we can close the integration contour either in the upper or the lower complex plane because the modulus of the half-circle integral behaves as $R^{-2+1} \to 0$ for $R \to \infty$, where $|z| = R$ is the radius of the half circle along which the contour is being closed. At the singularity, at $\omega = i\eta$, we have

$$\frac{\eta}{(\omega + i\eta)(\omega - i\eta)} \approx \frac{\eta}{2i\eta(\omega - i\eta)} \qquad (3.249)$$

and hence

$$\operatorname*{Res}_{\omega = i\eta} \frac{\eta}{(\omega + i\eta)(\omega - i\eta)} = \frac{1}{2i}. \qquad (3.250)$$

The result is

$$\int_{-\infty}^{\infty} d\omega \, \frac{\eta}{\omega^2 + \eta^2} = 2\pi i \, \frac{1}{2i} = \pi. \qquad (3.251)$$

Reversing the argument, for closing the contour in the lower half of the complex plane, we have

$$\int_{-\infty}^{\infty} d\omega \, \frac{\eta}{\omega^2 + \eta^2} = (-2\pi i) \left(-\frac{1}{2i} \right) = \pi. \qquad (3.252)$$

The first minus sign is due to the reversed sense of revolution around the singularity. The second minus sign is due to an additional sign reversal of the residue at $\omega = -i\eta$.

3.3.6 *Boundary Conditions on the Infinite Part of a Long Strip*

We now consider again a long strip, as in Sec. 3.3.4, but this time, the boundary condition is imposed on the infinite part of the long strip rather than the short (finite) part. Our boundary conditions are thus given as

$$\Phi(0, y) = \Phi(x, 0) = 0, \qquad \Phi(x, b) = f(x), \qquad 0 < x < a \to \infty, \qquad 0 < y < b. \qquad (3.253)$$

The boundary condition is given on the semi-infinite strip $y = b$. In this case, we switch from a summation to an integration. The expansion functions are labeled by a continuous variable, say k, for which the following relations are useful. In view of the boundary condition $\Phi(0, y) = \Phi(x, 0) = 0$, we choose the *ansatz*

$$\Phi(x, y) = \int_0^\infty dk\, \alpha(k) \sin(k x) \frac{\sinh(k y)}{\sinh(k b)}, \tag{3.254}$$

so that

$$\Phi(x, b) = f(x) = \int_0^\infty dk\, \alpha(k) \sin(k x). \tag{3.255}$$

It is nontrivial to show that

$$\alpha(k) = \frac{2}{\pi} \int_0^\infty dx\, f(x) \sin(k x). \tag{3.256}$$

First, one proves the following relation,

$$
\begin{aligned}
\int_0^\infty dx \sin(k'x) \sin(kx) &= \int_0^\infty dx\, \frac{1}{2i}\left[e^{ik'x} - e^{-ik'x}\right] \frac{1}{2i}\left[e^{ikx} - e^{-ikx}\right] \\
&= -\frac{1}{4} \int_0^\infty \left[\left(e^{ik'x}e^{ikx} + e^{-ik'x}e^{-ikx}\right)\right. \\
&\qquad \left. - \left(e^{ik'x}e^{-ikx} + e^{-ik'x}e^{+ikx}\right)\right] dx \\
&= -\frac{2\pi}{4} \frac{1}{2\pi} \int_{-\infty}^\infty \left(e^{ik'x}e^{ikx} - e^{ik'x}e^{-ikx}\right) dx \\
&= \frac{\pi}{2}\left[\delta(k' - k) - \delta(k' + k)\right]. \tag{3.257}
\end{aligned}
$$

We verify that

$$
\begin{aligned}
\frac{2}{\pi} \int_0^\infty dx\, f(x) \sin(k x) &= \frac{2}{\pi} \int_0^\infty dk'\, \alpha(k') \left[\int_0^\infty dx \sin(k' x) \sin(k x)\right] \\
&= \frac{2}{\pi} \int_0^\infty dk'\, \alpha(k') \frac{\pi}{2}\left[\delta(k' - k) + \delta(k' + k)\right] dk = \alpha(k'). \\
& \tag{3.258}
\end{aligned}
$$

For the boundary condition $\Phi(x, b) = f(x)$, we now choose

$$f(x) = \Phi_0. \tag{3.259}$$

As we shall see later, this example can be used in order to illustrate the properties of complex integrals. The Fourier-sine transform of $f(x) = \Phi_0$ is given by

$$\alpha(k) = \frac{2\Phi_0}{\pi} \int_0^\infty dx \sin(k x) = \frac{2\Phi_0}{\pi 2i} \int_0^\infty dx \left[e^{ikx} - e^{-ikx}\right].$$

In the sense of Riemann, this improper integral is formally divergent. We thus introduce the convergent factor $\exp(-\epsilon x)$ in the integrand, where $\epsilon > 0$ is real and small. Under this assumption, we can assign a finite value to the expansion coefficient $\alpha(k)$,

$$
\alpha(k) = \frac{2\,\Phi_0}{2\pi i} \lim_{\epsilon \to 0^+} \lim_{b \to +\infty} \left[\int_0^b e^{-(\epsilon - ik)x} \mathrm{d}x - \int_0^b e^{-(\epsilon + ik)x} \mathrm{d}x \right]
$$

$$
= \frac{2\,\Phi_0}{2\pi i} \lim_{\epsilon \to 0} \left(\frac{1}{\epsilon - ik} - \frac{1}{\epsilon + ik} \right) = \frac{2\,\Phi_0}{\pi k}. \tag{3.260}
$$

Using this result for $\alpha(k)$, the potential can be written as a convergent integral,

$$
\Phi(x, y) = \frac{2\,\Phi_0}{\pi} \int_0^\infty \mathrm{d}k \, \frac{\sin(kx)}{k} \frac{\sinh(ky)}{\sinh(kb)} = \frac{\Phi_0}{\pi} \int_{-\infty}^\infty \mathrm{d}k \, \frac{\sin(kx)}{k} \frac{\sinh(ky)}{\sinh(kb)},
$$
$$
\tag{3.261}
$$

where we notice the change in the integration limits in the second step. Integrals from $-\infty$ to ∞ are naturally prone to an evaluation by the method of residues, as one can close the contour from $+\infty$ to $-\infty$ along a circle either above or below the real axis, preferably along a large circle whose radius tends to infinity. We shall evaluate the integral using the theory of residues. If the integrand falls off sufficiently rapidly for large modulus of the argument, then the contribution of the large circle closing the contour is zero. Closing the contour, one then picks up the contributions from the pole terms only.

As it stands, the integral (3.261) has no poles along the real k axis. In particular, for $k \to 0$, we have the finite limit $\sin(kx)/k \to x$. However, pole terms are encountered as we use the formula

$$
\sin(kx) = \frac{\exp(ikx) - \exp(-ikx)}{2i}. \tag{3.262}
$$

In performing the k integral, we thus have two alternatives for encircling the poles on the real k axis. We can either shift the contour infinitesimally below the real k axis, letting $k = k - i\epsilon$ and later let $\epsilon \to 0^+$, or we can shift the contour infinitesimally above the k axis, letting $k = k + i\epsilon$ and later calculate the limit $\epsilon \to 0^+$. Yet, we have to make a decision and then stick to it. In the end, both alternatives must lead to equivalent results. The notion is that we later split the integrand into terms for which we can close the contour along either side of the real axis. We write

$$
\Phi(x, y) = \frac{\Phi_0}{2\pi i} \int_{-\infty}^\infty \frac{\exp(ikx) - \exp(-ikx)}{k - i\epsilon} \frac{\sinh(ky)}{\sinh(kb)} \mathrm{d}k
$$

$$
= \frac{\Phi_0}{2\pi i} \int_{C_+} \mathrm{d}k \frac{\exp(ikx)}{k - i\epsilon} \frac{\sinh(ky)}{\sinh(kb)} - \frac{\Phi_0}{2\pi i} \int_{C_-} \mathrm{d}k \frac{\exp(-ikx)}{k - i\epsilon} \frac{\sinh(ky)}{\sinh(kb)}. \tag{3.263}
$$

The contour will be closed along a half-circle of large radius in the upper complex half-plane (denoted as C_+ in the first integral), which leads to exponential damping for the term proportional to $\exp(ikx)$. The contour C_- involves a large half-circle in the lower complex half plane and suppresses the term $\exp(-ikx)$. The contour C_-

is traversed in the mathematically negative direction, leading to an overall minus sign for the residue from the contour closed in the lower half plane.

We now appeal to the discussion in Sec. 3.3.5 and attempt to identify the poles in the integrand in Eq. (3.263). Notably, we shall use the method outlined in Eqs. (3.248)–(3.252)] for the evaluation of relevant indefinite integrals. The poles come from three locations. The first of these is *(i)* from the $1/(k-i\epsilon)$ factor at $k = i\epsilon$ with $\epsilon \to 0^+$. There are *(ii)* poles at $kb = i\pi n$ from the $\sinh(kb)$ term (contour C_+) and *(iii)* poles at $kb = -i\pi n$ from the $\sinh(kb)$ term (contour C_-). The ensemble of the poles thus is a equidistant string of singularities along the imaginary axis, extending over all $kb = i\pi m$ with $m \in \mathbb{Z}_0$. Cauchy's residue theorem, as discussed in Sec. 3.3.5, is thus brought to bear on the subject. The single pole of the first category at $k = i\epsilon$ leads to the following residue

$$\operatorname*{Res}_{k=i\epsilon^+}\left(\frac{\Phi_0}{2\pi i}\frac{\exp(ikx)}{k-i\epsilon}\frac{\sinh(ky)}{\sinh(kb)}\right) = \lim_{\epsilon\to 0^+}\left(\frac{\Phi_0}{2\pi i}\frac{\exp(i\cdot i\epsilon x)}{1}\frac{\sinh(i\epsilon y)}{\sinh(i\epsilon b)}\right)$$

$$= \frac{\Phi_0}{2\pi i}\frac{i\epsilon y}{i\epsilon b} = \frac{\Phi_0\,y}{2\pi i b}. \tag{3.264}$$

The contribution to the potential $\Phi(x,y)$ is

$$\Phi_{\mathrm{I}}(x,y) = 2\pi i \operatorname*{Res}_{k=i\epsilon^+}\left(\frac{\Phi_0}{2\pi i}\frac{\exp(ikx)}{k-i\epsilon}\frac{\sinh(ky)}{\sinh(kb)}\right) = \frac{\Phi_0\,y}{b}. \tag{3.265}$$

For the poles of the second category, we have $kb = n\pi i$ and investigate the $\sinh(kb)$ term. Indeed, we have $\sinh(kb) = 0$ at $kb = n\pi i$ for the integrand in the upper half plane $(n \in \mathbb{N})$. The hyperbolic sine can be expanded near the pole, as follows,

$$\frac{1}{\sinh(kb)} = \frac{1}{\sinh(in\pi + (kb - in\pi))}$$

$$\approx \frac{1}{\sinh(in\pi) + \left(\dfrac{d}{d\xi}\sinh(\xi)\Big|_{\xi=in\pi}\right)(kb - in\pi)}$$

$$= \frac{1}{\cosh(i\pi n)\,(kb - in\pi)} = \frac{1}{b\,(-1)^n\,(k - in\pi/b)}. \tag{3.266}$$

This corresponds to a simple pole at $k = n\pi i/b$, with a coefficient as indicated. We therefore have for the residue,

$$\operatorname*{Res}_{k=n\pi i/b}\left(\frac{\Phi_0}{2\pi i}\frac{\exp(ikx)}{k-i\epsilon}\frac{\sinh(ky)}{\sinh(kb)}\right) = \frac{\Phi_0}{2\pi i}(-1)^n\frac{\exp(i\cdot in\pi x/b)}{b(n\pi i/b)}\sinh\left(\frac{n\pi i y}{b}\right)$$

$$= \frac{\Phi_0}{2\pi i}(-1)^n\frac{\exp(-n\pi x/b)}{n\pi i}\,i\sin\left(\frac{n\pi y}{b}\right)$$

$$= \frac{\Phi_0}{2\pi i}(-1)^n\frac{\exp(-n\pi x/b)}{n\pi}\sin\left(\frac{n\pi y}{b}\right). \tag{3.267}$$

The contribution to the potential $\Phi(x,y)$ therefore is

$$\Phi_{II,n}(x,y) = \Phi_0 \, (-1)^n \, \frac{\exp(-n\pi x/b)}{n\pi} \sin\left(\frac{n\pi y}{b}\right). \tag{3.268}$$

For the poles of the third category, at $k b = -n\pi i$ with $n \in \mathbb{N}$, we expand

$$\frac{1}{\sinh(k b)} = \frac{1}{\sinh(-in\pi + [k b + in\pi])} \approx \frac{1}{b(-1)^n[k + in\pi/b]}, \tag{3.269}$$

finding a simple pole at $k = n\pi i/b$. We therefore have for the residue,

$$\operatorname*{Res}_{k=-n\pi i/b}\left(-\frac{\Phi_0}{2\pi i}\frac{\exp(-ikx)}{k - i\epsilon}\frac{\sinh(k y)}{\sinh(k b)}\right) = -\frac{\Phi_0}{2\pi i}(-1)^n \frac{\exp(-n\pi x/b)}{b(-n\pi i/b)}\sinh\left(-\frac{n\pi i y}{b}\right)$$

$$= -\frac{\Phi_0}{2\pi i}(-1)^n\frac{\exp(-n\pi x/b)}{n\pi}\sin\left(\frac{n\pi y}{b}\right). \tag{3.270}$$

We encircle these poles in the clockwise direction, which is the mathematically negative direction. The third category of residues therefore is

$$\Phi_{III,n}(x,y) = \Phi_0 \, (-1)^n \, \frac{\exp(-n\pi x/b)}{n\pi} \sin\left(\frac{n\pi y}{b}\right). \tag{3.271}$$

The potential is obtained from the sum over all the residues

$$\Phi(x,y) = \Phi_I(x,y) + \sum_{n=1}^{\infty} \left(\Phi_{II,n}(x,y) + \Phi_{III,n}(x,y)\right)$$

$$= \frac{\Phi_0\, y}{b} + \frac{2\Phi_0}{\pi}\sum_{n=1}^{\infty}\frac{(-1)^n}{n}\exp\left(-\frac{n\pi x}{b}\right)\sin\left(\frac{n\pi y}{b}\right)$$

$$= \frac{\Phi_0\, y}{b} + \frac{\Phi_0}{\pi i}\sum_{n=1}^{\infty}\frac{(-1)^n}{n}\left\{\exp\left(-\frac{n\pi x}{b}+i\frac{n\pi y}{b}\right) - \exp\left(-\frac{n\pi x}{b}-i\frac{n\pi y}{b}\right)\right\}$$

$$= \frac{\Phi_0\, y}{b} + \frac{\Phi_0}{\pi i}\sum_{n=1}^{\infty}\frac{(-1)^{n+1}}{n}\left[\exp\left(-\frac{\pi x}{b}-i\frac{\pi y}{b}\right)\right]^n$$

$$\quad - \frac{\Phi_0}{\pi i}\sum_{n=1}^{\infty}\frac{(-1)^{n+1}}{n}\left[\exp\left(-\frac{\pi x}{b}+i\frac{\pi y}{b}\right)\right]^n$$

$$= \frac{\Phi_0\, y}{b} + \frac{\Phi_0}{\pi i}\ln\left[1+\exp\left(-\frac{\pi x}{b}-i\frac{\pi y}{b}\right)\right] - \frac{\Phi_0}{\pi i}\ln\left[1+\exp\left(-\frac{\pi x}{b}+i\frac{\pi y}{b}\right)\right]$$

$$= \frac{\Phi_0\, y}{b} + \frac{\Phi_0}{\pi i}\ln\left(\frac{\exp\left(\frac{\pi x}{b}\right)+\exp\left(-i\frac{\pi y}{b}\right)}{\exp\left(\frac{\pi x}{b}\right)+\exp\left(i\frac{\pi y}{b}\right)}\right). \tag{3.272}$$

We recall that according to Eqs. (3.253) and (3.259), the boundary conditions fulfilled by this solution are

$$\Phi(0,y) = \Phi(x,0) = 0, \qquad \Phi(x,b) = \Phi_0, \qquad 0 < x < \infty. \tag{3.273}$$

Convergence at the point $x = 0$, $y = b$ is nonuniform as is evident from Fig. 3.7. The nonuniform convergence can be verified explicitly. On the one hand, we have for

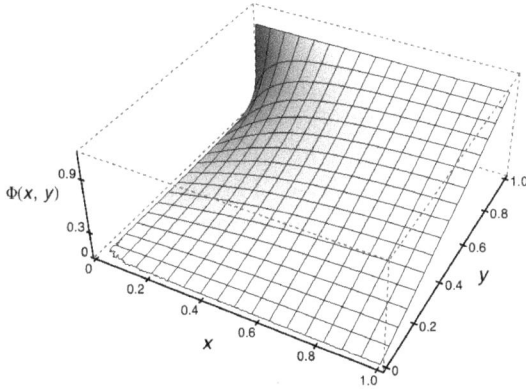

Fig. 3.7   Plot of the function $\Phi(x, y)$ as defined in Eq. (3.272) for $\Phi_0 = 1$ and $a = 1$.

the approach along the y axis,

$$
\begin{aligned}
\lim_{\epsilon \to 0^+} \Phi\,(0, b - \epsilon) &= \lim_{\epsilon \to 0^+} \left\{ \frac{\Phi_0\,(b - \epsilon)}{b} + \frac{\Phi_0}{\pi i} \ln \left(\frac{1 + \exp\left(-i\pi \frac{b-\epsilon}{b}\right)}{1 + \exp\left(i\pi \frac{b-\epsilon}{b}\right)} \right) \right\} \\
&= \lim_{\epsilon \to 0^+} \left\{ \Phi_0 + \frac{\Phi_0}{\pi i} \ln \left[\exp\left(-i\pi \frac{b - \epsilon}{b}\right) \frac{1 + \exp\left(i\pi \frac{b-\epsilon}{b}\right)}{1 + \exp\left(i\pi \frac{b-\epsilon}{b}\right)} \right] \right\} \\
&= \lim_{\epsilon \to 0^+} \left\{ \Phi_0 + \frac{\Phi_0}{\pi i}\,(-i\pi) \right\} = 0\,.
\end{aligned}
\tag{3.274}
$$

The approach parallel to the x axis for $y = b$ leads to

$$
\lim_{\epsilon \to 0^+} \Phi\,(\epsilon, b) = \lim_{\epsilon \to 0^+} \left\{ \frac{\Phi_0\, b}{b} + \frac{\Phi_0}{\pi i} \ln \left(\frac{\exp\left(\frac{\pi \epsilon}{b}\right) - 1}{\exp\left(\frac{\pi \epsilon}{b}\right) - 1} \right) \right\} = \lim_{\epsilon \to 0^+} \left\{ \frac{\Phi_0\, y}{b} + \frac{\Phi_0}{\pi i} \ln(1) \right\} = \Phi_0\,.
\tag{3.275}
$$

The potential is discontinuous at $(x, y) = (0, b)$; the boundary condition itself is discontinuous at this point. The discontinuity persists even though the result (3.272) is expressible in closed analytic form. The sum of the result

$$
\Phi(x, y) = \frac{\Phi_0\, y}{b} + \frac{\Phi_0}{\pi i} \ln \left(\frac{\exp\left(\frac{\pi x}{b}\right) + \exp\left(-i\frac{\pi y}{b}\right)}{\exp\left(\frac{\pi x}{b}\right) + \exp\left(i\frac{\pi y}{b}\right)} \right)
\tag{3.276}
$$

and the result given in Eq. (3.225) reads as $\Phi_0\, y/b$, which also is a solution of the Laplace equation, and fulfills appropriate boundary conditions.

3.3.7   Cylinders and Zeros of Bessel Functions

In Secs. 2.4.2 and 2.4.3, we have discussed a number of useful properties of Bessel functions. Equipped with this general knowledge on Bessel functions, we can now turn our attention to the application of the general solution of the Laplace equation in cylindrical coordinates, given in Eq. (2.142), to various example problems of cylindrical symmetry.

We start with a grounded outer rim of a cylinder of radius a, i.e., $\Phi = 0$ for $\rho = a$. In this case, the solutions are Bessel functions of the first kind satisfying the condition $J_m(k_{mn}\rho) = 0$ for $\rho = a$. This implies that we must choose $x_{mn} = k_{mn} a$ to be a zero of the mth Bessel function. Furthermore, we must discard solutions that involve Bessel Y functions as these are singular at the cylinder axis $\rho = 0$. This requirement determines values of $k = k_{mn}$, that is, k_{mn} with $n = 1, 2, \ldots$, where $x_{mn} = k_{mn} a$ is the nth root of the Bessel function of order m (see Table 2.1). The general solution to the Laplace equation $\vec{\nabla}^2 \Phi(\rho, \varphi, z) = 0$ with a vanishing potential at $\rho = a$ is a specialization of Eq. (2.142) and reads

$$\Phi(\rho,\varphi,z) = \sum_{m=-\infty}^{\infty} \sum_{n=1}^{\infty} a_{mn} J_m(k_{mn}\,\rho)\, e^{im\varphi}\, e^{\pm k_{mn}\, z}\,, \qquad \Phi(\rho = a, \varphi, z) = 0\,. \quad (3.277)$$

The condition to be fulfilled by the allowed wave numbers k_{mn} is

$$J_m(k_{mn}\, a) = 0\,, \qquad k_{mn} = \frac{x_{mn}}{a}\,, \quad (3.278)$$

where x_{mn} is the nth zero of the mth-order Bessel function.

The following boundary conditions

$$\Phi(\rho = a, \varphi, z) = 0\,, \qquad \Phi(\rho, \varphi, z = \pm L/2) = \Phi_0\,, \quad (3.279)$$

describe a zero potential on the outer rim of the cylinder, and constant potential on the endcaps. The potential is symmetric under $z \leftrightarrow -z$, and the *ansatz* (3.277) thus is modified to read

$$\Phi(\rho,\varphi,z) = \sum_{m=-\infty}^{\infty} \sum_{n=1}^{\infty} a_{mn} J_m(k_{mn}\,\rho)\, e^{im\varphi} \cosh(k_{mn}\, z)\,. \quad (3.280)$$

The area element on the cylinder endcaps is $dA = d\varphi\, d\rho\, \rho$ with integration domains $\varphi \in (0, 2\pi)$ and $\rho \in (0, a)$. Hence, the expansion coefficients are given by the condition

$$\int_0^a d\rho\, \rho \int_0^{2\pi} d\varphi\, \Phi(\rho, \varphi, \pm L/2)\, e^{-im'\varphi}\, J_{m'}(k_{m'n'}\,\rho)$$

$$= \Phi_0 \int_0^a d\rho\, \rho \int_0^{2\pi} d\varphi\, e^{-im'\varphi}\, J_{m'}(k_{m'n'}\,\rho)\,. \quad (3.281)$$

However, the determination of the expansion coefficients can be simplified somewhat, using the azimuthal symmetry of the potential. The only nonvanishing term is the one with $m = 0$, and Eq. (3.280) is thus simplified to

$$\Phi(\rho, \varphi, z) = \sum_{n=1}^{\infty} a_n J_0(k_{0n}\,\rho) \cosh(k_{0n}\, z)\,. \quad (3.282)$$

We now use the formula

$$\int_0^a d\rho\, \rho\, J_0(\xi\,\rho) = \frac{a}{\xi}\, J_1(\xi\, a)\,, \quad (3.283)$$

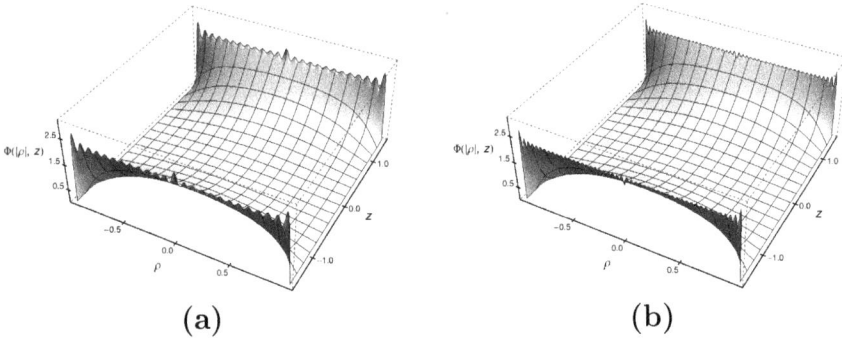

(a) (b)

Fig. 3.8 Plot of the partial sums up to terms with $n = 25$ and $n = 50$ of the expression (3.288) which describes a zero potential on the outer rim of a cylinder, but constant potential on the cylinder endcaps. The parameters are chosen as $a = 1$, $L = 3$ and $\Phi_0 = 2.3$. Slow convergence (oscillatory behavior, so-called Gibbs phenomenon) is clearly visible near the boundaries of the cylinder. Note that the potential $\Phi(\rho, z)$ has azimuthal symmetry and is independent of φ. The plot of $\Phi(|\rho|, z)$ with $-a < \rho < a$ corresponds to a plot from one outer rim of the cylinder to the other.

which can be shown as follows. One first differentiates the expression $x J_1(x)$ with the help of Eq. (2.149b)

$$\frac{\mathrm{d}}{\mathrm{d}x}\left[x J_1(x)\right] = J_1(x) + \frac{x}{2}\left[J_0(x) - J_2(x)\right]. \tag{3.284}$$

The recursion relation (2.149a) leads to

$$J_2(x) = -J_0(x) + \frac{2}{x} J_1(x). \tag{3.285}$$

Plugging this relation into the right-hand side of (3.284), one has

$$\frac{\mathrm{d}}{\mathrm{d}x}\left[x J_1(x)\right] = x J_0(x), \tag{3.286}$$

and a trivial scale transformation leads to (3.283). Finally, with the help of (3.283) and the orthogonality condition (2.172), one verifies that

$$a_n = \frac{2\Phi_0}{k_{0n}\, a}\, \frac{1}{\cosh(k_{0n} L/2)\, J_1(k_{0n} a)}. \tag{3.287}$$

The solution to the boundary-value problem (3.279) thus reads as

$$\Phi(\rho, \varphi, z) = \frac{2\Phi_0}{a} \sum_{n=1}^{\infty} \frac{1}{k_{n0}} \frac{J_0(k_{0n}\, \rho)}{J_1(k_{0n}\, a)} \frac{\cosh(k_{0n}\, z)}{\cosh(k_{0n} L/2)}. \tag{3.288}$$

A plot of the partial sums of this expansion up to $n = 5$ and $n = 50$, is given in Fig. 3.8.

Another question is how to modify the calculation when the potential vanishes on the cylinder endcaps at $z = \pm L/2$. In this case, we must choose linear combinations of the functions $\exp(k z)$ that vanish at $z = \pm L/2$. The only way to achieve this goal is to choose k as a complex rather than real number. We thus form linear combinations proportional to sine and cosine functions. Provided the potential is regular at the cylinder axis $\rho = 0$, we again discard the Neumann functions and write

$$\Phi(\rho, \varphi, z) = \sum_{m=-\infty}^{\infty} \sum_{n=0}^{\infty} a_{mn} \, J_m(\mathrm{i} \, k_n^{(1)} \, \rho) \, \mathrm{e}^{\mathrm{i}m\varphi} \cos(k_n^{(1)} \, z)$$

$$+ \sum_{m=-\infty}^{\infty} \sum_{n=1}^{\infty} b_{mn} \, J_m(\mathrm{i} \, k_n^{(2)} \, \rho) \, \mathrm{e}^{\mathrm{i}m\varphi} \sin(k_n^{(2)} \, z) \,, \tag{3.289a}$$

$$k_n^{(1)} = \frac{(2n+1)\pi}{L} \,, \qquad k_n^{(2)} = \frac{2n\pi}{L} \,, \qquad \Phi(\rho, \varphi, z = \pm L/2) = 0 \,. \tag{3.289b}$$

Observe that the sum over n goes through $n = 0, 2, \ldots, \infty$ in the first sum, whereas it starts from unity in the second sum. A convenient reformulation can be achieved in terms of the modified Bessel functions [5, 7–9],

$$I_m(x) = \mathrm{i}^{-m} \, J_m(\mathrm{i}x) \,. \tag{3.290}$$

The asymptotic behavior of the Bessel I functions is given as

$$I_m(x) \sim J_m(x) \sim \frac{1}{\Gamma(m+1)} \left(\frac{x}{2}\right)^m \,, \qquad x \to 0 \,, \tag{3.291a}$$

$$I_m(x) \sim \frac{1}{(2\pi x)^{1/2}} \exp(x) \,, \qquad x \to \infty \,. \tag{3.291b}$$

The general solution (3.289) to the homogeneous equation becomes

$$\Phi(\rho, \varphi, z) = \sum_{m=-\infty}^{\infty} \sum_{n=1}^{\infty} a_{mn} \, I_m(k_n^{(1)} \, \rho) \, \mathrm{e}^{\mathrm{i}m\varphi} \cos(k_n^{(1)} \, z)$$

$$+ \sum_{m=-\infty}^{\infty} \sum_{n=0}^{\infty} b_{mn} \, I_m(k_n^{(2)} \, \rho) \, \mathrm{e}^{\mathrm{i}m\varphi} \sin(k_n^{(2)} \, z) \,. \tag{3.292}$$

The sine terms are asymmetric under $z \leftrightarrow -z$, whereas the cosine terms are symmetric.

We choose as boundary conditions a zero potential on the endcaps of the cylinder, and constant potential on the outer rim,

$$\Phi(\rho = a, \varphi, z) = \Phi_0 \,, \qquad \Phi(\rho, \varphi, z = \pm L/2) = 0 \,. \tag{3.293}$$

The potential is symmetric under $z \leftrightarrow -z$. Hence,

$$\Phi(\rho, \varphi, z) = \sum_{m=-\infty}^{\infty} \sum_{n=1}^{\infty} a_{mn} \, I_m(k_n^{(1)} \, \rho) \, \mathrm{e}^{\mathrm{i}m\varphi} \cos(k_n^{(1)} \, z) \,, \tag{3.294}$$

where we recall that $k_n^{(1)} = (2n+1)\pi/L$. The infinitesimal area element on the outer rim of the cylinder is $\mathrm{d}A = a \, \mathrm{d}\varphi \, \mathrm{d}z$ with the integration domains $\varphi \in (0, 2\pi)$ and $z \in (-L/2, L/2)$. Hence, the expansion coefficients a_{mn} can be determined based on

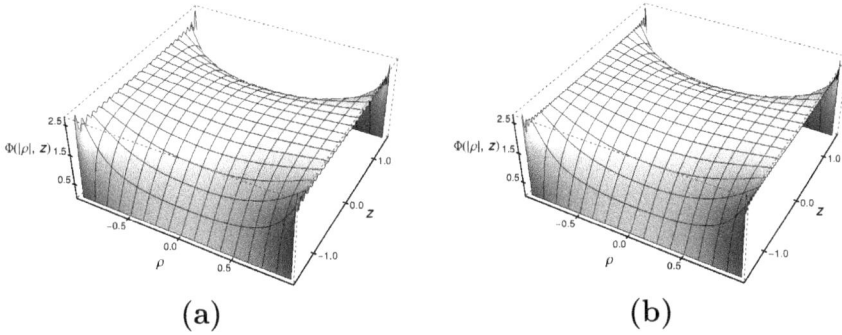

(a)   (b)

Fig. 3.9   Plot of the partial sums up to terms with $n = 5$ and $n = 50$ of the expression (3.298) which describes a zero potential on the cylinder endcaps, but constant potential on the outer rim of the cylinder. The parameters are $a = 1$, $L = 3$ and $\Phi_0 = 2.3$. Note that the potenial $\Phi(\rho, z)$ has azimuthal symmetry and is independent of φ. Our plot of $\Phi(|\rho|, z)$ with $-a < \rho < a$ corresponds to a plot from one outer rim of the cylinder to the other.

the conditions that

$$
\int_0^{2\pi} \mathrm{d}\varphi \int_{-L/2}^{L/2} \mathrm{d}z\, \Phi(a, \varphi, z)\, \mathrm{e}^{-\mathrm{i}m'\varphi} \cos(k_{n'}^{(1)} z) = \Phi_0 \int_0^{2\pi} \mathrm{d}\varphi \int_{-L/2}^{L/2} \mathrm{d}z\, \mathrm{e}^{-\mathrm{i}m'\varphi} \cos(k_{n'}^{(1)} z),
$$

(3.295a)

$$
\int_0^{2\pi} \mathrm{d}\varphi \int_{-L/2}^{L/2} \mathrm{d}z\, \Phi(a, \varphi, z)\, \mathrm{e}^{-\mathrm{i}m'\varphi} \sin(k_{n'}^{(2)} z) = \Phi_0 \int_0^{2\pi} \mathrm{d}\varphi \int_{-L/2}^{L/2} \mathrm{d}z\, \mathrm{e}^{-\mathrm{i}m'\varphi} \sin(k_{n'}^{(2)} z).
$$

(3.295b)

However, again, there is a more direct way in view of the simplicity of the boundary condition which implies azimuthal symmetry. The only contributing term is the one with $m = 0$, and we have

$$
\Phi(\rho, \varphi, z) = \sum_{n=1}^{\infty} a_n I_0(k_n^{(1)} \rho) \cos(k_{0n}^{(1)} z).
$$

(3.296)

One verifies that

$$
a_n = \frac{4(-1)^n \Phi_0}{(2n+1)\pi\, I_0(k_n^{(1)} a)}.
$$

(3.297)

The solution to the boundary-value problem (3.293) thus reads as (see Fig. 3.9)

$$
\Phi(\rho, \varphi, z) = \frac{4\Phi_0}{\pi} \sum_{n=0}^{\infty} \frac{(-1)^n}{(2n+1)} \frac{I_0(k_n^{(1)} \rho)}{I_0(k_n^{(1)} a)} \cos(k_n^{(1)} z).
$$

(3.298)

A few final remarks are in order. A general cylindrical problem with boundary conditions on the cylinder boundaries (both outer rim as well as endcaps) can be separated into one with a vanishing solution on the cylinder endcaps and one with a vanishing solution on the outer cylinder boundaries. One adjusts the coefficients in Eq. (3.277) so that the boundary conditions on the endcaps are fulfilled, and then, one adjusts the coefficients in Eq. (3.289) so that the boundary conditions on the outer rim of the cylinder are fulfilled. A general boundary value problem on the cylinder thus involves both types of solutions and can be broken into pieces according to the above analysis.

3.4 Laplace Equation and Dirichlet Green Functions

3.4.1 *Dirichlet Green Function for Spherical Shells*

In Sec. 3.2, we have considered approximate solutions to the Laplace equation using the variational principle. In Sec. 3.3, the solution of the Laplace equation using series expansions was treated. Now, we shall turn our attention to the third method listed in Sec. 3.1.4, namely, the Dirichlet Green function. Let us consider grounded concentric spheres, with radii $a < b$. As discussed in Sec. 3.1.3, the Dirichlet Green function fulfills

$$\vec{\nabla}^2 G_D\left(\vec{r},\vec{r}'\right)\big|_{\vec{r},\,\vec{r}'\in V} = -\frac{1}{\epsilon_0}\delta^{(3)}(\vec{r}-\vec{r}'), \qquad G_D\left(\vec{r},\vec{r}'\right)\big|_{\vec{r}\in V,\vec{r}'\in\partial V} = 0. \qquad (3.299)$$

In order to illustrate the Green function techniques, we consider systems which lie in a spherical region $a \leq r \leq b$ bounded by two concentric conducting spherical shells of radii a and b. In this case, we assume that the potentials (rather than the field strengths) on the conductors are specified and the Dirichlet Green function (rather than the Neumann Green Function) is required.

In view of the spherical symmetry, we follow the approach outlined in Sec. 2.3.2 and use the following *ansatz* for the Dirichlet Green function,

$$G_D\left(\vec{r},\vec{r}'\right) = \sum_{\ell=0}^{\infty}\sum_{m=-\ell}^{\ell} g_\ell\left(r,r'\right) Y_{\ell m}\left(\theta,\varphi\right) Y_{\ell m}^*\left(\theta',\varphi'\right). \qquad (3.300)$$

We proceed as in the case of the ordinary Green function, i.e., we exclude the value at $r = r'$ and concatenate two solutions to the homogeneous equation, one for $a \leq r < r'$ and the other for $r' < r \leq b$. Just as in the case of the free Green function (Sec. 2.3.2), we can combine the solution regular at the origin, and the one regular at infinity, in such a way that they fulfill the homogeneous equation everywhere expect at the cusp $r = r'$. We then adjust the coefficients such as to ensure that $g_\ell\left(r,r'\right) = 0$ at the surfaces of the conductor. A suitable combination is

$$g_\ell\left(r,r'\right) = \alpha_\ell\left(r'\right)\left(\frac{r^\ell}{r'^{\,\ell+1}} - \frac{a^{2\ell+1}}{\left(r\,r'\right)^{\ell+1}}\right), \qquad a \leq r < r', \qquad (3.301\text{a})$$

$$g_\ell\left(r,r'\right) = \beta_\ell\left(r'\right)\left(-\frac{\left(r\,r'\right)^\ell}{b^{2\ell+1}} + \frac{r'^\ell}{r^{\ell+1}}\right), \qquad r' < r \leq b. \qquad (3.301\text{b})$$

For $a \to 0$ and $b \to \infty$, both of these equations reproduce the structure encountered for the free Green function. The *ansatz* (3.301) is justified by the idea that the radial terms of the free Green function are supplemented by a product of regular terms (at the origin, or at infinity) of products of solutions of the homogeneous equations, symmetric under $r \leftrightarrow r'$. Furthermore, in view of the boundary conditions imposed on the Green function, we need to have $g_\ell (r, r') = 0$ at $r = a$ and at $r = b$. Continuity of $g_\ell (r, r')$ at $r = r'$ requires that

$$\alpha_\ell (r') \left(\frac{1}{r'} - \frac{a^{2\ell+1}}{(r')^{2\ell+2}} \right) - \beta_\ell (r') \left(-\frac{(r')^{2\ell}}{b^{2\ell+1}} + \frac{1}{r'} \right) = 0 . \tag{3.302}$$

In order to satisfy the differential equation

$$\vec{\nabla}^2 G_D (\vec{r}, \vec{r}') = -\frac{1}{\epsilon_0} \delta(\vec{r} - \vec{r}') , \tag{3.303}$$

one needs to integrate Eq. (3.301) from $r = r' - \epsilon$ to $r = r' + \epsilon$.

This procedure results in the following equation,

$$\alpha_\ell (r') \left(\frac{\ell}{r'} + \frac{(\ell+1) a^{2\ell+1}}{r'^{2\ell+2}} \right) + \beta_\ell (r') \left(\frac{\ell r'^{2\ell}}{b^{2\ell+1}} + \frac{(\ell+1)}{r'} \right) = \frac{1}{\epsilon_0 r'} . \tag{3.304}$$

The two Eqs. (3.302) and (3.304) can be solved for $\alpha_\ell(r')$ and $\beta_\ell(r')$,

$$\alpha_\ell (r') = -\frac{1}{\epsilon_0} \frac{1}{2\ell + 1} \frac{1 - (r'/b)^{2\ell+1}}{1 - (a/b)^{2\ell+1}} , \qquad \beta_\ell (r') = -\frac{1}{\epsilon_0} \frac{1}{2\ell + 1} \frac{1 - (a/r')^{2\ell+1}}{1 - (a/b)^{2\ell+1}} . \tag{3.305}$$

The overall result for the Green function is

$$G_D (\vec{r}, \vec{r}') = \frac{1}{\epsilon_0} \sum_{\ell=0}^{\infty} \frac{\left(b^{2\ell+1} - r_>^{2\ell+1} \right) \left(r_<^{2\ell+1} - a^{2\ell+1} \right)}{(2\ell + 1) \left(b^{2\ell+1} - a^{2\ell+1} \right) (r\, r')^{\ell+1}} \sum_{m=-\ell}^{\ell} Y_{\ell m} (\theta, \varphi) \, Y_{\ell m} (\theta', \varphi')^* . \tag{3.306}$$

The Green function vanishes on the two concentric spheres at $r = a$ and at $r = b$. Two remarks are in order here. We have considered a system with an inner shell of radius a and an outer shell of radius b, both held at a constant zero potential. If we set $a = 0$, then this gives the Dirichlet Green function for the system confined to lie inside a spherical shell of radius b. If we let $b \to \infty$, then this is the Dirichlet Green function for the region outside a spherical shell of radius a.

3.4.2 *Boundary Condition of Dipole Symmetry*

We now turn to the application of the Dirichlet Green function to a boundary-value problem. We recall that, using the Dirichlet Green function and a suitable Dirichlet boundary condition $\Phi(\vec{r})$ for ∂V, and vanishing charge distribution, we can evaluate the potential $\Phi(\vec{r}')$ using Eq. (3.27), which we recall for convenience,

$$\Phi_D (\vec{r}') = -\epsilon_0 \int_{\partial V} d\vec{A} \cdot \vec{\nabla} G_D (\vec{r}, \vec{r}') \cdot \Phi (\vec{r}) . \tag{3.307}$$

Let us formulate a boundary condition which has a dipole structure on the inner sphere, whereas the outer sphere remains grounded,

$$\Phi(a, \theta, \varphi) = f_a(\theta) = \Phi_0 \cos \theta, \qquad \Phi(b, \theta, \varphi) = f_b(\theta) = 0. \tag{3.308}$$

The task then is to find the electrostatic potential everywhere. Using the symmetry property $G(\vec{r}, \vec{r}') = G(\vec{r}', \vec{r})$ of the Green function, the solution can be written as

$$\Phi(r, \theta, \varphi) = a^2 \epsilon_0 \int d\Omega' \, \Phi_0 \cos \theta' \, \frac{\partial}{\partial r'} G_D(\vec{r}, \vec{r}') \Big|_{r'=a}. \tag{3.309}$$

Using the orthogonality of the Legendre polynomials which are "hidden" in the spherical harmonics, this reduces to

$$\Phi(r, \theta, \varphi) = \frac{a^2}{3} \Phi_0 \cos \theta \, \frac{\partial}{\partial r'} \left(\frac{[b^3 - r^3] [r'^3 - a^3]}{(b^3 - a^3)(r r')^2} \right) \Big|_{r'=a}$$

$$= \Phi_0 \cos \theta \, \frac{a^2}{r^2} \frac{b^3 - r^3}{b^3 - a^3}, \tag{3.310}$$

which interpolates between zero (at $r = b$) and $\Phi_0 \cos \theta$ (at $r = a$). The concept of the multipole decomposition, outlined in Sec. 2.3.5, has been used with advantage in the analysis.

3.4.3   *Verification Using Series Expansion*

In order to establish contact with the series expansion method (3.3), it is useful to rederive the result given in Eq. (3.310) using the multipole decomposition, with a suitable projection of the multipole components. Azimuthal symmetry implies that in Eq. (3.170), we should only select the spherical harmonics with $\ell = 0$. Alternatively, in view of Eq. (2.53), we may formulate the angular part as being proportional to the Legendre polynomial $P_\ell(\cos \theta)$, to obtain

$$\Phi(r, \theta) = \sum_{\ell=0}^{\infty} \left(\alpha_\ell r^\ell + \beta_\ell r^{-\ell-1} \right) P_\ell(\cos \theta), \quad \Phi(a, \theta) = f_a(\theta), \quad \Phi(b, \theta) = f_b(\theta). \tag{3.311}$$

A rather straightforward calculation shows that the α_ℓ and β_ℓ are given by

$$\alpha_\ell = \frac{a^{\ell+1} I_\ell(a) - b^{\ell+1} I_\ell(b)}{a^{2\ell+1} - b^{2\ell+1}}, \tag{3.312a}$$

$$\beta_\ell = -(a b)^{\ell+1} \frac{b^\ell I_\ell(a) - a^\ell I_\ell(b)}{a^{2\ell+1} - b^{2\ell+1}}, \tag{3.312b}$$

where the overlap integrals $I_\ell(a)$ involve the boundary conditions $f_a(\theta)$ and $f_b(\theta)$,

$$I_\ell(a) = \frac{2\ell + 1}{2} \int_0^\pi f_a(\theta) P_\ell(\cos \theta) \sin \theta \, d\theta, \tag{3.313a}$$

$$I_\ell(b) = \frac{2\ell + 1}{2} \int_0^\pi f_b(\theta) P_\ell(\cos \theta) \sin \theta \, d\theta. \tag{3.313b}$$

Using the boundary conditions (3.308), we obtain $I_\ell(b) = 0$ for all ℓ, and the only nonvanishing α_ℓ and β_ℓ coefficients are found for $\ell = 1$,

$$I_1(a) = \Phi_0, \qquad \alpha_1 = \Phi_0 \frac{a^2}{a^3 - b^3}, \qquad \beta_1 = \Phi_0 \frac{a^2 b^3}{a^3 - b^3}. \tag{3.314}$$

Plugging these coefficients into Eq. (3.311), we confirm that

$$\Phi(r, \theta, \varphi) = \Phi_0 \cos\theta \, \frac{a^2}{r^2} \frac{b^3 - r^3}{b^3 - a^3}, \tag{3.315}$$

and thus, the validity of the result given in Eq. (3.310).

3.5 Poisson Equation and Dirichlet Green Functions

3.5.1 *Source Terms with Boundary Conditions*

In Sec. 3.1.3, we had developed the formalism for the treatment of boundary-value problems (of the Dirichlet and Neumann type) with nontrivial source terms and nontrivial boundary conditions. Given a Dirichlet boundary condition $\Phi(\vec{r}) = f(\vec{r})$ for $r \in \partial V$, we recall that the solution is given by Eq. (3.21),

$$\Phi_D(\vec{r}') = \int_V d^3 r \, G_D(\vec{r}, \vec{r}') \, \rho(\vec{r}) - \epsilon_0 \int_{\partial V} d\vec{S} \cdot \vec{\nabla} G_D(\vec{r}, \vec{r}') \, f(\vec{r}), \tag{3.316}$$

where the defining relations for the Dirichlet Green function are [see Eq. (3.20)],

$$\vec{\nabla}^2 G_D(\vec{r}, \vec{r}') \big|_{\vec{r}, \vec{r}' \in V} = -\frac{1}{\epsilon_0} \delta^{(3)}(\vec{r} - \vec{r}'), \qquad G_D(\vec{r}, \vec{r}') \big|_{\vec{r} \in V, \vec{r}' \in \partial V} = 0. \tag{3.317}$$

If the charge distribution vanishes, $\rho(\vec{r}) = 0$, then the solution to the boundary-value problem is given by the second term on the right-hand side of (3.316). By contrast, if the potential is supposed to vanish on the boundary, then the solution is given by the first term on the right-hand side of (3.316).

A hybrid approach can be taken to problems which involve both nontrivial boundary conditions as well as nontrivial source terms. Namely, one can split Dirichlet boundary-value problems into two parts, the first of which is given by the charge distribution and needs to be integrated using the Dirichlet Green function, and the second of which is given by the boundary condition and can be integrated using a series expansion, using the variational principle or with the Dirichlet Green function itself. In the following, we shall consider the first term on the right-hand side of (3.316), and its integration using the Dirichlet Green function.

3.5.2 *Sources and Fields in a Spherical Shell*

We consider the Dirichlet Green function for concentric spheres with radii $a < b$ [see Eq. (3.306)],

$$G_D(\vec{r}, \vec{r}') = \frac{1}{\epsilon_0} \sum_{\ell=0}^{\infty} \frac{\left(b^{2\ell+1} - r_>^{2\ell+1}\right)\left(r_<^{2\ell+1} - a^{2\ell+1}\right)}{(2\ell+1)\left(b^{2\ell+1} - a^{2\ell+1}\right)(r\,r')^{\ell+1}} \sum_{m=-\ell}^{\ell} Y_{\ell m}(\theta, \varphi) \, Y_{\ell m}(\theta', \varphi')^*. \tag{3.318}$$

For concentric spheres with radii $a < b$, we consider a uniformly charged annulus (ring) with total charge Q_0, located in the plane $\theta = \pi/2$ (xy plane), i.e., stretched between the two concentric spherical shells. This corresponds to a nontrivial source term in Eq. (3.316); we recall that nontrivial boundary conditions have been treated using two alternative approaches in Sec. 3.4. The potential in the interior region between the concentric spheres, due to the uniformly charged annulus, can be calculated with the help of the Dirichlet Green function (3.318), as follows,

$$\Phi_D(\vec{r}) = \int_V d^3 r' \, G_D(\vec{r}, \vec{r}') \, \rho(\vec{r}') . \tag{3.319}$$

The (constant) surface charge density on the ring (annulus) between the spheres is

$$\sigma_0 = \frac{Q_0}{\pi \left(b^2 - a^2 \right)} , \tag{3.320}$$

where Q_0 is the total charge on the annulus [see also Fig. 3.10(a)]. The volume charge density on the annulus therefore is

$$\rho(\vec{r}') = \sigma_0 \, \delta(z') = \sigma_0 \, \delta(r' \cos\theta') = \frac{\sigma_0}{r'} \, \delta(\cos\theta') , \qquad a \leq r \leq b, \tag{3.321}$$

where the latter equality is valid for purposes of the integration over θ', in which case r' can be assumed to be a constant. The boundary conditions, which are automatically implemented by the use of the Dirichlet Green function, read as $\Phi(r = a, \theta, \varphi) = \Phi(r = b, \theta, \varphi) = 0$. We use Eq. (3.319) and write

$$
\begin{aligned}
\Phi(r, \theta, \varphi) &= \int_{a \leq r' \leq b} d^3 r' \, G_D(\vec{r}, \vec{r}') \, \rho(\vec{r}') \\
&= \frac{1}{\epsilon_0} \int_{a \leq r' \leq b} d^3 r' \sum_{\ell=0}^{\infty} \frac{\left(b^{2\ell+1} - r_>^{2\ell+1} \right) \left(r_<^{2\ell+1} - a^{2\ell+1} \right)}{(2\ell+1) \left(b^{2\ell+1} - a^{2\ell+1} \right) (r \, r')^{\ell+1}} \\
&\quad \times \sum_{m=-\ell}^{\ell} Y_{\ell m}(\theta, \varphi) \, Y_{\ell m}(\theta', \varphi')^* \, \sigma_0 \, \delta(r' \cos\theta') \\
&= \frac{1}{\epsilon_0} \int_{a \leq r' \leq b} dr' \, r'^2 \, (2\pi) \, d\cos\theta' \sum_{\ell=0}^{\infty} \frac{\left(b^{2\ell+1} - r_>^{2\ell+1} \right) \left(r_<^{2\ell+1} - a^{2\ell+1} \right)}{(2\ell+1) \left(b^{2\ell+1} - a^{2\ell+1} \right) (r \, r')^{\ell+1}} \\
&\quad \times \left(\sqrt{\frac{2\ell+1}{4\pi}} \right)^2 P_\ell(\cos\theta) \, P_\ell(\cos\theta') \, \frac{\sigma_0}{r'} \, \delta(\cos\theta') \\
&= \frac{\sigma_0}{2 \, \epsilon_0} \sum_{\ell=0}^{\infty} \frac{P_\ell(\cos\theta) \, P_\ell(0)}{b^{2\ell+1} - a^{2\ell+1}} \int_a^b dr' \, r' \, \frac{\left(b^{2\ell+1} - r_>^{2\ell+1} \right) \left(r_<^{2\ell+1} - a^{2\ell+1} \right)}{(r \, r')^{\ell+1}} .
\end{aligned}
\tag{3.322}
$$

For the Legendre polynomials of zero argument, one can use the formula

$$P_\ell(0) = \frac{\sqrt{\pi}}{\Gamma(\frac{1}{2} - \frac{1}{2}\ell) \, \Gamma(1 + \frac{1}{2}\ell)} = \cos\left(\frac{\pi\ell}{2} \right) \frac{\Gamma(\frac{1}{2} + \frac{1}{2}\ell)}{\sqrt{\pi} \, \Gamma(1 + \frac{1}{2}\ell)} . \tag{3.323}$$

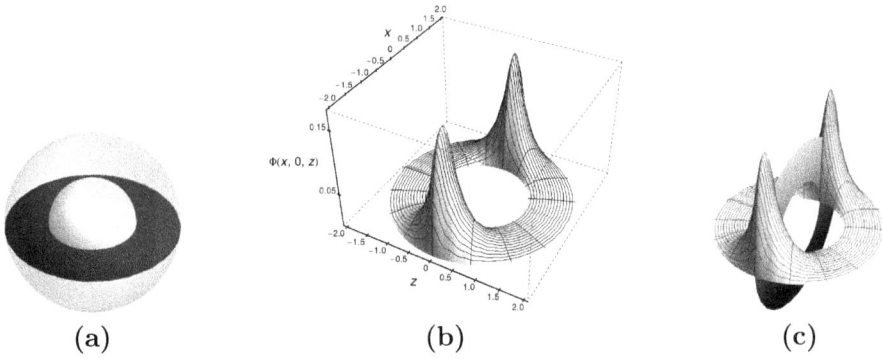

Fig. 3.10 Panel (a) shows the schematic arrangement consisting of the two concentric spheres, with the annulus centered in the $z = 0$ plane. Panel (b) shows a plot of the potential (3.326) in the plane of azimuth angle $\varphi = 0$, i.e., in the $y = 0$ plane. The potential has a peak at $z = 0$, as outlined in the plot. Panel (c) shows the position of the annulus in plot (b), confirming that the potential peaks where the annulus is. The problem is symmetric with respect to the azimuth angle, i.e., symmetric with respect to rotations about the z axis.

The integral over r' can be carried out for the first few ℓ by an explicit calculation, where we use the general identity

$$\int_a^b dr'\, f(r_<, r_>) = \int_a^r dr'\, f(r', r) + \int_r^b dr'\, f(r, r').\qquad(3.324)$$

The result for the first two novanishing terms (monopole and quadrupole) is

$$\Phi(r, \theta, \varphi) = \frac{\sigma_0}{\epsilon_0} \frac{(a - r)(r - b)}{4r}$$
$$- \frac{5\sigma_0}{32\epsilon_0} \frac{b^4(b - r)r^4 + a^5(b^4 - r^4) + a^4(r^5 - b^5)}{(b^5 - a^5)r^3}(3\cos^2\theta - 1) + \dots\,.$$

$$(3.325)$$

The general result consists of the even multipoles,

$$\Phi(r, \theta, \varphi) = \frac{\sigma_0}{\epsilon_0} \sum_{\ell=0}^{\infty} \cos\left(\frac{\pi\ell}{2}\right) \frac{(\ell + \frac{1}{2})\Gamma(\frac{1}{2} + \frac{1}{2}\ell)}{\sqrt{\pi}\,\Gamma(1 + \frac{1}{2}\ell)} P_\ell(\cos\theta)$$
$$\times \frac{(br)^{\ell+1}(b^\ell r - br^\ell) + a^{2\ell+1}(b^{\ell+2} - r^{\ell+2}) + a^{\ell+2}(r^{2\ell+1} - b^{2\ell+1})}{r^{\ell+1}(b^{2\ell+1} - a^{2\ell+1})(\ell^2 + \ell - 2)}\,.$$

$$(3.326)$$

Odd ℓ are eliminated under the sum by the factor $\cos(\pi\ell/2)$. The plot in Fig. 3.10 shows the sum of the first 40 multipoles in the expression (3.326) for the solution of the charged annulus inside the two concentric, grounded, spherical shells. It is given in the xz plane. The annulus lies in the plane $z = 0$ and is located where the potential is large. As we cannot realistically plot the potential as a function of the three coordinates x, y and z, we have chosen the x and z coordinates. The potential is rotationally symmetric about the z axis.

3.5.3 *Induced Charge Distributions on the Boundaries*

In Sec. 3.5.2, we have considered a situation where a nonvanishing charge distribution leads to a nontrivial potential in the interior region between the two spherical shells. The boundary condition for the potential, namely, the assumption that $\Phi(\vec{r}) = 0$ for $r = a$ and $r = b$, is automatically implemented through the use of the Dirichlet Green function. Given the charge density on the annulus as in Eq. (3.321), Dirichlet boundary conditions on the two concentric spheres actually imply that compensating charge must be brought onto the two concentric spheres in order to annihilate the electrostatic potential there.

The charge on the inner surface of the outer sphere can be calculated as follows ($r_> = r$, and $r_< = r'$). We invoke a Gaussian cylinder surrounding the surface, and take into account that the outer layer is grounded, i.e., held at zero potential. Hence, the electric field outside of the concentric sphere vanishes. Furthermore, this implies that the contribution to the surface integral (through the Gaussian cylinder) comes exclusively from the "inner" region where the surface normal is $-\hat{e}_r$. So, there isn't a "factor $1/2$ coming in from the other cap of the Gaussian cylinder" as one might otherwise suggest. We simply have $\sigma(\vec{r}) = -\epsilon_0\, E_r$ where E_r is the normal component of the electric field (which is pointing in the negative radial direction for the outer sphere),

$$
Q_b = \int_{r=b} d\Omega\, r^2\, \sigma(\vec{r}) = \int_{r=b} d\Omega\, r^2\, (-\epsilon_0\, E_r) = b^2\, \epsilon_0 \int d\Omega\, \left.\frac{\partial \Phi\,(r,\theta,\varphi)}{\partial r}\right|_{r=b}
$$

$$
= \frac{\sigma_0\, b^2}{2}\, (2\pi) \sum_{\ell=0}^{\infty} \int_{-1}^{1} d\cos\theta\, \frac{P_\ell\,(\cos\theta)\, P_\ell\,(0)}{b^{2\ell+1} - a^{2\ell+1}}
$$

$$
\times \frac{\partial}{\partial r} \int_{a}^{b} dr'\, r'\, \left.\frac{\left(b^{2\ell+1} - r^{2\ell+1}\right)\left(r'^{2\ell+1} - a^{2\ell+1}\right)}{(r\, r')^{\ell+1}}\right|_{r=b}
$$

$$
= \frac{\sigma_0\, b^2}{2}\, (2\pi) \int_{-1}^{1} d\cos\theta\, \frac{P_0(\cos\theta)\, P_0\,(0)}{b-a}\, \frac{\partial}{\partial r} \int_{a}^{b} dr'\, \frac{b-r}{r}\, r'\, \left.\frac{r'-a}{r'}\right|_{r=b}
$$

$$
= \frac{\sigma_0\, b^2}{2}\, (2\pi)\, \frac{2}{b-a} \left(\left.\frac{\partial}{\partial r}\, \frac{b-r}{r}\right|_{r=b}\right) \int_{a}^{b} dr'\, (r'-a)
$$

$$
= \frac{\sigma_0\, b^2}{2}\, (2\pi)\, (2)\, \frac{1}{b-a} \left(-\frac{b}{b^2}\right) \int_{a}^{b} dr'\, (r'-a) = -\pi\sigma_0\, b\, (b-a)\,, \tag{3.327}
$$

which has the correct physical dimension. In the derivation, we have started from the integral representation in the last line of Eq. (3.322). On the inner spherical shell ($r = a$), we have $\sigma(\vec{r}) = \epsilon_0\, E_r$ because the normal component of the electric field is pointing in the positive radial direction. Furthermore, on the outer surface of the inner sphere, we have $r_> = r'$ and $r_< = r$, and the surface charge therefore is given by

$$
Q_a = \int_{r=a} d\Omega\, r^2\, \sigma(\vec{r}) = \int_{r=a} d\Omega\, r^2\, (+\epsilon_0\, E_r) = -a^2\,\epsilon_0 \int d\Omega\, \frac{\partial \Phi\,(r,\theta,\varphi)}{\partial r}\bigg|_{r=a}
$$

$$
= -\frac{\sigma_0\, a^2}{2}\, (2\pi) \sum_{\ell=0}^{\infty} \int_{-1}^{1} d\cos\theta\, \frac{P_\ell\,(\cos\theta)\, P_\ell\,(0)}{b^{2\ell+1} - a^{2\ell+1}}
$$

$$
\times \frac{\partial}{\partial r} \int_a^b r'\, \frac{\left(b^{2\ell+1} - r'^{2\ell+1}\right)\left(r^{2\ell+1} - a^{2\ell+1}\right)}{(r\, r')^{\ell+1}}\, dr'\bigg|_{r=a}
$$

$$
= -\frac{\sigma_0\, a^2}{2}\, (2\pi) \int_{-1}^{1} \left(\frac{P_0\,(\cos\theta)\, P_0\,(0)}{b-a}\, d\cos\theta\right) \frac{\partial}{\partial r} \int_a^b r'\, \frac{b-r'}{r'}\, \frac{r-a}{r}\, dr'\bigg|_{r=b}
$$

$$
= -\frac{\sigma_0\, a^2}{2}\, (2\pi)\, \frac{2}{b-a} \left(\frac{\partial}{\partial r}\, \frac{r-a}{r}\bigg|_{r=a}\right) \int_a^b (b-r')\, dr'
$$

$$
= -\frac{\sigma_0\, a^2}{2}\, (2\pi)\, (2)\, \frac{1}{b-a}\, \left(\frac{a}{a^2}\right) \int_a^b (b-r')\, dr' = -\pi\sigma_0\, a\, (b-a)\,. \tag{3.328}
$$

The two charges Q_b and Q_a both have the same sign but are manifestly different; the ratio is b/a. This is in line with intuition because of the following consideration: Namely, if the induced charge density on both concentric spheres were the same, then the ratio would be equal to the ratio of the surface areas, i.e., b^2/a^2. However, the average distance to the surface elements of the inner concentric sphere is smaller, and because the potential goes with the inverse distance, the change in the ratio from b^2/a^2 to b/a is in line with intuition. In the derivation, we have used the orthogonality of the Legendre polynomials in the form

$$
\int_{-1}^{1} P_\ell\,(x)\, P_\ell\,(0)\, dx = P_\ell\,(0) \int_{-1}^{1} P_\ell\,(x)\, dx = P_\ell\,(0) \int_{-1}^{1} P_\ell\,(x)\, P_0(x)\, dx
$$

$$
= P_\ell\,(0)\, \frac{2\,\delta_{\ell 0}}{\ell+1} = 2\,\delta_{\ell 0}\,, \tag{3.329}
$$

where $P_0(0) = P_0(x) = 1$. The orthogonality property (2.48) has also been employed.

3.6 Exercises

• **Exercise 3.1**: For $\vec{r}' \in \partial V$, Eq. (3.21) reduces to

$$
\Phi_D\,(\vec{r}')\bigg|_{\vec{r}' \in \partial V} = \int_{\partial V} d\vec{A} \cdot \vec{\nabla} G_D\,(\vec{r},\vec{r}')\, f\,(\vec{r})\,. \tag{3.330}
$$

By applying the divergence theorem to the right-hand side, show that the right-hand side is equal to $f\,(\vec{r}')$, tantamount to the Dirichlet boundary condition.

- **Exercise 3.2**: For $\vec{r}' \in \partial V$, Eq. (3.26) can be written as

$$\vec{\nabla}_{\vec{r}'} \Phi_N \left(\vec{r}' \right) \big|_{\vec{r}' \in \partial V} = \int_V d^3 r \vec{\nabla}_{\vec{r}'} G_N \left(\vec{r}, \vec{r}' \right) \rho(\vec{r}) + \epsilon_0 \int_{\partial V} \left(\vec{\nabla}_{\vec{r}'} d\vec{A} \cdot \vec{\nabla}_{\vec{r}} \Phi(\vec{r}) G_N \left(\vec{r}, \vec{r}' \right) \right).$$
(3.331)

Show that the right-hand side of Eq. (3.331) is indeed equal to $\vec{g}(\vec{r})$, where $\vec{g}(\vec{r}')$ is defined as in Eq. (3.15), by applying the divergence theorem to the third term on the right-hand side of Eq. (3.331).

- **Exercise 3.3**: For a functional of the form

$$S[f] = \int dx\, F(f(x), f'(x)),$$
(3.332)

show that

$$\frac{\delta S}{\delta f(x)} = \frac{\partial F}{\partial f(x)} - \frac{d}{dx} \frac{\partial F}{\partial f'(x)}$$
(3.333)

where you treat the partial differentiations as if f and f' were independent variables. Hint: In carrying out the variations, you still need to take into account that a change in $f \to f + \delta f$ implies a concomitant change in $f' \to f' + \delta f'$.

- **Exercise 3.4**: Consider the three-dimensional generalization of Eq. (3.45),

$$\frac{\delta S}{\delta f(\vec{r})} = \lim_{\epsilon \to 0} \int_V d^3 r' \frac{F\left(f(\vec{r}') + \epsilon \delta^{(3)}(\vec{r}' - \vec{r}) \right) - F\left(f(\vec{r}') \right)}{\epsilon}.$$
(3.334)

Show that, for

$$S[f(\vec{r})] = \frac{1}{2} \int_V d^3 r' \left[\vec{\nabla} f(\vec{r}') \right]^2,$$
(3.335)

one has

$$\frac{\delta S}{\delta f(\vec{r})} = -\vec{\nabla}^2 f(\vec{r}).$$
(3.336)

- **Exercise 3.5**: For the functional

$$S[\Phi] = \int_V d^3 r \left(\vec{\nabla} \Phi(\vec{r}) \right)^2,$$
(3.337)

show that

$$\frac{\delta^2 S}{\partial \Phi(\vec{r}) \partial \Phi(\vec{r}')} = -2\delta^{(3)}(\vec{r} - \vec{r}') \vec{\nabla}^2.$$
(3.338)

Verify the result in two ways, *(i)* with recourse to the definition (3.45), suitably generalized to a three-dimensional integral, *(ii)* by writing the first functional derivative as an integral, and then reading off the second functional derivative "under the integral sign", as in Eq. (3.44).

- **Exercise 3.6**: Carry out the generalization of Eqs. (3.71)–(3.73) to three dimensions explicitly. For an infinite integration volume, consult Chaps. 4 and 6 for an introduction to Fourier transforms.

• **Exercise 3.7**: Write the following *ansatz* for the potential energy stored in the potential in a long coaxial cylinder,

$$E = \frac{\epsilon_0}{2} (2\pi) L \int_b^c d\rho\, \rho \left| \frac{\partial w}{\partial \rho} \right|^2, \tag{3.339}$$

where $w = w(\rho)$ is your trial function for the potential, and L is the cylinder length. This *ansatz* is inspired by Eq. (3.105) upon dropping the dependence on φ and z of the potential, which is irrelevant in the limit of large L. By variation of this function using the Euler–Lagrange equations, obtain the exact form given in Eq. (3.117).

• **Exercise 3.8**: Show Eq. (3.124) by expressing the unit vector $\hat{n}$ in Cartesian coordinates, and using the definition (1.16) of the gradient operator.

• **Exercise 3.9**: Calculate the surface area of an ellipsoid given by the equation

$$F(\vec{r}) = (x^2 + y^2)/a^2 + z^2/c^2 - 1 = 0 \tag{3.340}$$

using Eq. (3.128).

• **Exercise 3.10**: Calculate the potential on the symmetry axis due to a uniform charge density σ_0 on the surface of an ellipsoid given by Eq. (3.340). Employ Eq. (3.128).

• **Exercise 3.11**: Exercise: Verify the result (3.147) by a numerical integration, choosing convenient numerical values of the parameters (proposal: $b = 0.2\,\text{cm}$, $\rho = 1.0\,\text{cm}$).

• **Exercise 3.12**: Treat the following boundary-value problem for a cube with volume $0 < x < a$, $0 < y < a$, and $0 < z < a$, whose upper side at $z = a$ is at a potential Φ_0, with grounded five other surfaces (zero potential on five of the six surfaces). Hint: you may want to choose the *ansatz*

$$\Phi(x, y, z) = \sum_{mn} C_{nm} \sin\left(\frac{n\pi x}{a}\right) \sin\left(\frac{m\pi y}{a}\right) \sinh\left(\sqrt{\left(\frac{n\pi}{a}\right)^2 + \left(\frac{m\pi}{a}\right)^2}\, z\right). \tag{3.341}$$

Now set $\Phi = \Phi_0$ for $z = a$, resulting in the condition

$$\int_0^a dx \int_0^a dy\, \Phi(x, y, a) \sin\left(\frac{n'\pi x}{a}\right) \sin\left(\frac{m'\pi y}{a}\right)$$
$$= \Phi_0 \int_0^a dx \int_0^a dy \sin\left(\frac{n'\pi x}{a}\right) \sin\left(\frac{m'\pi y}{a}\right). \tag{3.342}$$

• **Exercise 3.13**: Repeat the calculation in Sec. 3.3.4 for the boundary condition

$$g(y) = \Phi_0 \left(\frac{y}{b}\right)^2. \tag{3.343}$$

• **Exercise 3.14**: Show that (maybe with the help of computer algebra)

$$\frac{1}{6!} \frac{d^6}{dz^6} \left\{ z^7 \left(\frac{\sin(z)}{z^6 \cos(z)}\right) \right\} \bigg|_{z=0} = \frac{2}{15}. \tag{3.344}$$

Thus, calculate the residue

$$\operatorname*{Res}_{z=0}\left[\tan(z)\, z^{-6}\right].\tag{3.345}$$

• **Exercise 3.15**: Show that

$$\operatorname*{Res}_{z=0}\left(\frac{1}{z^4}\exp(z)\right)=\frac{1}{6},\tag{3.346a}$$

$$\operatorname*{Res}_{z=0}\left(\frac{1}{z^5}\exp(z)\right)=\frac{1}{24},\tag{3.346b}$$

$$\operatorname*{Res}_{z=0}\left(\frac{1}{z^5}\frac{\exp(iz)}{\sin(z)}\right)=0,\tag{3.346c}$$

$$\operatorname*{Res}_{z=0}\left(\frac{1}{z^4}\exp(z)\cos(z)\right)=-\frac{1}{3},\tag{3.346d}$$

$$\operatorname*{Res}_{z=0}\left(\frac{1}{z^5}\exp(z)\cos(z)\right)=-\frac{1}{6}.\tag{3.346e}$$

• **Exercise 3.16**: Repeat the derivation outlined in Eqs. (3.264)–(3.272) upon an alteration of the pole prescription in Eq. (3.263), i.e., $1/k \to 1/(k+i\epsilon)$ instead of $1/k \to 1/(k-i\epsilon)$.

• **Exercise 3.17**: Fill in the missing intermediate steps in the derivation of Eq. (3.288) for the boundary conditions given in Eq. (3.279). Then, generalize the treatment for the following boundary-value problem:

$$\Phi(\rho=a,\varphi,z)=0,\qquad \Phi(\rho,\varphi,z=\pm L/2)=\pm\Phi_0,\tag{3.347}$$

i.e., for an upper, positively charged endcap, and a lower, negatively charged endcap.

• **Exercise 3.18**: Fill in the missing intermediate steps in the derivation of Eq. (3.298) for the boundary conditions given in Eq. (3.293). Then, generalize the treatment for the following boundary-value problem:

$$\Phi(\rho=a,\varphi,z)=\Phi_0\frac{2z}{L},\qquad \Phi(\rho,\varphi,z=\pm L/2)=0,\tag{3.348}$$

i.e., for grounded endcaps, and a nonvanishing potential on the outer rim of the cylinder, which is odd under $z \leftrightarrow -z$.

• **Exercise 3.19**: Derive (3.304) based on (3.302) and ideas outlined in Eqs. (2.81)–(2.90). **Hint**: Consider Eq. (2.86), in the form

$$r'^2\left(\frac{\partial}{\partial r}G_\ell(r,r')\Big|_{r=r'+\varepsilon}-\frac{\partial}{\partial r}G_\ell(r,r')\Big|_{r=r'-\varepsilon}\right)=-\frac{1}{\epsilon_0},\tag{3.349}$$

• **Exercise 3.20**: Fill in the missing intermediate steps in the derivation of Eq. (3.310), starting from Eq. (3.308). In particular, clarify why the expression $\vec{\nabla}G_D(\vec{r},\vec{r}')\cdot d\vec{A}$ can be replaced by $[(\partial/\partial r)G_D(\vec{r},\vec{r}')]\,dA$. Generalize the solution for the boundary-value problem

$$\Phi(a,\theta,\varphi)=f_a(\theta)=0,\qquad \Phi(b,\theta,\varphi)=f_b(\theta)=\Phi_0\,\cos\theta.\tag{3.350}$$

• **Exercise 3.21**: Fill in the missing steps in the derivation extending from Eq. (3.311) to (3.315). Hint: Use the orthogonality relation (2.48) and Eq. (2.44).

• **Exercise 3.22**: Repeat the analysis in Secs. 3.4.2 and 3.4.3 for the boundary conditions

$$\Phi\left(a,\theta,\varphi\right) = f_a(\theta) = \Phi_0\left(3\cos^2\theta - 1\right), \qquad \Phi\left(b,\theta,\varphi\right) = f_b(\theta) = 0. \qquad (3.351)$$

• **Exercise 3.23**: Derive Eq. (3.326), using as input all other formulas presented in Sec. 3.5.1.

Chapter 4

Green Functions of Electrodynamics

4.1 Green Function for the Wave Equation

4.1.1 *Integral Representation of the Green Function*

In contrast to electrostatics, electrodynamics describes time-dependent sources, vector potentials and fields. Of prime importance are electromagnetic wave phenomena. In consequence, it is natural to start the discussion of electrodynamics by considering the wave equation, first, in the context of a model problem. Namely, we assume that a signal function $\psi(\vec{r}, t)$ is generated by sources $F(\vec{r}, t)$ and has the general form

$$\left(\frac{1}{c^2} \frac{\partial^2}{\partial t^2} - \vec{\nabla}^2 \right) \psi(\vec{r}, t) = \frac{1}{\epsilon_0} F(\vec{r}, t) . \tag{4.1}$$

One approach to the solution of this equation is to work with the space-time Fourier transforms of the fields and sources. The space-time functions are related to the corresponding functions in Fourier space, referred to as $\psi(\vec{k}, \omega)$ and $F(\vec{k}, \omega)$,

$$\psi(\vec{r}, t) = \int \frac{d^3 k}{(2\pi)^3} \int \frac{d\omega}{2\pi} \psi(\vec{k}, \omega) \exp\left(i\vec{k} \cdot \vec{r} - i\omega t \right) , \tag{4.2a}$$

$$F(\vec{r}, t) = \int \frac{d^3 k}{(2\pi)^3} \int \frac{d\omega}{2\pi} F(\vec{k}, \omega) \exp\left(i\vec{k} \cdot \vec{r} - i\omega t \right) . \tag{4.2b}$$

Here, we denote the original function as well as its Fourier transform by the same symbol, as already done in Sec. 2.2. The wave vector is $\vec{k}$ and the angular frequency is denoted as ω. The Fourier backtransformation is given as

$$\psi(\vec{k}, \omega) = \int d^3 r \int dt\, \psi(\vec{r}, t) \exp\left(-i\vec{k} \cdot \vec{r} + i\omega t \right) , \tag{4.3a}$$

$$F(\vec{k}, \omega) = \int d^3 r \int dt\, F(\vec{r}, t) \exp\left(-i\vec{k} \cdot \vec{r} + i\omega t \right) , \tag{4.3b}$$

where we use the relationship

$$\int_{-\infty}^{\infty} \frac{dv}{2\pi} e^{-iuv} = \delta(u) . \tag{4.4}$$

The Fourier transform of the wave function fulfills the Fourier transformed version
of Eq. (4.1),

$$\int \frac{d^3k}{(2\pi)^3} \int \frac{d\omega}{2\pi} \left(-\vec{\nabla}^2 + \frac{1}{c^2} \frac{\partial^2}{\partial t^2} \right) \psi(\vec{k}, \omega) \, e^{i\vec{k}\cdot\vec{r} - i\omega t}$$

$$= \frac{1}{\epsilon_0} \int \frac{d^3k}{(2\pi)^3} \int \frac{d\omega}{2\pi} F(\vec{k}, \omega) \, e^{i\vec{k}\cdot\vec{r} - i\omega t} \,. \tag{4.5}$$

Because the differential operators only act on the exponential terms, the equation
is converted into an algebraic equation,

$$\int \frac{d^3k}{(2\pi)^3} \int \frac{d\omega}{2\pi} \left\{ \left(\vec{k}^2 - \frac{\omega^2}{c^2} \right) \psi(\vec{k}, \omega) - \frac{1}{\epsilon_0} F(\vec{k}, \omega) \right\} e^{i\vec{k}\cdot\vec{r} - i\omega t} = 0 \,. \tag{4.6}$$

Since each Fourier component $\exp(i\vec{k}\cdot\vec{r} - i\omega t)$ is linearly independent, all the Fourier
coefficients must be zero. In other words, if a function vanishes, then its Fourier
transform vanishes, and vice versa. The solution of the algebraic equation fulfilled
by the Fourier transform therefore is

$$\psi(\vec{k}, \omega) = \frac{1}{\epsilon_0} \frac{F(\vec{k}, \omega)}{\vec{k}^2 - (\omega/c)^2} \,. \tag{4.7}$$

The problem of calculating $\psi(\vec{r}, t)$ is now reduced to taking the inverse Fourier
transform of the expression $F(\vec{k}, \omega)/[\vec{k}^2 - (\omega/c)^2]$. There is at least one difficulty.
The integrand has singularities at $\omega = \pm c|\vec{k}|$, which need to be handled when evalu-
ating the ω integral. The different choices for the contour will later be understood
in terms of the different choices for the boundary conditions imposed on the Green
function. In the case of electrostatics, the Green function $G(\vec{r}, \vec{r}') = 1/(4\pi\epsilon_0 |\vec{r} - \vec{r}'|)$
can be modified to fulfill different boundary conditions (in space), by adding to the
Green function a solution of the homogeneous equation. Likewise, here, we will
see that the Green function can fulfill the defining equation and still fulfill different
boundary conditions in space and time.

The Green function $G(\vec{r} - \vec{r}', t - t')$ with its Fourier transform $G(\vec{k}, \omega)$ solves
Eq. (4.1) with the source term given by

$$F(\vec{r}, t) = \delta^{(3)}(\vec{r} - \vec{r}') \, \delta(t - t') = \int \frac{d^3k}{(2\pi)^3} \int \frac{d\omega}{2\pi} e^{i\vec{k}\cdot(\vec{r} - \vec{r}') - i\omega(t - t')} \,, \tag{4.8}$$

i.e., the Green function in coordinate space fulfills

$$\left(\frac{1}{c^2} \frac{\partial^2}{\partial t^2} - \vec{\nabla}^2 \right) G(\vec{r} - \vec{r}', t - t') = \frac{1}{\epsilon_0} \delta^{(3)}(\vec{r} - \vec{r}') \, \delta(t - t') \,, \tag{4.9}$$

where $\vec{\nabla} = \partial/\partial\vec{r}$ is a differential operator with respect to $\vec{r}$ (not $\vec{r}'$). In the case of
translation invariance, the dependence of the Green function on $\vec{r}'$ is absorbed in
the notation

$$G(\vec{r} - \vec{r}', t - t') \equiv G(\vec{r}, t, \vec{r}', t') \,, \tag{4.10}$$

where the latter form explicitly indicates the dependence on the four independent
arguments. One can find the Fourier transform $G(\vec{k}, \omega)$ by applying the operator

$(1/c^2)\partial^2/\partial t^2 - \vec{\nabla}^2$ under the integral sign, to the Fourier decomposition. This operation is analogous to the calculation that leads to Eqs. (4.6) and (4.7), and results in the identity

$$\int \frac{d^3 k}{(2\pi)^3} \frac{d\omega}{2\pi} \left\{ \left(\vec{k}^2 - \frac{\omega^2}{c^2} \right) G(\vec{k}, \omega) - \frac{1}{\epsilon_0} \right\} e^{i\vec{k}\cdot(\vec{r}-\vec{r}')-i\omega(t-t')} = 0, \qquad (4.11)$$

which must be valid for all Fourier components separately. In Fourier space, the Green function therefore reads as

$$G(\vec{k}, \omega) = \frac{1}{\epsilon_0} \frac{1}{\vec{k}^2 - (\omega/c)^2} . \qquad (4.12)$$

Furthermore, in terms of space-time coordinates, the Green function for the wave equation is found by Fourier backtransformation

$$G(\vec{r} - \vec{r}', t - t') = \frac{1}{\epsilon_0} \int \frac{d^3 k}{(2\pi)^3} \int \frac{d\omega}{2\pi} \frac{e^{i\vec{k}\cdot(\vec{r}-\vec{r}')-i\omega(t-t')}}{\vec{k}^2 - (\omega/c)^2} . \qquad (4.13)$$

The integrand has singularities at $k = \pm\omega/c$.

As already anticipated in the discussion following Eq. (4.7), we must invoke Cauchy's residue theorem in evaluating the pole contributions in Eq. (4.13) (in the ω integral), after a suitable choice of the integration contour. In view of the identity

$$\frac{1}{\vec{k}^2 - \omega^2/c^2} = \frac{c}{2|\vec{k}|} \left(\frac{\omega - c|\vec{k}|}{\omega^2 - (c\vec{k})^2} - \frac{\omega + c|\vec{k}|}{\omega^2 - (c\vec{k})^2} \right) = \frac{c}{2|\vec{k}|} \left(\frac{1}{\omega + c|\vec{k}|} - \frac{1}{\omega - c|\vec{k}|} \right), \qquad (4.14)$$

we write the Green function as follows,

$$G(\vec{r}-\vec{r}', t-t') = \int_{-\infty}^{\infty} \frac{d^3 k}{(2\pi)^3} \frac{c\, e^{i\vec{k}\cdot(\vec{r}-\vec{r}')}}{2|\vec{k}|\,\epsilon_0} \int_{-\infty}^{\infty} \frac{d\omega}{2\pi} e^{-i\omega(t-t')} \left(\frac{1}{\omega + c|\vec{k}|} - \frac{1}{\omega - c|\vec{k}|} \right). \qquad (4.15)$$

The ω integration over the interval $\omega \in (-\infty, +\infty)$ cannot be properly defined in the Riemannian sense; the integration contour has to be deformed into the complex plane in order for the residue theorem to be applicable. The integrand has two first-order poles along the real ω axis. We write G as follows,

$$G(\vec{r} - \vec{r}', t - t') = \int \frac{d^3 k}{(2\pi)^3} f(\vec{k}) P_C(t - t', |\vec{k}|), \qquad f(\vec{k}) = \frac{c\, e^{i\vec{k}\cdot(\vec{r}-\vec{r}')}}{2|\vec{k}|\epsilon_0}, \qquad (4.16)$$

where the following characteristic frequency integral needs to be studied,

$$P_C(t - t', |\vec{k}|) = \int_{-\infty}^{\infty} \frac{d\omega}{2\pi} e^{-i\omega(t-t')} \left(\frac{1}{\omega + c|\vec{k}|} - \frac{1}{\omega - c|\vec{k}|} \right)$$

$$\rightarrow \oint_C \frac{d\omega}{2\pi} e^{-i\omega(t-t')} \left(\frac{1}{\omega + c|\vec{k}|} - \frac{1}{\omega - c|\vec{k}|} \right). \qquad (4.17)$$

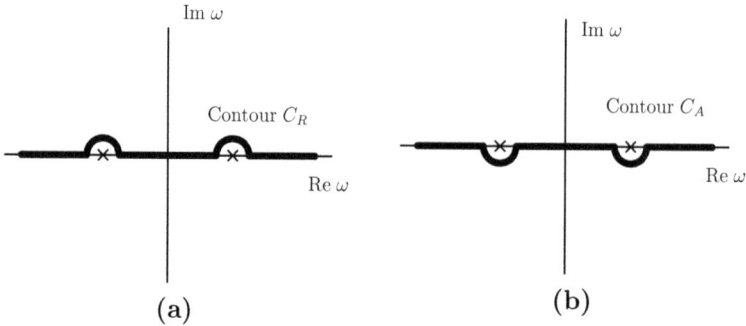

(a)   **(b)**

Fig. 4.1   The ω integration contours C_R and C_A in the complex plane are illustrated. Panel (a) shows the contour that gives the retarded Green function, while panel (b) leads to the advanced Green function. The contour C_R used in evaluating the integral in Eq. (4.17) encircles the poles from above the real axis. For $t < t'$, we have to close the contour C_R in the upper complex half plane, and the result for the integral (4.17) is zero; the retarded Green function vanishes for $t < t'$. By contrast, the contour C_A is below the real axis. So, if C_A is used, then for $t > t'$, the contour C_A is completed below the real axis and the advanced Green function vanishes.

Here, C is a contour that is infinitesimally displaced from the poles. We will discuss its explicit form below; there is some freedom of choice. Indeed, the value of $P_C(t - t', |\vec{k}|)$ depends on the choice of the contour. Integrating ω over the interval $(-\infty, \infty)$, one encounters poles at $\omega = -c|\vec{k}|$ and $\omega = +c|\vec{k}|$. The question then is how these poles should be handled. Indeed, all reasonable ways of treating the poles on the initial integration contour lead to Green functions that solve the defining Eq. (4.9). One may, for example, treat the pole terms using a principal-value integration, which is equivalent to the arithmetic mean of the results obtained using the "upper" and "lower" contours, which encircle the poles infinitesimally above and below the real axis. Alternatively, one may either encircle both of the poles infinitesimally above or below the real axis, or one above and one below. All Green functions thus obtained fulfill the defining equation (4.9). In order to apply the residue theorem, we must close the contour C in Eq. (4.17). The poles have been infinitesimally displaced from the real axis, and the contour initially extends from $\omega = -\infty$ to $\omega = \infty$; we need to close it alongside a semi-circle at $|\omega| = \infty$, with exponential damping in the upper or lower half plane to ensure the convergence of the integral. In our derivations, we thus assume that the integral along the infinite semi-circle S at $|\omega| = \infty$,

$$Q_C(t - t', |\vec{k}|) = \int_S \frac{d\omega}{2\pi} e^{-i\omega(t-t')} \left(\frac{1}{\omega + c|\vec{k}|} - \frac{1}{\omega - c|\vec{k}|} \right) \tag{4.18}$$

vanishes because of the exponential suppression of the integrand for large $|\omega|$. As we shall see, the way in which the contour has to be closed may depend on the sign of $t - t'$. Details are discussed in the following. Possible choices for the contour C are given in Fig. 4.1 and Fig. 4.2.

4.1.2 Why So Many Green Functions?

In the following considerations, we shall encounter the retarded, advanced, and FEYNMAN[1] Green functions. All of these solve the basic, defining equation for the Green function, given in Eq. (4.9), which we recall for convenience,

$$\left(\frac{1}{c^2} \frac{\partial^2}{\partial t^2} - \vec{\nabla}^2 \right) G\left(\vec{r} - \vec{r}', t - t' \right) = \frac{1}{\epsilon_0} \delta^{(3)} \left(\vec{r} - \vec{r}' \right) \delta \left(t - t' \right). \tag{4.19}$$

However, the Green functions differ in their causal behavior and in their analytic structure.

In terms of a classical interpretation, let us first consider the case where the Green function describes a string. The retarded Green function $G_R(x, t, x', t')$ gives the response of the string (initially at rest) to a "unit of momentum" (Dirac-δ disturbance) applied to the string at a point in time t' at a point x' along the string, with the response being recorded at space-time point (x, t) with $t > t'$. The advanced Green function $G_A(x, t, x', t')$ gives the initial field configuration of the string, designed such that a "unit of momentum" applied at (x', t') causes it to come to rest. The initial configuration is described at the space-time points (x, t) with $t < t'$. The advanced Green function "propagates into the past", the retarded Green function "propagates into the future".

In electrodynamics, the retarded Green function describes outgoing waves [scalar and vector potentials $\Phi(\vec{r}, t)$ and $\vec{A}(\vec{r}, t)$], generated by the sources $\rho(\vec{r}', t')$ and $\vec{J}(\vec{r}', t')$, with $t > t'$, i.e., the electromagnetic potentials and fields are generated by a unit disturbance (Dirac-δ function) at $(\vec{r}', t')$, with the fields being zero for $t < t'$. The advanced Green function in electrodynamics describes incoming waves, i.e., it describes the scalar and vector potentials $\Phi(\vec{r}, t)$ and $\vec{A}(\vec{r}, t)$] which converge to a unit disturbance at $(\vec{r}', t')$, with $t < t'$, with the fields being zero for $t > t'$.

The Feynman Green function was originally devised by STUECKELBERG[2] and Feynman. The Feynman Green function essentially relies on a complex formalism, in the sense of a Fourier transform of the source configuration. Positive-frequency components of the source configuration are interpreted as sources for outgoing waves, which propagate into the future, whereas negative-frequency components are interpreted as sinks of incoming waves, which propagate into the past. This enables one to describe both field quanta creation as well as field quanta annihilation processes with one and the same Green function; one thus summarizes a number of different time orderings in the expansion of the time-dependent perturbation theory, into one and the same Green function [20].

The necessity of introducing yet a third Green function, namely, the Feynman Green function comes from field theory, notably, quantum electrodynamics. Roughly speaking, in field theory, in order to describe processes which involve the time evolution of the quantum fields, one needs to consider the so-called time-

---

[1] Richard Phillips Feynman (1918–1988)
[2] Count Ernst Carl Gerlach von Stueckelberg (1905–1984)

ordered T product of field operators [see Eqs. (3.124) and (3.125) of Ref. [20]],

$$\langle 0|T\,A^\mu(x)\,A^\nu(x')|0\rangle = \Theta(t-t')\,\langle 0|A^\mu(x)\,A^\nu(x')|0\rangle + \Theta(t'-t)\,\langle 0|A^\mu(x')\,A^\nu(x)|0\rangle\,,$$
$$(4.20)$$

where Θ is the Heaviside step function, $x = (t,\vec{r})$ is a space-time coordinate four-vector, and $A^\mu(x)$ is a four-vector potential operator ($\mu,\nu = 0,\dots,3$). The vacuum state is denoted as $|0\rangle$. The Green function

$$G_F^{\mu\nu}(x-x') = g^{\mu\nu}\,G_F(x-x') = -\mathrm{i}\,\langle 0|T\,A^\mu(x)\,A^\nu(x')|0\rangle \qquad (4.21)$$

contains both retarded [$\propto \Theta(t-t')$] as well as advanced [$\propto \Theta(t'-t)$] components and is proportional to the Feynman propagator G_F. The space-time metric is $g^{\mu\nu} = \mathrm{diag}(1,-1,-1,-1)$.

A final point: The three Green functions (retarded, advanced, and Feynman) differ by a solution of the homogeneous equation, i.e.,

$$\left(\frac{1}{c^2}\frac{\partial^2}{\partial t^2} - \vec{\nabla}^2\right)\left(G_F(\vec{r}-\vec{r}',t-t') - G_R(\vec{r}-\vec{r}',t-t')\right) = 0\,, \qquad (4.22)$$

and the same relation holds for $G_F - G_A$, and $G_R - G_A$. In our electrostatic analogy the different Green functions would correspond to different solutions of the defining equation (2.66) of the electrostatic Green function, which we recall as

$$\vec{\nabla}^2 G(\vec{r}-\vec{r}') = -\frac{1}{\epsilon_0}\,\delta^{(3)}(\vec{r}-\vec{r}')\,, \qquad (4.23)$$

with different boundary conditions, implemented for the "boundaries" of space-time which can include the half-spaces $t > t'$ and $t' < t$. Pursuing the analogy further, one may remark that the Green function of electrostatics has the representation $-1/\vec{k}^2$ in wave number space, but in coordinate space, one may add a solution of the homogeneous equation $\vec{\nabla}^2 G(\vec{r}-\vec{r}') = 0$, which leads to the Dirichlet and Neumann Green functions. Analogous considerations are true for the Green function of electrodynamics, with $-1/\vec{k}^2$ being replaced by $(\epsilon_0\,(\vec{k}^2 - (\omega/c)^2))^{-1}$ [see Eq. (4.12)].

4.1.3   *Retarded and Advanced Green Function*

We recall once more that according to Eqs. (4.16) and (4.17), the Green function is given as

$$G(\vec{r}-\vec{r}',t-t') = \int\frac{\mathrm{d}^3 k}{(2\pi)^3}\,f(\vec{k})\,P_C(t-t',|\vec{k}|)\,, \qquad f(\vec{k}) = \frac{c}{2|\vec{k}|\epsilon_0}\,\mathrm{e}^{\mathrm{i}\vec{k}\cdot(\vec{r}-\vec{r}')}\,. \quad (4.24)$$

We recall the following characteristic frequency integral, defined in Eq. (4.17),

$$P_C(t-t',|\vec{k}|) = \int_C\frac{\mathrm{d}\omega}{2\pi}\,\mathrm{e}^{-\mathrm{i}\omega(t-t')}\left(\frac{1}{\omega+c|\vec{k}|} - \frac{1}{\omega-c|\vec{k}|}\right)\,. \qquad (4.25)$$

The contour C consists two elements, the first being equal to the real axis, with an appropriate choice regarding the poles of the integrand, the second being the semi-circle, with infinite radius, in either the upper half complex ω plane or in the lower complex ω half plane. Unlike the way in which the pole terms are encircled, the choice of the semi-circle actually is determined by the requirement for the integral to converge to a finite value.

Together with the choice of the integration contour for the pole terms, this requirement determines the behavior of the Green function for positive and negative time difference $t - t'$. Let us consider the term in round brackets in the integrand of Eq. (4.25), which causes the integrand to vanish, on either semi-circle, as $|\omega|^{-2}$ for large $|\omega|$ provided the exponential in the numerator is well behaved. In order to analyze the behavior on the semi-circle, we set $\omega = \mathrm{Re}(\omega) + \mathrm{i}\,\mathrm{Im}(\omega)$. The absolute value of the exponential in the integrand of Eq. (4.25) is determined by the relation

$$\left| \exp\left[-\mathrm{i}\omega\,(t - t') \right] \right| = \exp\left[\mathrm{Im}(\omega)\,(t - t') \right]. \tag{4.26}$$

If $t - t' > 0$ and $\mathrm{Im}(\omega) \to +\infty$, then this expression diverges. However, if $t - t' > 0$ and $\mathrm{Im}(\omega) \to -\infty$, then this expression vanishes exponentially. Thus, for $t > t'$, the path used in evaluating (4.17) must be closed in the lower half of the complex ω plane. Conversely, for $t < t'$, the path must be closed along a semi-circle in the upper half of the complex ω plane. The first path gives a nonvanishing contribution for the retarded Green function (contour C_R, see Fig. 4.1), whereas the second yields a nonvanishing contribution for the advanced Green function (contour C_A). For completeness, we should point out that two obvious further possibilities for distorting the path exist which encircle the poles on opposite sides of the real axis; these are not shown in Fig. 4.1.

Both integration paths outlined in Fig. 4.1 have their justification, and both results for the Green function are valid solutions of the defining equation (4.9), i.e., they lead to solutions of the inhomogeneous differential equation

$$\left(\frac{1}{c^2} \frac{\partial^2}{\partial t^2} - \vec{\nabla}^2 \right) G\left(\vec{r} - \vec{r}', t - t' \right) = \frac{1}{\epsilon_0} \delta^{(3)}\left(\vec{r} - \vec{r}' \right) \delta\left(t - t' \right). \tag{4.27}$$

The different solutions fulfill different boundary conditions, which is natural for a second-order partial differential equation.

The results obtained for $P_C(t - t', |\vec{k}|)$ along the different contours differ in the boundary conditions fulfilled by the corresponding Green functions. For the Dirichlet Green function in electrostatics, the boundary conditions are given along two-dimensional submanifolds (surfaces) of three-dimensional space. The Dirichlet Green function is required to vanish whenever one of the two arguments $\vec{r}$ or $\vec{r}'$ is on the defining surface. For the time-dependent Green function, boundary conditions are defined on submanifolds of four-dimensional space-time. Intuitively, one would associate boundary conditions with limiting cases of the variables, e.g., in the infinite past or the infinite future. In all cases considered, we have so far defined our Green function to vanish in the infinite past or future, or, for large distances.

Theoretically, we could have added to any of our Green functions a solution to the homogeneous equation to the Green function (e.g., a constant term) that changes the behavior as $t, t' \to \pm\infty$ or $|\vec{r}|, |\vec{r}'| \to \infty$. However, we choose not to do so. Or, we could have added a solution of the homogeneous equation, proportional to a plane and uniform wave $\cos(\vec{k} \cdot \vec{r} - \omega t)$. This wave does not vanish for $|t - t'| \to \infty$, and $|\vec{r} - \vec{r}'| \to \infty$. Indeed, the difference of the retarded and advanced Green functions is just a solution of the homogeneous equation.

Let us now evaluate both the retarded as well as the advanced Green function, explicitly. We consider the ω or angular frequency part of the integrand

$$P_C(t - t', |\vec{k}|) = \oint_{\text{closed}} \frac{d\omega}{2\pi} e^{-i\omega(t-t')} \left(\frac{1}{\omega + c|\vec{k}|} - \frac{1}{\omega - c|\vec{k}|} \right), \tag{4.28}$$

where C is the path of integration. If the path C_R for the retarded Green function in Fig. 4.1 is used and $t < t'$, then we must close the integration path in the upper half of the complex plane, and the result for P_C is zero. However, if C_R is used and $t > t'$, then the poles are encircled in the clockwise (mathematically negative) direction, and we obtain

$$P_{C_R}\left(t - t', |\vec{k}|\right) = -\frac{2\pi i}{2\pi} \Theta(t - t') \left[\operatorname*{Res}_{\omega = -ck} \left(\frac{e^{-i\omega(t-t')}}{\omega + c|\vec{k}|} \right) - \operatorname*{Res}_{\omega = +ck} \left(\frac{e^{-i\omega(t-t')}}{\omega - c|\vec{k}|} \right) \right]$$

$$= -i\,\Theta(t - t') \left(e^{-i(-ck)(t-t')} - e^{-i(+ck)(t-t')} \right)$$

$$= -i\,\Theta(t - t') \left(e^{ick(t-t')} - e^{-ick(t-t')} \right)$$

$$= 2\,\Theta(t - t')\, \sin(ck(t - t')). \tag{4.29}$$

The function $P_{C_R}(t - t', |\vec{k}|)$ is nonvanishing only for $t > t'$; this is manifest in the step function. By looking at the integration contour, one might superficially assume that the pole of the integrand is only half-encircled. However, the continuation of the integration contour into the lower complex plane actually makes it a full pole. The factor $1/(2\pi)$ of the ω integration thus is compensated by the factor 2π from the residue theorem.

On account of Eqs. (4.16) and (4.29), we obtain the retarded Green function G_R,

$$G_R(\vec{r} - \vec{r}', t - t') = \int \frac{d^3k}{(2\pi)^3} f(\vec{k}) P_{C_R}(t - t', |\vec{k}|)$$

$$= \Theta(t - t') \frac{c}{\epsilon_0} \int \frac{d^3k}{(2\pi)^3} \frac{e^{i\vec{k} \cdot (\vec{r} - \vec{r}')}}{|\vec{k}|} \sin(ck(t - t')), \tag{4.30}$$

where we recall the definition of $f(\vec{k})$ given in Eq. (4.16). Now we do the k integration in spherical $\vec{k}$ space, assuming (without loss of generality) that $\vec{r} - \vec{r}'$ points along the $\hat{e}_z$ direction $[u_k = \cos(\theta_k)]$:

$$G_R = \Theta(t - t') \frac{c}{\epsilon_0 (8\pi^3)} \int_0^\infty dk \, k^2 \int_0^{2\pi} d\varphi_k \int_{-1}^1 du_k \frac{\exp(ik|\vec{r} - \vec{r}'| u_k)}{|\vec{k}|} \sin(ck(t - t'))$$

$$= \Theta(t - t') \frac{c(2\pi)}{\epsilon_0 (8\pi^3)} \int_0^\infty dk \, k^2 \int_{-1}^1 du_k \frac{\exp(ik|\vec{r} - \vec{r}'| u_k)}{|\vec{k}|} \sin(ck(t - t'))$$

$$= \Theta(t - t') \frac{c}{4\pi^2 \epsilon_0} \int_0^\infty dk \, k^2 \frac{\exp(ik|\vec{r} - \vec{r}'|) - \exp(-ik|\vec{r} - \vec{r}'|)}{ik^2|\vec{r} - \vec{r}'|} \sin(ck(t - t'))$$

$$= \Theta(t - t') \frac{c}{4\pi^2 \epsilon_0 i |\vec{r} - \vec{r}'|} \int_0^\infty dk \left[e^{ik|\vec{r} - \vec{r}'|} \sin(ck(t - t')) \right.$$

$$\left. + e^{i(-k)|\vec{r} - \vec{r}'|} \sin(c(-k)(t - t')) \right]. \tag{4.31}$$

In view of the symmetry under $k \leftrightarrow -k$, we can extend the integration domain to the entire real axis,

$$G_R = \Theta(t - t') \frac{c}{4\pi^2 \epsilon_0 i |\vec{r} - \vec{r}'|} \int_{-\infty}^\infty dk \exp(ik|\vec{r} - \vec{r}'|) \sin(ck(t - t'))$$

$$= \Theta(t - t') \frac{c}{4\pi^2 \epsilon_0 i |\vec{r} - \vec{r}'|} \int_{-\infty}^\infty dk \left(\frac{1}{2i}\right) \exp(ik|\vec{r} - \vec{r}'|) \left[e^{ick(t - t')} - e^{-ick(t - t')} \right]$$

$$= \Theta(t - t') \left(-\frac{c}{8\pi^2 \epsilon_0 |\vec{r} - \vec{r}'|}\right) \int_{-\infty}^\infty dk \, e^{ik|\vec{r} - \vec{r}'|} \left[e^{ick(t - t')} - e^{-ick(t - t')} \right]$$

$$= \Theta(t - t') \left(-\frac{c}{8\pi^2 \epsilon_0 |\vec{r} - \vec{r}'|}\right) \int_{-\infty}^\infty dk \left[e^{ik(|\vec{r} - \vec{r}'| + c(t - t'))} - e^{ik(|\vec{r} - \vec{r}'| - c(t - t'))} \right]. \tag{4.32}$$

The k integration is now trivial in view of the formula

$$\int_{-\infty}^\infty dk \, e^{ikx} = 2\pi \delta(x). \tag{4.33}$$

The retarded Green function thus is given as

$$G_R(\vec{r} - \vec{r}', t - t') = -\frac{c}{4\pi\epsilon_0} \frac{1}{|\vec{r} - \vec{r}'|} \Theta(t - t') \left\{ \delta\Big(|\vec{r} - \vec{r}'| + c(t - t')\Big) - \delta\Big(|\vec{r} - \vec{r}'| - c(t - t')\Big) \right\}, \tag{4.34}$$

which can be simplified to read

$$G_R(\vec{r} - \vec{r}', t - t') = \frac{c}{4\pi\epsilon_0} \frac{1}{|\vec{r} - \vec{r}'|} \Theta(t - t') \left\{ \delta\Big(|\vec{r} - \vec{r}'| - c(t - t')\Big) - \delta\Big(|\vec{r} - \vec{r}'| + c(t - t')\Big) \right\}. \tag{4.35}$$

One might otherwise argue that it should be possible to discard the last term in curly brackets in Eq. (4.35), because the expression $|\vec{r} - \vec{r}'| + c(t - t')$ cannot vanish if $t - t' > 0$, which in turn is necessary for the step function to be nonvanishing. One might thus think that the Green function should be expressible as

$$G_R(\vec{r} - \vec{r}', t - t') \approx \frac{c}{4\pi \epsilon_0} \frac{\Theta(t - t')}{|\vec{r} - \vec{r}'|} \delta\left(|\vec{r} - \vec{r}'| - c(t - t')\right). \tag{4.36}$$

In many cases, it is indeed possible to discard the second term in curly brackets in Eq. (4.35), as done in Ref. [21]. However, as argued in Ref. [22], the approximation (4.36) can only be used if no partial integrations are to be carried out, i.e., if we have manifestly $t - t' > 0$ for all contributing source terms in a calculation. The deeper reason is the following: In any representation of the Dirac-δ distribution, e.g., formulated in terms of normalized Gaussians whose width tends to zero, the step function will assume values of unity within the width of the representation of the Dirac-δ function when $|\vec{r} - \vec{r}'| \to 0^+$ and $c(t - t') \to 0^+$. In other words, for any representation of the Dirac-δ distribution in terms of Gaussians with a small width, the Dirac-δ function $\delta(|\vec{r} - \vec{r}'| + c(t - t'))$ will start to "ramp up" already when $|\vec{r} - \vec{r}'| + c(t - t')$ assumes slightly positive, nonvanishing values $(t - t' > 0)$, which lie in the domain where $\Theta(t - t') > 0$. One thus cannot discard any term in Eq. (4.35) if one would like to carry out partial integrations (with respect to time or space) correctly. Such partial integrations are useful in the investigation of the coupling of the sources to the potentials and fields in classical electrodynamics [22].

It is sometimes useful to note that G_R can be written as

$$\begin{aligned}
G_R(\vec{r} - \vec{r}', t - t') = & \frac{c}{4\pi\epsilon_0} \frac{1}{|\vec{r} - \vec{r}'|} \Theta(t - t')\, \delta\left(|\vec{r} - \vec{r}'| - c(t - t')\right) \\
& + \frac{(-c)}{4\pi\epsilon_0} \frac{1}{|\vec{r} - \vec{r}'|} \Theta(t - t')\, \delta\left(|\vec{r} - \vec{r}'| - (-c)(t - t')\right),
\end{aligned} \tag{4.37}$$

i.e., as the sum of two terms, the second of which is derived from the first by a change in the sign of the speed of light. Likewise, the approximation

$$G_R(\vec{r} - \vec{r}', t - t') \approx \frac{c}{2\pi\epsilon_0} \Theta(t - t')\, \delta\left((\vec{r} - \vec{r}')^2 - c^2(t - t')^2\right) \tag{4.38}$$

is valid provided we can discard the second term in curly brackets in Eq. (4.35). Indeed, if the only contributing zero of the argument of the Dirac-δ is the one at

$$|\vec{r} - \vec{r}'| - c(t - t') = 0, \qquad\qquad t - t' > 0, \tag{4.39}$$

then in view of Eq. (1.34),

$$\delta\left((\vec{r} - \vec{r}')^2 - c^2(t - t')^2\right) = \frac{\delta\left(|\vec{r} - \vec{r}'| - c(t - t')\right)}{2|\vec{r} - \vec{r}'|}, \qquad t - t' > 0. \tag{4.40}$$

Under these assumptions, the approximations (4.38) and (4.36) remain valid.

The advanced Green function obtained using the path C_A (see Fig. 4.1) is similar to the retarded Green function but is nonvanishing only for $t < t'$. Note that one must repeat all the steps leading to the expression (4.35) for the retarded Green

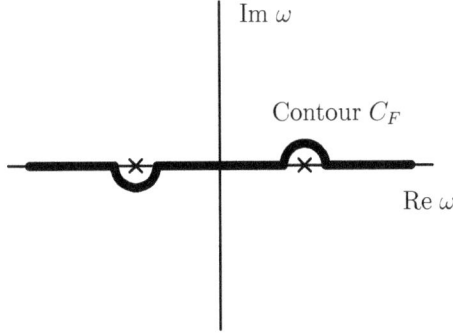

Fig. 4.2   The Feynman contour C_F corresponds to the time-ordered product of photon field operators in field theory.

function but use the contour C_A, rather than C_R. The net result of this calculation is that the advanced Green function is obtained from Eq. (4.34) by the substitution $t - t' \to t' - t$,

$$G_A\left(\vec{r} - \vec{r}', t - t'\right) = \frac{c}{4\pi\epsilon_0} \frac{1}{|\vec{r} - \vec{r}'|} \Theta(t' - t) \left\{ \delta\left(|\vec{r} - \vec{r}'| + c\left(t - t'\right)\right) - \delta\left(|\vec{r} - \vec{r}'| - c\left(t - t'\right)\right) \right\}.$$
(4.41)

This can be rewritten as

$$G_A\left(\vec{r} - \vec{r}', t - t'\right) = \frac{c}{4\pi\epsilon_0} \frac{1}{|\vec{r} - \vec{r}'|} \Theta(t' - t) \left\{ \delta\left(|\vec{r} - \vec{r}'| - c\left(t' - t\right)\right) - \delta\left(|\vec{r} - \vec{r}'| + c\left(t' - t\right)\right) \right\}.$$
(4.42)

Under appropriate assumptions (manifestly $t - t' < 0$, and no partial integrations) we can approximate the advanced Green function as

$$G_A\left(\vec{r} - \vec{r}', t - t'\right) \approx \frac{c}{4\pi\epsilon_0} \frac{1}{|\vec{r} - \vec{r}'|} \Theta\left(t' - t\right) \delta\left(|\vec{r} - \vec{r}'| + c\left(t - t'\right)\right),$$
(4.43)

or as

$$G_A\left(\vec{r} - \vec{r}', t - t'\right) \approx \frac{c}{2\pi\epsilon_0} \Theta\left(t' - t\right) \delta\left((\vec{r} - \vec{r}')^2 - c^2\left(t - t'\right)^2\right),$$
(4.44)

where the latter form is valid if we neglect the overlap region of the Θ function with the chosen representation of the Dirac-δ; this approximation is generally valid unless partial integrations have to be taken [22].

4.1.4   *Feynman Contour Green Function*

We recall the definition of the P_C function, which for the Feynman contour as given in Fig. 4.2 reads

$$P_{C_F}\left(t - t', |\vec{k}|\right) = \oint_{C_F} \frac{d\omega}{2\pi} e^{-i\omega(t-t')} \left(\frac{1}{\omega + c|\vec{k}|} - \frac{1}{\omega - c|\vec{k}|} \right).$$
(4.45)

For $t-t' > 0$, the Feynman integration contour has to be closed in the mathematically negative sense, and the pole at $\omega = c|\vec{k}|$ is relevant. For $t - t' < 0$, the contour is closed in the upper complex half plane, and the pole at $\omega = -c|\vec{k}|$ is relevant. It is encircled in the mathematically positive sense. Alternatively, we could include infinitesimal imaginary contributions into the denominators,

$$P_{C_F}(t-t',|\vec{k}|) = \int_{-\infty}^{\infty} \frac{d\omega}{2\pi} e^{-i\omega(t-t')} \left(\frac{1}{\omega + c|\vec{k}| - i\epsilon} - \frac{1}{\omega - c|\vec{k}| + i\epsilon} \right). \tag{4.46}$$

So,

$$P_{C_F}\left(t-t',|\vec{k}|\right) = -\frac{(-2\pi i)}{2\pi} \Theta(t-t') \operatorname*{Res}_{\omega=+ck} \left(\frac{e^{-i\omega(t-t')}}{\omega - c|\vec{k}|} \right)$$

$$+ \frac{2\pi i}{2\pi} \Theta(t'-t) \operatorname*{Res}_{\omega=-ck} \left(\frac{e^{-i\omega(t-t')}}{\omega + c|\vec{k}|} \right)$$

$$= i\,\Theta(t-t')\, e^{-ick(t-t')} + i\,\Theta(t'-t)\, e^{ick(t-t')}. \tag{4.47}$$

The second term equals the first under the replacement $t-t' \to t'-t$. The Feynman Green function G_F is thus obtained as

$$G_F(\vec{r}-\vec{r}',t-t') = \int \frac{d^3k}{(2\pi)^3} \frac{c\, e^{i\vec{k}\cdot(\vec{r}-\vec{r}')}}{2\epsilon_0|\vec{k}|} \left(i\Theta(t-t')\, e^{-ick(t-t')} + i\Theta(t'-t)\, e^{ick(t-t')} \right). \tag{4.48}$$

Now we do the k integration in spherical $\vec{k}$ space, assuming, without loss of generality, that $\vec{r}-\vec{r}'$ is aligned along the $\hat{e}_z$ direction $[u_k = \cos(\theta_k)]$. The Feynman Green function G_F consists of two terms, the first of which reads

$$G_F^{(1)} = i\,\Theta(t-t')\frac{c}{2\,\epsilon_0\,(8\pi^3)} \int_0^\infty dk\, k^2 \int_0^{2\pi} d\varphi_k \int_{-1}^1 du_k \frac{\exp\left(i\,k|\vec{r}-\vec{r}'|\,u_k\right)}{|\vec{k}|} e^{-ick(t-t')}$$

$$= i\,\Theta(t-t')\frac{c\,(2\pi)}{2\,\epsilon_0\,(8\pi^3)} \int_0^\infty dk\, k^2 \int_{-1}^1 du_k \frac{\exp\left(i\,k|\vec{r}-\vec{r}'|\,u_k\right)}{|\vec{k}|} e^{-ick(t-t')}$$

$$= i\,\Theta(t-t')\frac{c}{8\pi^2\epsilon_0} \int_0^\infty dk\, k^2 \frac{\exp\left(i\,k\,|\vec{r}-\vec{r}'|\right) - \exp\left(-i\,k|\vec{r}-\vec{r}'|\right)}{i\,k^2|\vec{r}-\vec{r}'|} e^{-ick(t-t')}$$

$$= i\,\Theta(t-t')\frac{c}{8\pi^2\epsilon_0\,i\,|\vec{r}-\vec{r}'|} \int_0^\infty dk\, \left[e^{i\,k\,|\vec{r}-\vec{r}'|-i\,ck\,(t-t')} - e^{-i\,k\,|\vec{r}-\vec{r}'|-ick\,(t-t')} \right]$$

$$= i\,\Theta(t-t')\frac{c}{8\pi^2\epsilon_0\,i\,|\vec{r}-\vec{r}'|} \int_0^\infty dk\, \left[e^{i\,k\,(|\vec{r}-\vec{r}'|-c(t-t')+i\epsilon)} - e^{-i\,k\,(|\vec{r}-\vec{r}'|+c\,(t-t')-i\epsilon)} \right]. \tag{4.49}$$

In the last step, we have introduced infinitesimal convergent factors $\pm i\epsilon$, which enables us to carry out the k integration. The limit $\epsilon \to 0$, after performing the integrations, is understood here and in the following unless stated otherwise. We finally obtain

$$G_F^{(1)} = i\, \Theta(t-t')\, \frac{c}{8\pi^2\epsilon_0\, i\, |\vec{r}-\vec{r}'|} \left(\frac{1}{-i\,(|\vec{r}-\vec{r}'|-c(t-t')+i\epsilon)} \right.$$

$$\left. - \frac{1}{i\,(|\vec{r}-\vec{r}'|+c(t-t')-i\epsilon)} \right)$$

$$= i\, \Theta(t-t')\, \frac{c}{8\pi^2\epsilon_0\, |\vec{r}-\vec{r}'|} \left(\frac{1}{|\vec{r}-\vec{r}'|-c(t-t')+i\epsilon} + \frac{1}{|\vec{r}-\vec{r}'|+c(t-t')-i\epsilon} \right)$$

$$= i\, \Theta(t-t')\, \frac{c}{4\pi^2\epsilon_0}\, \frac{1}{(\vec{r}-\vec{r}')^2 - c^2\,(t-t')^2 + i\eta}$$

$$= -i\, \Theta(t-t')\, \frac{c}{4\pi^2\epsilon_0}\, \frac{1}{c^2\,(t-t')^2 - (\vec{r}-\vec{r}')^2 - i\eta}\,. \tag{4.50}$$

Here $\eta = 2\epsilon c\,(t-t') > 0$ is an infinitesimal quantity. The second term in the Feynman Green function G_F is analogously obtained as

$$G_F^{(2)} = i\, \Theta(t'-t)\, \frac{c}{2\,\epsilon_0\,(8\pi^3)} \int_0^\infty dk\, k^2 \int_0^{2\pi} d\varphi_k \int_{-1}^1 du_k\, \frac{\exp\left(i\,k|\vec{r}-\vec{r}'|\,u_k\right)}{|\vec{k}|}\, e^{ick(t-t')}$$

$$= i\, \Theta(t'-t)\, \frac{c}{8\pi^2\epsilon_0} \int_0^\infty dk\, k^2\, \frac{\exp\left(i\,k|\vec{r}-\vec{r}'|\right) - \exp\left(-i\,k|\vec{r}-\vec{r}'|\right)}{i\,k^2|\vec{r}-\vec{r}'|}\, e^{ick(t-t')}$$

$$= i\, \Theta(t'-t)\, \frac{c}{8\pi^2\epsilon_0\, i\, |\vec{r}-\vec{r}'|} \int_0^\infty dk\, \left[e^{i\,k\left(|\vec{r}-\vec{r}'|+c(t-t')+i\epsilon\right)} - e^{-i\,k\left(|\vec{r}-\vec{r}'|-c(t-t')-i\epsilon\right)} \right]$$

$$= i\, \Theta(t'-t)\, \frac{c}{8\pi^2\epsilon_0\, |\vec{r}-\vec{r}'|} \left(\frac{1}{|\vec{r}-\vec{r}'|+c(t-t')+i\epsilon} + \frac{1}{|\vec{r}-\vec{r}'|-c(t-t')-i\epsilon} \right)$$

$$= -i\, \Theta(t'-t)\, \frac{c}{4\pi^2\epsilon_0}\, \frac{1}{c^2\,(t-t')^2 - (\vec{r}-\vec{r}')^2 - i\eta'}\,, \tag{4.51}$$

where $\eta' = 2\epsilon c\,(t'-t) > 0$ in this case, because of the prefactor $\Theta(t'-t)$. Adding the two contributions for $G_F = G_F^{(1)} + G_F^{(2)}$ and interpolating trivially for $t = t'$, we obtain the following result for the Feynman propagator,

$$G_F(\vec{r}-\vec{r}', t-t') = -i\, \frac{c}{4\pi^2\epsilon_0}\, \frac{1}{c^2\,(t-t')^2 - (\vec{r}-\vec{r}')^2 - i\epsilon}\,. \tag{4.52}$$

Here, we have substituted $\eta \to \epsilon$ for the infinitesimal quantity in the denominator. Two important alternative forms of this propagator are as follows. First, because of the prescription

$$\frac{1}{g-i\epsilon} = (\text{P.V.})\,\frac{1}{g} + i\pi\,\delta(g)\,, \tag{4.53}$$

we have

$$G_F(\vec{r} - \vec{r}', t - t') = -\,\mathrm{i}\,\frac{c}{4\pi^2\epsilon_0}\,(\text{P.V.})\,\frac{1}{c^2\,(t-t')^2 - (\vec{r} - \vec{r}')^2}$$

$$+\frac{c}{4\pi\epsilon_0}\,\delta(c^2\,(t-t')^2 - (\vec{r} - \vec{r}')^2)\,. \tag{4.54}$$

Here, (P.V.) denotes the principal value. Another representation is

$$G_F(\vec{r} - \vec{r}', t - t') = -\mathrm{i}\,\frac{c}{8\pi^2\epsilon_0\,|\vec{r} - \vec{r}'|}\left(\frac{1}{c\,(t-t') - |\vec{r} - \vec{r}'| - \mathrm{i}\,\epsilon} - \frac{1}{c\,(t-t') + |\vec{r} - \vec{r}'| + \mathrm{i}\,\epsilon}\right)\,. \tag{4.55}$$

The poles are at

$$c\,(t-t') = |\vec{r} - \vec{r}'| + \mathrm{i}\,\epsilon\,, \qquad c\,(t-t') = -|\vec{r} - \vec{r}'| - \mathrm{i}\,\epsilon\,. \tag{4.56}$$

We consider the Fourier transformation with respect to $\tau = t - t'$ of the Feynman Green function for real rather than complex ω,

$$G_F(\vec{r} - \vec{r}', \omega) = -\,\mathrm{i}\int_{-\infty}^{\infty}\mathrm{d}\tau\,e^{\mathrm{i}\omega\tau}\,\frac{1}{8\pi^2\epsilon_0|\vec{r} - \vec{r}'|}\left(\frac{1}{\tau - |\vec{r} - \vec{r}'|/c - \mathrm{i}\,\epsilon} - \frac{1}{\tau + |\vec{r} - \vec{r}'|/c + \mathrm{i}\,\epsilon}\right)$$

$$= \begin{cases} -\mathrm{i}\,2\pi\,\mathrm{i}\,e^{\mathrm{i}\omega|\vec{r} - \vec{r}'|/c}\,\dfrac{1}{8\pi^2\epsilon_0\,|\vec{r} - \vec{r}'|} & (\omega > 0) \\[2ex] -(-\mathrm{i})\,(-2\pi\,\mathrm{i})\,e^{-\mathrm{i}\omega|\vec{r} - \vec{r}'|/c}\,\dfrac{1}{8\pi^2\epsilon_0\,|\vec{r} - \vec{r}'|} & (\omega < 0) \end{cases} = \frac{e^{\mathrm{i}\,\omega\,\mathrm{sgn}(\omega)|\vec{r} - \vec{r}'|/c}}{4\pi\epsilon_0\,|\vec{r} - \vec{r}'|}\,. \tag{4.57}$$

Here, $\mathrm{sgn}(x)$ is the sign function, which equals $+1$ for $x > 0$ and -1 for $x < 0$. For complex ω, the Fourier integral would diverge either at $\tau = -\infty$ or at $\tau = +\infty$ and could not be directly calculated. The result (4.57) can be understood as follows: There are two poles in the τ integration, namely, at

$$\tau = \frac{|\vec{r} - \vec{r}'|}{c} + \mathrm{i}\,\epsilon\,, \qquad \tau = -\frac{|\vec{r} - \vec{r}'|}{c} - \mathrm{i}\,\epsilon\,. \tag{4.58}$$

If $\omega > 0$, then the τ integral needs to be closed in the upper half plane, i.e., the pole at $\tau = |\vec{r} - \vec{r}'|/c + \mathrm{i}\,\epsilon$ contributes. Conversely, if $\omega < 0$, then the τ integral needs to be closed in the lower half plane, i.e., the pole at $\tau = -|\vec{r} - \vec{r}'|/c - \mathrm{i}\,\epsilon$ contributes. For real ω, we have in view of $\omega\,\mathrm{sgn}(\omega) = |\omega|$,

$$G_F(\vec{r} - \vec{r}', \omega) = \frac{e^{\mathrm{i}\,|\omega|\,|\vec{r} - \vec{r}'|/c}}{4\pi\epsilon_0\,|\vec{r} - \vec{r}'|}\,. \tag{4.59}$$

The question now is how to analytically continue the modulus function for manifestly complex ω. First of all, for real ω, we can write

$$\omega\,\mathrm{sgn}(\omega) = |\omega| = \sqrt{\omega^2}\,, \tag{4.60}$$

and this relation gives us a hint about how the analytic continuation needs to be done. The boundary condition in frequency space translates into the condition that

the contribution of frequencies of very large complex modulus to the Feynman propagator needs to vanish. This boundary condition can be implemented as follows.
We first supply an infinitesimal imaginary part under the square root,

$$\sqrt{\omega^2} \to \sqrt{\omega^2 + i\epsilon}. \tag{4.61}$$

This is compatible with the requirement that $\sqrt{\omega^2} = |\omega|$ for real ω, no matter whether
we lay the branch cut of the square root along the positive or the negative real axis.
Then,

$$G_F(\vec{r} - \vec{r}', \omega) = \frac{\exp\left(i\sqrt{\omega^2 + i\epsilon}\,\dfrac{|\vec{r} - \vec{r}'|}{c}\right)}{4\pi\epsilon_0 |\vec{r} - \vec{r}'|} = \frac{\exp\left(-b\dfrac{|\vec{r} - \vec{r}'|}{c}\right)}{4\pi\epsilon_0 |\vec{r} - \vec{r}'|}, \tag{4.62}$$

$$b = -i\sqrt{\omega^2 + i\epsilon}, \qquad \mathrm{Re}(b) \overset{!}{\geq} 0, \qquad \mathrm{Im}\sqrt{\omega^2 + i\epsilon} \overset{!}{\geq} 0. \tag{4.63}$$

This condition is fulfilled if we choose the branch cut of the square root function
along the positive real axis. The argument of the square root function vanishes for
$\omega = \pm\sqrt{\epsilon}\exp\left(-\frac{i\pi}{4}\right)$. If we define the branch cut of the square root function to be
along the positive real axis, then branch cuts, as a function of ω, extend to $\omega \to \pm\infty$,
from the branch points, and the latter possibility is the one that allows us to use
the Feynman contour for the ω integration in the usual manner.

Having specified the infinitesimal imaginary part as $\sqrt{\omega^2 + i\epsilon}$, and defining the
branch cut to be along the positive real axis, we may deform the Feynman contour
to the entire real axis, and we have the Fourier backtransformation,

$$G_F(\vec{r} - \vec{r}', \tau) = \int_{-\infty}^{\infty} \frac{d\omega}{2\pi} \frac{\exp\left(i\sqrt{\omega^2 + i\epsilon}\,\dfrac{|\vec{r} - \vec{r}'|}{c} - i\omega\tau\right)}{4\pi\epsilon_0 |\vec{r} - \vec{r}'|}$$

$$= \int_{0}^{\infty} \frac{d\omega}{2\pi} \frac{e^{\left(i\frac{|\vec{r} - \vec{r}'|}{c} - i\tau - \eta\right)\omega}}{4\pi\epsilon_0 |\vec{r} - \vec{r}'|} + \int_{-\infty}^{0} \frac{d\omega}{2\pi} \frac{e^{\left(-i\frac{|\vec{r} - \vec{r}'|}{c} - i\tau + \eta\right)\omega}}{4\pi\epsilon_0 |\vec{r} - \vec{r}'|}$$

$$= \frac{1}{8\pi^2\epsilon_0 |\vec{r} - \vec{r}'|} \left(-\frac{1}{i\dfrac{|\vec{r} - \vec{r}'|}{c} - i\tau - \eta} - \frac{1}{i\dfrac{|\vec{r} - \vec{r}'|}{c} + i\tau - \eta} \right)$$

$$= \frac{1}{8\pi^2\epsilon_0 |\vec{r} - \vec{r}'|} \left(-\frac{ic}{-|\vec{r} - \vec{r}'| + c\tau - i\eta} - \frac{ic}{-|\vec{r} - \vec{r}'| - c\tau - i\eta} \right)$$

$$= -i\frac{c}{4\pi^2\epsilon_0} \frac{1}{c^2\tau^2 - (\vec{r} - \vec{r}')^2 - i\eta}, \tag{4.64}$$

where η in the second line is a convergent factor. The limit $\eta \to 0$ is again understood in intermediate steps after all integrations have been carried out; η is

an auxiliary infinitesimal parameter which we could have labeled as ϵ. A priori, one has in the limit $\epsilon \to 0$, simply $\sqrt{\omega^2 + i\epsilon} \to |\omega|$, but one can justify the occurrence of the convergent factor based on the expansion

$$\sqrt{\omega^2 + i\epsilon} = |\omega| + \frac{i\epsilon}{2|\omega|} + \mathcal{O}(\epsilon^2), \qquad (4.65)$$

which leads to an infinitesimal exponential suppression factor, in accordance with Eq. (4.63).

Alternatively, for $\tau > 0$, we can evaluate the Fourier backtransformation by evaluating the cut of the photon propagator, along the positive real axis, by bending the Feynman contour into a segment C_{up}, infinitesimally above the real axis, extending from zero to $\omega = \infty \times \exp(i\epsilon)$, and a segment C_{down}, extending from $\omega = \infty \times \exp(-i\epsilon)$, to the origin of the ω plane,

$$G_F(\vec{r} - \vec{r}', \tau > 0) = \int_{C_{\text{up}}} \frac{d\omega}{2\pi} \frac{e^{i\omega \frac{|\vec{r}-\vec{r}'|}{c} - i\omega\tau - \epsilon\omega}}{4\pi\epsilon_0 |\vec{r} - \vec{r}'|} - \int_{C_{\text{down}}} \frac{d\omega}{2\pi} \frac{e^{-i\omega \frac{|\vec{r}-\vec{r}'|}{c} - i\omega\tau - \epsilon\omega}}{4\pi\epsilon_0 |\vec{r} - \vec{r}'|}$$

$$= \int_0^\infty \frac{d\omega}{2\pi} \frac{e^{\left(i\frac{|\vec{r}-\vec{r}'|}{c} - i\tau - \epsilon\right)\omega}}{4\pi\epsilon_0 |\vec{r} - \vec{r}'|} - \int_0^\infty \frac{d\omega}{2\pi} \frac{e^{\left(-i\frac{|\vec{r}-\vec{r}'|}{c} - i\tau - \epsilon\right)\omega}}{4\pi\epsilon_0 |\vec{r} - \vec{r}'|}$$

$$= \frac{1}{8\pi^2\epsilon_0 |\vec{r} - \vec{r}'|} \left(-\frac{ic}{-|\vec{r} - \vec{r}'| + c\tau - i\epsilon} + \frac{ic}{|\vec{r} - \vec{r}'| + c\tau - i\epsilon} \right)$$

$$= -i \frac{c}{4\pi^2\epsilon_0} \frac{1}{c^2\tau^2 - (\vec{r} - \vec{r}')^2 - i\epsilon \operatorname{sgn}(\tau)}$$

$$= -i \frac{c}{4\pi^2\epsilon_0} \frac{1}{c^2\tau^2 - (\vec{r} - \vec{r}')^2 - i\epsilon}, \qquad (4.66)$$

where the last transformation holds because we have assumed $\tau > 0$ in the first place.

Let us also carry out a full Fourier transformation with respect to space and time. With the definition of the Feynman propagator, we have

$$G_F(\vec{r} - \vec{r}', \omega) = \frac{\exp\left(i\sqrt{\omega^2 + i\epsilon} \frac{|\vec{r} - \vec{r}'|}{c}\right)}{4\pi\epsilon_0 |\vec{r} - \vec{r}'|} \qquad (4.67)$$

and with $\vec{\rho} = \vec{r} - \vec{r}'$,

$$G_F(\vec{k}, \omega) = \int d^3(\vec{r} - \vec{r}') \, G_F(\vec{r} - \vec{r}', \omega) \, e^{-i\vec{k}\cdot(\vec{r}-\vec{r}')}$$

$$= \int d^3\rho \, G_F(\vec{\rho}, \omega) \, e^{-i\vec{k}\cdot\vec{\rho}}, \qquad (4.68)$$

with two obvious variable substitutions. Assuming, without loss of generality, that the $\vec{k}$ vector is parallel to the z axis, we can perform the calculation as follows,

$$G_F(\vec{k}, \omega) = \int d^3r \, G_F(\vec{r}, \omega) \, e^{-i\vec{k}\cdot\vec{r}}$$

$$= 2\pi \int_0^\infty dr \, r^2 \int_{-1}^1 du \, \frac{\exp\left(i\sqrt{\omega^2 + i\epsilon}\,\frac{r}{c}\right)}{4\pi\epsilon_0 r} \, e^{-ikru}$$

$$= \frac{1}{\epsilon_0} \int_0^\infty dr \, \exp\left(i\sqrt{\omega^2 + i\epsilon}\,\frac{r}{c}\right) \frac{\sin(kr)}{k} = \frac{1}{\epsilon_0} \frac{1}{k^2 - \omega^2/c^2 - i\epsilon}. \quad (4.69)$$

The poles of the Fourier transform of the Feynman Green function are thus seen to be at $\omega = c|\vec{k}| - i\epsilon$ and at $\omega = -c|\vec{k}| + i\epsilon$, consistent with Fig. 4.2.

4.1.5 *Summary: Green Functions of Electrodynamics*

A summary of relevant formulas for the retarded, advanced, and Feynman Green functions appears to be in order. All of these Green functions fulfill Eq. (4.9), which reads

$$\left(\frac{1}{c^2}\frac{\partial^2}{\partial t^2} - \vec{\nabla}^2\right) G(\vec{r} - \vec{r}', t - t') = \frac{1}{\epsilon_0} \delta^{(3)}(\vec{r} - \vec{r}') \, \delta(t - t'). \quad (4.70)$$

According to Eq. (4.42), the advanced Green function G_A reads

$$G_A(\vec{r} - \vec{r}', t - t') = \frac{c}{4\pi\epsilon_0} \frac{1}{|\vec{r} - \vec{r}'|} \Theta(t'-t) \left\{ \delta\left(|\vec{r}-\vec{r}'| - c(t'-t)\right) - \delta\left(|\vec{r}-\vec{r}'| + c(t'-t)\right) \right\}. \quad (4.71)$$

Its Fourier transform with respect to time is

$$G_A(\vec{r} - \vec{r}', \omega) = \frac{e^{-i\omega|\vec{r}-\vec{r}'|/c}}{4\pi\epsilon_0|\vec{r} - \vec{r}'|}, \quad (4.72)$$

and the full Fourier transform into frequency-wave-number space is

$$G_A(\vec{k}, \omega) = \frac{c}{2|\vec{k}|\epsilon_0} \left(\frac{1}{\omega + c|\vec{k}| - i\epsilon} - \frac{1}{\omega - c|\vec{k}| - i\epsilon}\right) = \frac{1}{\epsilon_0} \frac{1}{k^2 - \omega^2/c^2 - i\epsilon\,\mathrm{sgn}(\omega)}. \quad (4.73)$$

This result is immediately obvious from Fig. 4.1. The retarded Green Function G_R can be found in Eq. (4.35),

$$G_R(\vec{r} - \vec{r}', t - t') = \frac{c}{4\pi\epsilon_0} \frac{1}{|\vec{r} - \vec{r}'|} \Theta(t-t') \left\{ \delta\left(|\vec{r}-\vec{r}'| - c(t-t')\right) - \delta\left(|\vec{r}-\vec{r}'| + c(t-t')\right) \right\}. \quad (4.74)$$

Its Fourier transform is

$$G_R(\vec{r} - \vec{r}', \omega) = \frac{e^{i\omega|\vec{r}-\vec{r}'|/c}}{4\pi\epsilon_0|\vec{r} - \vec{r}'|}. \quad (4.75)$$

The full Fourier transform into frequency-wave-number space is

$$G_R(\vec{k},\omega) = \frac{c}{2|\vec{k}|\,\epsilon_0} \left(\frac{1}{\omega + c|\vec{k}| + i\epsilon} - \frac{1}{\omega - c|\vec{k}| + i\epsilon} \right) = \frac{1}{\epsilon_0} \frac{1}{\vec{k}^2 - \omega^2/c^2 + i\epsilon\,\mathrm{sgn}(\omega)}\,. \quad (4.76)$$

The Feynman propagator G_F [see Eq. (4.52)] reads

$$G_F(\vec{r} - \vec{r}', t - t') = -i\,\frac{c}{4\pi^2\epsilon_0}\,\frac{1}{c^2\,(t - t')^2 - (\vec{r} - \vec{r}')^2 - i\epsilon}\,, \quad (4.77)$$

and its Fourier transform is

$$G_F(\vec{r} - \vec{r}', \omega) = \frac{\exp\left(i\,\sqrt{\omega^2 + i\epsilon}\,\dfrac{|\vec{r} - \vec{r}'|}{c} \right)}{4\pi\epsilon_0\,|\vec{r} - \vec{r}'|}\,, \qquad \mathrm{Im}\sqrt{\omega^2 + i\epsilon} > 0\,. \quad (4.78)$$

The latter condition clarifies how the branch cuts of the square root should be defined. The full Fourier transform of the Feynman Green function is

$$G_F(\vec{k},\omega) = \frac{c}{2|\vec{k}|\,\epsilon_0} \left(\frac{1}{\omega + c|\vec{k}| - i\epsilon} - \frac{1}{\omega - c|\vec{k}| + i\epsilon} \right) = \frac{1}{\epsilon_0} \frac{1}{\vec{k}^2 - \omega^2/c^2 - i\epsilon}\,. \quad (4.79)$$

The Feynman Green function is important in field-theoretical calculations [20, 23].

4.2 Action-at-a-Distance and Coulomb Gauge

4.2.1 *Potentials and Sources*

It is generally acknowledged that some mystery surrounds the so-called action-at-a-distance solution for the scalar potential which can be obtained in the Coulomb, or radiation, gauge. Due to the gauge condition in radiation gauge, the divergence of the vector potential vanishes, or, expressed differently, the vector potential is equal to its own transverse component [see Eq. (1.87)],

$$\vec{\nabla} \cdot \vec{A}_C\,(\vec{r},t) = 0\,, \qquad \vec{A}_C\,(\vec{r},t) = \vec{A}_{C\perp}\,(\vec{r},t)\,. \quad (4.80)$$

In the following, we distinguish the scalar and vector potentials by the subscripts C for Coulomb gauge, and L for Lorenz gauge. In Coulomb gauge, the coupling to the sources is governed by the following equations [see Eq. (1.92)],

$$\vec{\nabla}^2 \Phi_C\,(\vec{r},t) = -\frac{1}{\epsilon_0}\rho\,(\vec{r},t)\,, \quad (4.81a)$$

$$\left(\frac{1}{c^2} \frac{\partial^2}{\partial t^2} - \vec{\nabla}^2 \right) \vec{A}_{C\perp}\,(\vec{r},t) = \mu_0\,\vec{J}_\perp\,(\vec{r},t)\,, \quad (4.81b)$$

$$\epsilon_0 \frac{\partial}{\partial t}\vec{\nabla}\Phi_C\,(\vec{r},t) = \vec{J}_\parallel(\vec{r},t)\,. \quad (4.81c)$$

We recall that the longitudinal and transverse components of a general vector field are analyzed in Eqs. (1.48a) and (1.48b). Equation (4.81a) couples the electrostatic potential to the source; it has the instantaneous, action-at-a-distance solution given in Eq. (1.98), which we recall for convenience,

$$\Phi_C(\vec{r},t) = \frac{1}{4\pi\epsilon_0} \int d^3r' \frac{1}{|\vec{r}-\vec{r}'|} \rho(\vec{r}',t) + \Phi_{\text{hom}}(\vec{r},t), \qquad (4.82)$$

where $\Phi_{\text{hom}}(\vec{r},t)$ is a solution to the homogeneous equation. The instantaneous coupling of the electrostatic potential to the charge density gives rise to a number of concerns, which we have already discussed near the end of Chap. 1.2.5. Our task is to show that the instantaneous character of the solution (1.98) does not lead to a contradiction with respect to the causality principle; an electromagnetic signal cannot travel faster than light and the fields (as opposed to the potentials) should be manifestly retarded.

The Lorenz condition for the scalar potential Φ and the vector potential $\vec{A}$ has been given in Eq. (1.70),

$$\vec{\nabla} \cdot \vec{A}_L(\vec{r},t) + \frac{1}{c^2}\frac{\partial}{\partial t}\Phi_L(\vec{r},t) = 0. \qquad (4.83)$$

In Lorenz gauge, the scalar and vector potentials are coupled to the sources by inhomogeneous wave equations [see Eq. (1.77)],

$$\left(\frac{1}{c^2}\frac{\partial^2}{\partial t^2} - \vec{\nabla}^2\right)\Phi_L(\vec{r},t) = \frac{1}{\epsilon_0}\rho(\vec{r},t), \qquad (4.84a)$$

$$\left(\frac{1}{c^2}\frac{\partial^2}{\partial t^2} - \vec{\nabla}^2\right)\vec{A}_L(\vec{r},t) = \mu_0\vec{J}(\vec{r},t). \qquad (4.84b)$$

We here provide two perspectives on the problem. The first perspective (see Sec. 4.2.2) relies on the fact that the solution of Eq. (4.81a) might otherwise indicate that the scalar potential in Coulomb gauge is non-retarded. However, once charges move, currents are generated in view of the continuity equation. The seemingly instantaneous scalar potential in Coulomb gauge is connected to the longitudinal part of the current density by Eq. (4.81c). It is instructive to observe that Eq. (4.81c) cannot alone provide the solution of the problem; its divergence simply is the time derivative of Eq. (4.81a). However, Eq. (4.81c) shows that the situation is not so easy: We cannot argue that the seemingly instantaneous Coulomb interaction automatically gives rise to instantaneous fields; there is the additional condition (4.81c) which has to be fulfilled by the scalar potential. We shall see that the vector potential, in the Coulomb gauge, receives an additional contribution as compared to the Lorenz gauge. The additional term corresponds to the longitudinal part of the current density which has to be subtracted in order to obtain Eq. (4.81b). With the help of Eq. (4.81c), we are finally able to show that the supplementary term in the vector potential, in the Coulomb gauge, cancels the instantaneous contribution to the electric field and leads to manifestly retarded expressions.

Expressed differently, the additional constraint (4.81c) implies that the action-at-a-distance solution (4.82) is not universally a valid solution; it does not

automatically fulfill Eq. (4.81c). Indeed, we shall see that the homogeneous term
in Eq. (4.82) plays a crucial role in showing causality, together with Eq. (4.81c).

The second perspective (see Sec. 4.2.3) addresses the fact that the instantaneous
interaction integral can alternatively be written as a retarded integral, but with a
different source term, in accordance with arguments presented in Ref. [24]. Finally,
in Sec. 4.2.4 (third perspective), we find an explicitly retarded expression for the
Coulomb-gauge scalar potential, which fulfills both Eqs. (4.81a) as well as (4.81c).

4.2.2   *Cancellation of the Instantaneous Term*

According to Eq. (1.59), the electric field (in Coulomb gauge) is given as

$$\vec{E}(\vec{r},t) = -\vec{\nabla}\Phi_C(\vec{r},t) - \frac{\partial}{\partial t}\vec{A}_{C\perp}(\vec{r},t), \tag{4.85}$$

and therefore

$$\vec{E}_\parallel(\vec{r},t) = -\vec{\nabla}\Phi_C(\vec{r},t), \qquad \vec{E}_\perp(\vec{r},t) = -\frac{\partial}{\partial t}\vec{A}_{C\perp}(\vec{r},t). \tag{4.86}$$

These components fulfill, explicitly, $\vec{\nabla} \times \vec{E}_\parallel(\vec{r},t) = -(\vec{\nabla} \times \vec{\nabla})\,\Phi_C(\vec{r},t) = \vec{0}$, and
$\vec{\nabla} \cdot \vec{E}_\perp(\vec{r},t) = -\frac{\partial}{\partial t}\vec{\nabla} \cdot \vec{A}_{C\perp}(\vec{r},t) = 0$. Therefore, the full longitudinal component
of the electric field is given by $\vec{E}_\parallel(\vec{r},t) = -\vec{\nabla}\Phi_C(\vec{r},t)$, and the transverse component
in Coulomb gauge is purely given by the time derivative of the transverse vector
potential, without any "admixture" from the scalar potential. In Coulomb gauge,
the longitudinal component of the electric field is calculated as

$$\vec{E}_\parallel(\vec{r},t) = -\vec{\nabla}\Phi_C(\vec{r},t) = -\frac{1}{4\pi\epsilon_0}\vec{\nabla}\int d^3r' \frac{1}{|\vec{r}-\vec{r}'|}\rho(\vec{r}',t) - \vec{\nabla}\Phi_{\mathrm{hom}}(\vec{r},t), \tag{4.87}$$

where Φ_{hom} is the solution to the homogeneous equation, adjusted so that
Eq. (4.81c) is fulfilled.

The retarded Green function (4.35) enters the solutions of the Lorenz-gauge
couplings given in Eqs. (4.84a) and (4.84b),

$$\Phi_L(\vec{r},t) = \int d^3r' \, dt' \, G_R(\vec{r}-\vec{r}',t-t')\,\rho(\vec{r}',t'), \tag{4.88a}$$

$$\vec{A}_L(\vec{r},t) = \frac{1}{c^2}\int d^3r' \, dt' \, G_R(\vec{r}-\vec{r}',t-t')\,\vec{J}(\vec{r}',t'). \tag{4.88b}$$

As has been stressed in Sec. 1.2.4, the potentials are not uniquely defined even
within the family of potentials that fulfill the Coulomb gauge condition; a gauge
re-transformation within the Coulomb gauge is possible according to Eq. (1.93). A
permissible way to proceed is to use the retarded Green function to solve Eq. (4.81b),
and to find the vector potential $\vec{A}_C(\vec{r},t) = \vec{A}_{C\perp}(\vec{r},t)$. The solution $\vec{A}_C(\vec{r},t)$ can be

written in terms of the Lorenz-gauge expression $\vec{A}_L(\vec{r}, t)$ and the supplementary term $\vec{A}_S(\vec{r}, t)$. We have,

$$\vec{A}_C(\vec{r}, t) = \frac{1}{c^2} \int d^3 r' \int dt'\, G_R(\vec{r} - \vec{r}', t - t')\, \vec{J}_\perp(\vec{r}', t') = \vec{A}_L(\vec{r}, t) + \vec{A}_S(\vec{r}, t),$$
$$(4.89a)$$

$$\vec{A}_L(\vec{r}, t) = \frac{1}{c^2} \int d^3 r' \int dt'\, G_R(\vec{r} - \vec{r}', t - t')\, \vec{J}(\vec{r}', t'), \qquad (4.89b)$$

$$\vec{A}_S(\vec{r}, t) = -\frac{1}{c^2} \int d^3 r' \int dt'\, G_R(\vec{r} - \vec{r}', t - t')\, \vec{J}_\parallel(\vec{r}', t'), \qquad (4.89c)$$

where we have used the identity $\vec{J}_\perp(\vec{r}', t') = \vec{J}(\vec{r}', t') - \vec{J}_\parallel(\vec{r}', t')$. The supplementary term $\vec{A}_S(\vec{r}, t)$ is relevant to the Coulomb gauge and reads

$$\vec{A}_S(\vec{r}, t) = -\frac{1}{c^2} \int d^3 r' \int dt'\, G_R(\vec{r} - \vec{r}', t - t')\, \vec{J}_\parallel(\vec{r}', t'). \qquad (4.90)$$

The time derivative of the supplementary term $\vec{A}_S(\vec{r}, t)$ contributes a supplementary term $\vec{E}_S$ to the electric field,

$$\vec{E}_S(\vec{r}, t) = -\frac{\partial}{\partial t} \vec{A}_S(\vec{r}, t) = \frac{1}{c^2} \int d^3 r' \int dt' \left[\frac{\partial}{\partial t} G_R(\vec{r} - \vec{r}', t - t') \right] \vec{J}_\parallel(\vec{r}', t')$$

$$= -\frac{1}{c^2} \int d^3 r' \int dt' \left[\frac{\partial}{\partial t'} G_R(\vec{r} - \vec{r}', t - t') \right] \vec{J}_\parallel(\vec{r}', t')$$

$$= \frac{1}{c^2} \int d^3 r' \int dt'\, G_R(\vec{r} - \vec{r}', t - t') \left[\frac{\partial}{\partial t'} \vec{J}_\parallel(\vec{r}', t') \right], \qquad (4.91)$$

where we have first transformed $\partial/\partial t \to -\partial/\partial t'$ and then used integration by parts to move the derivative on the current. Using Eq. (4.81c), this can be rewritten in terms of the potential as

$$\vec{E}_S(\vec{r}, t) = \epsilon_0 \int d^3 r' \int dt'\, G_R(\vec{r} - \vec{r}', t - t')\, \vec{\nabla}' \left[\frac{1}{c^2} \frac{\partial^2}{\partial t'^2} \Phi_C(\vec{r}', t') \right]$$

$$= \epsilon_0 \vec{\nabla} \int d^3 r' \int dt'\, G_R(\vec{r}, t, \vec{r}', t') \left[\left(\frac{1}{c^2} \frac{\partial^2}{\partial t'^2} - \vec{\nabla}'^2 \right) + \vec{\nabla}'^2 \right] \Phi_C(\vec{r}', t')$$

$$= \epsilon_0 \vec{\nabla} \int d^3 r' \int dt' \left[\left(\frac{1}{c^2} \frac{\partial^2}{\partial t^2} - \vec{\nabla}^2 \right) G_R(\vec{r}, t, \vec{r}', t') \right] \Phi_C(\vec{r}', t')$$

$$+ \epsilon_0 \vec{\nabla} \int d^3 r' \int dt'\, G_R(\vec{r}, t, \vec{r}', t')\, \vec{\nabla}'^2 \Phi_C(\vec{r}', t'). \qquad (4.92)$$

For the first term, we use partial integration twice, and we also take advantage of the symmetry properties of the retarded Green function. With Eqs. (4.81a) and (4.9), we find

$$
\vec{E}_S(\vec{r},t) = \epsilon_0 \vec{\nabla} \int \mathrm{d}^3 r' \int \mathrm{d}t' \, \frac{1}{\epsilon_0} \delta^{(3)}(\vec{r}-\vec{r}') \, \delta(t-t') \, \Phi_C(\vec{r}',t')
$$
$$
- \epsilon_0 \vec{\nabla} \int \mathrm{d}^3 r' \int \mathrm{d}t' \, G_R(\vec{r},t,\vec{r}',t') \, \frac{1}{\epsilon_0} \rho(\vec{r}',t')
$$
$$
= \vec{\nabla} \Phi_C(\vec{r},t) - \vec{\nabla} \int \mathrm{d}^3 r' \int \mathrm{d}t' \, G_R(\vec{r},t,\vec{r}',t') \, \rho(\vec{r}',t')
$$
$$
= \vec{\nabla} \Phi_C(\vec{r},t) - \vec{\nabla} \Phi_L(\vec{r},t) \,. \tag{4.93}
$$

In Coulomb gauge, the supplementary term $\vec{E}_S = -\partial_t \vec{A}_S$ due to the time derivative of the supplementary vector potential cancels the gradient of the Coulomb gauge scalar potential and adds the Lorenz gauge gradient of the scalar potential. Comparing the formulas for the electric field in Coulomb and Lorenz gauge, the following identity follows immediately,

$$
\vec{E}_C(\vec{r},t) = -\vec{\nabla}\Phi_C(\vec{r},t) - \frac{\partial}{\partial t}\vec{A}_C(\vec{r},t) = -\vec{\nabla}\Phi_C(\vec{r},t) - \frac{\partial}{\partial t}\left(\vec{A}_L(\vec{r},t) + \vec{A}_S(\vec{r},t)\right)
$$
$$
= -\vec{\nabla}\Phi_C(\vec{r},t) - \frac{\partial}{\partial t}\vec{A}_L(\vec{r},t) + \left(\vec{\nabla}\Phi_C(\vec{r},t) - \vec{\nabla}\Phi_L(\vec{r},t)\right)
$$
$$
= -\vec{\nabla}\Phi_L(\vec{r},t) - \frac{\partial}{\partial t}\vec{A}_L(\vec{r},t) = \vec{E}_L(\vec{r},t) \,. \tag{4.94}
$$

We have temporarily denoted the "Coulomb gauge" electric field as $\vec{E}_C(\vec{r},t)$ and the "Lorenz gauge" electric field as $\vec{E}_L(\vec{r},t)$, even if both are actually equal due to gauge invariance, as shown.

Let us briefly summarize: In Coulomb gauge, there is an additional term $\vec{A}_S$ in the vector potential which is generated by the negative of the longitudinal component of the current density. The (negative of the) time derivative of the supplementary term in the vector potential yields an additional contribution to the electric field, in Coulomb gauge. The additional term in the electric field can be transformed into two parts, the first of which cancels the seemingly instantaneous electric field contribution in Coulomb gauge, obtained from the Coulomb-gauge electric potential, and the second yields the same result (the retarded one) as the gradient of the electric potential Φ_L in Lorenz gauge. In the end, the action-at-a-distance integral cancels, and the gauge invariance of the electric field is shown [Eq. (4.94)]. The overall conclusion is that in Coulomb gauge, in view of the condition (4.81c), the homogeneous solution Φ_{hom} in Eq. (4.82) has to be chosen so that the contribution of the action-at-a-distance term to the electric field cancels.

4.2.3 *Longitudinal Electric Field as a Retarded Integral*

We recall once more Eq. (4.87), which for the longitudinal component of the electric field reads as follows,

$$
\vec{E}_\parallel(\vec{r},t) = -\vec{\nabla}\Phi_C(\vec{r},t) = -\frac{1}{4\pi\epsilon_0}\vec{\nabla}\int \mathrm{d}^3 r' \, \frac{1}{|\vec{r}-\vec{r}'|}\rho(\vec{r}',t) - \vec{\nabla}\Phi_{\mathrm{hom}}(\vec{r},t) \,. \tag{4.95}
$$

The first term in this expression has an action-at-a-distance form, which could in principle lead to a contradiction with respect to the causality principle, were it not for the additional constraint (4.81c), which implies the necessity of adding a suitable solution of the homogeneous equation. We recall that the decomposition of the electric field into longitudinal and transverse components is unique; an instantaneous character of the longitudinal component also would have disastrous consequences. We should thus investigate if, taking into account Eq. (4.81c), the longitudinal component of the electric field can alternatively be written as a manifestly retarded integral, for which we guess the form [24]

$$\vec{E}_{\parallel}(\vec{r},t) = - \int \mathrm{d}^3 r' \int \mathrm{d}t' \, G_R(\vec{r}-\vec{r}',t-t') \left(\frac{1}{c^2} \frac{\partial}{\partial t'} \vec{J}_{\parallel}(\vec{r}',t') + \vec{\nabla}' \rho(\vec{r}',t') \right). \quad (4.96)$$

In this expression, we use Eq. (4.81c) in order to substitute for $\vec{J}_{\parallel}$ and Eq. (4.81a) in order to substitute for $\rho(\vec{r}',t')$,

$$\vec{E}_{\parallel}(\vec{r},t) = - \int \mathrm{d}^3 r' \int \mathrm{d}t' G_R(\vec{r}-\vec{r}',t-t') \left[\frac{1}{c^2} \frac{\partial}{\partial t'} \left(\epsilon_0 \vec{\nabla}' \frac{\partial}{\partial t'} \Phi_C(\vec{r}',t') \right) \right.$$

$$\left. + \vec{\nabla}' \left(-\epsilon_0 \vec{\nabla}'^2 \Phi_C(\vec{r}',t') \right) \right]$$

$$= - \int \mathrm{d}^3 r' \int \mathrm{d}t' \, G_R(\vec{r},t,\vec{r}',t') \left(\frac{1}{c^2} \frac{\partial^2}{\partial t'^2} - \vec{\nabla}'^2 \right) \epsilon_0 \vec{\nabla}' \Phi_C(\vec{r}',t'). \quad (4.97)$$

A double partial integration and use of Eq. (4.9) leads to the relation

$$\vec{E}_{\parallel}(\vec{r},t) = - \int \mathrm{d}^3 r' \int \mathrm{d}t' \, \delta^{(3)}(\vec{r}-\vec{r}') \, \delta(t-t') \, \vec{\nabla}' \Phi_C(\vec{r}',t) = -\vec{\nabla} \Phi_C(\vec{r},t), \quad (4.98)$$

which was to be shown.

In summary, we have demonstrated that the instantaneous integral for the longitudinal part of the electric field (4.87) can be rewritten as an integral involving the manifestly retarded Green function, with a nonstandard source term that does not only involve the charge density but also the longitudinal part of the current density. In the derivation, we have used Eq. (4.81c) which relates the charge density to the longitudinal part of the current density in Coulomb gauge.

4.2.4 Coulomb-Gauge Scalar Potential as a Retarded Integral

The last step in the analysis of the Coulomb gauge entails the calculation of the scalar potential, which is tantamount to finding an explicit expression for the homogeneous term Φ_{hom} in Eq. (4.82). We start from Eq. (4.96), which we recall,

$$\vec{E}_{\parallel}(\vec{r},t) = - \int \mathrm{d}^3 r' \int \mathrm{d}t' \, G_R(\vec{r}-\vec{r}',t-t') \left(\frac{1}{c^2} \frac{\partial}{\partial t'} \vec{J}_{\parallel}(\vec{r}',t') + \vec{\nabla}' \rho(\vec{r}',t') \right). \quad (4.99)$$

Here, in view of Eq. (1.48a), we can write the longitudinal component of the current density as the gradient of a scalar field $\mathcal{J}(\vec{r}',t')$,

$$\vec{J}_{\parallel}(\vec{r}',t') = \vec{\nabla}' \mathcal{J}(\vec{r}',t'). \quad (4.100)$$

So, the longitudinal component of the electric field becomes

$$\vec{E}_{\parallel}(\vec{r},t) = -\int d^3r' \int dt' \, G_R(\vec{r}-\vec{r}',t-t') \, \vec{\nabla}' \left(\frac{1}{c^2} \frac{\partial}{\partial t'} \mathcal{J}(\vec{r}',t') + \rho(\vec{r}',t') \right)$$

$$= \int d^3r' \int dt' \, \vec{\nabla}' G_R(\vec{r}-\vec{r}',t-t') \left(\frac{1}{c^2} \frac{\partial}{\partial t'} \mathcal{J}(\vec{r}',t') + \rho(\vec{r}',t') \right)$$

$$= -\vec{\nabla} \int d^3r' \int dt' \, G_R(\vec{r}-\vec{r}',t-t') \left(\frac{1}{c^2} \frac{\partial}{\partial t'} \mathcal{J}(\vec{r}',t') + \rho(\vec{r}',t') \right). \quad (4.101)$$

Because $\vec{E}_{\parallel}(\vec{r},t) = -\vec{\nabla}\Phi_C(\vec{r},t)$, a valid *ansatz* for the scalar potential is

$$\Phi_C(\vec{r},t) = \int dt' \int d^3r' \, G_R(\vec{r}-\vec{r}',t-t') \left(\frac{1}{c^2} \frac{\partial}{\partial t'} \mathcal{J}(\vec{r}',t') + \rho(\vec{r}',t') \right), \quad (4.102)$$

where $\mathcal{J}$ is given in Eq. (4.100).

The decisive step is to show that our *ansatz* (4.102) fulfills Eq. (4.81c),

$$\epsilon_0 \frac{\partial}{\partial t} \vec{\nabla}\Phi_C(\vec{r},t) = \vec{J}_{\parallel}(\vec{r},t). \quad (4.103)$$

The proof is relatively straightforward. One first integrates by parts,

$$\epsilon_0 \frac{\partial}{\partial t} \vec{\nabla}\Phi_C(\vec{r},t) = \epsilon_0 \frac{\partial}{\partial t} \vec{\nabla} \int dt' \int d^3r' G_R(\vec{r}-\vec{r}',t-t') \left(\frac{1}{c^2} \frac{\partial}{\partial t'} \mathcal{J}(\vec{r}',t') + \rho(\vec{r}',t') \right)$$

$$= \epsilon_0 \int dt' \int d^3r' \, G_R(\vec{r}-\vec{r}',t-t')$$

$$\times \left(\frac{1}{c^2} \frac{\partial^2}{\partial t'^2} \vec{\nabla}' \mathcal{J}(\vec{r}',t') + \frac{\partial}{\partial t'} \vec{\nabla}' \rho(\vec{r}',t') \right). \quad (4.104)$$

Use of Eq. (4.102) and of the continuity equation leads to

$$\epsilon_0 \, \partial_t \vec{\nabla}\Phi_C(\vec{r},t) = \epsilon_0 \int dt' \int d^3r' G_R(\vec{r}-\vec{r}',t-t') \left(\frac{1}{c^2} \frac{\partial^2}{\partial t'^2} \vec{J}_{\parallel}(\vec{r}',t') + \vec{\nabla}' \frac{\partial}{\partial t'} \rho(\vec{r}',t') \right)$$

$$= \epsilon_0 \int dt' \int d^3r' \, G_R(\vec{r}-\vec{r}',t-t')$$

$$\times \left(\frac{1}{c^2} \frac{\partial^2}{\partial t'^2} \vec{J}_{\parallel}(\vec{r}',t') + \vec{\nabla}' \left(-\vec{\nabla}' \cdot \vec{J}_{\parallel}(\vec{r}',t') \right) \right). \quad (4.105)$$

This can be summarized as follows,

$$\epsilon_0 \, \partial_t \vec{\nabla}\Phi_C(\vec{r},t) = \epsilon_0 \int dt' \int d^3r' \, G_R(\vec{r}-\vec{r}',t-t') \left(\frac{1}{c^2} \frac{\partial^2}{\partial t'^2} - \vec{\nabla}'^2 \right) \vec{J}_{\parallel}(\vec{r}',t'). \quad (4.106)$$

After a double partial integration, and use of Eq. (4.9), one can finally show that Eq. (4.102) fulfills Eq. (4.103). Because Eq. (4.102) is manifestly retarded, we have explicitly shown that the particular form of the scalar potential in Coulomb gauge, given in Eq. (4.102), does not lead to a contradiction with respect to the causality principle; the additional constraint (4.81c) ensures that the scalar potential is manifestly retarded.

4.3 Exercises

• **Exercise 4.1**: Show that all contour choices C_R, C_A and C_F shown in Figs. 4.1 and 4.2 lead to valid representations of Green functions that fulfill Eq. (4.9). Hint: Apply the differential operator $(1/c^2)\partial^2/\partial t^2 - \vec{\nabla}^2$ under the integral sign, invoking a Fourier representation of the Green function.

• **Exercise 4.2**: Generalize the steps outlined in Eqs. (4.29)–(4.35), for the advanced as opposed to the retarded Green function, using the contour C_A given in Fig. 4.1. Thus, verify the results given in Eqs. (4.41)–(4.44).

• **Exercise 4.3**: Generalize Eq. (4.66) to the case $\tau < 0$. How would you have to deform the integration contour in this case in order to identify the cut (in the complex plane) of the photon propagator?

• **Exercise 4.4**: Pick some $\psi(\vec{r}, t) = \cos(\omega_0 t)/(\vec{r}^2 + a^2)^2$ and $\psi(\vec{r}, t) = \delta(t - t_0)/(\vec{r}^2 + a^2)^2$ and propagate these wave packets using (a) the retarded (b) the advanced, and (c) the Feynman Green function. The dynamical equation is (4.1).

• **Exercise 4.5**: Show that the supplementary term in the vector potential, defined in Eqs. (4.89a) and (4.90), can be written as a gradient vector and therefore does not affect the result for the magnetic field, which thus is the same in Coulomb and Lorenz gauges.

Chapter 5

Paradigmatic Calculations in Electrodynamics

5.1 Overview

5.1.1 *General Considerations*

In electrodynamics as opposed to electrostatics, the emphasis is on the radiation emitted by moving charge distributions, i.e., current distributions. In practically important cases, the source oscillates at a specific frequency. Yet, the radiated fields are needed at specific points in space, and one can assume that the radiation is emitted continuously. In that case, it makes sense to transform the Green function to the mixed frequency-coordinate representation. In this representation, the defining equation for the Green function becomes the so-called Helmholtz equation. For classical fields, the radiation pattern from oscillating sources is described by the retarded Green function, which has been given in Eq. (4.75) in the mixed frequency-coordinate representation (see Sec. 4.1.5). In a typical case, the oscillating source (antenna) is a dipole. However, higher-order multipole radiation can also be important. It is thus imperative to expand the Green function of the Helmholtz equation (which is equal to the retarded Green function in the mixed coordinate-frequency representation) into multipoles. Appropriate integrations then lead to the relevant formulas for the radiation pattern.

Finally, when calculating antenna problems, it is usually assumed that the antenna (oscillating charge distribution) remains static during the emission process, i.e., it does not move. In Secs. 2.3.4 and 2.3.5, we encountered the multipole decomposition of the electrostatic Green function, which led to the multipole decomposition of the electrostatic potential generated by a static charge distribution [Eq. (2.99)]. The Helmholtz Green function fulfills a defining equation with a scalar structure [see Eq. (5.19)]. In consequence, the angular momentum decomposition of the Helmholtz Green function given in Eq. (5.72) involves the spherical harmonics, which are scalar (not tensor) quantities. Indeed, in antenna problems, the equation which relates the scalar potential to the oscillating charge distribution (5.77) retains a scalar structure. By contrast, the equation which relates the current density to the vector potential [Eq. (5.78)] has a vector structure, and the Green function, strictly speaking, therefore is promoted to a tensor Green function. This is not obvious from Eq. (5.78), but becomes more evident as one considers Eq. (5.158). The tensor structure of the Helmholtz Green function necessitates an analysis of the angular algebra which goes beyond the scalar structure of the scalar Helmholtz Green function; the "spin" of the photon (the vector structure of the vector potential) needs to be integrated into the formalism. This endeavor proceeds via the introduction of vector spherical harmonics (Sec. 5.4.2).

By contrast, the emission of radiation by a moving point charge constitutes a complementary problem where the charge distribution is trivial but the radiation is due to the motion of the charge relative to the observer. The scalar and vector potentials due to a moving point charge are otherwise known as the LIÉNARD[1]– WIECHERT[2] potentials, which can be given in Lorenz and Coulomb gauges. It is instructive to go through the alternative derivations because the final formulas reveal a concrete realization of the cancellation mechanism originally described in Sec. 4.2.

5.1.2 *Wave Equation and Green Functions*

We consider how general solutions to the wave equation can be obtained using the retarded Green function. We integrate the source term directly (no partial integrations) and may thus use the approximate simplified version of the retarded Green function already given in Eq. (4.36). This is admissible because no additional partial integrations need to be carried out. We recall the wave equation (4.1) with sources, which has the general form

$$\left(\frac{1}{c^2}\frac{\partial^2}{\partial t^2} - \vec{\nabla}^2\right)\Psi(\vec{r},t) = \frac{1}{\epsilon_0}F(\vec{r},t)\,. \tag{5.1}$$

Here, Ψ is the signal generated by the source F. In the approximate form (4.36), the retarded Green function is given by

$$G_R(\vec{r}-\vec{r}',t-t') \approx \frac{c}{4\pi\epsilon_0}\frac{1}{|\vec{r}-\vec{r}'|}\,\Theta(t-t')\,\delta(|\vec{r}-\vec{r}'|-c(t-t'))\,. \tag{5.2}$$

Using the retarded Green function and an arbitrary solution $\Psi_{\text{hom}}(\vec{r},t)$ to the homogeneous wave equation,

$$\left(\frac{1}{c^2}\frac{\partial^2}{\partial t^2} - \vec{\nabla}^2\right)\Psi_{\text{hom}}(\vec{r},t) = 0\,, \tag{5.3}$$

we obtain the general solution to the inhomogeneous wave equation as follows,

$$\Psi(\vec{r},t) = \Psi_{\text{hom}}(\vec{r},t) + \int G_R(\vec{r}-\vec{r}',t-t')\,F(\vec{r}',t')\,\mathrm{d}^3r'\,\mathrm{d}t'$$

$$= \Psi_{\text{hom}}(\vec{r},t) + \frac{1}{4\pi\epsilon_0}\int_{\mathbb{R}^3}\mathrm{d}^3r'\int_{-\infty}^{t}\mathrm{d}t'\,\frac{1}{|\vec{r}-\vec{r}'|}\,\delta\left(t'-\left[t-\frac{|\vec{r}-\vec{r}'|}{c}\right]\right)F(\vec{r}',t')$$

$$= \Psi_{\text{hom}}(\vec{r},t) + \frac{1}{4\pi\epsilon_0}\int\mathrm{d}^3r'\,\frac{1}{|\vec{r}-\vec{r}'|}\,F\left(\vec{r}',t-\frac{|\vec{r}-\vec{r}'|}{c}\right)\,. \tag{5.4}$$

---

[1] Alfred–Marie Liénard (1869–1958)
[2] Emil Johann Wiechert (1861–1928)

Here, the expression

$$t_{\text{ret}} = t - \frac{|\vec{r} - \vec{r}'|}{c} < t \qquad (5.5)$$

is the retarded time, which expresses the fact that the electromagnetic perturbation propagates at the speed of light. Because we have $t_{\text{ret}} < t$ for $\vec{r} \neq \vec{r}'$, the integration over t' from $-\infty$ to t always gives a nonvanishing result.

The same steps, applied to the advanced Green function, lead to

$$\Psi(\vec{r}, t) = \Psi_{\text{hom}}(\vec{r}, t) + \int G_A(\vec{r} - \vec{r}', t - t') \, F(\vec{r}', t') \, \mathrm{d}^3 r' \, \mathrm{d}t'$$

$$= \Psi_{\text{hom}}(\vec{r}, t) + \frac{1}{4\pi\epsilon_0} \int \mathrm{d}^3 r' \, \frac{1}{|\vec{r} - \vec{r}'|} \, F\left(\vec{r}', t + \frac{|\vec{r} - \vec{r}'|}{c}\right), \qquad (5.6)$$

where

$$t_{\text{adv}} = t + \frac{|\vec{r} - \vec{r}'|}{c} > t \qquad (5.7)$$

is the advanced time; this solution describes an electromagnetic signal (or a signal of a different physical origin) that propagates from the future into the past.

5.2 Helmholtz Equation

5.2.1 *Helmholtz Equation and Green Function*

It is rather straightforward to apply the formalism developed in Sec. 5.1.2 to electromagnetic radiation phenomena. A simple radiating system consists of a localized charge density $\rho(\vec{r}, t)$ and a localized current density $\vec{J}(\vec{r}, t)$. The system is considered to be localized if its dimensions are small compared to the wavelength of the radiation. We consider radiating sources in a vacuum and begin with the equations for the vector and scalar potentials in the Lorenz gauge [Eq. (1.77)],

$$\left(\frac{1}{c^2} \frac{\partial^2}{\partial t^2} - \vec{\nabla}^2\right) \Phi(\vec{r}, t) = \frac{1}{\epsilon_0} \, \rho(\vec{r}, t) , \qquad (5.8)$$

$$\left(\frac{1}{c^2} \frac{\partial^2}{\partial t^2} - \vec{\nabla}^2\right) \vec{A}(\vec{r}, t) = \mu_0 \, \vec{J}(\vec{r}, t) . \qquad (5.9)$$

The second of these equations can be written as

$$\left(\frac{1}{c^2} \frac{\partial^2}{\partial t^2} - \vec{\nabla}^2\right) c \vec{A}(\vec{r}, t) = \frac{\mu_0}{c^2} \frac{1}{c} \vec{J}(\vec{r}, t) = \frac{1}{\epsilon_0} \left(\frac{1}{c} \vec{J}(\vec{r}, t)\right) . \qquad (5.10)$$

The latter form is suggested by the relativistic formalism; indeed, one can order the scalar and vector potentials, and the charge and current densities, into 4-vectors as

$$\left(\Phi, c\vec{A}\right) \qquad \text{and} \qquad \left(\rho, \frac{1}{c} \vec{J}\right) . \qquad (5.11)$$

The consistency of the physical units of the components of the first 4-vector can be verified immediately, as follows: The electric field has the dimension of the expression $\vec{\nabla}\Phi$, the magnetic induction field has the dimension of $\vec{\nabla} \times \vec{A}$. For traveling plane waves, one has $|\vec{E}| \sim c|\vec{B}|$, and so, in terms of physical units, the factor c in the four-vector $(\Phi, c\vec{A})$ is explained. The charge density has units of $[\rho] = [Q]/[r]^3 = [Q]/([t] \, [r]^2) \, ([t]/[r]) = [\vec{J}]/[c]$, which explains the units for the components of the second 4-vector given in Eq. (5.11).

We integrate the wave equation as before,

$$\Phi\left(\vec{r},t\right) = \Phi_{\text{hom}}\left(\vec{r},t\right) + \int \mathrm{d}^3 r' \mathrm{d}t' G_R\left(\vec{r} - \vec{r}', t - t'\right) \rho\left(\vec{r}', t'\right) \tag{5.12a}$$

$$c\vec{A}\left(\vec{r},t\right) = c\vec{A}_{\text{hom}}\left(\vec{r},t\right) + \frac{1}{c} \int \mathrm{d}^3 r' \mathrm{d}t' G_R\left(\vec{r} - \vec{r}', t - t'\right) \vec{J}\left(\vec{r}', t'\right). \tag{5.12b}$$

Here, Φ_{hom} and $\vec{A}_{\text{hom}}$ are solutions to the homogeneous wave equation. Using the identity

$$\delta\left(|\vec{r} - \vec{r}| - c\left(t - t'\right)\right) = \frac{1}{c}\,\delta\left(t - t' - \frac{|\vec{r} - \vec{r}'|}{c}\right), \tag{5.13}$$

we may carry out t' integration, and find

$$\Phi\left(\vec{r},t\right) = \Phi_{\text{hom}}\left(\vec{r},t\right) + \frac{1}{4\pi\epsilon_0} \int \mathrm{d}^3 r' \frac{1}{|\vec{r} - \vec{r}'|} \rho\left(\vec{r}', t - \frac{|\vec{r} - \vec{r}'|}{c}\right), \tag{5.14a}$$

$$c\vec{A}\left(\vec{r},t\right) = c\vec{A}_{\text{hom}}\left(\vec{r},t\right) + \frac{1}{4\pi\epsilon_0 c} \int \mathrm{d}^3 r' \frac{1}{|\vec{r} - \vec{r}'|} \vec{J}\left(\vec{r}', t - \frac{|\vec{r} - \vec{r}'|}{c}\right). \tag{5.14b}$$

The latter equation can of course be rewritten as

$$\vec{A}\left(\vec{r},t\right) = \vec{A}_{\text{hom}}\left(\vec{r},t\right) + \int \mathrm{d}^3 r' \underbrace{\left[\frac{1}{\epsilon_0 c^2}\right]}_{=\mu_0} \vec{J}\left(\vec{r}', t - \frac{|\vec{r} - \vec{r}'|}{c}\right) \frac{1}{4\pi|\vec{r} - \vec{r}'|}, \tag{5.15}$$

and we see that the vacuum permeability μ_0 naturally appears in the calculation. The equation fulfilled by the retarded Green function,

$$\left(\frac{1}{c^2}\frac{\partial^2}{\partial t^2} - \vec{\nabla}^2\right) G_R\left(\vec{r} - \vec{r}', t - t'\right) = \frac{1}{\epsilon_0}\delta^{(3)}\left(\vec{r} - \vec{r}'\right)\delta(t - t'), \tag{5.16}$$

transforms to Fourier space as follows,

$$\left(-\frac{\omega^2}{c^2} - \vec{\nabla}^2\right) G_R\left(\vec{r} - \vec{r}', \omega\right) = \frac{1}{\epsilon_0}\delta\left(\vec{r} - \vec{r}'\right). \tag{5.17}$$

Setting $k = \omega/c$, and defining

$$G_R\left(k, \vec{r} - \vec{r}'\right) \equiv G_R\left(\vec{r} - \vec{r}', \omega = ck\right), \tag{5.18}$$

we obtain the (inhomogeneous) Helmholtz equation

$$\left(\vec{\nabla}^2 + k^2\right) G_R(k, \vec{r} - \vec{r}') = -\frac{1}{\epsilon_0}\delta\left(\vec{r} - \vec{r}'\right). \tag{5.19}$$

The "natural parameter" of the Helmholtz equation is the wave number k, not the frequency ω. The frequency argument in the expression $G_R(\vec{r}-\vec{r}', \omega)$ is generated by the Fourier transform of $G_R(\vec{r} - \vec{r}', t - t')$; it is thus natural to write ω in the second argument slot. However, when k becomes a mere parameter of the (inhomogeneous)

Helmholtz equation (5.19), then it is natural to write it as the first argument of the Green function. The Helmholtz equation is solved by the Green function given in Eq. (4.75),

$$G_R(k, \vec{r} - \vec{r}') = \frac{e^{i\,k\,|\vec{r} - \vec{r}'|}}{4\pi\epsilon_0\,|\vec{r} - \vec{r}'|}. \tag{5.20}$$

The electrostatic Green function G defined in Eq. (2.22) is recovered as the static limit of the electrodynamic Green function,

$$\lim_{\omega\to 0} G_R(\vec{r} - \vec{r}', \omega) = \lim_{k\to 0} G_R(k, \vec{r} - \vec{r}') = \frac{1}{4\pi\epsilon_0|\vec{r} - \vec{r}'|} = G(\vec{r}, \vec{r}'). \tag{5.21}$$

In Fourier space, we have by convolution,

$$\Phi(\vec{r}, \omega) = \int d^3r'\, G_R\left(\frac{\omega}{c}, \vec{r} - \vec{r}'\right) \rho(\vec{r}', \omega) = \int d^3r'\, \frac{e^{i\omega|\vec{r} - \vec{r}'|/c}}{4\pi\epsilon_0\,|\vec{r} - \vec{r}'|} \rho(\vec{r}', \omega), \tag{5.22a}$$

$$c\,\vec{A}(\vec{r}, \omega) = \frac{1}{c}\int d^3r'\, G_R\left(\frac{\omega}{c}, \vec{r} - \vec{r}'\right) \vec{J}(\vec{r}', \omega) = \frac{1}{c}\int d^3r'\, \frac{e^{i\omega|\vec{r} - \vec{r}'|/c}}{4\pi\epsilon_0\,|\vec{r} - \vec{r}'|} \vec{J}(\vec{r}', \omega). \tag{5.22b}$$

This calculation identifies Eqs. (5.22a) and (5.22b) as the Fourier transforms of Eqs. (5.12a) and (5.12b). Convolution in time is equivalent to multiplication in frequency space.

5.2.2 *Helmholtz Equation in Spherical Coordinates*

The theory of Green functions of the wave equation is connected with the Helmholtz equation (5.19), whose homogeneous form reads as follows,

$$\left(\vec{\nabla}^2 + k^2\right)\Phi(r, \theta, \varphi) = 0. \tag{5.23}$$

We eventually aim to expand the Helmholtz Green function (5.20) into multipole components, in the spherical basis. To this end, we first need to study the solution of the homogeneous Helmholtz equation in the spherical basis, before connecting them at the cusp, in order to calculate the multipole expansion of the Green function. Therefore, we need to study the solutions of the homogeneous Helmholtz equation, expressed in spherical coordinates r, θ, and φ,

$$\left(\vec{\nabla}^2 + k^2\right)\Phi(r, \theta, \varphi) = \left(\frac{1}{r}\frac{\partial^2}{\partial r^2} r + \frac{1}{r^2\sin\theta}\frac{\partial}{\partial\theta}\sin\theta\frac{\partial}{\partial\theta} + \frac{1}{r^2\sin\theta}\frac{\partial^2}{\partial\varphi^2} + k^2\right)\Phi(r, \theta, \varphi)$$

$$= \left(\frac{\partial^2}{\partial r^2} + \frac{2}{r}\frac{\partial}{\partial r} - \frac{\vec{L}^2}{r^2} + k^2\right)\Phi(r, \theta, \varphi) = 0, \tag{5.24}$$

with $|m| \le \ell = 0, 1, 2, \ldots$. The operator $\vec{L}^2$ has been defined in Eq. (2.31). The general solution of the homogeneous Helmholtz equation can be written as

$$\Phi(r, \theta, \varphi) = \sum_{\ell=0}^{\infty}\sum_{m=-\ell}^{\ell}\left(a_{\ell m}\, j_\ell(kr) + b_{\ell m}\, y_\ell(kr)\right) Y_{\ell m}(\theta, \varphi), \tag{5.25}$$

where the separation constants $a_{\ell m}$ and $b_{\ell m}$ can take arbitrary values, and each function in the set fulfills the Helmholtz equation, separately. The spherical Bessel and Neumann functions are written as j_ℓ and y_ℓ. One uses lowercase letters instead of the uppercase notation, which is used for the ordinary Bessel function J_ℓ and Y_ℓ discussed in Sec. 2.4.2. The spherical Bessel functions have been mentioned very briefly in Sec. 2.4.3 [see Eq. (2.179)]. We recall that they are defined as follows,

$$j_\ell(x) = \left(\frac{\pi}{2x}\right)^{1/2} J_{\ell+1/2}(x), \qquad y_\ell(x) = \left(\frac{\pi}{2x}\right)^{1/2} Y_{\ell+1/2}(x). \tag{5.26}$$

The defining differential equation for spherical Bessel functions is given as

$$\left(\frac{\partial^2}{\partial x^2} + \frac{2}{x}\frac{\partial}{\partial x} - \frac{\ell(\ell+1)}{x^2} + 1\right) j_\ell(x) = 0. \tag{5.27}$$

Spherical Bessel functions fulfill the following recursion relations,

$$j_{\ell-1}(x) + j_{\ell+1}(x) = \frac{2\ell+1}{x} j_\ell(x), \tag{5.28a}$$

$$\ell\, j_{\ell-1}(x) - (\ell+1)\, j_{\ell+1}(x) = (2\ell+1)\, j'_\ell(x), \tag{5.28b}$$

where $j'_\ell(x) = \partial j_\ell(x)/\partial x$ is the derivative with respect to the argument (not parameter). Alternatively, one can express the derivative as follows,

$$j'_\ell(x) = j_{\ell-1}(x) - \frac{\ell+1}{x} j_\ell(x) = -j_{\ell+1}(x) + \frac{\ell}{x} j_\ell(x)$$
$$= \frac{1}{2}\left(j_{\ell+1}(x) + j_{\ell-1}(x)\right) - \frac{1}{2x} j_\ell(x). \tag{5.29}$$

By a straightforward generalization of formulas given in Sec. 2.4.2, we can establish the following asymptotic behavior

$$j_\ell(x) \sim \frac{x^\ell}{(2\ell+1)!!}, \qquad y_\ell(x) \sim -\frac{(2\ell+1)!!}{x^{\ell+1}}, \qquad x \to 0, \tag{5.30}$$

$$j_\ell(x) \sim \frac{1}{x}\sin\left(x - \frac{\ell\pi}{2}\right), \qquad y_\ell(x) \sim -\frac{1}{x}\cos\left(x - \frac{\ell\pi}{2}\right), \qquad x \to \infty. \tag{5.31}$$

The double factorial fulfills the general relations (if ℓ is an integer)

$$(2\ell+1)!! = (2\ell+1)(2\ell-1)(2\ell-3)\ldots 5\cdot 3\cdot 1 = \frac{(2\ell+2)!}{2^{\ell+1}\,(\ell+1)!} = \frac{2^{\ell+1}\,\Gamma(\ell+3/2)}{\sqrt{\pi}}, \tag{5.32}$$

where the latter formula provides the generalization of the result (5.30) to non-integer order ℓ.

In order for Eq. (5.25) to represent a general solution to the Helmholtz equation, we have to demand that

$$\left(\frac{\partial^2}{\partial r^2} + \frac{2}{r}\frac{\partial}{\partial r} - \frac{\ell(\ell+1)}{r^2} + k^2\right) j_\ell(kr) = 0, \tag{5.33}$$

which is in fact equivalent to Eq. (5.27), under the specialization $x \to kr$.

Let us try to find a connection of the ordinary and spherical Bessel functions. Indeed, from Bessel's differential equation (2.144), we may infer that the function $Z = x^\alpha J_n(\beta x^\gamma)$ fulfills the relation

$$\frac{\partial^2}{\partial x^2} Z(x) + \frac{1 - 2\alpha}{x} \frac{\partial}{\partial x} Z(x) + \left(\beta^2 \gamma^2 x^{2\gamma - 2} + \frac{\alpha^2 - n^2 \gamma^2}{x^2} \right) Z(x) = 0. \tag{5.34}$$

For our case, we set $\gamma = 1$, $\alpha = -\frac{1}{2}$, $\beta = k$, and $n = \ell + 1/2$. Then,

$$\frac{\partial^2 Z}{\partial x^2} + \frac{2}{x} \frac{\partial Z}{\partial x} + \left(k^2 + \frac{\frac{1}{4} - (\ell + \frac{1}{2})^2}{x^2} \right) Z = 0, \tag{5.35}$$

which is trivially equivalent to

$$\frac{\partial^2 Z}{\partial x^2} + \frac{2}{x} \frac{\partial Z}{\partial x} + \left(k^2 - \frac{\ell(\ell + 1)}{x^2} \right) Z = 0, \qquad Z = Z(x) = \sqrt{\frac{\pi}{2x}} J_{\ell + 1/2}(k x). \tag{5.36}$$

This is just the desired Eq. (5.33). The complementary solution y_ℓ also fulfills the defining equation. Let us verify some asymptotic properties by way of example. For $\ell = 10$, we have

$$j_{\ell = 10}(x) \sim \frac{x^{10}}{13\,749\,310\,575}, \qquad x \to 0. \tag{5.37}$$

This is illustrated in Fig. 5.1(a). For intermediate, finite values of x, there are deviations from the asymptotic behavior [see Fig. 5.1(b)]. For large argument, we have

$$j_{\ell = 10}(x) \sim \frac{1}{x} \sin\left(x - \frac{10\,\pi}{2} \right), \qquad x \to \infty. \tag{5.38}$$

This is verified in Fig. 5.1(c). The spherical Bessel functions can be expressed in terms of trigonometric functions,

$$j_0(x) = \frac{\sin x}{x}, \qquad j_1(x) = \frac{\sin x}{x^2} - \frac{\cos x}{x}, \qquad j_2(x) = \left(\frac{3}{x^3} - \frac{1}{x} \right) \sin x - \frac{3}{x^2} \cos x, \tag{5.39a}$$

and we have for the Neumann functions,

$$y_0(x) = -\frac{\cos x}{x}, \qquad y_1(x) = -\frac{\cos x}{x^2} - \frac{\sin x}{x}, \qquad y_2(x) = \left(-\frac{3}{x^3} + \frac{1}{x} \right) \cos x - \frac{3}{x^2} \sin x. \tag{5.40a}$$

The Hankel function of the first and second kind are given as

$$h_\ell^{(1)}(x) = j_\ell(x) + i\, y_\ell(x), \qquad h_\ell^{(2)}(x) = j_\ell(x) - i\, y_\ell(x). \tag{5.41}$$

They can be written in terms of $\exp(ix)$ and powers of x. Based on Eqs. (5.39) and (5.40), we can easily show that

$$h_0^{(1)}(x) = -i \frac{e^{ix}}{x}, \qquad h_1^{(1)}(x) = -\frac{e^{ix}}{x} - i \frac{e^{ix}}{x^2}, \qquad h_2^{(1)}(x) = i \frac{e^{ix}}{x} - \frac{3 e^{ix}}{x^2} - i \frac{3 e^{ix}}{x^3}. \tag{5.42}$$

For large x,

$$h_\ell^{(1)}(x) \to -i \frac{e^{i(x - \ell\pi/2)}}{x} = (-i)^{\ell+1} \frac{e^{ix}}{x} \qquad x \to \infty, \tag{5.43}$$

i.e., the Hankel functions differ by a complex phase in the limit $x \to +\infty$.

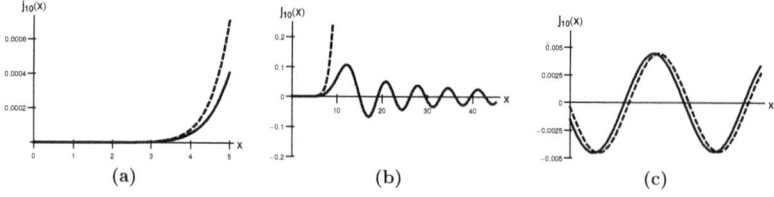

Fig. 5.1   Panel (a) illustrates the verification of Eq. (5.37), i.e., of the asymptotics of the spherical Bessel function $j_{10}(x)$ for $x \to 0$. In panel (b), we verify Eq. (5.37) on a larger scale; the oscillations of the Bessel function become evident. Finally, panel (c) verifies Eq. (5.38), in a regime where the asymptotic behavior is approximated by a trigonometric function, with a phase and a prefactor. In all panels, the solid curve is for the exact Bessel function, whereas the dashed curve is the approximation.

5.2.3   Radiation Green Function

We have come across the inhomogeneous Helmholtz equation (5.19),

$$\left(\vec{\nabla}^2 + k^2\right) G_R\left(k, \vec{r} - \vec{r}'\right) = -\frac{1}{\epsilon_0}\, \delta\left(\vec{r} - \vec{r}'\right), \tag{5.44}$$

and the Helmholtz Green function given in Eqs. (4.75) and (5.20),

$$G_R\left(k, \vec{r} - \vec{r}'\right) = \frac{\exp\left(i\,k\,|\vec{r} - \vec{r}'|\right)}{4\pi\epsilon_0\,|\vec{r} - \vec{r}'|}. \tag{5.45}$$

Then, using the relation $\vec{\nabla} r = \vec{r}/r$, it is instructive to independently verify that G_R fulfills the defining inhomogeneous Helmholtz equation. One needs to carefully keep track of all singular terms in order not to miss the Dirac-δ function,

$$\vec{\nabla}^2 \frac{\exp\left(i k\, r\right)}{r} = \vec{\nabla} \cdot \vec{\nabla} \frac{\exp\left(i k\, r\right)}{r} = \vec{\nabla} \cdot \left(\exp\left(i k r\right)\left[\frac{ik}{r}\vec{\nabla}r + \vec{\nabla}\frac{1}{r}\right]\right),$$

$$= \exp\left(i k\, r\right)\left(-\frac{k^2}{r}\vec{\nabla}r \cdot \vec{\nabla}r + ik\,\vec{\nabla}r \cdot \vec{\nabla}\left(\frac{1}{r}\right) + \frac{ik}{r}\vec{\nabla} \cdot \vec{\nabla}r + ik\vec{\nabla}r \cdot \vec{\nabla}\frac{1}{r} + \vec{\nabla}^2\frac{1}{r}\right)$$

$$= \exp\left(i k\, r\right)\left(-\frac{k^2}{r}\left(\frac{\vec{r}}{r} \cdot \frac{\vec{r}}{r}\right) + ik\,\frac{\vec{r}}{r} \cdot \left(\frac{-\vec{r}}{r^3}\right) + \frac{ik}{r}\left(\frac{2}{r}\right) + ik\left(\frac{\vec{r}}{r}\right) \cdot \left(\frac{-\vec{r}}{r^3}\right) + \vec{\nabla}^2\frac{1}{r}\right)$$

$$= \exp\left(i k\, r\right)\left(\frac{-k^2}{r} - \frac{ik}{r^2} + \frac{2ik}{r^2} - \frac{ik}{r^2} + \vec{\nabla}^2\frac{1}{r}\right)$$

$$= \exp\left(i k\, r\right)\left(\frac{-k^2}{r} + \vec{\nabla}^2\frac{1}{r}\right) = -\frac{k^2}{r}\exp\left(i k\, r\right) - 4\pi\delta^{(3)}\left(\vec{r} - \vec{r}'\right). \tag{5.46}$$

Thus, the validity of Eq. (5.45) is verified once more.

    The central result of the current derivation is the multipole expansion of the radiation Green function, which is analogous to the electrostatic case discussed in Sec. 2.3.5, but a little more involved. An appropriate *ansatz* for the Green function in spherical coordinates reads as follows,

$$G_R(k, \vec{r} - \vec{r}') = \frac{\exp\left(\mathrm{i}\,k\,|\vec{r} - \vec{r}'|\right)}{4\pi\epsilon_0\,|\vec{r} - \vec{r}'|} = \sum_{\ell=0}^{\infty} \sum_{m=-l}^{\ell} G_\ell(k, r, r')\, Y_{\ell m}(\theta, \varphi)\, Y_{\ell m}(\theta', \varphi')^* \,.$$

(5.47)

It is our task to find an appropriate expression for $G_\ell(k, r, r')$. We proceed as in Sec. 2.3.5, by recalling that the operator $\vec{\nabla}^2 + k^2$ acts onto a test function of the form $R(r)\,Y_{\ell m}(\hat{r})$ as follows,

$$\left(\vec{\nabla}^2 + k^2\right) R(r)\,Y_{\ell m}(\hat{r}) = \left(\frac{1}{r^2}\frac{\partial}{\partial r}r^2\frac{\partial}{\partial r} + k^2 - \frac{\ell(\ell+1)}{r^2}\right) R(r)\,Y_{\ell m}(\hat{r})\,,$$

(5.48)

The Dirac-δ function is expanded as follows,

$$\delta(\vec{r} - \vec{r}') = \frac{1}{r^2\,\sin\theta}\,\delta(r - r')\,\delta(\theta - \theta')\,\delta(\varphi - \varphi')\,.$$

(5.49)

The completeness relation for the sum over spherical harmonics reads

$$\sum_{\ell m} Y_{\ell m}(\hat{r})\,Y_{\ell m}^*(\hat{r}) = \frac{1}{\sin\theta}\,\delta(\theta - \theta')\,\delta(\varphi - \varphi')\,.$$

(5.50)

The radial Green function $G_\ell(k, r, r')$ as defined in Eq. (5.47) thus has to fulfill the following relation,

$$\frac{\partial}{\partial r}\left(r^2\frac{\partial}{\partial r}G_\ell(k, r, r')\right) + \left(k^2\,r^2 - \ell(\ell+1)\right) G_\ell(k, r, r') = -\frac{1}{\epsilon_0}\,\delta(r - r')\,.$$

(5.51)

Let us start with the "homogeneous domain" away from the peak of the Dirac-δ function. We shall assume that the radial part of the Green function has the functional form $G_\ell(r, r') = f_\ell(k\,r, k\,r')$ and seek solutions to the homogeneous equation

$$\frac{\partial}{\partial r}\left(r^2\frac{\partial}{\partial r}f_\ell(k\,r, k\,r')\right) + \left(k^2\,r^2 - \ell(\ell+1)\right) f_\ell(k\,r, k\,r') = 0\,.$$

(5.52)

The general solution is a linear combination of the two spherical Bessel functions,

$$f_\ell(k\,r, k\,r') = a_\ell(k\,r')\,j_\ell(k\,r) + b_\ell(k\,r')\,y_\ell(k\,r)\,.$$

(5.53)

We now appeal to Eq. (5.26) for the behavior of the Bessel and Neumann functions near the origin. Since the Green function is regular at $r = 0$, and in particular, for $r < r'$, we must choose the regular solution, i.e., the spherical Bessel function j_ℓ, in this domain,

$$f_\ell(k\,r, k\,r') = G_\ell(k, r, r') = a_\ell(k\,r')\,j_\ell(k\,r)\,, \qquad r < r'\,.$$

(5.54)

For $r > r'$, we certainly need a contribution of the Bessel y_ℓ which is not regular at the origin. However, if we wish to construct a Wronskian upon action of the radial differential operator, then we need to add a contribution from j_ℓ as well. A natural assumption is to make an *ansatz* for the solution G_ℓ to be a superposition of an incoming and an outgoing wave,

$$\propto \exp\left[-\mathrm{i}\left(k\,r - \frac{\ell\pi}{2}\right)\right] \text{ (incoming)}\,, \qquad \propto \exp\left[\mathrm{i}\left(k\,r - \frac{\ell\pi}{2}\right)\right] \text{ (outgoing)}\,.$$

(5.55)

Two considerations support the outgoing wave. The first is that the Green function "propagates", as it were, the point $\vec{r}'$ to the point $\vec{r}$. An incoming wave converging to the point $\vec{r}'$ should be propagated to point $\vec{r}$. In an eigenfunction decomposition of the Green function, we would use outgoing and incoming waves as given in Eq. (5.55). These considerations are valid for $r > r'$, in which case the wave is outgoing at point $\vec{r}$. From the analogy with the Green function for electrostatics, we also conjecture that the solution should fall off as r^{-1} for large r. We thus assume that

$$f_\ell(kr, kr') = G_\ell(k, r, r') = c_\ell(kr')\,[i\,j_\ell(kr) - y_\ell(kr)]\,, \qquad r > r'. \qquad (5.56)$$

Then, for $kr \gg 1$,

$$G_\ell(k, r, r') = c_\ell(kr')\,[i\,j_\ell(kr) - y_\ell(kr)] \longrightarrow c_\ell(kr')\,\frac{1}{kr}\,\exp\left[i\left(kr - \frac{\ell\pi}{2}\right)\right]. \qquad (5.57)$$

Continuity at $r = r'$ is automatically fulfilled if we write the radial component of the Green functions as a product, supplementing for $a_\ell(r')$ and $c_\ell(r')$ the respective "other" solution of the radial Helmholtz equation. The result of this operation is

$$G_\ell(r, r') = a_0(k)\,j_\ell(kr_<)\,[i\,j_\ell(kr_>) - y_\ell(kr_>)]\,, \qquad (5.58)$$

where $a_0(k)$ is a prefactor which can only depend on k, and remains to be determined. In order to convince ourselves of the continuity, we consider r' to be constant and vary r, starting at $r = 0$. As we increase r, the variable r assumes the role of $r_<$ until $r = r'$ (cusp). At the cusp, it does not matter which identification we make, because r and r' are equal. For $r > r'$, the identification of $r_<$ and $r_>$ is different, but we have gone through the cusp. In some sense, the cusp thus ensures the continuity.

The constant a_0 is determined from the discontinuity in the derivative of $G_\ell(k, r, r')$ at $r = r'$. Note that $G_\ell(k, r, r')$ is continuous at $r = r'$, but the derivative $\frac{d}{dr}G_\ell(r, r')$ is discontinuous. Only the region

$$r' - \epsilon \le r \le r' + \epsilon \qquad (5.59)$$

gives a non-zero contribution. Taking notice of the continuity of the Green function and of the discontinuity of its derivative at the cusp, we have

$$\lim_{\epsilon \to 0} \int_{r'-\epsilon}^{r'+\epsilon} \left[\frac{\partial}{\partial r}\left(r^2 \frac{\partial}{\partial r} G_\ell(k, r, r')\right) + \left(k^2 r^2 - \ell(\ell+1)\right) G_\ell((k, r, r')\right] dr$$

$$= -\frac{1}{\epsilon_0} \int_{r'-\epsilon}^{r'+\epsilon} \delta(r - r')\,dr. \qquad (5.60)$$

This implies that the derivative of the radial component of the Green function must have a discontinuity at the cusp, of magnitude [see Eq. (2.85)]

$$r^2 \frac{\partial}{\partial r} G_\ell(r, r')\bigg|_{r=r'+\varepsilon} - r^2 \frac{\partial}{\partial r} G_\ell(r, r')\bigg|_{r=r'-\varepsilon} = -\frac{1}{\epsilon_0}. \qquad (5.61)$$

We can now identify the places where the differentiations take place, immediately to the right and left of the cusp, based on the formulation of the radial Green function in terms of $r_<$ and $r_>$,

$$r^2 \frac{\partial}{\partial r} a_0(k) \, j_\ell \, (k \, r_<) \, [i j_\ell \, (k \, r_>) - y_\ell \, (k \, r_>)] \Big|_{r=r'-\epsilon}^{r=r'+\epsilon} = -\frac{1}{\epsilon_0} \, . \tag{5.62}$$

The differentiations are carried out as follows,

$$r^2 a_0(k) \left\{ j_\ell \, (k r') \left(\frac{\mathrm{d}}{\mathrm{d}r} \, [i j_\ell \, (k \, r) - y_\ell \, (k \, r)] \Big|_{r=r'+\epsilon} \right) \right.$$
$$\left. - \left(\frac{\mathrm{d}}{\mathrm{d}r} j_\ell \, (k r) \Big|_{r=r'-\epsilon} \right) [i j_\ell \, (k \, r') - y_\ell \, (k \, r')] \right\} = -\frac{1}{\epsilon_0} \, . \tag{5.63}$$

Finally, we evaluate the resulting expression in the limit $\epsilon \to 0$, which implies $r = r'$, and obtain

$$r^2 a_0(k) \left(j_\ell(kr) \frac{\mathrm{d}}{\mathrm{d}r} \, [i j_\ell(kr) - y_\ell(kr)] - \left(\frac{\mathrm{d}}{\mathrm{d}r} j_\ell(kr) \right) [i j_\ell(kr') - y_\ell(kr')] \right) = -\frac{1}{\epsilon_0} \, . \tag{5.64}$$

The term proportional to the mixed product of Bessel functions and derivatives cancels, and we have

$$a_0(k) \, r^2 \left(-j_\ell \, (k \, r) \frac{\mathrm{d}}{\mathrm{d}r} y_\ell \, (k \, r) + y_\ell \, (k \, r) \frac{\mathrm{d}}{\mathrm{d}r} j_\ell \, (k \, r) \right) = -\frac{1}{\epsilon_0} \, . \tag{5.65}$$

From the differential equation (5.33) fulfilled by the j_ℓ and y_ℓ, we can easily derive that

$$\frac{\mathrm{d}}{\mathrm{d}r} \left[r^2 \left(-j_\ell \, (k \, r) \frac{\mathrm{d}}{\mathrm{d}r} y_\ell \, (k \, r) + y_\ell \, (k \, r) \frac{\mathrm{d}}{\mathrm{d}r} j_\ell \, (k \, r) \right) \right] = 0 \, , \tag{5.66}$$

and so the Wronskian $-j_\ell \, y_\ell' + y_\ell \, j_\ell'$ can be evaluated for any argument. In particular, we can use the form for the two spherical Bessel functions as $r' \to 0$,

$$a_0(k) \, r^2 \left\{ \frac{-(k \, r)^\ell}{(2\ell + 1)!!} \left[\frac{\mathrm{d}}{\mathrm{d}r} \left(-\frac{(2\ell - 1)!!}{(k \, r)^{\ell+1}} \right) \right] + \left(-\frac{(2\ell - 1)!!}{(k \, r)^{\ell+1}} \right) \left[\frac{\mathrm{d}}{\mathrm{d}r} \frac{(k \, r)^\ell}{(2\ell + 1)!!} \right] \right\} = -\frac{1}{\epsilon_0} \, , \tag{5.67}$$

Carrying out the differentiation, we find that

$$a_0(k) \, r^2 \left[\frac{(k \, r)^\ell}{(2\ell + 1)} \left(\frac{-(\ell + 1) \, k}{(k \, r)^{\ell+2}} \right) + \left(-\frac{1}{(k \, r)^{\ell+1}} \right) \left(\frac{\ell k \, (k \, r)^{\ell-1}}{(2\ell + 1)} \right) \right] = -\frac{1}{\epsilon_0} \, . \tag{5.68}$$

Assembling all factors, we finally obtain the relation

$$a_0(k) \, r^2 \left[\frac{-(\ell + 1) k - \ell k}{(k r)^2 \, (2\ell + 1)} \right] = a_0(k) \left(-\frac{1}{k} \right) = -\frac{1}{\epsilon_0} \, . \tag{5.69}$$

The explicit expression for the overall coefficient thus reads as

$$a_0(k) = \frac{k}{\epsilon_0} \, , \tag{5.70}$$

and the final form for $G_\ell(k, r, r')$ is

$$G_\ell(k, r, r') = \frac{1}{\epsilon_0} k\, j_\ell(k r_<) \left[i\, j_\ell(k r_>) - y_\ell(k r_>) \right] = \frac{1}{\epsilon_0} i\, k\, j_\ell(k r_<) \, h_\ell^{(1)}(k r_>) \,,$$

(5.71)

where the Hankel functions have been defined in Eq. (5.41). We conclude that in spherical coordinates, the Green function has the following angular momentum decomposition,

$$G_R(k, \vec{r} - \vec{r}') = \frac{\exp\left(i\, k\, |\vec{r} - \vec{r}'|\right)}{4\pi\epsilon_0 |\vec{r} - \vec{r}'|}$$

$$= \frac{i\, k}{\epsilon_0} \sum_{\ell=0}^{\infty} \sum_{m=-\ell}^{\ell} j_\ell(k r_<) \, h_\ell^{(1)}(k r_>) \, Y_{\ell m}(\theta, \varphi) \, Y_{\ell m}^*(\theta', \varphi') \,.$$

(5.72)

By contrast, we recall the angular momentum decomposition [Eqs. (2.25) and (2.91)] of the Green function of electrostatics,

$$G(\vec{r}, \vec{r}') = \frac{1}{4\pi\epsilon_0 |\vec{r} - \vec{r}'|} = \frac{1}{\epsilon_0} \sum_{\ell=0}^{\infty} \sum_{m=-\ell}^{\ell} \frac{1}{2\ell + 1} \frac{r_<^\ell}{r_>^{\ell+1}} Y_{\ell m}(\theta, \varphi) \, Y_{\ell m}^*(\theta', \varphi') \,.$$

(5.73)

The matching is successful if

$$\lim_{k \to 0} G_R(k, \vec{r} - \vec{r}') = G(\vec{r}, \vec{r}') \,, \qquad \lim_{k \to 0} i\, k\, j_\ell(k r_<) \, h_\ell^{(1)}(k r_>) = \frac{1}{2\ell + 1} \frac{r_<^\ell}{r_>^{\ell+1}} \,.$$

(5.74)

The asymptotic relations (5.30) lead to the following expansion,

$$i\, k\, j_\ell(k r_<) \, h_\ell^{(1)}(k r_>) = i\, k\, j_\ell(k r_<) \left(j_\ell(k r_<) + i\, y_\ell(k r_>) \right)$$

$$\overset{k \to 0}{=} - k\, j_\ell(k r_<) \, y_\ell(k r_>)$$

$$\overset{k \to 0}{=} - k\, \frac{(k r_<)^\ell}{(2\ell + 1)!!} \left(-\frac{(2\ell - 1)!!}{(k r_>)^{\ell+1}} \right) = \frac{1}{2\ell + 1} \frac{r_<^\ell}{r_>^{\ell+1}} \,,$$

(5.75)

confirming Eq. (5.74).

5.3 Localized Harmonically Oscillating Sources

5.3.1 *Basic Formulas and Multipole Expansion*

We consider sources which oscillate at a fixed angular frequency ω,

$$\rho(\vec{r}, t) = \rho_0(\vec{r}) \, \exp\left(-i\, \omega\, t\right) \,, \qquad \vec{J}(\vec{r}, t) = \vec{J}_0(\vec{r}) \, \exp\left(-i\, \omega\, t\right) \,.$$

(5.76)

The real part of these source functions and of the potentials are the physical parameters, but it is computationally useful to add the imaginary part, in order to be able to restrict the discussion to a single Fourier component. In view of Eq. (5.22b), the scalar and vector potentials generated by this current source are given as

$$\Phi(\vec{r}, t) = e^{-i\omega t} \int d^3 r' \, G_R\left(\frac{\omega}{c}, \vec{r} - \vec{r}'\right) \rho_0(\vec{r}') \,,$$

(5.77)

$$\vec{A}(\vec{r}, t) = e^{-i\omega t} \int d^3 r' \, \frac{1}{c^2} G_R\left(\frac{\omega}{c}, \vec{r} - \vec{r}'\right) \vec{J}_0(\vec{r}') \,.$$

(5.78)

It is natural to assume that the vector potentials and fields have the same harmonic dependence,

$$\vec{A}(\vec{r},t) = \vec{A}_0(\vec{r})\,\exp(-i\omega t)\,, \quad \vec{B}(\vec{r},t) = \vec{B}_0(\vec{r})\,\exp(-i\omega t)\,,$$

$$\vec{E}(\vec{r},t) = \vec{E}_0(\vec{r})\,\exp(-i\omega t)\,. \tag{5.79}$$

The spatial variation of the vector potential is contained in the quantity $\vec{A}_0(\vec{r})$ which only depends on space (but not on time),

$$\vec{A}_0(\vec{r}) = \int \mathrm{d}^3 r'\, \frac{\exp(i\omega|\vec{r}-\vec{r}'|/c)}{4\pi\epsilon_0 c^2 |\vec{r}-\vec{r}'|}\, \vec{J}_0(\vec{r}')\,. \tag{5.80}$$

The magnetic induction field is given by

$$\vec{B}_0(\vec{r}) = \vec{\nabla} \times \vec{A}_0(\vec{r})\,. \tag{5.81}$$

At the observation point, which is far away from the source, the current density vanishes, $\vec{J}_0(\vec{r}) = \vec{0}$, and the electric field is given by the Ampere–Maxwell law

$$\vec{\nabla} \times \vec{B}_0(\vec{r},t) = \frac{1}{c^2}\frac{\partial}{\partial t}\vec{E}_0(\vec{r},t) = -i\frac{\omega}{c^2}\,\vec{E}_0(\vec{r},t)\,. \tag{5.82}$$

The spatial dependence of the electric field thus reads as

$$\vec{E}_0(\vec{r}) = \frac{ic}{\omega}\left(\vec{\nabla}\times\vec{B}_0(\vec{r})\right) = \frac{ic}{\omega}\,\vec{\nabla}\times\left(\vec{\nabla}\times\vec{A}_0(\vec{r})\right)\,. \tag{5.83}$$

Equations (5.81) and (5.83) imply that the knowledge of the vector potential $\vec{A}_0(\vec{r})$ is sufficient for the calculation of both $\vec{E}_0(\vec{r})$ and $\vec{B}_0(\vec{r})$ in the source-free region; it is not necessary to consider the scalar potential (5.77) at all.

We have already restricted our sources configurations (the domain where $\vec{J}_0$ is nonvanishing) to be localized with a characteristic dimension, d, satisfying

$$d \ll \frac{c}{\omega} = \frac{\lambda}{2\pi}\,, \tag{5.84}$$

i.e., with a spatial dimension much less than the wavelength of the radiation. This restriction only pertains to the dimension d of the sources configuration, not to the distance r from the source itself. The exponential term in the integrand for the vector potential suggests that the potential will have a drastically different spatial dependence depending on the range of $\vec{r}$. The condition $d \ll r \ll c/\omega$ is relevant to the near field or static zone, less than a wavelength away, while the condition $r \gg c/\omega$ characterizes the far field or radiation zone, many wavelengths away. The vector potential, given by Eq. (5.78), can be evaluated using the angular momentum decomposition (5.72) for the Helmholtz Green function, which we recall for convenience,

$$G_R(k,\vec{r}-\vec{r}') = \frac{ik}{\epsilon_0}\sum_{\ell=0}^{\infty}\sum_{m=-\ell}^{\ell} j_\ell(kr_<)\,h_\ell^{(1)}(kr_<)\,Y_{\ell m}(\theta,\varphi)\,Y_{\ell m}^*(\theta',\varphi')\,. \tag{5.85}$$

In view of Eq. (5.78), the vector potential $\vec{A}_0(\vec{r})$ can be decomposed into multipoles as follows,

$$\vec{A}_0(\vec{r}) = \frac{ik}{\epsilon_0 c^2}\sum_{\ell m} Y_{\ell m}(\theta,\varphi)\int \mathrm{d}^3 r'\,\vec{J}_0(\vec{r}')j_\ell(kr_<)h_\ell^{(1)}(kr_>)Y_{\ell m}^*(\theta',\varphi')\,. \tag{5.86}$$

We here introduce the notation $\sum_{\ell m}$ for the sum $\sum_{\ell=0}^{\infty} \sum_{m=-\ell}^{\ell}$; it will be used in the following. This expansion ignores, in some sense, the intrinsic angular momentum of the current density $\vec{J}_0(\vec{r}\,')$. The current density $\vec{J}_0(\vec{r}\,')$ transforms as a vector, just like the spherical harmonics $Y_{\ell m}(\theta, \varphi)$. A more systematic expansion is found if one adds the angular momentum inherent to the vector field to the angular momentum of the spherical harmonic. As it stands, in order to obtain a valid expansion in Eq. (5.86) and use the orthogonality properties of the spherical harmonics, one has to expand each individual Cartesian component of the vector-valued current density $\vec{J}_0(\vec{r}\,')$ into spherical harmonics. This procedure mixes Cartesian and spherical coordinate systems and is somewhat unsystematic. A better way will be described in Sec. 5.4; but Eq. (5.86) is good enough for our purposes, for the time being. In fact, we shall first discuss dipole radiation based on Eq. (5.86), before generalizing to arbitrary multipole orders using the tensor Green function.

If the source current is localized near $r' \approx 0$, we can set $r_< = r'$ and $r_> = r$, and the radiated vector potential reads

$$\vec{A}_0(\vec{r}) = \frac{ik}{\epsilon_0 c^2} \sum_{\ell m} h_\ell^{(1)}(kr)\, Y_{\ell m}(\theta, \varphi) \int \mathrm{d}^3 r'\, \vec{J}_0(\vec{r}\,')\, j_\ell(kr')\, Y_{\ell m}^*(\theta', \varphi') \qquad (5.87)$$

If the wavelength emitted by the radiating source is much larger than the characteristic length scale d of the charge distribution, then we have $kd \ll 1$. Furthermore, in this case, the argument kr' of the Bessel function $j_\ell(kr')$ fulfills $kr' < kd \ll 1$, and the spherical Bessel function can be approximated with its asymptotic form near $r' = 0$, that is, $j_\ell(z) \sim z^\ell/(2\ell+1)!!$. One then obtains

$$\vec{A}_0(\vec{r}) \approx \frac{ik}{\epsilon_0 c^2} \sum_{\ell=0}^{\infty} \sum_{m=-\ell}^{\ell} \frac{k^\ell h_\ell^{(1)}(kr)}{(2\ell+1)!!}\, Y_{\ell m}(\theta, \varphi) \int \mathrm{d}r'\, r'^{\ell+2} \int \mathrm{d}\Omega'\, \vec{J}_0(\vec{r}\,')\, Y_{\ell m}^*(\theta', \varphi')\,.$$

$$(5.88)$$

We define the near-field region to be the spatial region closer than a wavelength away from the antenna, but still large compared to the spatial extent of the charge distribution,

$$d < r \ll \frac{c}{\omega}\,, \qquad kd \ll kr \ll 1\,, \qquad h_\ell^{(1)}(kr) \approx +i\, y_\ell(kr) \approx -i\, \frac{(2\ell-1)!!}{(kr)^{\ell+1}}\,, \quad (5.89a)$$

$$\vec{A}_0(\vec{r}) \approx \frac{1}{\epsilon_0 c^2} \sum_{\ell=0}^{\infty} \sum_{m=-\ell}^{\ell} \frac{Y_{\ell m}(\theta, \varphi)}{(2\ell+1)!!\, r^{\ell+1}} \int \mathrm{d}r'\, r'^{\ell+2} \int \mathrm{d}\Omega'\, \vec{J}_0(\vec{r}\,')\, Y_{\ell m}^*(\theta', \varphi')\,.$$

$$(5.89b)$$

So, in the near-field region, the double factorial $(2\ell-1)!!$ cancels, and we have a very compact expression for a long-wavelength emitter. In the far zone, still for a long-wavelength emitter, we have in view of Eq. (5.43),

$$kd \ll 1\,, \qquad kr \gg 1\,, \qquad h_\ell^{(1)}(kr) \approx \frac{e^{i(kr-\ell\pi/2)}}{ikr}\,, \qquad (5.90a)$$

$$\vec{A}_0(\vec{r}) \approx \sum_{\ell=0}^{\infty} \sum_{m=-\ell}^{\ell} \frac{k^\ell e^{i(kr-\pi\ell/2)}\, Y_{\ell m}(\theta, \varphi)}{\epsilon_0 c^2 r\, (2\ell+1)!!} \int \mathrm{d}r'\, r'^{\ell+2} \int \mathrm{d}\Omega'\, \vec{J}_0(\vec{r}\,')\, Y_{\ell m}^*(\theta', \varphi')\,.$$

$$(5.90b)$$

By dimensional analysis, the source integral is proportional to

$$\int \mathrm{d}r'\, r'^{\ell+2} \int \mathrm{d}\Omega'\, \vec{J}_0\left(\vec{r}'\right) Y_{\ell m}^*\left(\theta', \varphi'\right) \sim j_0\, d^{\ell+3}\,, \tag{5.91}$$

where j_0 is a characteristic scale of the current density. The expansion in multipoles thus is seen to be an expansion in powers of the parameter $(k\,d)^\ell$, where $k = \omega\, c$, and can be evaluated term by term. The lowest nonvanishing term then gives the dominant contribution. In evaluating the source integral (5.91), one has to multiply each of the three spatial components of $\vec{J}_0\left(\vec{r}'\right)$ by the spherical harmonic $Y_{\ell m}^*\left(\theta', \varphi'\right)$ and evaluate the overlap integral over the entire solid angle $\mathrm{d}\Omega'$.

5.3.2 *Asymptotic Limits of Dipole Radiation*

The lowest nonvanishing contribution to electromagnetic radiation comes from an oscillating dipole and is referred to as dipole radiation. By contrast, the integral over $\vec{J}_0\left(\vec{r}'\right)$ for $\ell = 0$ and $m = 0$ in Eq. (5.91) projects out only the spherically symmetric part of the $\vec{J}_0\left(r', \theta', \varphi'\right)$,

$$\vec{D} = \int \vec{J}_0\left(\vec{r}'\right) r'^2\, Y_{00}\left(\theta', \varphi'\right)^* \mathrm{d}r'\mathrm{d}\Omega' = \frac{1}{\sqrt{4\pi}} \int \vec{J}_0\left(\vec{r}'\right) \mathrm{d}^3 r'\,, \tag{5.92}$$

If $\vec{J}_0\left(\vec{r}'\right)$ were a scalar, then this integral would naturally be referred as a monopole radiation integral. Somehow, we have to relate the vector-valued space integral over $\vec{J}_0\left(\vec{r}'\right)$ to the total oscillating dipole moment. Charge conservation helps. In the mixed frequency-coordinate representation, the charge conservation condition reads as

$$\vec{\nabla} \cdot \vec{J}_0\left(\vec{r}\right) = -\left(-\mathrm{i}\omega\rho_0\left(\vec{r}\right)\right) = \mathrm{i}\omega\rho_0\left(\vec{r}\right)\,. \tag{5.93}$$

This suggests the following trick in order to convert the integral over the current distribution into an integral over the charge distribution: just multiply the current density by a coordinate and calculate the divergence. The result consists of two terms, one of which reproduces the current density, and the second yields the charge density by virtue of current conservation. The integral over the total divergence vanishes, and the charge density is obtained in the final result.

This program is implemented as follows. The integral over the current density is converted to one over the charge density using the identity

$$\vec{\nabla} \cdot \left(x_m\, \vec{J}_0(\vec{r})\right) = \sum_{n=1}^{3} \frac{\partial}{\partial x_n}\left[x_m\, J_{0,n}\left(\vec{r}\right)\right] = \sum_{n=1}^{3} J_{0,n}\left(\vec{r}\right)\delta_{nm} + x_m\, \vec{\nabla} \cdot \vec{J}_0\left(\vec{r}\right)$$

$$= J_{0,m}\left(\vec{r}\right) + \mathrm{i}\omega\, x_m\, \rho_0\left(\vec{r}\right)\,, \tag{5.94}$$

where $J_{0,m}\left(\vec{r}\right)$ refers to the mth Cartesian component of $\vec{J}_0(\vec{r})$. So,

$$\vec{D} = \frac{1}{\sqrt{4\pi}} \int \vec{J}_0\left(\vec{r}'\right) \mathrm{d}^3 r' = -\mathrm{i}\frac{\omega}{\sqrt{4\pi}} \int \sum_{m=1}^{3} \hat{e}_m\, x_m'\, \rho_0\left(\vec{r}'\right) \mathrm{d}^3 r'$$

$$= -\mathrm{i}\frac{\omega}{\sqrt{4\pi}} \int \vec{r}'\, \rho_0\left(\vec{r}'\right) \mathrm{d}^3 r' = -\mathrm{i}\frac{\omega}{\sqrt{4\pi}}\, \vec{p}_0\,, \qquad \vec{p}_0 = \int \vec{r}'\, \rho_0\left(\vec{r}'\right) \mathrm{d}^3 r'\,. \tag{5.95}$$

This last integral is the electric dipole moment of the charge distribution, i.e., $\vec{p}_0$. We recall Eq. (5.90); for the dipole case, this formula is exact for a localized, long-wavelength source with $kd \ll 1$, because the Hankel function $h_0^{(1)}(kr)$ consists of only one term [see Eq. (5.42). Then, the $\ell = 0$ component amounts to

$$\vec{A}_0(\vec{r}) = \frac{k}{\epsilon_0 c^2} \frac{\exp(ikr)}{kr} \frac{1}{\sqrt{4\pi}} \left(\int \vec{J}_0(\vec{r}') \, r'^2 \, \frac{1}{\sqrt{4\pi}} dr' d\Omega' \right)$$

$$= \frac{k}{4\pi\epsilon_0 c^2} \frac{\exp(ikr)}{kr} \left(-i\frac{\omega}{4\pi} \vec{p}_0 \right) = -i\frac{k \vec{p}_0}{4\pi\epsilon_0 c} . \tag{5.96}$$

The magnetic induction generated by an oscillating electric dipole is

$$\vec{B}_0(\vec{r}) = \vec{\nabla} \times \vec{A}_0(\vec{r}) = -i\frac{k}{4\pi\epsilon_0 c} \vec{\nabla} \times \left[\vec{p}_0 \left(\frac{\exp(ikr)}{r} \right) \right] = i\frac{k \vec{p}_0}{4\pi\epsilon_0 c} \times \vec{\nabla} \left(\frac{\exp(ikr)}{r} \right)$$

$$= i\frac{k \vec{p}_0}{4\pi\epsilon_0 c} \times \left(\frac{d}{dr} \frac{\exp(ikr)}{r} \right) \vec{\nabla} r = -i\frac{k}{4\pi\epsilon_0 c} \left(\frac{\vec{r}}{r} \times \vec{p}_0 \right) \frac{\exp(ikr)}{r} \left(ik - \frac{1}{r} \right)$$

$$= \frac{k^2 \exp(ikr)}{4\pi\epsilon_0 cr} \left(1 - \frac{1}{ikr} \right) (\hat{r} \times \vec{p}_0) , \tag{5.97}$$

where $\hat{r} = \vec{r}/r$. The leading term, for large r, is the magnetic field in the radiation zone,

$$\vec{B}_0(\vec{r}) \sim \frac{k^2 \exp(ikr)}{4\pi\epsilon_0 cr} (\hat{r} \times \vec{p}_0) , \qquad kr \to \infty . \tag{5.98}$$

We can use Eq. (5.83) in order to calculate the electric field,

$$\vec{E}_0(\vec{r}) = \frac{ic^2}{\omega} \left(\vec{\nabla} \times \vec{B}_0(\vec{r}) \right) = i\frac{c}{k} \left(\vec{\nabla} \times \vec{B}_0(\vec{r}) \right) = \frac{ik}{4\pi\epsilon_0} \vec{\nabla} \times \left[\left(\frac{\vec{r}}{r} \times \vec{p}_0 \right) f(r) \right] ,$$

$$f(r) = \frac{\exp(ikr)}{r} \left(1 - \frac{1}{ikr} \right) . \tag{5.99}$$

The leading term, for large r, is generated by the gradient operator pulling down a factor $\vec{k}$ from the exponential, and can be written as

$$\vec{E}_0(\vec{r}) \approx \frac{ik}{4\pi\epsilon_0} (\vec{\nabla} r) \times \left(\frac{\vec{r}}{r} \times \vec{p}_0 \right) \frac{d}{dr} f(r) \approx \frac{ik}{4\pi\epsilon_0} \left[\frac{\vec{r}}{r} \times \left(\frac{\vec{r}}{r} \times \vec{p}_0 \right) \right] \frac{ik \exp(ikr)}{r}$$

$$= \frac{k^2 \exp(ikr)}{4\pi\epsilon_0 r} \left[(\hat{r} \times \vec{p}_0) \times \hat{r} \right] . \tag{5.100}$$

So, the electric field in the radiation zone is given as

$$\vec{E}_0(\vec{r}) \sim \frac{k^2 \exp(ikr)}{4\pi\epsilon_0 r} \left[(\hat{r} \times \vec{p}_0) \times \hat{r} \right] , \qquad kr \to \infty . \tag{5.101}$$

Together with the corresponding asymptotic formula for the magnetic field given in Eq. (5.98), one infers that

$$\vec{E}_0(\vec{r}) \sim c\vec{B}_0(\vec{r}) \times \hat{r} , \qquad kr \to \infty . \tag{5.102}$$

In the radiation zone, the electric field, the magnetic field as well as the position unit vector $\hat{r} \parallel \vec{E}_0 \times \vec{B}_0$ form a right-handed system.

It is very instructive to calculate the energy density radiated by the oscillating dipole. To this end, we need the Poynting vector $\vec{S}(\vec{r}, t)$. From the electric and magnetic fields $\vec{E}(\vec{r}, t)$ and $\vec{B}(\vec{r}, t)$ (which are functions of the spatial coordinates and of time), the instantaneous Poynting vector $\vec{S}(\vec{r}, t)$ is calculated as follows,

$$\vec{S}(\vec{r}, t) = \frac{1}{\mu_0} \vec{E}(\vec{r}, t) \times \vec{B}(\vec{r}, t) . \tag{5.103}$$

With the electromagnetic energy density

$$u(\vec{r}, t) = \frac{1}{2\mu_0} \vec{B}^2(\vec{r}, t) + \frac{\epsilon_0}{2} \vec{E}^2(\vec{r}, t) , \tag{5.104}$$

which measures the amount of energy stored in the electromagnetic field per unit volume, we have the Poynting theorem (1.106)

$$\vec{\nabla} \cdot \vec{S}(\vec{r}, t) + \frac{\partial}{\partial t} u(\vec{r}, t) + \vec{E}(\vec{r}, t) \cdot \vec{J}(\vec{r}, t) = 0 . \tag{5.105}$$

Energy leaving the reference volume per unit time is obtained by integrating $\int_V \vec{\nabla} \cdot \vec{S}(\vec{r}, t) \mathrm{d}^3 r = \int_{\partial V} \vec{S}(\vec{r}, t) \cdot \mathrm{d}\vec{A}$. The Poynting theorem states that energy leaving the reference volume, as described by the integral of the Poynting vector over the surface normal, plus the work done on the charge in the reference volume by the electric field, is accompanied by a loss in field energy within the reference volume. For an oscillatory field described by a frequency component $\exp(-i\omega t)$, the physically relevant quantity is the real part which is proportional to $\mathrm{Re}[\exp(-i\omega t)] = \cos(\omega t)$. The average value of $\cos(\omega t)$ over a period of oscillation is just equal to $1/2$. So, for a component of angular frequency ω described by Eq. (5.79), the average Poynting vector $\langle \vec{S}(\vec{r}) \rangle$ (over a period of oscillation) is given as

$$\langle \vec{S}(\vec{r}) \rangle = \frac{1}{2\mu_0} \vec{E}_0(\vec{r}) \times \vec{B}_0^*(\vec{r}) . \tag{5.106}$$

The vector identity $\vec{a} \times (\vec{b} \times \vec{c}) = \vec{b} \times (\vec{a} \cdot \vec{c}) - \vec{c} \times (\vec{a} \cdot \vec{b})$ can be used to calculate this expression,

$$\langle \vec{S}(\vec{r}) \rangle = \frac{1}{2\mu_0} \vec{E}_0(\vec{r}) \times \vec{B}_0^*(\vec{r}) = \frac{1}{2\mu_0} \frac{k^2}{4\pi\epsilon_0} \frac{k^2}{4\pi\epsilon_0 c} \frac{1}{r^2} [(\hat{r} \times \vec{p}_0) \times \hat{r}] \times (\hat{r} \times \vec{p}_0)$$

$$= -\frac{c}{2\mu_0 c^2} \frac{k^4}{(4\pi\epsilon_0)^2} \frac{1}{r^2} (\hat{r} \times \vec{p}_0) \times [(\hat{r} \times \vec{p}_0) \times \hat{r}] = \frac{\epsilon_0 c}{2} \frac{k^4}{(4\pi\epsilon_0)^2} \frac{1}{r^2} \hat{r} (\hat{r} \times \vec{p}_0)^2$$

$$= \frac{\epsilon_0 c}{2} \frac{k^4}{(4\pi\epsilon_0)^2} \frac{1}{r^2} \hat{r} \left(\vec{p}_0^2 - (\hat{r} \cdot \vec{p}_0)^2 \right) = \frac{\epsilon_0 c}{2} \frac{k^4 \vec{p}_0^2}{(4\pi\epsilon_0)^2} \frac{1}{r^2} \hat{r} \left(1 - \cos\theta^2 \right) , \tag{5.107}$$

where $\cos\theta = \hat{r} \cdot \hat{p}_0$. The angular distribution of the average radiated power $\mathrm{d}P_{\mathrm{avg}}$ per area $\mathrm{d}A$ is

$$\frac{\mathrm{d}P_{\mathrm{avg}}(\Omega)}{\mathrm{d}A} = \hat{r} \cdot \langle S(\vec{r}) \rangle = \frac{\epsilon_0 c}{2} \frac{k^4 \vec{p}_0^2}{(4\pi\epsilon_0)^2} \frac{1}{r^2} \left(1 - \cos\theta^2 \right) . \tag{5.108}$$

The average intensity radiated parallel to the dipole vector is zero. The time averaged power $\langle P \rangle$ radiated by an oscillating electric dipole is obtained as

$$\langle P \rangle = \frac{\epsilon_0 c}{2} \frac{k^4 \vec{p}_0^2}{(4\pi\epsilon_0)^2} \int \frac{1}{r^2} \left(1 - \cos^2\theta \right) \left(r^2 \, \mathrm{d}\Omega \right) = \frac{c}{3} \frac{k^4 \vec{p}_0^2}{4\pi\epsilon_0} . \tag{5.109}$$

The radiated power is proportional to the fourth power of the frequency of the oscillation.

5.3.3 Exact Expression for the Radiating Dipole

Our exact result for the magnetic field generated by oscillating dipole (5.97) is given as

$$\vec{B}_0(\vec{r}) = \frac{k^2 \exp(ik\,r)}{4\pi\epsilon_0\, c\, r} \left(1 - \frac{1}{ik\,r}\right)(\hat{r} \times \vec{p}_0)\,. \tag{5.110}$$

In Eq. (5.101), the electric field is obtained as

$$\vec{E}_0(\vec{r}) = \frac{ic^2}{\omega}\left(\vec{\nabla} \times \vec{B}_0(\vec{r})\right) \sim \frac{k^2 \exp(ik\,r)}{4\pi\epsilon_0\, r}\left[(\hat{r} \times \vec{p}_0) \times \hat{r}\right]\,, \qquad k\,r \to \infty\,. \tag{5.111}$$

The exact expression for $\vec{E}_0(\vec{r})$ remains to be derived. Certainly,

$$\begin{aligned}
\vec{E}_0(\vec{r}) &= \frac{ic}{k}\frac{k^2}{4\pi\epsilon_0\, c}\left(\vec{\nabla} \times \left(\frac{\vec{r}}{r} \times \vec{p}_0\right)\frac{\exp(ik\,r)}{r}\left(1 - \frac{1}{ik\,r}\right)\right) \\
&= \frac{ik}{4\pi\epsilon_0}\left(\vec{\nabla} \times (\vec{r} \times \vec{p}_0)\frac{\exp(ik\,r)}{r^2}\left(1 - \frac{1}{ik\,r}\right)\right) \\
&= \frac{ik}{4\pi\epsilon_0}(\vec{\nabla}r) \times \left(\frac{\vec{r}}{r} \times \vec{p}_0\right) r\frac{d}{dr}\left(\frac{\exp(ik\,r)}{r^2}\left(1 - \frac{1}{ik\,r}\right)\right) \\
&\quad + \frac{ik}{4\pi\epsilon_0}\left(\frac{\exp(ik\,r)}{r^2}\left(1 - \frac{1}{ik\,r}\right)\right)\left(\vec{\nabla} \times (\vec{r} \times \vec{p}_0)\right)\,. \tag{5.112}
\end{aligned}$$

According to the rule that $\vec{a} \times (\vec{b} \times \vec{c}) = \vec{b}(\vec{a}\cdot\vec{c}) - \vec{c}(\vec{a}\cdot\vec{b})$,

$$\left(\vec{\nabla} \times (\vec{r} \times \vec{p}_0)\right) = \left(\vec{\nabla}\cdot\vec{p}_0\right)\vec{r} - \vec{p}_0\left(\vec{\nabla}\cdot\vec{r}\right) = \vec{p}_0 - 3\,\vec{p}_0 = -2\vec{p}_0\,.$$

We have taken into account the fact that the $\vec{\nabla}$ operator acts on everything to the right. So,

$$\begin{aligned}
\vec{E}_0(\vec{r}) &= \frac{ik}{4\pi\epsilon_0}\left[\frac{\vec{r}}{r} \times \left(\frac{\vec{r}}{r} \times \vec{p}_0\right)\right]\left[r\frac{d}{dr}\left(\frac{\exp(ik\,r)}{r^2}\left(1 - \frac{1}{ik\,r}\right)\right)\right] \\
&\quad + \frac{ik}{4\pi\epsilon_0}\left(\frac{\exp(ik\,r)}{r^2}\left(1 - \frac{1}{ik\,r}\right)\right)(-2\vec{p}_0) \\
&= \frac{ik}{4\pi\epsilon_0}[\hat{r} \times (\hat{r} \times \vec{p}_0)]\left(\frac{ik}{r} - \frac{3}{r^2} - \frac{3i}{k\,r^3}\right)\exp(ik\,r) + \frac{ik}{4\pi\epsilon_0}\left(\frac{\exp(ik\,r)}{r^2}\left(1 - \frac{1}{ik\,r}\right)\right)(-2\vec{p}_0) \\
&= \frac{ik}{4\pi\epsilon_0}[\hat{r} \times (\hat{r} \times \vec{p}_0)]\frac{ik}{r}\exp(ik\,r) + \frac{ik}{4\pi\epsilon_0}[\hat{r} \times (\hat{r} \times \vec{p}_0)]\left(-\frac{3}{r^2} - \frac{3i}{k\,r^3}\right)\exp(ik\,r) \\
&\quad + \frac{ik}{4\pi\epsilon_0}\left(\frac{\exp(ik\,r)}{r^2}\left(1 - \frac{1}{ik\,r}\right)\right)(-2\vec{p}_0)\,. \tag{5.113}
\end{aligned}$$

Finally,

$$\vec{E}_0(\vec{r}) = \frac{k^2}{4\pi\epsilon_0} \left[(\hat{r} \times \vec{p}_0) \times \hat{r}\right] \frac{\exp(ik\,r)}{r} + \frac{1}{4\pi\epsilon_0} \left[\hat{r}(\hat{r} \cdot \vec{p}_0 - \vec{p}_0)\right] \left(-\frac{3ik}{r^2} + \frac{3}{r^3}\right) \exp(ik\,r)$$

$$+ \frac{1}{4\pi\epsilon_0} \left(\frac{\exp(ik\,r)}{r^2}\left(ik - \frac{1}{r}\right)\right)(-2\vec{p}_0) \,. \tag{5.114}$$

The first term is the leading term. The remaining terms have the structure of a quadrupole term component when projected onto the dipole vector $\vec{p}_0$,

$$\vec{E}_0(\vec{r}) = \frac{k^2}{4\pi\epsilon_0} \left[(\hat{r} \times \vec{p}_0) \times \hat{r}\right] \frac{\exp(ik\,r)}{r} + \frac{1}{4\pi\epsilon_0} \left[3\hat{r}(\hat{r} \cdot \vec{p}_0 - 3\vec{p}_0)\right] \left(\frac{1}{r^3} - \frac{ik}{r^2}\right) \exp(ik\,r)$$

$$+ \frac{1}{4\pi\epsilon_0} \exp(ik\,r) \left(-\frac{ik}{r^2} + \frac{1}{r^3}\right)(+2\vec{p}_0)$$

$$= \frac{k^2}{4\pi\epsilon_0} \left[(\hat{r} \times \vec{p}_0) \times \hat{r}\right] \frac{\exp(ik\,r)}{r} + \frac{1}{4\pi\epsilon_0} \left[3\hat{r}(\hat{r} \cdot \vec{p}_0 - 3\vec{p}_0)\right] \left(\frac{1}{r^3} - \frac{ik}{r^2}\right) \exp(ik\,r)$$

$$+ \frac{1}{4\pi\epsilon_0} (2\vec{p}_0) \left(\frac{1}{r^3} - \frac{ik}{r^2}\right) \exp(ik\,r)$$

$$= \frac{k^2}{4\pi\epsilon_0} \left[(\hat{r} \times \vec{p}_0) \times \hat{r}\right] \frac{\exp(ik\,r)}{r} + \frac{1}{4\pi\epsilon_0} \left[3\hat{r}(\hat{r} \cdot \vec{p}_0 - \vec{p}_0)\right] \left(\frac{1}{r^3} - \frac{ik}{r^2}\right) \exp(ik\,r) \,. \tag{5.115}$$

Finally, we have found the exact result for the electric field radiated from the dipole, valid for all r,

$$\vec{E}_0(\vec{r}) = \frac{k^2 \exp(ik\,r)}{4\pi\epsilon_0 \, r} \left[(\hat{r} \times \vec{p}_0) \times \hat{r}\right] - \frac{ik \exp(ik\,r)}{4\pi\epsilon_0 \, r^2} \left(1 - \frac{1}{ikr}\right) \left[3\hat{r}(\hat{r} \cdot \vec{p}_0 - \vec{p}_0)\right] \,. \tag{5.116}$$

In the near field, $kr \ll 1$, we can replace $\exp(ik\,r) \to 1$ and find for the magnetic field in the near zone,

$$\vec{B}_0(\vec{r}) \sim \frac{ik}{4\pi\epsilon_0 \, cr^2} \, (\hat{r} \times \vec{p}_0) \,, \qquad k\,r \to 0 \,, \tag{5.117}$$

and for the electric field in the near zone,

$$\vec{E}_0(\vec{r}) \sim \frac{1}{4\pi\epsilon_0 \, r^3} \left[3\hat{r}(\hat{r} \cdot \vec{p}_0 - \vec{p}_0)\right] \,, \qquad k\,r \to 0 \,. \tag{5.118}$$

Even for dipole radiation, the derivation of the exact result (5.116) for the electric field is a nontrivial exercise, and it is helpful to include all intermediate steps.

5.4 Tensor Green Function

5.4.1 *Clebsch–Gordan Coefficients: Motivation*

Angular momentum algebra is connected to the so-called Clebsch–Gordan or vector addition coefficients. These coefficients are of rather universal applicability over wide ranges of physical theory. One uses them in order to find the expansion coefficients of a tensor of higher rank as it is composed out of elements of tensors of lower rank, hence the name "vector" addition coefficients (while "tensor" addition might otherwise be a more precise formulation).

One of the most important symmetries in physics concerns the group of rotations, or, the special orthogonal group in three dimensions, called SO(3). We know that the scalar product $\vec{u} \cdot \vec{v}$ of two vectors is invariant under rotations. The vector product $\vec{u} \times \vec{v}$ transforms as a (pseudo-)vector itself. We recall that a pseudo-vector, in contrast to a vector, conserves its sign under parity, $\vec{u} \times \vec{v} \to (-\vec{u}) \times (-\vec{v})$. Vectors do not mix with scalars under rotations. The vectors are of rank one, whereas scalars are tensors of rank zero. Furthermore, from the tensor product of two vectors, we can extract a third quantity which is a quadrupole tensor of rank two, which also does not mix with tensors of different rank under rotations. From the tensor product of two vectors, one may thus extract tensors of rank zero, one, and two. The components of these "constructed" tensors are linear combinations of products of the components of the vectors. The corresponding formalism, in the spherical basis, involves the vector addition, or Clebsch–Gordan, coefficients.

The procedure is illustrated most effectively by way of an example. We consider a second-rank tensor from the tensor product of two vectors,

$$\mathbb{T} = \vec{u} \otimes \vec{v} = \begin{pmatrix} u_x v_x & u_x v_y & u_x v_z \\ u_y v_x & u_y v_y & u_y v_z \\ u_z v_x & u_z v_y & u_z v_z \end{pmatrix}. \tag{5.119}$$

We can decompose $\mathbb{T}$ as follows,

$$\mathbb{T} = \mathbb{T}|_{\ell=0} + \mathbb{T}|_{\ell=1} + \mathbb{T}|_{\ell=2}, \tag{5.120}$$

$$\mathbb{T}|_{\ell=0} = \frac{\mathrm{trc}(\mathbb{T})}{3} \mathbb{1}_{3\times3}, \qquad \mathrm{trc}(\mathbb{T}) = u_x v_x + u_y v_y + u_z v_z, \tag{5.121}$$

$$\mathbb{T}|_{\ell=1} = \frac{1}{2} \left(\mathbb{T} - \mathbb{T}^T\right) = \begin{pmatrix} 0 & \frac{1}{2}\left(u_x v_y - u_y v_x\right) & -\frac{1}{2}\left(u_z v_x - u_x v_z\right) \\ -\frac{1}{2}\left(u_x v_y - u_y v_x\right) & 0 & \frac{1}{2}\left(u_y v_z - u_z v_y\right) \\ \frac{1}{2}\left(u_z v_x - u_x v_z\right) & -\frac{1}{2}\left(u_y v_z - u_z v_y\right) & 0 \end{pmatrix}. \tag{5.122}$$

The ($\ell = 0$)-component is invariant under rotations. The entries of the matrix $\mathbb{T}|_{\ell=1}$ can be identified as the components of the vector product of $\vec{u}$ and $\vec{v}$, with details to be discussed below. Finally, the ($\ell = 2$)-component reads as

$$\mathbb{T}|_{\ell=2} = \frac{1}{2}\left(\mathbb{T} + \mathbb{T}^T\right) - \frac{\mathrm{trc}(\mathbb{T})}{3}\mathbb{1}_{3\times3} \tag{5.123}$$

$$= \begin{pmatrix} \dfrac{2u_x v_x - u_y v_y - u_z v_z}{3} & \dfrac{u_x v_y + u_y v_x}{2} & \dfrac{u_x v_z + u_z v_x}{2} \\ \dfrac{u_x v_y + u_y v_x}{2} & \dfrac{2u_y v_y - u_x v_x - u_z v_z}{3} & \dfrac{u_y v_z + u_z v_y}{2} \\ \dfrac{u_x v_z + u_z v_x}{2} & \dfrac{u_y v_z + u_z v_y}{2} & \dfrac{2u_z v_z - u_x v_x - u_y v_y}{3} \end{pmatrix}.$$

As already anticipated, and with reference to Eq. (1.23), we can order the non-vanishing components of the antisymmetric tensor $\mathbb{T}|_{\ell=1}$ into a vector, namely, the vector product of $\vec{u}$ and $\vec{v}$,

$$\vec{u} \times \vec{v} = \begin{pmatrix} u_y\, v_z - u_z\, v_y \\ u_z\, v_x - u_x\, v_z \\ u_x\, v_y - u_y\, v_x \end{pmatrix}, \qquad (\vec{u} \times \vec{v})_i = \epsilon_{ijk}\, u_j\, v_k, \tag{5.124}$$

where a summation over j and k is understood by the Einstein summation convention [see Eq. (1.45)]. The fact that $\mathbb{T}|_{\ell=1}$ transforms as a (pseudo-)vector shows that it is possible to construct a vector whose components themselves are products of vector components. In Cartesian components, the "coupling coefficients" can directly be read off from Eq. (5.124), in the sense that the components of a tensor of rank one (of the vector product) are obtained by multiplying the products $u_j\, v_k$ of the components of the vectors $\vec{u}$ and $\vec{v}$ by the coupling coefficient ϵ_{ijk}.

However, the canonical formalism employs the spherical basis. We recall from Eq. (2.30) that the z component of the angular momentum operator has a particularly simple form in spherical coordinates [see Eq. (2.30c)],

$$L_z = -\mathrm{i}\left(x\, \frac{\partial}{\partial y} - y\, \frac{\partial}{\partial x} \right) = -\mathrm{i}\, \frac{\partial}{\partial \varphi}. \tag{5.125}$$

The spherical basis of vector components is chosen to generate an explicit φ dependence of the form $\exp(\mathrm{i}m\varphi)$, i.e., it consists of eigenfunctions of L_z. In the spherical basis, the components are denoted as x_{+1}, x_0 and x_{-1}; they read as follows,

$$x_{+1} = -\frac{1}{\sqrt{2}}(x + \mathrm{i}y) = -\frac{1}{\sqrt{2}}|\vec{r}|\sin\theta\, \mathrm{e}^{\mathrm{i}\varphi} = \sqrt{\frac{4\pi}{3}}|\vec{r}|\, Y_{11}(\theta, \varphi), \tag{5.126a}$$

$$x_0 = z = |\vec{r}|\cos\theta = \sqrt{\frac{4\pi}{3}}|\vec{r}|\, Y_{10}(\theta, \varphi), \tag{5.126b}$$

$$x_{-1} = \frac{1}{\sqrt{2}}(x - \mathrm{i}y) = \frac{1}{\sqrt{2}}|\vec{r}|\sin\theta\, \mathrm{e}^{-\mathrm{i}\varphi} = \sqrt{\frac{4\pi}{3}}|\vec{r}|\, Y_{1-1}(\theta, \varphi), \tag{5.126c}$$

where we have allowed ourselves the luxury to write $r = |\vec{r}| = \sqrt{x^2 + y^2 + z^2}$ explicitly. The emergence of phase factors of the form $\exp(\mathrm{i}m\,\varphi)$ with $m = -1, 0, 1$ is evident. The spherical components are complemented by spherical basis vectors as follows,

$$\vec{e}_{+1} = -\frac{1}{\sqrt{2}}(\hat{e}_x + \mathrm{i}\hat{e}_y), \qquad \vec{e}_0 = \hat{e}_z, \qquad \vec{e}_{-1} = \frac{1}{\sqrt{2}}(\hat{e}_x - \mathrm{i}\hat{e}_y). \tag{5.127}$$

The coordinate vector can easily be expanded into the spherical basis,

$$\vec{r} = \sum_{q=-1}^{1} x_q\, \vec{e}_q^{\,*} = \sum_{q=-1}^{1} (-1)^q\, x_{-q}\, \vec{e}_q, \qquad \vec{e}_q^{\,*} = (-1)^q\, \vec{e}_{-q}. \tag{5.128}$$

The spherical basis vectors are normalized as follows,

$$\vec{e}_q \cdot \vec{e}_{q'} = (-1)^q\, \delta_{q\,-q'}, \qquad \vec{e}_q \cdot \vec{e}_{q'}^{\,*} = \delta_{q\,q'}. \tag{5.129}$$

The sum over q runs over the indices $q = -1, 0, 1$. For absolute clarity, we should mention that the latter Kronecker symbol in Eq. (5.128) is to be understood as $\delta_{q,-q'}$, i.e., q has to be equal to $-q'$ in order for the Kronecker symbol to be equal to unity rather than zero. Components can be extracted by calculating the scalar product of the coordinate vector $\vec{r}$ with a spherical basis vector,

$$\vec{r} \cdot \vec{e}_q = \sum_{q'=-1}^{1} x_{q'} \, \vec{e}_{q'}^{\,*} \cdot \vec{e}_q = \sum_{q'=-1}^{1} x_{q'} \, \delta_{q'\,q} = x_q \,. \tag{5.130}$$

The paradigm of the vector coupling, or Clebsch–Gordan, coefficients, is that one obtains a tensor component of magnetic quantum number m for a tensor of rank j by coupling two tensors of rank j_1 and j_2 as follows,

$$v(jm) = \sum_{m_1=-j_1}^{j_1} \sum_{m_2=-j_2}^{j_2} C_{j_1 m_1 j_2 m_2}^{jm} \, u(j_1 \, m_1) \, u(j_2 \, m_2) \,. \tag{5.131}$$

Here, $u(j_1 \, m_1)$ and $u(j_2 \, m_2)$ are two distinct vectors. In our example, we couple two tensors of rank one (the vectors $\vec{u}$ and $\vec{v}$ with spherical components u_{+1}, u_0 and u_{-1}, as well as v_{+1}, v_0 and v_{-1}), to a tensor of rank one, which is the vector product $\vec{w} = \vec{u} \times \vec{v}$. So, we have $j_1 = j_2 = j = 1$ for our example, which is given by Eq. (5.124) in Cartesian coordinates. The Clebsch–Gordan coefficients are tabulated and nowadays implemented in most modern computer algebra systems. Using tabulated values, one finds

$$w_q = \sum_{q'q''} C_{1q'\,1q''}^{1q} \, u_{q'} \, v_{q''} \,, \tag{5.132a}$$

$$w_{+1} = \frac{u_{+1} v_0 - u_0 v_{+1}}{\sqrt{2}} = \left[\frac{\mathrm{i}}{\sqrt{2}}\right] \left\{-\frac{1}{\sqrt{2}}\left[(\vec{u} \times \vec{v})_x + \mathrm{i}\,(\vec{u} \times \vec{v})_y\right]\right\}, \tag{5.132b}$$

$$w_0 = \frac{u_{+1} v_{-1} - u_{-1} v_{+1}}{\sqrt{2}} = \left[\frac{\mathrm{i}}{\sqrt{2}}\right] (\vec{u} \times \vec{v})_z \,, \tag{5.132c}$$

$$w_{-1} = \frac{u_0 v_{-1} - u_{-1} v_0}{\sqrt{2}} = \left[\frac{\mathrm{i}}{\sqrt{2}}\right] \left\{\frac{1}{\sqrt{2}}\left[(\vec{u} \times \vec{v})_x - \mathrm{i}\,(\vec{u} \times \vec{v})_y\right]\right\}, \tag{5.132d}$$

where the subscripts x, y, z denote the Cartesian components of the vector product, i.e., $(u \times v)_z = u_x v_y - u_y v_x$ and further by cyclic permutation. A comparison of Eq. (5.124) with Eq. (5.126) shows that the components of $\vec{w}$ are equal to those obtained by writing $\vec{u} \times \vec{v}$ in the spherical basis, up to a prefactor $\mathrm{i}/\sqrt{2}$, i.e.,

$$\vec{w} = \sum_{q=-1}^{1} (-1)^q \, w_{-q} \, \vec{e}_q = \frac{\mathrm{i}}{\sqrt{2}} \, (\vec{u} \times \vec{v}) \,. \tag{5.133}$$

One can form two further linear combinations of the spherical basis vectors $\vec{e}_q$ which are of interest,

$$\vec{e}_q \times \vec{e}_{q'} = \mathrm{i}\sqrt{2} \sum_{\lambda=-1}^{1} C_{1q1q'}^{1\lambda} \, \vec{e}_\lambda \,, \qquad \vec{r} = -\sqrt{3} \sum_{q=-1}^{1} \sum_{q'=-1}^{1} C_{1q1q'}^{00} \, x_q \, \vec{e}_{q'} \,. \tag{5.134}$$

The first of these illustrates that the vector product of two basis vectors in the spherical basis again is a vector; the second clarifies that the coordinate vector $\vec{r}$

actually is a scalar under rotations, obtained as the scalar combination (tensor of rank zero, component number zero) composed out of the spherical coordinates x_q and the spherical basis vectors $\vec{e}_q$. Of course, the vector composed of the components x_q is not a scalar. However, the scalar product of the vector composed of the x_q with the vector composed of the $\vec{e}_q$ constitutes a physical vector $\vec{r}$ which does not change just upon a change of the reference frame [see Eq. (5.128)]. Upon rotation, this scalar product is equal to the coordinate vector obtained using the new coordinates x'_q, but multiplied with the rotated basis vectors $\vec{e}'_q$, which leaves the physical coordinate vector $\vec{r}$ invariant. This corresponds to a passive interpretation of the rotation.

Let us now investigate the ($\ell = 2$)-component given in Eq. (5.123). It is symmetric and traceless and has five independent components. The counting works as follows: We have five independent components, because the matrix is symmetric and traceless. This leaves three off-diagonal and two diagonal components to be determined; the third component on the diagonal is fixed by the condition trc $(\mathbb{T}|_{\ell=2}) = 0$. A scalar ($\ell = 0$) has one component, a vector ($\ell = 1$) has three magnetic components x_m (which depend on φ as $\exp(im\varphi)$ with $m = -1, 0, 1$), and a quadrupole tensor has five magnetic components [which depend on φ as $\exp(im\varphi)$ with $m = -2, -1, 0, 1, 2$]. The generalization calls for $2\ell + 1$ magnetic components for a tensor of rank ℓ. Again, using tabulated values for Clebsch–Gordan coefficients of the form $C^{2q}_{1q'\,1q''}$, with $q', q'' = -1, 0, 1$ and $q = -2, -1, 0, 1, 2$, we have

$$t_q = \sum_{q'q''} C^{2q}_{1q'\,1q''}\, u_{q'}\, v_{q''}\,, \tag{5.135a}$$

$$t_{+2} = u_{+1}\, v_{+1}\,, \qquad t_{+1} = \frac{u_{+1}\, v_0 + u_0\, v_{+1}}{\sqrt{2}}\,, \tag{5.135b}$$

$$t_0 = \frac{3\, u_0\, v_0 - \vec{u}\cdot\vec{v}}{\sqrt{6}}\,, \tag{5.135c}$$

$$t_{-1} = \frac{u_{-1}\, v_0 + u_0\, v_{-1}}{\sqrt{2}}\,, \qquad t_{-2} = u_{-1}\, v_{-1}\,, \tag{5.135d}$$

where the spherical components u_{-1}, u_0 and u_{+1} are defined in Eq. (5.126). If the operators $\vec{L}_1$ and $\vec{L}_2$ address the vectors $\vec{u}$ and $\vec{v}$ separately, then the functions t_q are eigenfunctions of the operator $\vec{L}^2 = (\vec{L}_1 + \vec{L}_2)^2$ with eigenvalue $\vec{L}^2 \to \ell(\ell+1) = 6$ and of the z component $L_z = L_{1z} + L_{2z}$ with an eigenvalue q of L_z.

5.4.2 Vector Additions and Vector Spherical Harmonics

Vector spherical harmonics are obtained upon adding the "spin" of the photon, namely, the spherical basis vectors which are used in the expansion of the vector potential, to the spherical harmonics which represent the "orbital" angular momentum of the photon. Photons (light particles) are spin-1 objects. The spherical basis vectors are components of a tensor of rank one. To this tensor we add, vectorially, the orbital angular momentum ℓ of the photon, as manifest in the spherical harmonic. Hence, in view of Eq. (5.131), the vector spherical harmonic is given as

$$\vec{Y}^{\ell}_{j\mu}(\hat{r}) = \sum_{m=-\ell}^{\ell} \sum_{q=-1}^{1} C^{j\mu}_{\ell m\, 1q}\, Y_{\ell m}(\theta,\varphi)\, \vec{e}_q \,. \tag{5.136}$$

Because neither the total angular momentum quantum number j nor the orbital angular momentum ℓ can be negative, the vector spherical harmonics with $j = -1$ and $\ell = -1$ vanish; this observation comes in handy in regard to a number of summations discussed in the following. From Eq. (5.127), we recall the basis vectors in the spherical basis,

$$\vec{e}_{+1} = -\frac{1}{\sqrt{2}} \left(\hat{e}_x + \mathrm{i}\,\hat{e}_y \right), \qquad \vec{e}_0 = \hat{e}_z\,, \qquad \vec{e}_{-1} = \frac{1}{\sqrt{2}} \left(\hat{e}_x - \mathrm{i}\,\hat{e}_y \right). \tag{5.137}$$

The Clebsch–Gordan coefficients $C^{j\mu}_{\ell m\, 1q}$ assemble a tensor of angular symmetry $j\mu$ from a spherical harmonic of angular symmetry ℓm and a spherical basis vector of angular symmetry $1q$. We prefer the above notation for the vector spherical harmonic; the magnetic projection μ may assume values from $-j$ to j. The superscript ℓ reminds us of the orbital momentum which was used in the construction of the vector spherical harmonic. The spin of the photon (equal to one) is added to the orbital angular momentum; hence the total angular momentum j can differ from ℓ by at most unity; otherwise the vector spherical harmonic vanishes.

The spin operators of the photon are given by the matrices $\$_i$ $(i = 1, 2, 3)$,

$$(\$_k)_{ij} = -\mathrm{i}\,\epsilon_{kij}\,, \tag{5.138}$$

where ϵ_{ijk} is the Levi–Cività tensor [see Eq. (1.23)]. The explicit representation for $k = 1, 2, 3$ reads as

$$\$_1 = \begin{pmatrix} 0 & 0 & 0 \\ 0 & 0 & -\mathrm{i} \\ 0 & \mathrm{i} & 0 \end{pmatrix} \qquad \$_2 = \begin{pmatrix} 0 & 0 & \mathrm{i} \\ 0 & 0 & 0 \\ -\mathrm{i} & 0 & 0 \end{pmatrix}, \qquad \$_3 = \begin{pmatrix} 0 & -\mathrm{i} & 0 \\ \mathrm{i} & 0 & 0 \\ 0 & 0 & 0 \end{pmatrix}. \tag{5.139}$$

The matrix $\mathbb{M}$ given above is identified as the lower right 2×2 submatrix of $\mathbb{M}_1$, and also as the upper left 2×2 submatrix of $\mathbb{M}_3$. These matrices and the corresponding components of the angular momentum vector fulfill the algebraic relations

$$[\,\$_i, \$_j\,] = \mathrm{i}\,\epsilon_{ijk}\, \$_k\,, \tag{5.140}$$

where the Einstein summation convention is used for the sum over $k = 1, 2, 3$ on the right-hand side. The vector square of the $\$$ matrices is

$$\vec{\$}^2 = \begin{pmatrix} 2 & 0 & 0 \\ 0 & 2 & 0 \\ 0 & 0 & 2 \end{pmatrix} = S\,(S+1)\,\mathbb{1}_{3\times3}\,, \qquad S = 1\,, \tag{5.141}$$

demonstrating that the photon is a spin-1 particle (with $S = 1$). The total angular momentum operator of the photon is given by

$$\vec{J} = \vec{L}\,\mathbb{1}_{3\times3} + \vec{\$}\,, \tag{5.142}$$

where we are pedantic in multiplying the orbital angular momentum operator by the three-dimensional unit matrix, implying that, say, the z component $L_z \vec{A} = L_z \mathbb{1} \cdot \vec{A}$ acts on the entire vector $\vec{A}$, i.e., on all of the components of $\vec{A}$ separately. The z component of $\vec{J}$ acts on a vector-valued function as $J_z \vec{V}(\theta, \varphi) = L_z \vec{V}(\theta, \varphi) + \$_z \cdot \vec{V}(\theta, \varphi)$. The vector spherical harmonics have the properties

$$\vec{J}^2 \vec{Y}^\ell_{j\mu}(\theta, \varphi) = j(j+1) \vec{Y}^\ell_{j\mu}(\theta, \varphi), \tag{5.143a}$$

$$\vec{L}^2 \vec{Y}^\ell_{j\mu}(\theta, \varphi) = \ell(\ell+1) \vec{Y}^\ell_{j\mu}(\theta, \varphi), \tag{5.143b}$$

$$J_z \vec{Y}^\ell_{j\mu}(\theta, \varphi) = \mu \vec{Y}^\ell_{j\mu}(\theta, \varphi). \tag{5.143c}$$

The orthonormality relations are

$$\int d\Omega \, \vec{Y}^{\ell\,*}_{j\mu}(\theta, \varphi) \cdot \vec{Y}^{\ell'}_{j'\mu'}(\theta, \varphi) = \delta_{jj'} \, \delta_{\ell\ell'} \, \delta_{\mu\mu'}, \tag{5.144a}$$

$$\sum_{\mu=-j}^{j} \int d\Omega \, \vec{Y}^\ell_{j\mu}(\theta, \varphi) \otimes \vec{Y}^{\ell\,*}_{j\mu}(\theta, \varphi) = \frac{2j+1}{3} \mathbb{1}_{3\times 3}, \tag{5.144b}$$

where we assume that the vector spherical harmonic is nonvanishing, i.e., $|j - \ell| \leq 1$. In the second sum, j and ℓ are held constant; they just have to be the same for both vector spherical harmonics but are not summed over. For given j, there are three possible values of ℓ, namely, $\ell = j - 1, j, j + 1$; summing over these, one obtains a factor $(2j+1)$ instead of $(2j+1)/3$ on the right-hand side.

There is also a completeness relation,

$$\sum_{j=\ell-1}^{\ell+1} \sum_{\mu=-j}^{j} \vec{Y}^\ell_{j\mu}(\theta, \varphi) \otimes \vec{Y}^{\ell\,*}_{j\mu}(\theta', \varphi') = \sum_{m=-\ell}^{\ell} Y_{\ell m}(\theta, \varphi) Y^*_{\ell m}(\theta', \varphi') \mathbb{1}_{3\times 3}, \tag{5.145}$$

which implies that

$$\sum_{\ell=0}^{\infty} \sum_{j=\ell-1}^{\ell+1} \sum_{\mu=-j}^{j} \vec{Y}^\ell_{j\mu}(\theta, \varphi) \otimes \vec{Y}^{\ell\,*}_{j\mu}(\theta', \varphi') = \frac{1}{\sin\theta} \delta(\theta - \theta') \delta(\varphi - \varphi') \mathbb{1}_{3\times 3}. \tag{5.146}$$

The relation analogous to Eq. (2.56a) is

$$\vec{Y}^{\ell\,*}_{j\mu}(\theta, \varphi) = (-1)^{\ell+1+j} (-1)^\mu \vec{Y}^\ell_{j\,-\mu}(\theta, \varphi). \tag{5.147}$$

Explicit representations are given as follows,

$$\vec{Y}^j_{j\mu}(\theta, \varphi) = \frac{1}{\sqrt{j(j+1)}} \vec{L} Y_{j\mu}(\theta, \varphi), \tag{5.148a}$$

$$\vec{Y}^{j-1}_{j\mu}(\theta, \varphi) = \frac{1}{\sqrt{j(2j+1)}} \left(j\hat{r} + r\vec{\nabla} \right) Y_{j\mu}(\theta, \varphi)$$

$$= -\frac{1}{\sqrt{j(2j+1)}} \left(i\hat{r} \times \vec{L} - j\hat{r} \right) Y_{j\mu}(\theta, \varphi), \tag{5.148b}$$

$$\vec{Y}^{j+1}_{j\mu}(\theta, \varphi) = -\frac{1}{\sqrt{(j+1)(2j+1)}} \left((j+1)\hat{r} - r\vec{\nabla} \right) Y_{j\mu}(\theta, \varphi)$$

$$= -\frac{1}{\sqrt{(j+1)(2j+1)}} \left(i\hat{r} \times \vec{L} + (j+1)\hat{r} \right) Y_{j\mu}(\theta, \varphi). \tag{5.148c}$$

Here, μ can take on the values $\mu = -j, \ldots, j$. Again, we emphasize that vector spherical harmonics with $|j - \ell| > 1$ vanish.

The two equivalent representations of the vector spherical harmonics can be reconciled with each other on the basis of the operator identity

$$(\mathrm{i}\,\hat{r} \times \vec{L})\, f(\theta, \varphi) = -r\, \vec{\nabla} f(\theta, \varphi)\,, \tag{5.149}$$

which is valid for any test function f that only depends on the angular variables.

We can write a representation of the vector spherical harmonics in terms of Clebsch–Gordan coefficients. For example, we have in the case $\ell = j$,

$$\vec{Y}_{j\mu}^{j}(\theta, \varphi) = C_{j\,\mu-1\,1\,1}^{j\mu}\, \vec{e}_{+1}\, Y_{j\,\mu-1}(\theta, \varphi) + C_{j\,\mu\,1\,0}^{j\mu}\, \vec{e}_{0}\, Y_{j\,\mu}(\theta, \varphi) + C_{j\,\mu+1\,1\,-1}^{j\mu}\, \vec{e}_{-1}\, Y_{j\,\mu+1}(\theta, \varphi)\,. \tag{5.150}$$

Using known formulas for the Clebsch–Gordan coefficients, the vector spherical harmonics with $\ell = j$ find the following representation,

$$\vec{Y}_{j\mu}^{j}(\theta, \varphi) = -\sqrt{\frac{(j+\mu)\,(j-\mu+1)}{2j(j+1)}}\, Y_{j,\mu-1}(\theta, \varphi)\, \vec{e}_{+1} + \frac{\mu}{\sqrt{j(j+1)}}\, Y_{j,\mu}(\theta, \varphi)\, \vec{e}_{0}$$

$$+ \sqrt{\frac{(j-\mu)\,(j+\mu+1)}{2j(j+1)}}\, Y_{j,\mu+1}(\theta, \varphi)\, \vec{e}_{-1}\,. \tag{5.151a}$$

For the case $\ell = j - 1$, one has

$$\vec{Y}_{j\mu}^{j-1}(\theta, \varphi) = \sqrt{\frac{(j+\mu-1)\,(j+\mu)}{2j(2j-1)}}\, Y_{j-1,\mu-1}(\theta, \varphi)\, \vec{e}_{+1}$$

$$+ \sqrt{\frac{(j-\mu)\,(j+\mu)}{j(2j-1)}}\, Y_{j-1,\mu}(\theta, \varphi)\, \vec{e}_{0}$$

$$+ \sqrt{\frac{(j-\mu-1)\,(j-\mu)}{2j(2j-1)}}\, Y_{j-1,\mu+1}(\theta, \varphi)\, \vec{e}_{-1}\,. \tag{5.151b}$$

Finally, for the case $\ell = j + 1$, one has

$$\vec{Y}_{j\mu}^{j+1}(\theta, \varphi) = \sqrt{\frac{(j-\mu+1)\,(j-\mu+2)}{2(j+1)(2j+3)}}\, Y_{j+1,\mu-1}(\theta, \varphi)\, \vec{e}_{+1}$$

$$- \sqrt{\frac{(j-\mu+1)\,(j+\mu+1)}{(j+1)(2j+3)}}\, Y_{j+1,\mu}(\theta, \varphi)\, \vec{e}_{0}$$

$$+ \sqrt{\frac{(j+\mu+2)\,(j+\mu+1)}{2(j+1)(2j+3)}}\, Y_{j+1,\mu+1}(\theta, \varphi)\, \vec{e}_{-1}\,. \tag{5.151c}$$

Here, one easily discerns the addition of the magnetic quantum number, of the spherical harmonic and the spherical basis vector, to the total magnetic projection of the vector spherical harmonic.

5.4.3 *Scalar Helmholtz Green Function and Scalar Potential*

We recall that the Helmholtz Green function is given as follows [see Eq. (5.20)],

$$G_R(k, \vec{r} - \vec{r}') = \frac{\exp(ik|\vec{r} - \vec{r}'|)}{4\pi\epsilon_0 |\vec{r} - \vec{r}'|}. \tag{5.152}$$

This Green function couples the scalar potential to the source, according to Eq. (5.77),

$$\Phi(\vec{r}, t) = e^{-i\omega t} \Phi_0(\vec{r}), \qquad \Phi_0(\vec{r}) = \int d^3 r' \, G_R\left(\frac{\omega}{c}, \vec{r} - \vec{r}'\right) \rho_0(\vec{r}'). \tag{5.153}$$

For definiteness, we also recall the angular decomposition of the Green function for the Helmholtz equation according to Eq. (5.72),

$$G_R(k, \vec{r} - \vec{r}') = \frac{1}{\epsilon_0} \sum_{\ell,m} i k \, j_\ell(k \, r_<) \, h_\ell^{(1)}(k \, r_>) \, Y_{\ell m}(\theta, \varphi) \, Y_{\ell m}^*(\hat{r}'). \tag{5.154}$$

Outside of the charge distribution, the scalar potential is thus given by

$$\Phi_0(\vec{r}) = \frac{i k}{\epsilon_0} \sum_{\ell=0}^{\infty} \sum_{m=-\ell}^{\ell} p_{\ell m} \, h_\ell^{(1)}(k \, r) \, Y_{\ell m}(\theta, \varphi), \tag{5.155}$$

with

$$p_{\ell m} = \int d^3 r \, j_\ell(k \, r) \, \rho(\vec{r}) \, Y_{\ell m}^*(\theta, \varphi) \approx \frac{k^\ell}{(2\ell + 1)!!} \int d^3 r \, r^\ell \, \rho(\vec{r}) \, Y_{\ell m}^*(\theta, \varphi). \tag{5.156}$$

The $p_{\ell m}$ generalize the multipole components of a static charge distribution, defined according to Eq. (2.97), for a dynamical process, namely, the emission of radiation. In the notation, we suppress their dependence on the wave number k. Indeed, for $k \to 0$ (zero-frequency radiation), the leading term in the expansion of $p_{\ell m}$ according to Eq. (5.156) is proportional to $q_{\ell m}$, a fact which is obvious from Eq. (5.30).

5.4.4 *Tensor Helmholtz Green Function and Vector Potential*

The tensor Helmholtz Green function is obtained from the Helmholtz Green function by a multiplication with the unit matrix,

$$\mathbb{G}_R(k, \vec{r} - \vec{r}') = \mathbb{1}_{3\times3} \, G_R(k, \vec{r} - \vec{r}'). \tag{5.157}$$

It enters the equation that couples the vector potential to the source, Eq. (5.78), which we write as follows,

$$\vec{A}(\vec{r}, t) = e^{-i\omega t} \vec{A}_0(\vec{r}), \qquad \vec{A}_0(\vec{r}) = \int d^3 r' \, \frac{1}{c^2} \, \mathbb{G}_R\left(\frac{\omega}{c}, \vec{r} - \vec{r}'\right) \cdot \vec{J}_0(\vec{r}'). \tag{5.158}$$

Note the explicit matrix product of the tensor Green function and the current density, which differentiates Eq. (5.158) from (5.78).

In order to proceed with the analysis of the tensor Green function, we first need to write a tensorial decomposition. The unit matrix in Eq. (5.157), which describes the spin of the photon, needs to be incorporated into the analysis. Of course, the hope is that once we form the tensor product of all the vector spherical harmonics pertaining to the same orbital angular momentum ℓ, we would somehow recover the angular structure in Eq. (5.72), namely, $\sum_{\ell m} Y_{\ell m}(\theta, \varphi) Y_{\ell m}^*(\theta', \varphi')$, multiplied by the unit matrix $\mathbb{1}_{3\times 3}$. Indeed, we recall Eq. (5.145),

$$\sum_{j=\ell-1}^{\ell+1} \sum_{\mu=-j}^{j} \vec{Y}_{j\mu}^{\ell}(\theta, \varphi) \otimes \vec{Y}_{j\mu}^{\ell*}(\theta', \varphi') = \sum_{m=-\ell}^{\ell} Y_{\ell m}(\theta, \varphi) Y_{\ell m}^*(\theta', \varphi') \, \mathbb{1}_{3\times 3} \, . \tag{5.159}$$

For the case $\ell = 0$, we recall the observation reported in the text following Eq. (5.136). Essentially, in Eq. (5.145), for given ℓ, one sums over the possible values of j, namely, $j = \ell - 1, \ell, \ell + 1$, and then over the possible magnetic projections μ, and obtains an angular structure which is familiar from Eq. (5.72). Here, the tensor product is the one which transforms the two vectors $Y_{\ell m}(\theta, \varphi)$ and $Y_{\ell m}^*(\theta', \varphi')$ into a matrix; pedantically, one might otherwise have indicated the transpose of the latter vector.

The angular decomposition of the tensor Green function (5.72) for the vector Helmholtz equation can thus be given in terms of the vector spherical harmonics,

$$\mathbb{G}_R(k, \vec{r} - \vec{r}') = \frac{\exp(\mathrm{i}k|\vec{r} - \vec{r}'|)}{4\pi\epsilon_0 |\vec{r} - \vec{r}'|} \, \mathbb{1}_{3\times 3}$$

$$= \frac{\mathrm{i}k}{\epsilon_0} \sum_{j=0}^{\infty} \sum_{\mu=-j}^{j} \sum_{\ell=j-1}^{j+1} j_\ell(k r_<) h_\ell^{(1)}(k r_>) \vec{Y}_{j\mu}^{\ell}(\theta, \varphi) \otimes \vec{Y}_{j\mu}^{\ell*}(\theta', \varphi')$$

$$= \frac{\mathrm{i}k}{\epsilon_0} \sum_{j=0}^{\infty} \sum_{\mu=-j}^{j} \Big(j_{j-1}(k r_<) h_{j-1}^{(1)}(k r_>) \vec{Y}_{j\mu}^{j-1}(\theta, \varphi) \otimes \vec{Y}_{j\mu}^{j-1*}(\theta', \varphi')$$

$$+ j_j(k r_<) h_j^{(1)}(k r_>) \vec{Y}_{j\mu}^{j}(\theta, \varphi) \otimes \vec{Y}_{j\mu}^{j*}(\theta', \varphi')$$

$$+ j_{j+1}(k r_<) h_{j+1}^{(1)}(k r_>) \vec{Y}_{j\mu}^{j+1}(\theta, \varphi) \otimes \vec{Y}_{j\mu}^{j+1*}(\theta', \varphi') \Big) \, . \tag{5.160}$$

So, outside of the charge distribution, the vector potential is thus given by

$$\vec{A}_0(\vec{r}) = \frac{1}{c^2} \int \mathrm{d}^3 r' \, \mathbb{G}_R(k, \vec{r} - \vec{r}') \cdot \vec{J}_0(\vec{r}')$$

$$= \int \mathrm{d}^3 r' \, \frac{\exp(\mathrm{i}k|\vec{r} - \vec{r}'|)}{4\pi\epsilon_0 c^2 |\vec{r} - \vec{r}'|} \, \mathbb{1}_{3\times 3} \cdot \vec{J}_0(\vec{r}')$$

$$= \mathrm{i}\mu_0 k \sum_{j=0}^{\infty} \sum_{\mu=-j}^{j} \sum_{\ell=j-1}^{j+1} p_{j\mu}^{\ell} \, h_\ell^{(1)}(k r) \, \vec{Y}_{j\mu}^{\ell}(\theta, \varphi) \, , \tag{5.161}$$

with

$$p_{j\mu}^{\ell} = \int \mathrm{d}^3 r \, j_\ell(k r) \, \vec{J}_0(\vec{r}) \cdot \vec{Y}_{j\mu}^{\ell*}(\theta, \varphi) \approx \frac{k^\ell}{(2\ell+1)!!} \int \mathrm{d}^3 r \, r^\ell \, \vec{J}_0(\vec{r}) \cdot \vec{Y}_{j\mu}^{\ell*}(\theta, \varphi) \, . \tag{5.162}$$

The advantage of Eq. (5.161) over Eq. (5.87) lies in the fact that the vector structure of the radiated vector potential is resolved and the spin of the photon is incorporated into the formalism.

But we are not quite there yet. Namely, the decomposition (5.161) does not clearly separate the longitudinal and transverse components to the electric and magnetic fields generated by the vector potential. An alternative decomposition, which accomplishes a separation into electric and magnetic multipole radiation, reads as follows,

$$\mathbb{G}_R(k,\vec{r}-\vec{r}') = \frac{\mathrm{i}k}{\epsilon_0} \sum_{j=0}^{\infty} \sum_{\mu=-j}^{j} \left(\vec{M}_{j\mu}^{(1)}(k,\vec{r}_>) \otimes \vec{M}_{j\mu}^{(0)*}(k,\vec{r}_<) \right.$$

$$\left. + \vec{N}_{j\mu}^{(1)}(k,\vec{r}_>) \otimes \vec{N}_{j\mu}^{(0)*}(k,\vec{r}_<) + \vec{L}_{j\mu}^{(1)}(k,\vec{r}_>) \otimes \vec{L}_{j\mu}^{(0)*}(k,\vec{r}_<) \right). \quad (5.163)$$

From the above discussion, it is clear that ℓ takes the role otherwise taken by j, because we have added the spin of the photon to its orbital angular momentum. We shall later see that the angular dependence of the functions $M_{j\mu}^{(K)}$, $N_{j\mu}^{(K)}$, and $L_{j\mu}^{(K)}$, is given by vector spherical harmonics with a total angular momentum number j. The notation is to be explained in the following. For the magnetic (M), electric (N), and longitudinal (L) multipole moments, we have

$$\vec{M}_{j\mu}^{(K)}(k,\vec{r}) = \frac{1}{\sqrt{j(j+1)}} f_j^{(K)}(kr)\, \vec{L} Y_{j\mu}(\theta,\varphi), \quad (5.164a)$$

$$\vec{N}_{j\mu}^{(K)}(k,\vec{r}) = \frac{\mathrm{i}}{k} \vec{\nabla} \times \vec{M}_{j\mu}^{(K)}(k,\vec{r}), \quad (5.164b)$$

$$\vec{L}_{j\mu}^{(K)}(k,\vec{r}) = \frac{1}{k} \vec{\nabla} \left(f_j^{(K)}(kr)\, Y_{j\mu}(\theta,\varphi) \right), \quad (5.164c)$$

and conversely

$$\vec{M}_{j\mu}^{(K)}(k,\vec{r}) = -\frac{\mathrm{i}}{k} \vec{\nabla} \times \vec{N}_{j\mu}^{(K)}(k,\vec{r}). \quad (5.165)$$

The definition of $\vec{M}_{00}^{(K)}(k,\vec{r})$ needs to be clarified for Eq. (5.164). In principle, we are dividing zero by zero because the application of the $\vec{L}$ operator to Y_{00} leads to zero, but the prefactor $1/\sqrt{j(j+1)}$ has a zero in the denominator. The solution is to allow for an infinitesimal displacement of the angular momentum $j \to j+\varepsilon$ before applying the $\vec{L}$ operator and then letting $\varepsilon \to 0$ at the end of the calculation. This clarifies that

$$\vec{M}_{00}^{(K)}(k,\vec{r}) = \vec{N}_{00}^{(K)}(k,\vec{r}) = \vec{0}. \quad (5.166)$$

The $f_j^{(K)}(kr)$ with $K = -1,0,1$ are Bessel and Hankel functions, as follows,

$$f_j^{(0)}(kr) = j_j(kr), \qquad f_j^{(1)}(kr) = h_j^{(1)}(kr), \qquad f_j^{(-1)}(kr) = h_j^{(2)}(kr). \quad (5.167)$$

The vector multipole decomposition involves the following moments,

$$m_{j\mu} = \int \mathrm{d}^3 r'\, \vec{J}_0(\vec{r}') \cdot \vec{M}_{j\mu}^{(0)*}(\theta',\varphi'), \quad (5.168a)$$

$$n_{j\mu} = \int \mathrm{d}^3 r'\, \vec{J}_0(\vec{r}') \cdot \vec{N}_{j\mu}^{(0)*}(\theta',\varphi'), \quad (5.168b)$$

$$l_{j\mu} = \int \mathrm{d}^3 r'\, \vec{J}_0(\vec{r}') \cdot \vec{L}_{j\mu}^{(0)*}(\theta',\varphi'), \quad (5.168c)$$

whose dependence on k is suppressed. (This dependence is due to the $f_j^{(K)}$ functions defined in Eq. (5.167), which enter the $\vec{M}_{j\mu}^{(0)*}$, $\vec{N}_{j\mu}^{(0)*}$, and $\vec{L}_{j\mu}^{(0)*}$, according to Eq. (5.164).) We note that the curl operator in Eq. (5.168b) admixes Bessel functions j_{j-1} and j_{j+1} to j_j. The approximation

$$f_j^{(0)}(kr) = j_j(kr) \approx \frac{(kr)^j}{(2j+1)!!} \tag{5.169}$$

is otherwise applicable for a localized source, in much the same way as in Eqs. (5.156) and (5.162). Finally, the decomposition of the vector potential reads as

$$\vec{A}_0(\vec{r}) = ik\,\mu_0 \sum_{j=0}^{\infty} \sum_{\mu=-j}^{j} \left(m_{j\mu}\,\vec{M}_{j\mu}^{(1)}(k,\vec{r}) + n_{j\mu}\,\vec{N}_{j\mu}^{(1)}(k,\vec{r}) + l_{j\mu}\,\vec{L}_{j\mu}^{(1)}(k,\vec{r}) \right). \tag{5.170}$$

Here, the $m_{j\mu}$ are the magnetic multipole moments, the $n_{j\mu}$ are the electric multipole moments, and the $l_{j\mu}$ are longitudinal multipole moments which do not contribute to the fields.

A (perhaps) more systematic expansion writes the $\vec{M}_{j\mu}^{(K)}(k,\vec{r})$, $\vec{N}_{j\mu}^{(K)}(k,\vec{r})$, and $\vec{L}_{j\mu}^{(K)}(k,\vec{r})$ in terms of the vector spherical harmonics with $\ell = j-1, j, j+1$, but definite j,

$$\vec{M}_{j\mu}^{(K)}(k,\vec{r}) = f_j^{(K)}(kr)\,\vec{Y}_{j\mu}^j(\theta,\varphi), \tag{5.171a}$$

$$\vec{N}_{j\mu}^{(K)}(k,\vec{r}) = -\sqrt{\frac{j+1}{2j+1}}\,f_{j-1}^{(K)}(kr)\,\vec{Y}_{j\mu}^{j-1}(\theta,\varphi) + \sqrt{\frac{j}{2j+1}}\,f_{j+1}^{(K)}(kr)\,\vec{Y}_{j\mu}^{j+1}(\theta,\varphi), \tag{5.171b}$$

$$\vec{L}_{j\mu}^{(K)}(k,\vec{r}) = \sqrt{\frac{j}{2j+1}}\,f_{j-1}^{(K)}(kr)\,\vec{Y}_{j\mu}^{j-1}(\theta,\varphi) + \sqrt{\frac{j+1}{2j+1}}\,f_{j+1}^{(K)}(kr)\,\vec{Y}_{j\mu}^{j+1}(\theta,\varphi). \tag{5.171c}$$

From Eq. (5.171b), one might think that $\vec{N}_{00}^{(K)}(k,\vec{r})$ could incur a nonvanishing contribution proportional to $\vec{Y}_{00}^1(\theta,\varphi)$, but the prefactor of this term vanishes. Equation (5.171c) teaches us that the term proportional to $\vec{Y}_{00}^1(\theta,\varphi)$ is part of the longitudinal component of the vector potential, which does not contribute to the electric and magnetic fields. There is no such thing as a photon with vanishing total angular momentum $j = \mu = 0$. Yet another representation is as follows,

$$\vec{M}_{j\mu}^{(K)}(k,\vec{r}) = f_j^{(K)}(kr)\,\vec{Y}_{j\mu}^j(\theta,\varphi), \tag{5.172a}$$

$$\vec{N}_{j\mu}^{(K)}(k,\vec{r}) = \frac{1}{kr}\left\{ \frac{d[kr\,f_j^{(K)}(kr)]}{d(kr)}[i\hat{r}\times\vec{Y}_{j\mu}^j(\theta,\varphi)] - \hat{r}\sqrt{j(j+1)}f_j^{(K)}(kr)\,Y_{j\mu}(\theta,\varphi) \right\}, \tag{5.172b}$$

$$\vec{L}_{j\mu}^{(K)}(k,\vec{r}) = \hat{r}\,\frac{d[f_j^{(K)}(kr)]}{d(kr)}\,Y_{j\mu}(\theta,\varphi) - \sqrt{j(j+1)}\,\frac{1}{kr}\,f_j^{(K)}(kr)\,(i\hat{r}\times\vec{Y}_{j\mu}^j(\theta,\varphi)). \tag{5.172c}$$

Let us now try to interpret the contributions $\vec{M}_{j\mu}^{(K)}(k,\vec{r})$, $\vec{N}_{j\mu}^{(K)}(k,\vec{r})$, and $\vec{L}_{j\mu}^{(K)}(k,\vec{r})$ physically. From Eq. (5.164c), we infer that $\vec{L}_{j\mu}^{(K)}(k,\vec{r})$ is the "longitudinal" solution constructed by taking the gradient of a scalar solution of the Helmholtz equation. The magnetic and electric fields are obtained from the vector potential, via the equations

$$\vec{B}_0(\vec{r}) = \vec{\nabla} \times \vec{A}_0(\vec{r}), \qquad \vec{E}_0(\vec{r}) = \frac{ic^2}{\omega} \vec{\nabla} \times \vec{B}_0(\vec{r}), \qquad (5.173)$$

in the source-free region. Because $\vec{L}_{j\mu}^{(K)}(k,\vec{r})$ is a gradient of a scalar solution of the Helmholtz equation, its curl vanishes. Furthermore, in view of Eq. (5.173), the fields generated by the longitudinal components of the vector potential vanish. We note that in view of the relation $\vec{E}(\vec{r},t) = -\vec{\nabla}\Phi(\vec{r},t) - \partial_t\vec{A}(\vec{r},t)$, the electric field can alternatively be calculated as

$$\vec{E}_0(\vec{r}) = -\vec{\nabla}\Phi_0(\vec{r}) + i\omega\vec{A}_0(\vec{r}). \qquad (5.174)$$

We note that $\vec{L}_{j\mu}^{(K)}(k,\vec{r})$ is longitudinal, just like $\vec{\nabla}\Phi_0(\vec{r})$ (finally, it is just a gradient). We shall thus later have to show that in calculating $\vec{E}_0(\vec{r})$, the gradient of Φ_0 given in Eq. (5.155) actually cancels against the time derivative of the longitudinal contribution proportional to $i\omega\,\vec{L}_{j\mu}^{(K)}(k,\vec{r})$ from the time derivative of the vector potential.

Hence, we have

$$\vec{E}_0(\vec{r}) = i\omega\vec{A}_{0\perp}(\vec{r}), \qquad (5.175a)$$

$$\vec{A}_{0\perp}(\vec{r}) = ik\,\mu_0 \sum_{j=0}^{\infty} \sum_{\mu=-j}^{j} \left(m_{j\mu}\,\vec{M}_{j\mu}^{(1)}(k,\vec{r}) + n_{j\mu}\,\vec{N}_{j\mu}^{(1)}(k,\vec{r}) \right), \qquad (5.175b)$$

$$\vec{E}_0(\vec{r}) = -k^2\,c\,\mu_0 \sum_{j=0}^{\infty} \sum_{\mu=-j}^{j} \left(m_{j\mu}\,\vec{M}_{j\mu}^{(1)}(k,\vec{r}) + n_{j\mu}\,\vec{N}_{j\mu}^{(1)}(k,\vec{r}) \right). \qquad (5.175c)$$

From Eq. (5.164a), we infer that $\vec{M}_{j\mu}$ is the (normalized) elementary solution consisting of a Bessel function times $\vec{L}\,Y_{j\mu}(\theta,\varphi) \propto \vec{Y}_{j\mu}^{j}(\theta,\varphi)$. It is (by construction) purely transverse, because $\hat{r} \cdot \vec{Y}_{j\mu}^{j}(\theta,\varphi) \propto \vec{r} \cdot (\vec{r} \times \vec{\nabla})Y_{j\mu}(\theta,\varphi) = 0$. The general rationale is this: The $\vec{M}$ terms are the magnetic multipoles. We have $\vec{B} = \vec{\nabla} \times \vec{A}$ and $\vec{E} \propto \vec{\nabla} \times \vec{B}$. The electric field $\vec{E} \propto \vec{M}$ in this case is transverse, i.e., $\hat{r} \cdot \vec{E} = 0$. This is the right characteristic of a magnetic multipole.

In view of Eq. (5.164b), we infer that $\vec{N}_{j\mu}$ is the solution constructed by the taking the curl of $\vec{M}_{j\mu}$. The $\vec{N}_{j\mu}$ terms are the electric multipoles. Forming the curl of $\vec{N}$, one obtains an expression for $\vec{B}$ which is proportional to $\vec{M}_{j\mu}$, which evidently is transverse. In this particular case, one has $\vec{r} \cdot \vec{B} = 0$. The general result is as follows,

$$\vec{B}_0(\vec{r}) = k^2\,\mu_0 \sum_{j=0}^{\infty} \sum_{\mu=-j}^{j} \left(m_{j\mu}\,\vec{N}_{j\mu}^{(1)}(k,\vec{r}) - n_{j\mu}\,\vec{M}_{j\mu}^{(1)}(k,\vec{r}) \right), \qquad (5.175d)$$

as will be shown below. A transverse magnetic field is characteristic of an electric multipole. One observes, for the electric dipole, that the exact expression for the magnetic field, given in Eq. (5.97), is transverse, while the exact expression for the electric field, given in Eq. (5.116), is not transverse.

It is perhaps instructive to discuss the "construction principle" that leads to Eq. (5.170), i.e., the rationale behind the transformation from Eq. (5.161) to (5.170). For given j and μ, one first identifies the maximal longitudinal subcomponent and calls it $\vec{L}_{j\mu}^{(1)}(k,\vec{r})$. One then constructs the maximal transverse component whose scalar product with $\hat{r}$ vanishes, and identifies it as $\vec{M}_{j\mu}^{(1)}(k,\vec{r}) = 0$. The rest of the component of the vector potential with given j and μ finally is the electric multipole, called $\vec{N}_{j\mu}^{(1)}(k,\vec{r}) = 0$.

A remark is in order. For the magnetic multipoles, the corresponding term in $\vec{A}_0(\vec{r})$ [the term containing $\vec{M}_{j\mu}^{(1)}(k,\vec{r})$] is transverse, i.e., its scalar product with $\hat{r}$ vanishes. The electric field is parallel to the time derivative of the sum of the non-longitudinal components of $\vec{A}_0(\vec{r})$. The electric field involves the combination $m_{j\mu}\vec{M}_{j\mu}^{(1)}(k,\vec{r}) + n_{j\mu}\vec{N}_{j\mu}^{(1)}(k,\vec{r})$, so it is transverse for the magnetic multipoles. The magnetic induction field has the combination $m_{j\mu}\vec{N}_{j\mu}^{(1)}(k,\vec{r}) - n_{j\mu}\vec{M}_{j\mu}^{(1)}(k,\vec{r})$, so it is transverse for the electric multipoles. This observation generalizes the behavior found in Eqs. (5.97) and (5.116), for the exact expressions pertaining to the electric and magnetic fields radiated by a dipole (the magnetic field was found to be transverse).

5.5 Radiation and Angular Momenta

5.5.1 *Radiated Electric and Magnetic Fields*

We shall now try to verify Eq. (5.175d) explicitly. The magnetic radiated field is calculated as follows,

$$\vec{B}_0(\vec{r}) = \vec{\nabla} \times \vec{A}_0(\vec{r})$$

$$= i\,k\,\mu_0 \sum_{\ell m} \left(m_{j\mu}\,\vec{\nabla} \times \vec{M}_{j\mu}^{(1)}(k,\vec{r}) + n_{j\mu}\,\vec{\nabla} \times \vec{N}_{j\mu}^{(1)}(k,\vec{r}) + l_{j\mu}\,\vec{\nabla} \times \vec{L}_{j\mu}^{(1)}(k,\vec{r}) \right)$$

$$= i\,k\,\mu_0 \sum_{j\mu} \left(m_{j\mu}\left(-i\,k\,\vec{N}_{j\mu}^{(1)}(k,\vec{r})\right) + n_{j\mu}\left(i\,k\,\vec{M}_{j\mu}^{(1)}(k,\vec{r})\right) \right)$$

$$= k^2\,\mu_0 \sum_{j\mu} \left(m_{j\mu}\,\vec{N}_{j\mu}^{(1)}(k,\vec{r}) - n_{j\mu}\,\vec{M}_{j\mu}^{(1)}(k,\vec{r}) \right), \qquad (5.176)$$

where we have used Eqs. (5.164) and (5.165). In order to calculate the electric field, one observes that in a source-free region, the Ampere–Maxwell law implies that [see also Eq. (6.87)]

$$\vec{\nabla} \times \vec{B}_0(\vec{r}) = -\frac{i\omega}{c^2}\,\vec{E}_0(\vec{r}) \Rightarrow \vec{E}_0(\vec{r}) = \frac{ic^2}{\omega}\,\vec{\nabla} \times \vec{B}_0(\vec{r}). \qquad (5.177)$$

Using Eqs. (5.164b) and (5.165), this is evaluated as

$$
\vec{E}_0(\vec{r}) = \frac{\mathrm{i}c^2}{\omega} \vec{\nabla} \times \vec{B}_0(\vec{r})
$$

$$
= k^2 \mu_0 \left(\frac{\mathrm{i}c^2}{\omega}\right) \sum_{j\mu} \left(m_{j\mu} \vec{\nabla} \times \vec{N}_{j\mu}^{(1)}(k,\vec{r}) - n_{j\mu} \vec{\nabla} \times \vec{M}_{j\mu}^{(1)}(k,\vec{r})\right)
$$

$$
= \frac{\mathrm{i}k^2 \mu_0 c^2}{\omega} \sum_{j\mu} \left(m_{j\mu} \left(\mathrm{i}k\,\vec{M}_{j\mu}^{(1)}(k,\vec{r})\right) - n_{j\mu} \left(-\mathrm{i}k\,\vec{N}_{j\mu}^{(1)}(k,\vec{r})\right)\right)
$$

$$
= \frac{(\mathrm{i}k)\,\mathrm{i}k^2 \mu_0 c^2}{ck} \sum_{j\mu} \left(m_{j\mu}\,\vec{M}_{j\mu}^{(1)}(k,\vec{r}) + n_{j\mu}\,\vec{N}_{j\mu}^{(1)}(k,\vec{r})\right)
$$

$$
= -k^2 c \mu_0 \sum_{j\mu} \left(m_{j\mu}\,\vec{M}_{j\mu}^{(1)}(k,\vec{r}) + n_{j\mu}\,\vec{N}_{j\mu}^{(1)}(k,\vec{r})\right), \tag{5.178}
$$

confirming the result anticipated in Eq. (5.175c).

5.5.2 *Gauge Condition, Vector and Scalar Potentials*

We have defined the multipoles for the radiated scalar potential in Eq. (5.156); the longitudinal components of the vector potential have been discussed in Eqs. (5.168c) and (5.164c). In our discussion of the radiated electric field, we anticipated the cancellation of the longitudinal contributions, in between the gradient of the scalar potential and the time derivative of the vector potential, in Eqs. (5.174) and (5.175). In consequence, we anticipate a simple relation between the $p_{\ell m}$ coefficients from Eq. (5.156) and the $l_{j\mu}$ coefficients from Eq. (5.168c). The derivation of this relation is necessary in order to establish the fulfillment of the Lorenz gauge condition and will be discussed in the following. We remember that the equations used by us in order to relate the sources to the potentials, namely, Eqs. (5.153) and (5.158), precisely are the Lorenz gauge versions, equivalent to Eq. (1.77).

In the derivation, we shall use the representation (5.164c) for the $L_{j\mu}^{(K)}(k,\vec{r})$ functions. Furthermore, we have the charge conservation condition $\vec{\nabla} \cdot \vec{J}(\vec{r},t) = -\partial_t \rho(\vec{r},t)$ and so $\vec{\nabla} \cdot \vec{J}(\vec{r}) = \mathrm{i}\omega\rho(\vec{r})$. A simple partial integration then leads to the formulas

$$
l_{j\mu} = \int \mathrm{d}^3 r'\,\vec{J}(\vec{r}') \cdot \vec{L}_{j\mu}^{(0)*}(\theta',\varphi')
$$

$$
= \int \mathrm{d}^3 r'\,\vec{J}(\vec{r}') \cdot \vec{\nabla}' \left(\frac{1}{k} j_j(k r') Y_{j\mu}^*(\theta',\varphi')\right)
$$

$$
= -\int \mathrm{d}^3 r' \left[\vec{\nabla}' \cdot \vec{J}(\vec{r}')\right] \frac{1}{k} j_j(k r') Y_{j\mu}^*(\theta',\varphi')
$$

$$
= -\int \mathrm{d}^3 r'\,\mathrm{i}\omega\,\rho(\vec{r}') \frac{1}{k} j_j(k r') Y_{j\mu}(\theta',\varphi')
$$

$$
= -\mathrm{i}\frac{\omega}{k} \int \mathrm{d}^3 r'\,\rho(\vec{r}')\,j_j(k r') Y_{j\mu}(\theta',\varphi') = -\mathrm{i}c\,p_{j\mu}. \tag{5.179}
$$

With $p_{j\mu} = \int d^3r' \, \rho(\vec{r}') \, j_j(k\,r') \, Y^*_{j\mu}(\theta', \varphi')$, we thus verify the relation

$$l_{j\mu} = -\mathrm{i}c\,p_{j\mu}\,. \tag{5.180}$$

Let us recall the gauge condition (1.70) fulfilled by the potentials,

$$\vec{\nabla} \cdot \vec{A} + \frac{1}{c^2}\frac{\partial \Phi}{\partial t} = 0\,. \tag{5.181}$$

In view of the decompositions (5.155) and (5.170), we have for the scalar and vector potentials,

$$\Phi_0(\vec{r}) = \frac{\mathrm{i}k}{\epsilon_0} \sum_{j=0}^{\infty} \sum_{\mu=-j}^{j} p_{j\mu}\, h_j^{(1)}(k\,r)\, Y_{j\mu}(\theta, \varphi)\,, \tag{5.182}$$

and

$$\vec{A}_0(\vec{r}) = \mathrm{i}k\,\mu_0 \sum_{j=0}^{\infty} \sum_{\mu=-j}^{j} \left(m_{j\mu}\, \vec{M}_{j\mu}^{(1)}(k,\vec{r}) + n_{j\mu}\, \vec{N}_{j\mu}^{(1)}(k,\vec{r}) + l_{j\mu}\, \vec{L}_{j\mu}^{(1)}(k,\vec{r}) \right). \tag{5.183}$$

In the mixed frequency-coordinate representation, the Lorenz gauge condition reads as

$$\vec{\nabla} \cdot \vec{A}_0(\vec{r}) - \frac{\mathrm{i}\omega}{c^2}\Phi_0(\vec{r}) = 0\,. \tag{5.184}$$

Because the magnetic and electric multipole terms are divergence-free [see Eq. (5.175b)], the Lorenz gauge condition is equivalent to

$$\mathrm{i}k \sum_{j\mu} \left\{ \mu_0\, l_{j\mu}\, \vec{\nabla} \cdot \vec{L}_{j\mu}^{(1)}(k,\vec{r}) + \frac{(-\mathrm{i}\omega)}{\epsilon_0\, c^2}\, p_{j\mu}\, h_j^{(1)}(k\,r)\, Y_{j\mu}(\theta,\varphi) \right\} = 0\,, \tag{5.185}$$

which implies that

$$\mathrm{i}k \sum_{j\mu} \left\{ l_{j\mu}\, \vec{\nabla} \cdot \vec{L}_{j\mu}^{(1)}(k,\vec{r}) - \mathrm{i}\,k c\, p_{j\mu}\, h_j^{(1)}(k\,r)\, Y_{j\mu}(\theta,\varphi) \right\} = 0\,. \tag{5.186}$$

We use Eq. (5.164c), in the form

$$\vec{L}_{j\mu}^{(1)}(k,\vec{r}) = \frac{1}{k}\, \vec{\nabla} \left(h_j^{(1)}(k\,r)\, Y_{j\mu}(\theta,\varphi) \right), \tag{5.187}$$

$$\vec{\nabla} \cdot \vec{L}_{j\mu}^{(1)}(k,\vec{r}) = \frac{1}{k}\, \vec{\nabla}^2 \left(h_j^{(1)}(k\,r)\, Y_{j\mu}(\theta,\varphi) \right)$$

$$= -k\, h_j^{(1)}(k\,r)\, Y_{j\mu}(\theta,\varphi)\,. \tag{5.188}$$

Advantage has been taken of the fact that the function $h_j^{(1)}(k\,r)\, Y_{j\mu}(\theta,\varphi)$ fulfills the Helmholtz equation, i.e., $(\vec{\nabla}^2 + k^2) h_j^{(1)}(k\,r)\, Y_{j\mu}(\theta,\varphi) = 0$. Inserting Eq. (5.187) into (5.186), we see that the gauge condition is equivalent to the relation

$$\sum_{j\mu} (l_{j\mu} + icp_{j\mu})\, h_{j\mu}^{(1)}(kr)\, Y_{j\mu}(\theta,\varphi) = 0, \tag{5.189}$$

which is fulfilled in view of Eq. (5.180).

Using Eq. (5.180), we should now be able to verify the cancellation of the longitudinal contributions to the electric field, in the steps leading from Eq. (5.174) to (5.175a). To this end, we calculate

$$
\begin{aligned}
\vec{E}_0(\vec{r}) &= -\vec{\nabla}\Phi_0(\vec{r}) + i\omega \vec{A}_0(\vec{r}) \\[2mm]
&= -\vec{\nabla}\left(\frac{ik}{\epsilon_0} \sum_{j\mu} p_{j\mu}\, h_j^{(1)}(kr)\, Y_{\ell m}(\hat r) \right) \\[2mm]
&\quad + i\omega\left(ik\mu_0 \sum_{\ell m} \left\{ m_{j\mu}\, \vec{M}_{j\mu}^{(1)}(k,\vec r) + n_{j\mu}\, \vec{N}_{j\mu}^{(1)}(k,\vec r) + l_{j\mu}\, \vec{L}_{j\mu}^{(1)}(k,\vec r) \right\} \right) \\[2mm]
&= -k^2\mu_0 c \sum_{j\mu} \left\{ m_{j\mu}\, \vec{M}_{j\mu}^{(1)}(k,\vec r) + n_{j\mu}\, \vec{N}_{j\mu}^{(1)}(k,\vec r) \right\} + \vec{\mathcal{E}}(\vec r), \tag{5.190}
\end{aligned}
$$

where the first term is the result given in Eq. (5.175a) and the additional term $\vec{\mathcal{E}}(\vec r)$ must be shown to vanish,

$$
\begin{aligned}
\vec{\mathcal{E}}(\vec r) &= -\frac{ik}{\epsilon_0} \sum_{j\mu} \left\{ p_{j\mu}\, \vec{\nabla}\left(h_j^{(1)}(kr)\, Y_{j\mu}(\hat r) \right) + \left(-\frac{\epsilon_0}{ik} \right) i\omega(ik\mu_0) l_{j\mu}\, \vec{L}_{j\mu}^{(1)}(k,\vec r) \right\} \\[2mm]
&= -\frac{ik}{\epsilon_0} \sum_{j\mu} \left\{ p_{j\mu}\, \vec{\nabla}\left(h_j^{(1)}(kr)\, Y_{j\mu}(\hat r) \right) - ikc\mu_0 \epsilon_0\, l_{j\mu}\, \vec{L}_{j\mu}^{(1)}(k,\vec r) \right\} \\[2mm]
&= -\frac{ik^2}{\epsilon_0} \sum_{j\mu} \left\{ \left(p_{j\mu} - \frac{i}{c} l_{j\mu} \right) \vec{L}_{j\mu}^{(1)}(k,\vec r) \right\} = 0, \tag{5.191}
\end{aligned}
$$

again in view of Eq. (5.180). We have shown that the longitudinal components of the electric field cancel, as they should, in the source-free region. A few remarks are in order. The factor $[-\epsilon_0/(ik)]$ in the first line is simply introduced in order to cancel the first prefactor. Alternatively, one may observe that the extra term in the electric field, from the longitudinal components of the vector potential, simply reads as $-k^2 c\mu_0 \sum_{j\mu} l_{j\mu}\, \vec{L}_{j\mu}^{(1)}(k,\vec r)$, with the prefactor being fixed in Eq. (5.178).

5.5.3 *Representations of the Vector Spherical Harmonics*

In the following, we shall endeavor in rather difficult calculations. It is a good moment to rest and to write down the most important representations of the vector spherical harmonics, and of the multipole functions. We start from Eqs. (5.148a) and (5.151a),

$$\vec{Y}^j_{j\,\mu}(\theta,\varphi) = \frac{1}{\sqrt{j(j+1)}}\, \vec{L} Y_{j\mu}(\theta,\varphi),$$

$$= -\sqrt{\frac{(j+\mu)\,(j-\mu+1)}{2j(j+1)}}\, Y_{j,\mu-1}(\theta,\varphi)\,\vec{e}_{+1} + \frac{\mu}{\sqrt{j(j+1)}}\, Y_{j,\mu}(\theta,\varphi)\,\vec{e}_0$$

$$+ \sqrt{\frac{(j-\mu)\,(j+\mu+1)}{2j(j+1)}}\, Y_{j,\mu+1}(\theta,\varphi)\,\vec{e}_{-1}\,. \tag{5.192a}$$

Equations (5.148b) and (5.151b) state that

$$\vec{Y}^{j-1}_{j\,\mu}(\theta,\varphi) = \frac{1}{\sqrt{j\,(2j+1)}}\,\left(j\,\hat{r} + r\,\vec{\nabla}\right) Y_{j\mu}(\theta,\varphi)$$

$$= -\frac{1}{\sqrt{j\,(2j+1)}}\,\left(\mathrm{i}\hat{r}\times\vec{L} - j\,\hat{r}\right) Y_{j\mu}(\theta,\varphi),$$

$$= \sqrt{\frac{(j+\mu-1)\,(j+\mu)}{2j(2j-1)}}\, Y_{j-1,\mu-1}(\theta,\varphi)\,\vec{e}_{+1}$$

$$+ \sqrt{\frac{(j-\mu)\,(j+\mu)}{j(2j-1)}}\, Y_{j-1,\mu}(\theta,\varphi)\,\vec{e}_0$$

$$+ \sqrt{\frac{(j-\mu-1)\,(j-\mu)}{2j(2j-1)}}\, Y_{j-1,\mu+1}(\theta,\varphi)\,\vec{e}_{-1}\,, \tag{5.192b}$$

Finally, we have from Eq. (5.148c) and (5.151c),

$$\vec{Y}^{j+1}_{j\,\mu}(\theta,\varphi) = -\frac{1}{\sqrt{(j+1)\,(2j+1)}}\,\left((j+1)\,\hat{r} - r\,\vec{\nabla}\right) Y_{j\mu}(\theta,\varphi)$$

$$= -\frac{1}{\sqrt{(j+1)\,(2j+1)}}\,\left(\mathrm{i}\hat{r}\times\vec{L} + (j+1)\,\hat{r}\right) Y_{j\mu}(\theta,\varphi)$$

$$= \sqrt{\frac{(j-\mu+1)\,(j-\mu+2)}{2(j+1)(2j+3)}}\, Y_{j+1,\mu-1}(\theta,\varphi)\,\vec{e}_{+1}$$

$$- \sqrt{\frac{(j-\mu+1)\,(j+\mu+1)}{(j+1)(2j+3)}}\, Y_{j+1,\mu}(\theta,\varphi)\,\vec{e}_0$$

$$+ \sqrt{\frac{(j+\mu+2)\,(j+\mu+1)}{2(j+1)(2j+3)}}\, Y_{j+1,\mu+1}(\theta,\varphi)\,\vec{e}_{-1}\,. \tag{5.192c}$$

The spherical harmonics are given by Eq. (2.53), which we recall for convenience,

$$Y_{j\mu}(\theta, \varphi) = (-1)^{\mu} \left(\frac{2j+1}{4\pi} \frac{(j-\mu)!}{(j+\mu)!} \right)^{1/2} P_{\ell}^{|\mu|}(\cos\theta)\, \mathrm{e}^{\mathrm{i}\mu\varphi} . \tag{5.193}$$

We also summarize Eqs. (5.164), (5.171), and (5.172) as follows,

$$\vec{M}_{j\mu}^{(K)}(k, \vec{r}) = \frac{1}{\sqrt{j(j+1)}} \vec{L}\left(f_j^{(K)}(kr)\, Y_{j\mu}(\theta, \varphi) \right)$$

$$= f_j^{(K)}(kr)\, \vec{Y}_{j\mu}^{j}(\theta, \varphi) = -\frac{\mathrm{i}}{k} \vec{\nabla} \times \vec{N}_{j\mu}^{(K)}(k, \vec{r}), \tag{5.194a}$$

$$\vec{N}_{j\mu}^{(K)}(k, \vec{r}) = \frac{\mathrm{i}}{k} \vec{\nabla} \times \vec{M}_{j\mu}^{(K)}(k, \vec{r}),$$

$$= -\sqrt{\frac{j+1}{2j+1}} f_{j-1}^{(K)}(kr)\, \vec{Y}_{j\mu}^{j-1}(\theta, \varphi) + \sqrt{\frac{j}{2j+1}} f_{j+1}^{(K)}(kr)\, \vec{Y}_{j\mu}^{j+1}(\theta, \varphi),$$

$$= \frac{1}{kr} \left\{ \frac{\mathrm{d}[kr\, f_j^{(K)}(kr)]}{\mathrm{d}(kr)} [\mathrm{i}\hat{r} \times \vec{Y}_{j\mu}^{j}(\theta, \varphi)] - \hat{r}\sqrt{j(j+1)} f_j^{(K)}(kr)\, Y_{j\mu}(\theta, \varphi) \right\}, \tag{5.194b}$$

$$\vec{L}_{j\mu}^{(K)}(k, \vec{r}) = \frac{1}{k} \vec{\nabla}\left(f_j^{(K)}(kr)\, Y_{j\mu}(\theta, \varphi) \right)$$

$$= \sqrt{\frac{j}{2j+1}} f_{j-1}^{(K)}(kr)\, \vec{Y}_{j\mu}^{j-1}(\theta, \varphi) + \sqrt{\frac{j+1}{2j+1}} f_{j+1}^{(K)}(kr)\, \vec{Y}_{j\mu}^{j+1}(\theta, \varphi)$$

$$= \hat{r}\, \frac{\mathrm{d}[f_j^{(K)}(kr)]}{\mathrm{d}(kr)} Y_{j\mu}(\theta, \varphi) - \sqrt{j(j+1)} \frac{1}{kr} f_j^{(K)}(kr)\, (\mathrm{i}\hat{r} \times \vec{Y}_{j\mu}^{j}(\theta, \varphi)). \tag{5.194c}$$

These results will be needed in the analysis of the Poynting vector.

5.5.4 *Poynting Vector of the Radiation*

We start from the relations (5.175c) and (5.175d),

$$\vec{E}_0(\vec{r}) = -k^2 c\,\mu_0 \sum_{j=0}^{\infty} \sum_{\mu=-j}^{j} \left(m_{j\mu}\, \vec{M}_{j\mu}^{(1)}(k, \vec{r}) + n_{j\mu}\, \vec{N}_{j\mu}^{(1)}(k, \vec{r}) \right), \tag{5.195}$$

and

$$\vec{B}_0(\vec{r}) = k^2 \mu_0 \sum_{j=0}^{\infty} \sum_{\mu=-j}^{j} \left(m_{j\mu}\, \vec{N}_{j\mu}^{(1)}(k, \vec{r}) - n_{j\mu}\, \vec{M}_{j\mu}^{(1)}(k, \vec{r}) \right). \tag{5.196}$$

The Poynting vector, which carries the physical dimension of radiated power per area, is [see Eq. (5.106)]

$$\vec{S}(\vec{r},t) = \frac{1}{\mu_0}\,\vec{E}(\vec{r},t) \times \vec{B}(\vec{r},t), \qquad \vec{S}_0(\vec{r}) = \frac{1}{2\mu_0}\,\vec{E}_0(\vec{r}) \times \vec{B}_0^*(\vec{r}). \tag{5.197}$$

From Eq. (5.164a), we recall the relation

$$\vec{M}_{j\mu}^{(K)}(k,\vec{r}) = f_j^{(K)}(k r)\,\vec{Y}_{j\mu}^j(\theta,\varphi), \tag{5.198}$$

which implies that

$$\vec{M}_{j\mu}^{(1)}(k,\vec{r}) = f_j^{(1)}(k r)\,\vec{Y}_{j\mu}^j(\theta,\varphi) = h_j^{(1)}(k r)\,\vec{Y}_{j\mu}^j(\theta,\varphi). \tag{5.199}$$

In view of the asymptotics (for large $k r$)

$$h_j^{(1)}(k r) \to \frac{e^{i k r-(j+1)i\pi/2}}{k r}, \tag{5.200}$$

we have

$$\vec{M}_{j\mu}^{(1)}(k,\vec{r}) = h_j^{(1)}(k r)\,\vec{Y}_{j\mu}^j(\theta,\varphi) \to \frac{e^{i k r-(j+1)i\pi/2}}{k r}\,\vec{Y}_{j\mu}^j(\theta,\varphi) = (-i)^{j+1}\frac{e^{i k r}}{k r}\,\vec{Y}_{j\mu}^j(\theta,\varphi). \tag{5.201}$$

From Eq. (5.164b), we infer that

$$\vec{N}_{j\mu}^{(K)}(k,\vec{r}) = -\sqrt{\frac{j+1}{2j+1}}\,f_{j-1}^{(K)}(k r)\,\vec{Y}_{j\mu}^{j-1}(\theta,\varphi) + \sqrt{\frac{j}{2j+1}}\,f_{j+1}^{(K)}(k r)\,\vec{Y}_{j\mu}^{j+1}(\theta,\varphi). \tag{5.202}$$

This implies that for large r,

$$\vec{N}_{j\mu}^{(1)}(k,\vec{r}) = -\sqrt{\frac{j+1}{2j+1}}\,h_{j-1}^{(1)}(k r)\,\vec{Y}_{j\mu}^{j-1}(\theta,\varphi) + \sqrt{\frac{j}{2j+1}}\,h_{j+1}^{(1)}(k r)\,\vec{Y}_{j\mu}^{j+1}(\theta,\varphi)$$

$$\to -\sqrt{\frac{j+1}{2j+1}}\,(-i)^{(j-1)+1}\frac{e^{i k r}}{k r}\,\vec{Y}_{j\mu}^{j-1}(\theta,\varphi)$$

$$+ \sqrt{\frac{j}{2j+1}}\,(-i)^{(j+1)+1}\frac{e^{i k r}}{k r}\,\vec{Y}_{j\mu}^{j+1}(\theta,\varphi)$$

$$\to -(-i)^j\,\frac{e^{i k r}}{k r}\left(\sqrt{\frac{j+1}{2j+1}}\,\vec{Y}_{j\mu}^{j-1}(\theta,\varphi) + \sqrt{\frac{j}{2j+1}}\,\vec{Y}_{j\mu}^{j+1}(\theta,\varphi)\right). \tag{5.203}$$

We use the fact that $h_{j-1}^{(1)}$ and $h_{j+1}^{(1)}$ only differ by a phase $(-i)^2 = -1$ in the long-distance limit. From Eqs. (5.148b) and (5.148c), one may derive the relation

$$\sqrt{\frac{j+1}{2j+1}}\,\vec{Y}_{j\mu}^{j-1}(\theta,\varphi) + \sqrt{\frac{j}{2j+1}}\,\vec{Y}_{j\mu}^{j+1}(\theta,\varphi) = -i\left(\hat{r} \times \vec{Y}_{j\mu}^j(\theta,\varphi)\right). \tag{5.204}$$

The long-range asymptotics of $\vec{N}_{j\mu}^{(1)}(k, \vec{r})$ are thus given by

$$
\vec{N}_{j\mu}^{(1)}(k, \vec{r}) = -(-\mathrm{i})^j \frac{\mathrm{e}^{\mathrm{i}kr}}{kr} \left(\sqrt{\frac{j+1}{2j+1}} \, \vec{Y}_{j\mu}^{j-1}(\theta, \varphi) + \sqrt{\frac{j}{2j+1}} \, \vec{Y}_{j\mu}^{j+1}(\theta, \varphi) \right)
$$

$$
= -(-\mathrm{i})^j \frac{\mathrm{e}^{\mathrm{i}kr}}{kr} (-\mathrm{i}) \left(\hat{r} \times \vec{Y}_{j\mu}^{j}(\theta, \varphi) \right) = -(-\mathrm{i})^{j+1} \frac{\mathrm{e}^{\mathrm{i}kr}}{kr} \left(\hat{r} \times \vec{Y}_{j\mu}^{j}(\theta, \varphi) \right).
$$

$$(5.205)$$

We now use Eqs. (5.175c) and (5.175d) and the long-range asymptotic formulas,

$$
\vec{M}_{j\mu}^{(1)}(k, \vec{r}) \to (-\mathrm{i})^{j+1} \frac{\mathrm{e}^{\mathrm{i}kr}}{kr} \, \vec{Y}_{j\mu}^{j}(\theta, \varphi), \quad \vec{N}_{j\mu}^{(1)}(k, \vec{r}) \to -(-\mathrm{i})^{j+1} \frac{\mathrm{e}^{\mathrm{i}kr}}{kr} \left(\hat{r} \times \vec{Y}_{j\mu}^{j}(\theta, \varphi) \right).
$$

$$(5.206)$$

Hence,

$$
\vec{E}_0(\vec{r}) \to -k^2 \, c\,\mu_0 \, \frac{\mathrm{e}^{\mathrm{i}kr}}{kr} \sum_{j=0}^{\infty} \sum_{\mu=-j}^{j} (-\mathrm{i})^{j+1} \left(m_{j\mu} \, \vec{Y}_{j\mu}^{j}(\theta, \varphi) - n_{j\mu} \left(\hat{r} \times \vec{Y}_{j\mu}^{j}(\theta, \varphi) \right) \right),
$$

$$(5.207)$$

and

$$
\vec{B}_0(\vec{r}) \to -k^2 \, \mu_0 \, \frac{\mathrm{e}^{\mathrm{i}kr}}{kr} \sum_{j=0}^{\infty} \sum_{\mu=-j}^{j} (-\mathrm{i})^{j+1} \left(m_{j\mu} \left(\hat{r} \times \vec{Y}_{j\mu}^{j}(\theta, \varphi) \right) + n_{j\mu} \, \vec{Y}_{j\mu}^{j}(\theta, \varphi) \right).
$$

$$(5.208)$$

The Poynting vector evaluates to

$$
\vec{S}_0(\vec{r}) = \frac{1}{2\mu_0} \, \vec{E}_0(\vec{r}) \times \vec{B}_0^*(\vec{r})
$$

$$
= \frac{1}{2} \frac{k^4 \, c\,\mu_0}{(kr)^2} \sum_{j=0}^{\infty} \sum_{\mu=-j}^{j} (-\mathrm{i})^{j+1} \left(m_{j\mu} \, \vec{Y}_{j\mu}^{j}(\theta, \varphi) - n_{j\mu} \left(\hat{r} \times \vec{Y}_{j\mu}^{j}(\theta, \varphi) \right) \right)
$$

$$
\times \sum_{j'=0}^{\infty} \sum_{\mu'=-j'}^{j'} \mathrm{i}^{j'+1} \left(m_{j'\mu'}^* \left(\hat{r} \times \vec{Y}_{j'\mu'}^{j'\,*}(\theta, \varphi) \right) + n_{j'\mu'}^* \, \vec{Y}_{j'\mu'}^{j'\,*}(\theta, \varphi) \right)
$$

$$
= \frac{k^2 \, c\,\mu_0}{2\,r^2} \sum_{j=0}^{\infty} \sum_{\mu=-j}^{j} \sum_{j'=0}^{\infty} \sum_{\mu'=-j'}^{j'} \mathrm{i}^{j'-j} \left(m_{j\mu} \, \vec{Y}_{j\mu}^{j}(\theta, \varphi) - n_{j\mu} \left(\hat{r} \times \vec{Y}_{j\mu}^{j}(\theta, \varphi) \right) \right)
$$

$$
\times \left(m_{j'\mu'}^* \left(\hat{r} \times \vec{Y}_{j'\mu'}^{j'\,*}(\theta, \varphi) \right) + n_{j'\mu'}^* \, \vec{Y}_{j'\mu'}^{j'\,*}(\theta, \varphi) \right),
$$

$$(5.209)$$

which is a double-infinite sum. The total power radiated through a sphere of radius r is

$$
P = \int \mathrm{d}\Omega \, r^2 \, \hat{r} \cdot \vec{S}_0(\vec{r}) = T_1 + T_2,
$$

$$(5.210)$$

and considerable simplifications occur in its calculation. Here,

$$T_1 = \frac{k^2 c \mu_0}{2} \int d\Omega \, \hat{r} \cdot \sum_{jj'\mu\mu'} i^{j'-j} \left(m_{j\mu} m^*_{j'\mu'} \, \vec{Y}^j_{j\mu}(\theta,\varphi) \times \left(\hat{r} \times \vec{Y}^{j'\,*}_{j'\mu'}(\theta,\varphi) \right) \right.$$

$$\left. - n_{j\mu} n^*_{j'\mu'} \left(\hat{r} \times \vec{Y}^j_{j\mu}(\theta,\varphi) \right) \times \vec{Y}^{j'\,*}_{j'\mu'}(\theta,\varphi) \right),$$

$$T_2 = \frac{k^2 c \mu_0}{2} \int d\Omega \, \hat{r} \cdot \sum_{jj'\mu\mu'} i^{j'-j} \left(m_{j\mu} n^*_{j'\mu'} \, \vec{Y}^j_{j\mu}(\theta,\varphi) \times \vec{Y}^{j'\,*}_{j'\mu'}(\theta,\varphi) \right.$$

$$\left. - n_{j\mu} m^*_{j'\mu'} \left(\hat{r} \times \vec{Y}^j_{j\mu}(\theta,\varphi) \right) \times \left(\hat{r} \times \vec{Y}^{j'\,*}_{j'\mu'}(\theta,\varphi) \right) \right). \tag{5.211}$$

Now the angular integrals have to be evaluated. We recall Eq. (5.148a),

$$\vec{Y}^j_{j\mu}(\theta,\varphi) = \frac{1}{\sqrt{j(j+1)}} \, \vec{L} Y_{j\mu}(\theta,\varphi), \qquad \hat{r} \cdot \vec{Y}^j_{j\mu}(\theta,\varphi) = \frac{1}{\sqrt{j(j+1)}} \, \hat{r} \cdot \vec{L} Y_{j\mu}(\theta,\varphi) = 0, \tag{5.212}$$

where $\vec{L} = -i \, (\vec{r} \times \vec{\nabla})$. Hence,

$$\int d\Omega \, \hat{r} \cdot \vec{Y}^j_{j\mu}(\theta,\varphi) \times \left(\hat{r} \times \vec{Y}^{j'\,*}_{j'\mu'}(\theta,\varphi) \right)$$

$$= \int d\Omega \, \hat{r} \cdot \left[\hat{r} \left(\vec{Y}^j_{j\mu}(\theta,\varphi) \cdot \vec{Y}^{j'\,*}_{j'\mu'}(\theta,\varphi) \right) - \vec{Y}^{j'\,*}_{j'\mu'}(\theta,\varphi) \left(\hat{r} \cdot \vec{Y}^j_{j\mu}(\theta,\varphi) \right) \right]$$

$$= \int d\Omega \left[\vec{Y}^j_{j\mu}(\theta,\varphi) \cdot \vec{Y}^{j'\,*}_{j'\mu'}(\theta,\varphi) - \left(\hat{r} \cdot \vec{Y}^{j'\,*}_{j'\mu'}(\theta,\varphi) \right) \left(\hat{r} \cdot \vec{Y}^j_{j\mu}(\theta,\varphi) \right) \right]$$

$$= \int d\Omega \, \vec{Y}^j_{j\mu}(\theta,\varphi) \cdot \vec{Y}^{j'\,*}_{j'\mu'}(\theta,\varphi) = \delta_{jj'} \, \delta_{\mu\mu'}, \tag{5.213}$$

where we have used Eq. (5.212). Similarly, one shows that

$$\int d\Omega \, \hat{r} \cdot \left[\left(\hat{r} \times \vec{Y}^j_{j\mu}(\theta,\varphi) \right) \times \vec{Y}^{j'\,*}_{j'\mu'}(\theta,\varphi) \right] = -\delta_{jj'} \, \delta_{\mu\mu'}. \tag{5.214}$$

The two identities

$$\int d\Omega \, \hat{r} \cdot \left(\vec{Y}^j_{j\mu}(\theta,\varphi) \times \vec{Y}^{j'\,*}_{j'\mu'}(\theta,\varphi) \right) = 0, \tag{5.215}$$

$$\int d\Omega \, \hat{r} \cdot \left(\hat{r} \times \vec{Y}^j_{j\mu}(\theta,\varphi) \right) \times \left(\hat{r} \times \vec{Y}^{j'\,*}_{j'\mu'}(\theta,\varphi) \right) = 0, \tag{5.216}$$

are a little harder to show and are left as exercises. Armed with the results (5.213)–(5.216), we can show that $T_2 = 0$ and simplify Eq. (5.209) drastically,

$$P = \frac{k^2 c \mu_0}{2} \sum_{j\mu} \left(|m_{j\mu}|^2 + |n_{j\mu}|^2 \right). \tag{5.217}$$

5.5.5 *Half-Wave Antenna*

Let us carry out the multipole decomposition explicitly for an example problem. We consider a half-wave antenna with a current inside the antenna,

$$I = I_0 \cos\left(\frac{2\pi z}{\lambda} \right) e^{-i\omega t}, \tag{5.218}$$

where $-\lambda/4 < z < \lambda/4$. The length of the antenna therefore is exactly equal to half the wavelength, hence the name. In order to invoke our formalism, we first need to translate the current into a current density. A first guess might be

$$\vec{J}(\vec{r}) = \vec{J}_0(\vec{r})\, e^{-i\omega t}, \qquad \vec{J}_0(\vec{r}) = \hat{e}_z\, I_0\, \cos\left(\frac{2\pi z}{\lambda}\right) \delta(x)\,\delta(y), \qquad (5.219)$$

because the antenna is oriented parallel to the z axis. However, this expression leads to problems as we put in the formulas $x = r \sin\theta \cos\varphi$ and $y = r \sin\theta \sin\varphi$ for the coordinates. How are we supposed to evaluate the integrals over θ and φ if they enter the arguments of the Dirac-δ function in such an awkward way? We therefore use the apparent radial symmetry of the problem, with respect to the z axis, and write

$$\vec{J}_0(r,\theta,\varphi) = \hat{e}_z\, I_0\, \cos\left(\frac{2\pi r}{\lambda}\right) [2\delta(\theta) + 2\delta(\pi-\theta)]\, \frac{1}{2\pi\, r^2\, \sin(\theta)} \qquad (5.220)$$

We use the convention [see Eq. (1.31)]

$$\int_a^b \mathrm{d}x\, \delta(x-a)\, f(x) = \frac{1}{2}\, f(a), \qquad \int_a^b \mathrm{d}x\, \delta(x-b)\, f(x) = \frac{1}{2}\, f(b), \qquad (5.221)$$

for a Dirac-δ function "on the boundary". One may thus calculate the integral over the (partial) surface area of a sphere, centered at the origin, which encompasses the antenna. The infinitesimal area element is

$$\mathrm{d}A = R^2\, \sin\theta\, \mathrm{d}\theta\, \mathrm{d}\varphi. \qquad (5.222)$$

The surface integral, with the polar angle $\theta \in (0,\epsilon)$, samples points whose z coordinate is just $z = R$, i.e., points in the vicinity of the "pole" of the "sampling sphere". It evaluates to

$$\int \mathrm{d}A\, \vec{J}_0(R,\theta,\varphi) = R^2 \int_0^\epsilon \mathrm{d}\theta\, \sin\theta \int_0^{2\pi} \mathrm{d}\varphi\, \vec{J}_0(R,\theta,\varphi)$$

$$= \hat{e}_z\, I_0\, \cos\left(\frac{2\pi R}{\lambda}\right) \int_0^{2\pi} \mathrm{d}\varphi \int_0^\epsilon \mathrm{d}\theta\, \sin\theta\, R^2\, [2\delta(\theta)]\, \frac{1}{2\pi\, R^2\, \sin(\theta)}$$

$$= \hat{e}_z\, I_0\, \cos\left(\frac{2\pi R}{\lambda}\right) \int_0^\epsilon \mathrm{d}\theta\, [2\delta(\theta)] = \hat{e}_z\, I_0\, \cos\left(\frac{2\pi z}{\lambda}\right). \qquad (5.223)$$

Assembling the current density from terms near the "North" and the "South" pole, one has

$$\vec{J}_0(r,\theta,\varphi) = I_0\, \cos\left(\frac{2\pi r}{\lambda}\right) \frac{\delta(\theta) + \delta(\pi-\theta)}{\pi\, r^2\, \sin(\theta)}\, \Theta(\tfrac{1}{4}\lambda - r)\, \hat{e}_z. \qquad (5.224)$$

In order to calculate the magnetic multipoles using Eq. (5.164a), one recalls that according to Eq. (5.148a),

$$\vec{Y}_{j\mu}^{\,j}(\hat{r}) = \frac{1}{\sqrt{j(j+1)}}\, \vec{L}\, Y_{j\mu}(\hat{r}), \qquad (5.225)$$

and so

$$\hat{e}_z \cdot \vec{Y}_{j\mu}^j(\hat{r}) = \frac{1}{\sqrt{j(j+1)}} L_z Y_{j\mu}(\hat{r}) = \frac{1}{\sqrt{j(j+1)}} \mu Y_{j\mu}(\hat{r}). \tag{5.226}$$

Thus,

$$
\begin{aligned}
m_{j\mu} &= \int d^3r \, \vec{J}_0(\vec{r}) \cdot \vec{M}_{j\mu}^{(0)*}(\vec{r}) = \int d^3r \, j_0(kr) \, \vec{J}_0(\vec{r}) \cdot \vec{Y}_{j\mu}^{j*}(\theta,\varphi) \\
&= \int d\varphi \int d\theta \sin\theta \int dr \, r^2 \, j_0(kr) \, I_0 \cos(kr) \, \frac{\delta(\theta) + \delta(\pi - \theta)}{\pi r^2 \sin(\theta)} \hat{e}_z \cdot \vec{Y}_{j\mu}^{j*}(\theta,\varphi) \\
&= \frac{I_0}{2\pi} \int d\varphi \int dr \, j_0(kr) \, \cos(kr) \left[\hat{e}_z \cdot \vec{Y}_{j\mu}^{j*}(0,\varphi) + \hat{e}_z \cdot \vec{Y}_{j\mu}^{j*}(\pi,\varphi) \right] \\
&= \frac{I_0}{2\pi} \int d\varphi \int dr \, j_0(kr) \, \cos(kr) \, \frac{1}{\sqrt{j(j+1)}} \left[\mu Y_{j\mu}^*(0,\varphi) + \mu Y_{j\mu}^*(\pi,\varphi) \right].
\end{aligned}
\tag{5.227}
$$

From Eq. (2.53), one has

$$Y_{j\mu}(0,\varphi) = \delta_{\mu 0} \sqrt{\frac{2j+1}{4\pi}},$$

$$Y_{j\mu}(0,\varphi) = (-1)^j \delta_{\mu 0} \sqrt{\frac{2j+1}{4\pi}}. \tag{5.228}$$

Both results are independent of φ, as they should be on the pole caps of the unit sphere. The latter result follows from the former by the parity transformation $\theta \to \pi - \theta$, and $\varphi \to \varphi + \pi$ (parity transformation). In view of $\mu \, \delta_{\mu 0} = 0$, it follows that

$$m_{j\mu} = 0, \tag{5.229}$$

or in other words, all the magnetic multipoles vanish.

Now we turn our attention to the electric multipoles. From Eq. (5.148b), we have

$$
\begin{aligned}
\hat{e}_z \cdot \vec{Y}_{j\mu}^{j-1*}(\theta,\varphi) &= \sqrt{\frac{(j-\mu)(j+\mu)}{j(2j-1)}} Y_{j-1,\mu}(\theta,\varphi) \hat{e}_z \cdot \vec{e}_0^* \\
&= \sqrt{\frac{(j-\mu)(j+\mu)}{j(2j-1)}} Y_{j-1,\mu}(\theta,\varphi), \tag{5.230}
\end{aligned}
$$

because $\vec{e}_q \, \vec{e}_{q'}^* = \delta_{qq'}$. At the "North" pole of the unit sphere, one therefore has the relation,

$$
\begin{aligned}
\hat{e}_z \cdot \vec{Y}_{j\mu}^{j-1*}(0,\varphi) &= \sqrt{\frac{(j-\mu)(j+\mu)}{j(2j-1)}} Y_{j-1,\mu}(0,\varphi) \\
&= \sqrt{\frac{(j-\mu)(j+\mu)}{j(2j-1)}} \delta_{\mu 0} \sqrt{\frac{2(j-1)+1}{4\pi}} = \delta_{\mu 0} \sqrt{\frac{j}{4\pi}}, \tag{5.231}
\end{aligned}
$$

while at the "South" pole of the unit sphere, one has

$$\hat{e}_z \cdot \vec{Y}_{j\mu}^{j-1\,*}(\pi,\varphi) = \sqrt{\frac{(j-\mu)\,(j+\mu)}{j(2j-1)}}\, Y_{j-1,\mu}(\mu,\varphi)$$

$$= (-1)^{j-1} \sqrt{\frac{(j-\mu)\,(j+\mu)}{j(2j-1)}}\, \delta_{\mu 0} \sqrt{\frac{2(j-1)+1}{4\pi}}$$

$$= (-1)^{j-1}\, \delta_{\mu 0} \sqrt{\frac{j}{4\pi}}\,. \tag{5.232}$$

Let us also investigate, at the "North" pole of the unit sphere, using Eq. (5.148c)

$$\hat{e}_z \cdot \vec{Y}_{j\mu}^{j+1\,*}(0,\varphi) = -\sqrt{\frac{(j-\mu+1)\,(j+\mu+1)}{(j+1)(2j+3)}}\, Y_{j+1,\mu}(0,\varphi)$$

$$= -\sqrt{\frac{(j-\mu+1)\,(j+\mu+1)}{(j+1)(2j+3)}}\, \delta_{\mu 0} \sqrt{\frac{2(j+1)+1}{4\pi}}$$

$$= -\delta_{\mu 0} \sqrt{\frac{j+1}{4\pi}}\,. \tag{5.233}$$

Conversely, at the "South" pole of the unit sphere, one has

$$\hat{e}_z \cdot \vec{Y}_{j\mu}^{j+1\,*}(\pi,\varphi) = -\sqrt{\frac{(j-\mu+1)\,(j+\mu+1)}{(j+1)(2j+3)}}\, Y_{j+1,\mu}(\pi,\varphi) = -(-1)^{j-1}\,\delta_{\mu 0}\sqrt{\frac{j+1}{4\pi}}\,, \tag{5.234}$$

where the intermediate steps are left as an exercise to the reader. From Eq. (5.164b), we recall that

$$\vec{N}_{j\mu}^{(K)}(k,\vec{r}) = \sqrt{\frac{j+1}{2j+1}}\, f_{j-1}^{(K)}(k\,r)\, \vec{Y}_{j\mu}^{j-1}(\theta,\varphi) - \sqrt{\frac{j}{2j+1}}\, f_{j+1}^{(K)}(k\,r)\, \vec{Y}_{j\mu}^{j+1}(\theta,\varphi)\,. \tag{5.235}$$

We may now calculate the electric multipoles,

$$n_{j\mu} = \int \mathrm{d}^3 r\, \vec{J}_0(\vec{r}) \cdot \vec{N}_{j\mu}^{(0)\,*}(\vec{r}) = \int \mathrm{d}^3 r \left(j_{j-1}(k\,r) \sqrt{\frac{j+1}{2j+1}}\, \vec{J}(\vec{r}) \cdot \vec{Y}_{j\mu}^{j-1\,*}(\theta,\varphi) \right.$$

$$\left. - j_{j+1}(k\,r) \sqrt{\frac{j}{2j+1}}\, \vec{J}(\vec{r}) \cdot \vec{Y}_{j\mu}^{j+1\,*}(\theta,\varphi) \right)$$

$$= \frac{I_0}{2\pi} \int \mathrm{d}\varphi \int \mathrm{d}r\, r^2\, \cos(kr) \left(j_{j-1}(k\,r) \sqrt{\frac{j+1}{2j+1}} \left[\hat{e}_z \cdot \vec{Y}_{j\mu}^{j-1\,*}(0,\varphi) \right. \right.$$

$$\left. \left. + \hat{e}_z \cdot \vec{Y}_{j\mu}^{j-1\,*}(\pi,\varphi) \right] - j_{j+1}(k\,r) \sqrt{\frac{j}{2j+1}} \left[\hat{e}_z \cdot \vec{Y}_{j\mu}^{j+1\,*}(0,\varphi) + \hat{e}_z \cdot \vec{Y}_{j\mu}^{j+1\,*}(\pi,\varphi) \right] \right)\,. \tag{5.236}$$

With the help of the results (5.231), (5.232), (5.233) and (5.234), one simplifies this expression to the form

$$
n_{j\mu} = \frac{I_0}{2\pi} \int d\varphi \int dr \, \cos(kr) \left(j_{j-1}(kr) \sqrt{\frac{j+1}{2j+1}} \left[1 + (-1)^{j-1} \right] \delta_{\mu 0} \sqrt{\frac{j}{4\pi}} \right.
$$

$$
\left. - j_{j+1}(kr) \sqrt{\frac{j}{2j+1}} \left(- \left[1 + (-1)^{j-1} \right] \delta_{\mu 0} \sqrt{\frac{j+1}{4\pi}} \right) \right)
$$

$$
= I_0 \, \delta_{\mu 0} \sqrt{\frac{j(j+1)}{4\pi(2j+1)}} \left[1 + (-1)^{j-1} \right] \int dr \, \cos(kr) \left(j_{j-1}(kr) + j_{j+1}(kr) \right) .
$$

$$
(5.237)
$$

Recalling the recursion relation for spherical Bessel functions, given in Eq. (5.28), we find that the multipole moment is given as follows,

$$
n_{j\mu} = I_0 \, \delta_{\mu 0} \sqrt{\frac{j(j+1)}{4\pi(2j+1)}} \left[1 + (-1)^{j-1} \right] \int dr \, \cos(kr) \, \frac{2j+1}{kr} \, j_j(kr)
$$

$$
= - I_0 \, \delta_{\mu 0} \sqrt{\frac{j(j+1)(2j+1)}{4\pi}} \left[1 + (-1)^{j-1} \right] \int dr \left(- \frac{\cos(kr)}{kr} \right) j_j(kr)
$$

$$
= - I_0 \, \delta_{\mu 0} \sqrt{\frac{j(j+1)(2j+1)}{4\pi}} \left[1 + (-1)^{j-1} \right] \int_0^{\lambda/4} dr \, y_0(kr) \, j_j(kr) \qquad (5.238)
$$

where y_0 of course is the spherical Neumann function. The spherical Bessel and Neumann functions fulfill the differential equation (5.27). Hence, one can show that

$$
\int dx \, f_j(x) \, g_{j'}(x) = \frac{x^2}{j'(j'+1) - j(j+1)} \left[f_j(x) \, g'_{j'}(x) - f'_j(x) \, g_{j'}(x) \right] . \qquad (5.239)
$$

Now we set $f_j = y_0$ and $g_{j'} = j_j$, and have

$$
\int dx \, y_0(x) \, j_j(x) = \frac{x^2}{j(j+1)} \left(y_0(x) \, j'_j(x) - y'_0(x) \, j_j(x) \right) . \qquad (5.240)
$$

For the case of interest, we have in view of $k = 2\pi/\lambda$, as well as $y_0(\pi/2) = 0$, and $y'_0(\pi/2) = 2/\pi$,

$$
\int_0^{\lambda/4} dr \, y_0(kr) \, j_j(kr) = \frac{1}{k} \int_0^{\pi/2} dx \, y_0(x) \, j_j(x)
$$

$$
= - \frac{1}{k} \frac{(\pi/2)^2}{j(j+1)} \, y'_0(\pi/2) \, j_j(\pi/2) = - \frac{\pi}{2k} \frac{1}{j(j+1)} \, j_j(\pi/2) .
$$

$$
(5.241)
$$

Hence,

$$
n_{j\mu} = \frac{\pi I_0}{2k} \, \delta_{\mu 0} \sqrt{\frac{2j+1}{4\pi j(j+1)}} \left[1 + (-1)^{j-1} \right] j_j(\pi/2) . \qquad (5.242)
$$

The results (5.229) and (5.242) enter the relations (5.175c) and (5.175d), which in our case simplify because $m_{j\mu} = 0$ [see Eq. (5.229)],

$$\vec{E}_0(\vec{r}) = -k^2 c \mu_0 \sum_{j=0}^{\infty} n_{j0} \vec{N}_{j0}^{(1)}(k,\vec{r}), \qquad (5.243a)$$

$$\vec{B}_0(\vec{r}) = -k^2 \mu_0 \sum_{j=0}^{\infty} n_{j0} \vec{M}_{j0}^{(1)}(k,\vec{r}). \qquad (5.243b)$$

For the half-wave antenna, one can drop the $m_{j\mu}$ terms and restrict the sum over the $n_{j\mu}$ terms to those with $\mu = 0$.

We now intend to use Eq. (5.217) in order to evaluate the radiated power in the far field. It is instructive to rewrite Eq. (5.217) somewhat and to introduce the factor

$$\mu_0 c = \frac{\mu}{\sqrt{\epsilon_0 \mu_0}} = Z_0 \approx 377\,\Omega, \qquad (5.244)$$

which is commonly referred to as the vacuum impedance. The radiated power P can be written as a product of the vacuum impedance and the square of the current I_0, times some numerical factor,

$$P = \frac{k^2}{2} Z_0 \sum_{j\mu} \left\{ |m_{j\mu}|^2 + |n_{j\mu}|^2 \right\} = \frac{k^2}{2} Z_0 \sum_{j\mu} |n_{j\mu}|^2$$

$$= \frac{k^2}{2} Z_0 \sum_{j \text{ odd}} |n_{j0}|^2 = Z_0 I_0^2 \frac{\pi}{8} \sum_{j \text{ odd}} \frac{2j+1}{j(j+1)} \left[j_j(\pi/2) \right]^2. \qquad (5.245)$$

The sum

$$\mathcal{S} = \sum_{j \text{ odd}} \frac{2j+1}{j(j+1)} \left[j_j(\pi/2) \right]^2 = 0.246\,986 \qquad (5.246)$$

is rapidly converging as j is increased. The term with $j = 1$ is $0.246\,384$, implying that dipole radiation accounts for 99.76% of the radiated power. Equating the radiated power with the "resistance to radiation emission", one has

$$P = \frac{1}{2} R_{\text{rad}} I_0^2, \qquad R_{\text{rad}} = Z_0 \mathcal{S} \frac{\pi}{4} = 73.079\,\Omega. \qquad (5.247)$$

The physical interpretation is that the antenna acts just like a resistor, as it emits energy in the form of outgoing radiation. The power absorbed by the antenna is proportional to $Z_0 I_0^2$. For given ϵ_0, if the speed of light were infinitely fast, then in view of $\mu_0 \epsilon_0 = 1/c^2$, we would have $\mu_0 \to 0$ and $Z_0 \to 0$; it would be "easier" to emit radiation, or in other words, the emitted electromagnetic waves would carry less energy. With a grain of salt, we can remark that, because the impedance of the vacuum has an appreciable numerical value of $\approx 377\,\Omega$, radio communication works well.

5.5.6   Long-Wavelength Limit of the Dipole Term

The somewhat elegant formalism we have discussed obscures the leading contributions to the dipole terms, and to other, higher-order multipole terms; we shall now attempt to recover these terms, at least in the long-wavelength limit. In particular, we attempt to show that the dipole term is part of the expression $N_{1\mu}^{(1)}$, and is hidden in the contribution of $j_{j-1}(kr) = j_0(kr)$. It remains to verify the prefactor. Let us therefore investigate

$$n_{j\mu} = \int d^3r \, \vec{J}_0(\vec{r}) \cdot \vec{N}_{j\mu}^{(0)*}(k,\vec{r}) . \tag{5.248}$$

Now, in view of Eqs. (5.164a) and (5.164b), we have

$$\vec{N}_{j\mu}^{(0)}(k,\vec{r}) = -\frac{i}{k} \vec{\nabla} \times \vec{M}_{j\mu}^{(K)}(k,\vec{r})$$

$$= -\frac{1}{\sqrt{j(j+1)}} \frac{1}{k} \vec{\nabla} \times \left(\vec{r} \times \vec{\nabla}\right) j_j(kr) Y_{j\mu}^*(\theta,\varphi) . \tag{5.249}$$

The expression $\vec{\nabla} \times \left(\vec{r} \times \vec{\nabla}\right) j_j(kr) Y_{j\mu}(\theta,\varphi)$ is tricky; one can simplify it but must remember that the leftmost gradient operator acts on everything that follows. The result is that

$$\vec{\nabla} \times \left(\vec{r} \times \vec{\nabla}\right) f(\vec{r}) = \left[\vec{r}\,\vec{\nabla}^2 - \vec{\nabla}\frac{\partial}{\partial r} r\right] f(\vec{r}) , \tag{5.250}$$

and so

$$\vec{N}_{j\mu}^{(0)}(k,\vec{r}) = -\frac{1}{\sqrt{j(j+1)}} \frac{1}{k} \left[\vec{r}\,\vec{\nabla}^2 - \vec{\nabla}\left(\frac{\partial}{\partial r} r\right)\right] j_j(kr) Y_{j\mu}^*(\theta,\varphi) . \tag{5.251}$$

One uses Eq. (5.250), the fact that $j_j(kr) Y_{j\mu}(\theta,\varphi)$ satisfies the Helmholtz equation, integration by parts, and the continuity equation, to write

$$n_{j\mu} = -\frac{1}{\sqrt{j(j+1)}} \frac{1}{k} \int d^3r \, \vec{J}_0(\vec{r}) \cdot \left[\vec{r}\,\vec{\nabla}^2 - \vec{\nabla}\left(\frac{\partial}{\partial r} r\right)\right] j_j(kr) Y_{j\mu}^*(\theta,\varphi)$$

$$= -\frac{1}{\sqrt{j(j+1)}} \frac{1}{k} \int d^3r \, \vec{J}_0(\vec{r}) \cdot \left[-\vec{r}\,k^2 - \vec{\nabla}\left(\frac{\partial}{\partial r} r\right)\right] j_j(kr) Y_{j\mu}^*(\theta,\varphi)$$

$$= \frac{1}{\sqrt{j(j+1)}} \frac{1}{k} \int d^3r \left[k^2 j_j(kr)\vec{r} \cdot \vec{J}_0(\vec{r}) - \left(\frac{\partial}{\partial r} r j_j(kr)\right) \vec{\nabla} \cdot \vec{J}_0(\vec{r})\right] Y_{j\mu}^*(\theta,\varphi)$$

$$= \frac{1}{\sqrt{j(j+1)}} \int d^3r \left[k j_j(kr)\vec{r} \cdot \vec{J}_0(\vec{r}) - \left(\frac{\partial}{\partial r} r j_j(kr)\right)(ic\rho_0(\vec{r}))\right] Y_{j\mu}^*(\theta,\varphi) . \tag{5.252}$$

In the long-wavelength limit, i.e, small-k limit, the dominant term obviously is the second one, as it carries no explicit factor k. Using the long-wavelength (small-k)

asymptotics of the Bessel function, we have

$$
\begin{aligned}
n_{j\mu} &= -\frac{ic}{\sqrt{j(j+1)}} \int d^3r \left(\frac{\partial}{\partial r} r j_j(kr) \right) Y_{j\mu}^*(\theta,\varphi)\, \rho_0(\vec{r}) \\
&\approx -\frac{ic}{\sqrt{j(j+1)}} \int d^3r \left(\frac{\partial}{\partial r} \frac{k^j\, r^{j+1}}{(2j+1)!!} \right) Y_{j\mu}^*(\theta,\varphi)\, \rho_0(\vec{r}) \\
&\approx -\frac{i k^j\, c}{\sqrt{j(j+1)}} (j+1) \int d^3r \left(\frac{r^j}{(2j+1)!!} \right) Y_{j\mu}^*(\theta,\varphi)\, \rho_0(\vec{r}) \\
&\approx -i k^j\, c \sqrt{\frac{j+1}{j}} \frac{1}{(2j+1)!!} \int d^3r\, r^j\, Y_{j\mu}^*(\theta,\varphi)\, \rho_0(\vec{r}) \\
&\approx -\frac{ic}{(2j+1)!!} \sqrt{\frac{j+1}{j}}\, k^j\, q_{j\mu}^{(0)}.
\end{aligned}
\tag{5.253}
$$

Here, the $q_{j\mu}^{(0)}$ are the multipole moments defined in Eq. (2.97), as familiar from electrostatics, but this time evaluated using the spatial part $\rho_0(\vec{r})$ of the oscillating charge density $\rho(\vec{r},t) = \rho_0(\vec{r}) \exp(-i\omega t)$. Setting $j = 1$, one recovers the long-wavelength limit of the electric-dipole term.

5.6 Potentials due to Moving Charges in Different Gauges

5.6.1 *Moving Charges and Lorenz Gauge*

The cancellation of the instantaneous interaction in the calculation of the electric field in Lorenz gauge has already been discussed in Sec. 4.2, leaving the charge and current distributions general. It is very instructive and nontrivial to verify the general results on the basis of a concrete problem, namely, the potentials generated by a moving point charge. In classical electrodynamics, all electric and magnetic fields can in principle be described as a superposition of fields generated by infinitesimal moving point charges, so that the specialization to a moving point charge does not necessarily imply a loss of generality. Again, the intriguing problem is that in Coulomb gauge (radiation gauge), the scalar potential Φ can be written as an instantaneous integral over charges that are far away from the observation point. A change in a charge distribution light years away would thus lead to an instantaneous change in the scalar potential at an observation point on Earth. In order to address this latter question, we here not only calculate the scalar potential, but also the vector potential generated by a moving charge, in Coulomb gauge and in general form. This problem has independently attracted some interest [25].

The representation (4.35) of the retarded Green function in coordinate space contains two Dirac-δ functions. It is highly singular. Therefore, it is certainly sensible to verify the conclusions reached thus far on the basis of a concrete problem where the actual form of the retarded Green function needs to be used, and partial integrations are required. We thus consider the potentials generated by a moving

point charge (Liénard–Wiechert potentials). The charge and current densities for
the moving point charge are given as

$$\rho(\vec{r}, t) = q \, \delta^{(3)} \left(\vec{r} - \vec{R}(t) \right) , \tag{5.254a}$$

$$\vec{J}(\vec{r}, t) = q \, \delta^{(3)} \left(\vec{r} - \vec{R}(t) \right) \left[\frac{\mathrm{d}}{\mathrm{d}t} \vec{R}(t) \right] , \tag{5.254b}$$

where $\vec{R}(t)$ is the particle trajectory. If we keep only the first term in square brackets
in Eq. (4.35) and write [see Eq. (4.36)]

$$G_R(\vec{r}, t, \vec{r}', t') \approx \frac{c}{4\pi \, \epsilon_0} \frac{\Theta(t - t')}{|\vec{r} - \vec{r}'|} \delta \left(|\vec{r} - \vec{r}'| - c(t - t') \right) , \tag{5.255}$$

then the integration of the Lorenz-gauge potentials [see Eqs. (4.88a) and (4.88b)] is
almost trivial,

$$\Phi_L(\vec{r}, t) = \int \mathrm{d}^3 r' \, \mathrm{d}t' \, G_R(\vec{r}, t, \vec{r}', t') \, \rho(\vec{r}', t') , \tag{5.256a}$$

$$\vec{A}_L(\vec{r}, t) = \frac{1}{c^2} \int \mathrm{d}^3 r' \, \mathrm{d}t' \, G_R(\vec{r}, t, \vec{r}', t') \, \vec{J}(\vec{r}', t') . \tag{5.256b}$$

After the integration over $\mathrm{d}^3 r'$, one easily obtains

$$\Phi_L(\vec{r}, t) = \frac{q}{4\pi\epsilon_0} \int \mathrm{d}t' \, \Theta(t - t') \frac{1}{|\vec{r} - \vec{R}(t')|} \delta \left(t' - \left(t - \frac{|\vec{r} - \vec{R}(t')|}{c} \right) \right) , \tag{5.257a}$$

$$\vec{A}_L(\vec{r}, t) = \frac{q}{4\pi\epsilon_0 c^2} \int \mathrm{d}t' \, \Theta(t - t') \frac{\dot{\vec{R}}(t')}{|\vec{r} - \vec{R}(t')|} \delta \left(t' - \left(t - \frac{|\vec{r} - \vec{R}(t')|}{c} \right) \right) . \tag{5.257b}$$

The δ function peaks at

$$t' = t_{\mathrm{ret}} = t - \frac{|\vec{r} - \vec{R}(t')|}{c} = t - \frac{|\vec{r} - \vec{R}(t_{\mathrm{ret}})|}{c} < t , \tag{5.258}$$

or

$$t_{\mathrm{ret}} = t - \frac{|\vec{r} - \vec{R}(t_{\mathrm{ret}})|}{c} , \qquad c(t - t_{\mathrm{ret}})^2 = \left(\vec{r} - \vec{R}(t_{\mathrm{ret}}) \right)^2 , \tag{5.259}$$

so that the step function is always unity at the point where the Dirac-δ peaks. That
means that all points on the trajectory of the particle for which the retardation
condition is fulfilled, contribute to the integrals. Let us assume that the Dirac-δ
peaks only once, namely, at $t' = t_{\mathrm{ret}}$. The integration over $\mathrm{d}t'$ in Eq. (5.257) still
leads to a nontrivial Jacobian. Indeed, we have

$$\frac{\mathrm{d}}{\mathrm{d}t'} \left(t' - t + \frac{|\vec{r} - \vec{R}(t')|}{c} \right) = 1 - \frac{1}{c} \frac{\mathrm{d}\vec{R}(t')}{\mathrm{d}t'} \cdot \frac{\vec{r} - \vec{R}(t')}{|\vec{r} - \vec{R}(t')|} , \tag{5.260}$$

at $t' = t_{\text{ret}}$. So, the Liénard–Wiechert potentials in Lorenz gauge read

$$\Phi_L(\vec{r}, t) = \frac{q}{4\pi\epsilon_0} \frac{1}{|\vec{r} - \vec{R}(t_{\text{ret}})|} \left(1 - \frac{\dot{\vec{R}}(t_{\text{ret}})}{c} \cdot \frac{\vec{r} - \vec{R}(t_{\text{ret}})}{|\vec{r} - \vec{R}(t_{\text{ret}})|}\right)^{-1}, \tag{5.261a}$$

$$\vec{A}_L(\vec{r}, t) = \frac{q}{4\pi\epsilon_0 c^2} \frac{\dot{\vec{R}}(t_{\text{ret}})}{|\vec{r} - \vec{R}(t_{\text{ret}})|} \left(1 - \frac{\dot{\vec{R}}(t_{\text{ret}})}{c} \cdot \frac{\vec{r} - \vec{R}(t_{\text{ret}})}{|\vec{r} - \vec{R}(t_{\text{ret}})|}\right)^{-1}. \tag{5.261b}$$

With the definitions $\vec{\beta} \equiv \vec{\beta}(t_{\text{ret}}) = \frac{\dot{\vec{R}}(t_{\text{ret}})}{c}$, as well as $\hat{n} \equiv \frac{\vec{r} - \vec{R}(t_{\text{ret}})}{|\vec{r} - \vec{R}(t_{\text{ret}})|}$ and $\mathcal{R} \equiv |\vec{r} - \vec{R}(t_{\text{ret}})|$, we can write the scalar and vector potentials in the familiar form,

$$\Phi_L(\vec{r}, t) = \frac{q}{4\pi\epsilon_0} \frac{1}{\mathcal{R}\left(1 - \vec{\beta} \cdot \hat{n}\right)}, \qquad \vec{A}_L(\vec{r}, t) = \frac{q}{4\pi\epsilon_0 c^2} \frac{\vec{\beta}}{\mathcal{R}\left(1 - \vec{\beta} \cdot \hat{n}\right)}, \tag{5.262}$$

where we keep SI mksA units.

5.6.2 *Liénard–Wiechert Potentials in Coulomb Gauge*

The calculation of the Liénard–Wiechert potentials in Coulomb gauge crucially relies on a careful consideration of the longitudinal and transverse parts of the current density of a point charge. We recall the formulas (5.254a) and (5.254b) in the form

$$\rho(\vec{r}, t) = q\, \delta^{(3)}(\vec{r} - \vec{R}(t)), \tag{5.263a}$$

$$\vec{J}(\vec{r}, t) = q\, \delta^{(3)}(\vec{r} - \vec{R}(t)) \left[\frac{\mathrm{d}}{\mathrm{d}t} \vec{R}(t)\right] = q\, \delta^{(3)}(\vec{r} - \vec{R}(t))\, \dot{\vec{R}}(t). \tag{5.263b}$$

According to Eq. (1.48a), the longitudinal part of the current density can be extracted as follows,

$$\vec{J}_\parallel(\vec{r}, t) = -\frac{1}{4\pi} \vec{\nabla} \int \mathrm{d}^3 r' \frac{\vec{\nabla}' \cdot \vec{J}(\vec{r}', t)}{|\vec{r} - \vec{r}'|}. \tag{5.264}$$

The divergence of the current density is easily calculated:

$$\vec{\nabla} \cdot \vec{J}(\vec{r}, t) = q\, \dot{\vec{R}}(t) \cdot \vec{\nabla} \delta^{(3)}(\vec{r} - \vec{R}(t)). \tag{5.265}$$

We now calculate an integral, which we call for convenience $\mathcal{L}$ (its gradient is proportional to $\vec{J}_\parallel$)

$$\begin{aligned}
\mathcal{L} &= \int \mathrm{d}^3 r' \frac{\vec{\nabla}' \cdot \vec{J}(\vec{r}', t)}{|\vec{r} - \vec{r}'|} = \int \mathrm{d}^3 r' \frac{1}{|\vec{r} - \vec{r}'|} q\, \dot{\vec{R}}(t) \cdot \vec{\nabla}' \delta^{(3)}(\vec{r}' - \vec{R}(t)) \\
&= \int \mathrm{d}^3 r'\, q\, \dot{\vec{R}}(t) \cdot \left(-\vec{\nabla}' \frac{1}{|\vec{r} - \vec{r}'|}\right) \delta^{(3)}(\vec{r}' - \vec{R}(t)) \\
&= \dot{\vec{R}}(t) \cdot \vec{\nabla} \int \mathrm{d}^3 r' \frac{q}{|\vec{r} - \vec{r}'|} \delta^{(3)}(\vec{r}' - \vec{R}(t)) \\
&= q\, \dot{\vec{R}}(t) \cdot \vec{\nabla} \frac{1}{|\vec{r} - \vec{R}(t)|}.
\end{aligned} \tag{5.266}$$

The longitudinal component $\vec{J}_\parallel$ is finally calculated as follows,

$$\vec{J}_\parallel(\vec{r},t) = -\frac{1}{4\pi}\vec{\nabla}\mathcal{L} = -\frac{1}{4\pi}\vec{\nabla}\int \mathrm{d}^3 r' \frac{\vec{\nabla}'\cdot\vec{J}(\vec{r}',t)}{|\vec{r}-\vec{r}'|}$$

$$= -\frac{q}{4\pi}\vec{\nabla}\left(\dot{\vec{R}}(t)\cdot\vec{\nabla}\frac{1}{|\vec{r}-\vec{R}(t)|}\right) = -\frac{q}{4\pi}\left(\dot{\vec{R}}(t)\cdot\vec{\nabla}\right)\vec{\nabla}\frac{1}{|\vec{r}-\vec{R}(t)|}.$$
(5.267)

We can verify that $\vec{J}_\parallel$ carries the entire divergence of $\vec{J}$,

$$\vec{\nabla}\cdot\vec{J}_\parallel(\vec{r},t) = -\frac{q}{4\pi}\dot{\vec{R}}(t)\cdot\vec{\nabla}\,\vec{\nabla}^2\frac{1}{|\vec{r}-\vec{R}(t)|}$$

$$= q\,\dot{\vec{R}}(t)\cdot\vec{\nabla}\,\delta^{(3)}(\vec{r}-\vec{R}(t)) = \vec{\nabla}\cdot\vec{J}(\vec{r},t).$$
(5.268)

In principle, we could now simply calculate the transverse component $\vec{J}_\perp(\vec{r},t)$ as the difference of the total current density $\vec{J}(\vec{r},t)$ and the longitudinal component $\vec{J}_\parallel(\vec{r},t)$, namely, $\vec{J}_\perp(\vec{r},t) = \vec{J}(\vec{r},t) - \vec{J}_\parallel(\vec{r},t)$. However, it is more instructive to calculate the transverse component according to Eq. (1.48b),

$$\vec{J}_\perp(\vec{r},t) = \frac{1}{4\pi}\vec{\nabla}\times\int \mathrm{d}^3 r' \frac{\vec{\nabla}'\times\vec{J}(\vec{r}',t)}{|\vec{r}-\vec{r}'|}.$$
(5.269)

We calculate the vector-valued integral

$$\vec{\mathcal{K}} = \int \mathrm{d}^3 r' \frac{\vec{\nabla}'\times\vec{J}(\vec{r}',t)}{|\vec{r}-\vec{r}'|} = \int \mathrm{d}^3 r' \frac{1}{|\vec{r}-\vec{r}'|}\vec{\nabla}'\times\left(q\,\delta^{(3)}(\vec{r}'-\vec{R}(t))\,\dot{\vec{R}}(t)\right)$$

$$= -q\int \mathrm{d}^3 r' \frac{1}{|\vec{r}-\vec{r}'|}\dot{\vec{R}}(t)\times\vec{\nabla}'\delta^{(3)}(\vec{r}'-\vec{R}(t))$$

$$= q\int \mathrm{d}^3 r'\,\delta^{(3)}(\vec{r}'-\vec{R}(t))\left(\dot{\vec{R}}(t)\times\vec{\nabla}'\frac{1}{|\vec{r}-\vec{r}'|}\right)$$

$$= -q\,\dot{\vec{R}}(t)\times\vec{\nabla}\frac{1}{|\vec{r}-\vec{R}(t)|}.$$
(5.270)

The result for the transverse component of the current density therefore is

$$\vec{J}_\perp(\vec{r},t) = \frac{1}{4\pi}\vec{\nabla}\times\vec{\mathcal{K}} = \frac{1}{4\pi}q\,\dot{\vec{R}}(t)\times\vec{\nabla}\times\vec{\nabla}\frac{1}{|\vec{r}-\vec{R}(t)|}.$$
(5.271)

The transverse component finally is found as follows,

$$\vec{J}_\perp(\vec{r},t) = \frac{1}{4\pi}q\,\dot{\vec{R}}(t)\times\vec{\nabla}\times\vec{\nabla}\frac{1}{|\vec{r}-\vec{R}(t)|}$$

$$= \frac{q}{4\pi}\dot{\vec{R}}(t)\cdot\vec{\nabla}\,\vec{\nabla}\frac{1}{|\vec{r}-\vec{R}(t)|} - \frac{q}{4\pi}\dot{\vec{R}}(t)\,\vec{\nabla}^2\frac{1}{|\vec{r}-\vec{R}(t)|}$$

$$= \frac{q}{4\pi}\dot{\vec{R}}(t)\cdot\vec{\nabla}\,\vec{\nabla}\frac{1}{|\vec{r}-\vec{R}(t)|} + q\,\dot{\vec{R}}(t)\,\delta^{(3)}(\vec{r}-\vec{R}(t)) = -\vec{J}_\parallel(\vec{r},t) + \vec{J}(\vec{r},t),$$
(5.272)

verifying the identity $\vec{J}_\parallel(\vec{r},t) + \vec{J}_\perp(\vec{r},t) = \vec{J}(\vec{r},t)$.

The results can be summarized as follows,

$$\vec{J}_{\parallel}(\vec{r},t) = \vec{\nabla}\left(-\frac{q}{4\pi}\dot{\vec{R}}(t)\cdot\vec{\nabla}\frac{1}{|\vec{r}-\vec{R}(t)|}\right) = \vec{\nabla}\mathcal{J}(\vec{r},t), \qquad (5.273a)$$

$$\mathcal{J}(\vec{r},t) = -\frac{q}{4\pi}\dot{\vec{R}}(t)\cdot\vec{\nabla}\frac{1}{|\vec{r}-\vec{R}(t)|}, \qquad (5.273b)$$

$$\vec{J}_{\perp}(\vec{r},t) = \frac{1}{4\pi}q\dot{\vec{R}}(t)\times\vec{\nabla}\times\vec{\nabla}\frac{1}{|\vec{r}-\vec{R}(t)|}, \qquad (5.273c)$$

where we define the function $\mathcal{J}$ in Eq. (5.273b). The longitudinal and transverse components of $\vec{J}$ enter the formulas for the scalar and vector potentials in Coulomb gauge. We recall Eq. (4.102),

$$\Phi_C(\vec{r},t) = \int dt'\int d^3r'\, G_R(\vec{r}-\vec{r}',t-t')\left(\rho(\vec{r}',t') + \frac{1}{c^2}\frac{\partial}{\partial t'}\mathcal{J}(\vec{r}',t')\right). \qquad (5.274)$$

As before, the retarded Green function may be used in the approximation [see Eq. (4.36)],

$$G_R(\vec{r}-\vec{r}',t-t') \approx \frac{c}{4\pi\epsilon_0}\frac{\Theta(t-t')}{|\vec{r}-\vec{r}'|}\delta\left(|\vec{r}-\vec{r}'|-c(t-t')\right). \qquad (5.275)$$

So,

$$\Phi_C(\vec{r},t) = \Phi_L(\vec{r},t) - \frac{q}{4\pi c^2}\int dt'\int d^3r'\, G_R(\vec{r}-\vec{r}',t-t')\frac{\partial}{\partial t'}\dot{\vec{R}}(t')\cdot\vec{\nabla}'\frac{1}{|\vec{r}'-\vec{R}(t')|}$$

$$= \Phi_L(\vec{r},t) + \Phi_S(\vec{r},t), \qquad (5.276)$$

where Φ_L is the Lorenz-gauge expression (4.88a). No attempt is made here to simplify the expression found for the supplementary term Φ_S. We can also write the Coulomb-gauge Liénard–Wiechert vector potential as

$$\vec{A}_C(\vec{r},t) = \frac{1}{c^2}\int d^3r'\,dt'\, G_R(\vec{r}-\vec{r}',t-t')\,\vec{J}_{\perp}(\vec{r}',t')$$

$$= \frac{1}{c^2}\int d^3r'\,dt'\, G_R(\vec{r}-\vec{r}',t-t')\left(\vec{J}(\vec{r}',t')-\vec{J}_{\parallel}(\vec{r}',t')\right)$$

$$= \vec{A}_L(\vec{r},t) + \frac{q}{4\pi c^2}\int d^3r'\,dt'\, G_R(\vec{r}-\vec{r}',t-t')\left(\dot{\vec{R}}(t')\cdot\vec{\nabla}'\right)\left(\vec{\nabla}'\frac{1}{|\vec{r}'-\vec{R}(t')|}\right)$$

$$= \vec{A}_L(\vec{r},t) + \vec{A}_S(\vec{r},t), \qquad (5.277)$$

where $\vec{A}_L(\vec{r},t)$ is the Lorenz-gauge expression, given in Eq. (4.88b). It is instructive to write the supplementary term $\vec{A}_S(\vec{r},t)$ as a gradient, after an integration by parts and a change $\vec{\nabla}' \rightarrow -\vec{\nabla}$ when acting on the Green function,

$$\vec{A}_S(\vec{r},t) = \vec{\nabla} \frac{q}{4\pi c^2} \int d^3r' \, dt' \, G_R(\vec{r}-\vec{r}',t-t') \, \dot{\vec{R}}(t') \cdot \vec{\nabla}' \frac{1}{|\vec{r}'-\vec{R}(t')|} . \qquad (5.278)$$

Evidently, the curl of $\vec{A}_S$ vanishes and there is no additional contribution to the magnetic induction field. The supplementary electric field must necessarily vanish,

$$\vec{E}_S(\vec{r},t) = -\vec{\nabla}\Phi_S(\vec{r},t) - \frac{\partial}{\partial t}\vec{A}_S(\vec{r},t)$$

$$= \frac{q}{4\pi c^2} \int dt' \int d^3r' \, \vec{\nabla} G_R(\vec{r}-\vec{r}',t-t') \frac{\partial}{\partial t'}\dot{\vec{R}}(t') \cdot \vec{\nabla}'\frac{1}{|\vec{r}'-\vec{R}(t')|}$$

$$- \frac{q}{4\pi c^2} \int d^3r' \, dt' \, \frac{\partial}{\partial t} G_R(\vec{r}-\vec{r}',t-t') \, \dot{\vec{R}}(t') \cdot \vec{\nabla}' \, \vec{\nabla}'\frac{1}{|\vec{r}'-\vec{R}(t')|}$$

$$= -\frac{q}{4\pi c^2} \int dt' \int d^3r' G_R(\vec{r}-\vec{r}',t-t') \frac{\partial}{\partial t'}\dot{\vec{R}}(t') \cdot \vec{\nabla}' \, \vec{\nabla}'\frac{1}{|\vec{r}'-\vec{R}(t')|}$$

$$+ \frac{q}{4\pi c^2} \int d^3r' \, dt' \, G_R(\vec{r}-\vec{r}',t-t') \frac{\partial}{\partial t'}\dot{\vec{R}}(t') \cdot \vec{\nabla}' \, \vec{\nabla}'\frac{1}{|\vec{r}'-\vec{R}(t')|} = 0 ,$$

$$(5.279)$$

and it does, confirming the gauge invariance of the fields.

5.7   Exercises

• **Exercise 5.1**: Fill in the missing intermediate steps in Eq. (5.4). In particular, explain why the Heaviside step function $\Theta(t-t')$ does not figure in the final result.
• **Exercise 5.2**: Show that the function $Z(x) = x^\alpha \, J_n(\beta \, x^\gamma)$ fulfills

$$\frac{\partial^2 Z}{\partial x^2} + \frac{1-2\alpha}{x}\frac{\partial Z}{\partial x} + \left(\beta^2\gamma^2 x^{2\gamma-2} + \frac{\alpha^2 - n^2\gamma^2}{x^2}\right) Z = 0 . \qquad (5.280)$$

Hint: Start from the relations

$$Z(x) = x^\alpha \, J_n(\beta \, x^\gamma), \quad z = \beta \, x^\gamma, \quad x = \left(\frac{z}{\beta}\right)^{1/\gamma}, \qquad (5.281)$$

$$J_n(z) = x^{-\alpha} \, Z(x) = \left(\frac{z}{\beta}\right)^{-\alpha/\gamma} Z\left[\left(\frac{z}{\beta}\right)^{1/\gamma}\right] . \qquad (5.282)$$

Then, insert these into Bessel's differential equation which reads as

$$\left(\frac{\partial^2}{\partial z^2} + \frac{1}{z}\frac{\partial}{\partial z} + \frac{z^2 - n^2}{z^2}\right)J_n(z) = \left(\frac{\partial^2}{\partial z^2} + \frac{1}{z}\frac{\partial}{\partial z} + \frac{z^2 - n^2}{z^2}\right)\left\{\left(\frac{z}{\beta}\right)^{-\alpha/\gamma} Z\left[\left(\frac{z}{\beta}\right)^{1/\gamma}\right]\right\} = 0 .$$

$$(5.283)$$

In particular, carry out all differentiations with respect to z and then replace $z \rightarrow \beta x^\gamma$.

• **Exercise 5.3**: Tabulate the first few Hankel functions and convince yourself that these can be written in terms of $\exp(i\rho)$ and powers of ρ. Show that

$$h_0^{(1)}(\rho) = -i\,\frac{e^{i\rho}}{\rho}\,, \tag{5.284a}$$

$$h_1^{(1)}(\rho) = -\frac{e^{i\rho}}{\rho} - i\,\frac{e^{i\rho}}{\rho^2}\,, \tag{5.284b}$$

$$h_2^{(1)}(\rho) = \frac{i e^{i\rho}}{\rho} - \frac{3 e^{i\rho}}{\rho^2} - \frac{3i e^{i\rho}}{\rho^3}\,. \tag{5.284c}$$

Show, by an explicit calculation for the three example cases above, that as $\rho \to +\infty$,

$$h_\ell^{(1)}(\rho) \to -i\,\frac{e^{i(\rho - \ell\pi/2)}}{\rho} = (-i)^{\ell+1}\,\frac{e^{i\rho}}{\rho}\,, \qquad e^{i(-\ell\pi/2)} = \left(e^{-i\pi/2}\right)^\ell = (-i)^\ell\,, \qquad \rho \to \infty\,, \tag{5.285}$$

i.e., that the Hankel functions of different order only differ by a complex phase in the limit $\rho \to +\infty$. Then, repeat the above exercise for the Hankel functions of the second kind, $h_0^{(2)}(\rho)$, $h_1^{(2)}(\rho)$, $h_2^{(2)}(\rho)$.

• **Exercise 5.4**: Verify (i) the property (5.140) and (ii) Eq. (5.141) with explicit reference to the Levi–Cività tensor.

• **Exercise 5.5**: We saw that the action of the photon spin operator is given by the matrices $(\$_k)_{ij} = -i\,\epsilon_{ijk}$, where $\$_k$ is the kth Cartesian spin matrix. Verify that you can alternatively write the spin operator as $\vec{\$} = i\,\mathbb{1}_{3\times3}\times$, where $\times$ is the vector product. Hint: You first have to think about how to interpret the expression $\mathbb{1}_{3\times3}\times$.

• **Exercise 5.6**: Consider the two-dimensional Pauli matrices,

$$\sigma_1 = \begin{pmatrix} 0 & 1 \\ 1 & 0 \end{pmatrix}\,, \qquad \sigma_2 = \begin{pmatrix} 0 & -i \\ i & 0 \end{pmatrix} \qquad \sigma_3 = \begin{pmatrix} 1 & 0 \\ 0 & -1 \end{pmatrix}\,. \tag{5.286}$$

Verify that the spin operator matrices $\$_i = \sigma_i/2$ (with $i = 1,2,3$) fulfill an analogous relation as compared to Eq. (5.140), namely,

$$[\$_i, \$_j] = i\,\epsilon_{ijk}\$_k\,, \tag{5.287}$$

where the Einstein summation convention is used on the right-hand side [even if the indices are all lower indices, see Eq. (1.45)].

• **Exercise 5.7**: Verify that $\vec{M}_{j\mu}^{(K)}(k,\vec{r})$, $\vec{N}_{j\mu}^{(K)}(k,\vec{r})$, and $\vec{L}_{j\mu}^{(K)}(k,\vec{r})$ fulfill the Helmholtz equation

$$\left(\vec{\nabla}^2 + k^2\right)\vec{M}_{j\mu}^{(K)}(k,\vec{r}) = \left(\vec{\nabla}^2 + k^2\right)\vec{N}_{j\mu}^{(K)}(k,\vec{r}) = \left(\vec{\nabla}^2 + k^2\right)\vec{L}_{j\mu}^{(K)}(k,\vec{r}) = 0\,. \tag{5.288}$$

• **Exercise 5.8**: Show Eq. (5.149).

• **Exercise 5.9**: Verify the equivalence of the representations (5.164), (5.171), and (5.172). In the proof, one uses Eqs. (5.29) and (5.148).

• **Exercise 5.10**: Show that Eq. (5.178) can alternatively be written as

$$\vec{E}_0(\vec{r}) = -k^2 Z_0 \sum_{\ell m} \left(m_{\ell m}\,\vec{M}_{\ell m}^{(1)}(k,\vec{r}) + n_{\ell m}\,\vec{N}_{\ell m}^{(1)}(k,\vec{r}) \right)\,. \tag{5.289}$$

where $Z_0 = \sqrt{\mu_0/\epsilon_0}$ is the impedance of the vacuum, around $377\,\Omega$.

- **Exercise 5.11**: We have encountered the Clebsch–Gordan coefficients. Show that

$$s = \sum_{q'q''} C^{00}_{1q'\,1q''}\, u_{q'}\, v_{q''} = -\frac{1}{\sqrt{3}}\, \vec{u} \cdot \vec{v} \tag{5.290}$$

where the components u_q and v_q are given in the spherical basis, i.e., we would have for the reference vector $\vec{r} = \sum x_q\, \vec{e}^*_q$ [see Eq. (5.126)],

$$x_{+1} = -\frac{1}{\sqrt{2}}\,(x+iy)\,, \qquad x_0 = z\,, \qquad x_{-1} = \frac{1}{\sqrt{2}}\,(x-iy)\,, \tag{5.291}$$

with [see Eq. (5.127)]

$$\vec{e}_{+1} = -\frac{1}{\sqrt{2}}\,(\hat{e}_x + i\hat{e}_y)\,, \qquad \vec{e}_0 = \hat{e}_z\,, \qquad \vec{e}_{-1} = \frac{1}{\sqrt{2}}\,(\hat{e}_x - i\hat{e}_y)\,. \tag{5.292}$$

Which summation limits have to be used for the sums over q' and q''?

- **Exercise 5.12**: Consider the relations (5.148a) and (5.148c),

$$\vec{Y}^{\,j-1}_{j\mu}(\theta,\varphi) = -\frac{1}{\sqrt{j\,(2j+1)}}\,\left(i\hat{r} \times \vec{L} - j\,\hat{r}\right) Y_{j\mu}(\theta,\varphi)\,, \tag{5.293a}$$

$$\vec{Y}^{\,j+1}_{j\mu}(\theta,\varphi) = -\frac{1}{\sqrt{(j+1)\,(2j+1)}}\,\left(i\hat{r} \times \vec{L} + (j+1)\,\hat{r}\right) Y_{j\mu}(\theta,\varphi)\,. \tag{5.293b}$$

With the help of these relations, and the recursion relations for the Bessel functions, show that

$$\vec{N}^{(K)}_{j\mu}(k,\vec{r}) = \sqrt{\frac{j+1}{2j+1}}\, f^{(K)}_{j-1}(k\,r)\, \vec{Y}^{\,j-1}_{j\mu}(\theta,\varphi) - \sqrt{\frac{j}{2j+1}}\, f^{(K)}_{j+1}(k\,r)\, \vec{Y}^{\,j+1}_{j\mu}(\theta,\varphi)\,,$$

$$= \frac{1}{kr}\,\left\{ \frac{\mathrm{d}[kr\, f^{(K)}_j(k\,r)]}{\mathrm{d}(kr)}\,[i\hat{r} \times \vec{Y}^{\,j}_{j\mu}(\theta,\varphi)] - \hat{r}\sqrt{\ell(\ell+1)}\, f^{(K)}_j(k\,r)\, Y_{j\mu}(\theta,\varphi) \right\}\,, \tag{5.294}$$

and that

$$\vec{L}^{(K)}_{j\mu}(k,\vec{r}) = \sqrt{\frac{j}{2j+1}}\, f^{(K)}_{j-1}(k\,r)\, \vec{Y}^{\,j-1}_{j\mu}(\theta,\varphi) + \sqrt{\frac{j+1}{2j+1}}\, f^{(K)}_{j+1}(k\,r)\, \vec{Y}^{\,j+1}_{j\mu}(\theta,\varphi)$$

$$= \sqrt{j(j+1)}\,\frac{1}{k\,r}\, f^{(K)}_j(k\,r)\,(i\hat{r} \times \vec{Y}^{\,j}_{j\mu}(\theta,\varphi)) - \hat{r}\,\frac{\mathrm{d}[f^{(K)}_\ell(k\,r)]}{\mathrm{d}(k\,r)}\, Y_{j\mu}(\theta,\varphi)\,. \tag{5.295}$$

- **Exercise 5.13**: Show that

$$\sqrt{\frac{j+1}{2j+1}}\, \vec{Y}^{\,j-1}_{j\mu}(\theta,\varphi) + \sqrt{\frac{j}{2j+1}}\, \vec{Y}^{\,j+1}_{j\mu}(\theta,\varphi) = -i\,\left(\hat{r} \times \vec{Y}^{\,j}_{j\mu}(\theta,\varphi)\right)\,. \tag{5.296}$$

- **Exercise 5.14**: Show that

$$\vec{\nabla} \times \left(\vec{r} \times \vec{\nabla}\right) f(\vec{r}) = \left[\vec{r}\,\vec{\nabla}^2 - \vec{\nabla}\,\frac{\partial}{\partial r}\, r\right] f(\vec{r})\,, \tag{5.297}$$

for any test function $f(\vec{r})$.

• **Exercise 5.14**: We have shown that in the long-wavelength limit [see Eq. (5.253)],

$$n_{j\mu} \approx -\frac{ic}{(2j+1)!!} \sqrt{\frac{j+1}{j}} \, k^j \, q_{j\mu} \,. \tag{5.298}$$

Complete the calculation of the corresponding electric dipole ($j = 1$) contribution to $\vec{A}_0(\vec{r})$, based on the representation (5.170),

$$\vec{A}_0(\vec{r}) = ik\,\mu_0 \sum_{j=0}^{\infty} \sum_{\mu=-j}^{j} \left(m_{j\mu}\,\vec{M}_{j\mu}^{(1)}(k,\vec{r}) + n_{j\mu}\,\vec{N}_{j\mu}^{(1)}(k,\vec{r}) + l_{j\mu}\,\vec{L}_{j\mu}^{(1)}(k,\vec{r}) \right)$$

$$\to \sum_{j=0}^{\infty} \sum_{\mu=-j}^{j} n_{j\mu}\,\vec{N}_{j\mu}^{(1)}(k,\vec{r}) \,, \tag{5.299}$$

and verify that you obtain the familiar form of the electric dipole term.

• **Exercise 5.15**: Show that

$$\int dx\, f_\ell(x)\, g_{\ell'}(x) = \frac{x^2}{\ell'(\ell'+1) - \ell(\ell+1)} \left[f_\ell(x)\, g_{\ell'}'(x) - f_\ell'(x)\, g_{\ell'}(x) \right] , \tag{5.300}$$

with the help of the differential equation (5.27),

$$\left(\frac{\partial^2}{\partial x^2} + \frac{2}{x}\frac{\partial}{\partial x} - \frac{\ell(\ell+1)}{x^2} - 1 \right) f_\ell(x) = 0 \,, \tag{5.301}$$

where f_ℓ can be j_ℓ or y_ℓ, or a superposition (spherical Hankel function).

• **Exercise 5.16**: This exercise summarizes important formulas and can be used as a reference as well. We know that the coupling of the vector potential to the current density reads as [see Eq. (5.158)],

$$\vec{A}_0(\vec{r}) = \frac{1}{c^2} \int d^3r' \, \mathbb{G}_R(k, \vec{r} - \vec{r}') \cdot \vec{J}_0(\vec{r}') \,. \tag{5.302}$$

We had encountered three ways to expand the potential into multipole moments.

(i) The first expansion [Eq. (5.87)] reads as follows:

$$\mathbb{G}_R(\vec{r} - \vec{r}', k) = \frac{1}{\epsilon_0} \sum_{\ell=0}^{\infty} \sum_{m=-\ell}^{\ell} ik\,j_\ell(k\,r_<)\,h_\ell^{(1)}(k\,r_>)\,Y_{\ell m}(\theta, \varphi)\,Y_{\ell m}^*(\theta', \varphi')\,\mathbb{1}_{3\times 3} \,. \tag{5.303}$$

We define multipole moments as follows:

$$\vec{p}_{\ell m} = \int d^3r \, j_\ell(k\,r)\,\vec{J}_0(\vec{r})\,Y_{\ell m}^*(\theta, \varphi) \,. \tag{5.304}$$

The vector potential is obtained as follows,

$$\vec{A}_0(\vec{r}) = i\,\mu_0\,k \sum_{\ell=0}^{\infty} \sum_{m=-\ell}^{\ell} \vec{p}_{\ell m}\,h_\ell^{(1)}(k\,r)\,Y_{\ell m}(\theta, \varphi) \,. \tag{5.305}$$

(ii) The second expansion [Eq. (5.161)] involves the vector spherical harmonics,

$$\mathbb{G}_R(\vec{r} - \vec{r}', k) = \frac{ik}{\epsilon_0} \sum_{j=0}^{\infty} \sum_{\mu=-j}^{j} \sum_{\ell=j-1}^{j+1} j_\ell(k\,r_<)\,h_\ell^{(1)}(k\,r_>)\,\vec{Y}_{j\mu}^\ell(\theta, \varphi) \otimes \vec{Y}_{j\mu}^{\ell*}(\theta', \varphi') \,. \tag{5.306}$$

The multipole moments read as follows [see Eq. (5.162)],

$$p_{j\mu}^\ell = \int d^3r\, j_\ell(kr)\, \vec{J}_0(\vec{r}) \cdot \vec{Y}_{j\mu}^{\ell*}(\theta,\varphi) \approx \frac{k^\ell}{(2\ell+1)!!} \int d^3r\, r^\ell\, \vec{J}_0(\vec{r}) \cdot \vec{Y}_{j\mu}^{\ell*}(\theta,\varphi)\,. \quad (5.307)$$

The vector potential is given as follows,

$$\vec{A}_0(\vec{r}) = i\,\mu_0\, k \sum_{j=0}^{\infty} \sum_{\mu=-j}^{j} \sum_{\ell=j-1}^{j+1} p_{j\mu}^\ell\, h_\ell^{(1)}(kr)\, \vec{Y}_{j\mu}^\ell(\theta,\varphi)\,. \quad (5.308)$$

(iii) The third expansion [Eq. (5.170)] involves the following formula for the tensor Helmholtz Green function,

$$\mathbb{G}_R(k,\vec{r}-\vec{r}') = \frac{ik}{\epsilon_0} \sum_{j=0}^{\infty} \sum_{\mu=-j}^{j} \Big(\vec{M}_{j\mu}^{(1)}(k,\vec{r}_>) \otimes \vec{M}_{j\mu}^{(0)*}(k,\vec{r}_<)$$

$$+ \vec{N}_{j\mu}^{(1)}(k,\vec{r}_>) \otimes \vec{N}_{j\mu}^{(0)*}(k,\vec{r}_<) + \vec{L}_{j\mu}^{(1)}(k,\vec{r}_>) \otimes \vec{L}_{j\mu}^{(0)*}(k,\vec{r}_<) \Big)\,. \quad (5.309)$$

The multipole moments are [see Eqs. (5.168a), (5.168b) and (5.168c)]

$$m_{j\mu} = \int d^3r'\, \vec{J}_0(\vec{r}') \cdot \vec{M}_{j\mu}^{(0)*}(\theta',\varphi')\,, \quad (5.310a)$$

$$n_{j\mu} = \int d^3r'\, \vec{J}_0(\vec{r}') \cdot \vec{N}_{j\mu}^{(0)*}(\theta',\varphi')\,, \quad (5.310b)$$

$$l_{j\mu} = \int d^3r'\, \vec{J}_0(\vec{r}') \cdot \vec{L}_{j\mu}^{(0)*}(\theta',\varphi')\,. \quad (5.310c)$$

The vector potential is obtained as follows [see Eq. (5.170)],

$$\vec{A}_0(\vec{r}) = ik\,\mu_0 \sum_{j=0}^{\infty} \sum_{\mu=-j}^{j} \Big(m_{j\mu}\, \vec{M}_{j\mu}^{(1)}(k,\vec{r}) + n_{j\mu}\, \vec{N}_{j\mu}^{(1)}(k,\vec{r}) + l_{j\mu}\, \vec{L}_{j\mu}^{(1)}(k,\vec{r}) \Big)\,. \quad (5.311)$$

The magnetic, electric and longitudinal multipole functions read as follows [see Eqs. (5.164a), (5.164b) and (5.164c)],

$$\vec{M}_{j\mu}^{(K)}(k,\vec{r}) = \frac{1}{\sqrt{j(j+1)}}\, f_j^{(K)}(kr)\, \vec{L} Y_{j\mu}(\theta,\varphi)\,, \quad (5.312a)$$

$$\vec{N}_{j\mu}^{(K)}(k,\vec{r}) = \frac{i}{k}\, \vec{\nabla} \times \vec{M}_{j\mu}^{(K)}(k,\vec{r})\,, \quad (5.312b)$$

$$\vec{L}_{j\mu}^{(K)}(k,\vec{r}) = \frac{1}{k}\, \vec{\nabla}\Big(f_j^{(K)}(kr)\, Y_{j\mu}(\theta,\varphi) \Big)\,. \quad (5.312c)$$

Treat the following current distributions in all three variants of the multipole expansion.

(i) Assume that

$$\vec{J}_0(\vec{r}) = \frac{I_0}{a^2}\left(3\frac{z^2}{r^2}-1\right)\exp\left(-\frac{r^2}{a^2}\right)\hat{e}_z\,, \quad (5.313)$$

and use all three multipole formalisms. You may replace the Bessel functions by their leading asymptotic behavior for small argument. **Hint:** You might obtain

$$n_{10} = -\frac{\pi}{25\sqrt{6}} k^2 a^3 I_0, \qquad n_{30} = -\frac{\pi}{25}\sqrt{\frac{3}{7}} k^2 a^3 I_0, \qquad (5.314)$$

$$l_{10} = -\frac{\pi}{25\sqrt{3}} k^2 a^3 I_0, \qquad l_{30} = \frac{3\pi}{50\sqrt{7}} k^2 a^3 I_0. \qquad (5.315)$$

Do all the magnetic multipoles vanish, i.e., $m_{j\mu} = 0$?

(ii) Assume that

$$\vec{J}_0(\vec{r}) = \frac{I_0}{a^2}\left(3\frac{z^2}{r^2} - 1\right)\exp\left(-\frac{r^2}{a^2}\right)\vec{e}_{+1}. \qquad (5.316)$$

Hint: You might obtain

$$m_{21} = -\frac{\pi}{10\sqrt{10}} k^2 a^3 I_0, \qquad n_{11} = -\frac{\pi}{50\sqrt{6}} k^2 a^3 I_0, \qquad (5.317)$$

$$n_{31} = -\frac{\pi}{25}\sqrt{\frac{2}{7}} k^2 a^3 I_0, \qquad l_{11} = \frac{\pi}{50\sqrt{3}} k^2 a^3 I_0, \qquad (5.318)$$

$$l_{31} = \frac{\pi}{25}\sqrt{\frac{3}{14}} k^2 a^3 I_0. \qquad (5.319)$$

Do all the other multipoles vanish, i.e., $m_{j\mu} = 0$?

(iii) Now here is another current distribution,

$$\vec{J}_0(\vec{r}) = \frac{I_0}{a^2}\frac{z}{a}\left(\frac{y}{a}\hat{e}_x - \frac{x}{a}\hat{e}_y\right)\exp\left(-\frac{r^2}{a^2}\right), \qquad (5.320)$$

which needs to be analyzed in terms of the electric and magnetic multipoles. **Hint:** You might obtain

$$m_{20} = \mathrm{i}\frac{\pi}{4\sqrt{30}} k^2 a^3 I_0. \qquad (5.321)$$

Do all the electric and longitudinal multipoles vanish?

• **Exercise 5.17:** Consider the equation defining the retarded time, $t_{\mathrm{ret}} = t - |\vec{r} - \vec{R}(t_{\mathrm{ret}})|/c$, for uniform motion in only one dimension (along the x direction), i.e., for $\vec{r} = r\hat{e}_x$, $r > 0$, and $\vec{R}(t) = v_0 t\hat{e}_x$. We thus consider only observation points $\vec{r}$ lying in the x axis, reducing the problem to one dimension.

(i) Solve the defining equation for t_{ret}, replacing all vector quantities by their x components.

(ii) Calculate the scalar and vector potentials (Liénard–Wiechert potentials),

$$\Phi(\vec{r},t) = \frac{q}{4\pi\epsilon_0}\frac{1}{|\vec{r} - \vec{R}(t_{\mathrm{ret}})|}\left(1 - \frac{\dot{\vec{R}}(t_{\mathrm{ret}})}{c}\cdot\frac{\vec{r} - \vec{R}(t_{\mathrm{ret}})}{|\vec{r} - \vec{R}(t_{\mathrm{ret}})|}\right)^{-1}, \qquad (5.322a)$$

$$\vec{A}(\vec{r},t) = \frac{\Phi(\vec{r},t)}{c^2}\dot{\vec{R}}(t_{\mathrm{ret}}), \qquad (5.322b)$$

and show that for the particular case, the final formulas are consistent with "no retardation occurring", because

$$\Phi(\vec{r}, t) = \frac{q}{4\pi\epsilon_0} \frac{1}{r - v_0 t}, \qquad \vec{A}(\vec{r}, t) = \frac{q}{4\pi\epsilon_0 c^2} \frac{v_0}{r - v_0 t}. \qquad (5.323)$$

The change in the interaction distance $|\vec{r} - \vec{R}(t_{\mathrm{ret}})|$ by the retardation is exactly compensated by the Jacobian factor $\left(1 - \frac{\dot{\vec{R}}(t_{\mathrm{ret}})}{c} \cdot \frac{\vec{r} - \vec{R}(t_{\mathrm{ret}})}{|\vec{r} - \vec{R}(t_{\mathrm{ret}})|}\right)^{-1}$. Show your work and all intermediate steps.

• **Exercise 5.18**: Investigate the scalar and vector potentials (5.323), but this time for general $\vec{r}$, constant speed $\vec{v}_0$,

$$\vec{R}(t) = \vec{v}_0, \qquad \beta_0 = \frac{\vec{v}_0}{c}, \qquad \frac{\vec{R}_0 \cdot \vec{\beta}_0}{|\vec{R}_0| \beta_0} = \cos(\theta_0) = 1 - \frac{1}{2!}\theta_0^2 + \frac{1}{4!}\theta_0^2 + \mathcal{O}(\theta_0^6). \quad (5.324)$$

Show that the first retardation correction for small θ_0 reads as follows,

$$\Phi(\vec{r}, t) = \frac{q}{4\pi\epsilon_0} \frac{1}{|\vec{r} - \vec{v}_0 t|} \left(1 + \frac{\beta_0^2 \theta_0^2}{2} + \mathcal{O}(\theta_0^4)\right). \qquad (5.325)$$

Hint: The formulas $r - v_0 t_{\mathrm{ret}} = c(t - t_{\mathrm{ret}})$, with $t_{\mathrm{ret}} = (ct - r)/(c - v_0)$, may be helpful. Then, calculate the term of order θ_0^4.

• **Exercise 5.19**: Verify that $\vec{A}_C(\vec{r}, t)$ given in Eq. (5.277) is transverse, by an explicit calculation. Hint: This may be explicitly verified in various ways; for example, one may take the divergence of both sides of Eq. (5.277), then use the identity $\vec{\nabla} G_R(\vec{r} - \vec{r}', t - t') = -\vec{\nabla}' G_R(\vec{r} - \vec{r}', t - t')$, integrate by parts under the integral sign, and take into account that $\vec{J}_\parallel$ carries the entire divergence of $\vec{J}$, according to Eq. (5.268).

Chapter 6

Electrodynamics in Media

6.1 Overview

The purpose of this chapter is threefold. *(i)* We shall derive, based on a microscopic model, the phenomenological Maxwell equations which involve the so-called dielectric displacement $\vec{D}(\vec{r}, t)$ and the magnetic field $\vec{H}(\vec{r}, t)$. This allows us to give a definite physical interpretation of the dielectric displacement and reaction of the material. We shall also study the frequency ranges over which the phenomenological treatment is valid. *(ii)* We shall introduce Fourier transforms, and briefly discuss the Maxwell equations in the mixed coordinate-frequency representation. This also provides for an alternative viewpoint on the electromagnetic waves. *(iii)* We apply the wisdom gathered to the frequency-dependent dielectric constant $\epsilon_r(\omega)$ of both dilute materials as well as dense materials, and explore how causality dictates the Kramers–Kronig relations which have to be fulfilled by the dielectric constant.

In particular, three examples of possible functional forms of dielectric permittivities are discussed. First, based on a modification of the Sellmeier equation, we shall study the connection of the atomic polarizability and the so-called Drude model, which is relevant to the analysis of near-perfect conductors (see Sec. 6.4). The dielectric permittivity is analyzed for dense materials, in Sec. 6.4.4, with concomitant necessary modifications to the functional form describing the dielectric response function. Wave propagation in a dispersive medium (Sec. 6.5) is an excellent testbed for the study of advanced mathematical techniques like the method of steepest descent. Finally, the Kramers–Kronig relationships (Sec. 6.6) allow us to connect the real (dispersive) and imaginary (absorptive) parts of the dielectric response function, and add important tools in the study of the optical properties of a medium.

6.2 Microscopic and Macroscopic Equations

6.2.1 *Macroscopic Equations and Measurements*

We consider the derivation of the macroscopic (phenomenological) Maxwell equations which involve the dielectric displacement $\vec{D}(\vec{r}, t)$ and the magnetic field

$\vec{H}(\vec{r},t)$. To this end, it is instructive to first consider the response of materials to electromagnetic radiation in different frequency ranges, as this determines the physical conditions under which the phenomenological equations are valid. Let us consider a light frequency range

$$\nu \sim \frac{c}{a_0}, \qquad \lambda \sim a_0, \tag{6.1}$$

where ν is the frequency of the electromagnetic radiation, and λ the wavelength. The speed of light is $c = \lambda \nu$. The Bohr radius is

$$a_0 = \frac{\hbar}{\alpha\, m_e\, c} \approx 0.529\,\text{Å}, \tag{6.2}$$

where m_e is the electron mass, and λ is the wavelength. The wavelength range (6.1) is commonly used for x-ray crystallography. This is because in a dense crystal lattice, the distance among atoms is of the order of a Bohr radius, and therefore, incoming x-rays are collectively scattered by crystal planes. In a Laue photograph, a single crystalline sample is irradiated by a coherent x-ray source, which results in defined maxima of the scattered intensity (corresponding to the bright spots in the photograph). In a Debye–Scherrer photograph, a polycrystalline sample is irradiated by x-rays. The crystalline grains are oriented in all spatial directions, and therefore the Laue spots become concentric rings.

Note that Laue and Debye–Scherrer crystal diffraction could never have been discovered on the basis of the macroscopic Maxwell equations alone. Rather, the macroscopic Maxwell equations describe the perturbations to the average propagation velocity of light waves due to the presence of a medium, which is being felt by the traveling light wave as a uniform medium. For that approximation to be valid, the wavelength of the traveling wave must be much longer than the average structural distances in the sample,

$$\nu \ll \frac{c}{a_0}, \qquad \lambda \gg a_0. \tag{6.3}$$

Specifically, the Bragg condition $n\,\lambda = 2\,d\,\sin\theta$ governing the refracted waves from the crystal planes could never have been discovered on the basis of the phenomenological Maxwell equations (in order to derive the Bragg condition, one investigates the path difference of light rays emerging from crystal layers displaced by a distance d). The Bragg condition can be derived based on a consideration of the path difference of light rays emitted from different crystal planes, with θ being the incident angle of the x-ray with respect to the crystal plane, or, equivalently the angle of the normal to the crystal plane with the normal to the incident light "sheet".

In the frequency range

$$\nu \gg \frac{c}{a_0}, \qquad \lambda_\gamma \ll a_0, \tag{6.4}$$

the scattering takes place not from individual atoms, but from individual electrons inside the sample. We can even treat the electrons as stationary, the reason being that the momentum p of the incoming photon is much larger than the average momentum of a bound electron. According to the de Broglie formula $p = h/\lambda$, we can write for the frequency range (6.4),

$$p = \frac{h}{\lambda} = \frac{h\nu}{c} \gg \frac{h}{a_0} = 2\pi\alpha m_e c \sim \alpha m_e c. \tag{6.5}$$

The classical "speed" of a bound electron is of the order of αc, where $\alpha \approx 1/137$ is the fine-structure constant, and so $\alpha m_e c$ is a typical electron momentum in the bound state of an atom. In the frequency range (6.4), the photon momentum is much larger than $\alpha m_e c$, implying that the electron can be assumed to be stationary.

We thus expect to use the macroscopic equations and tabulated electric and magnetic susceptibilities only if the electromagnetic fields have wavelengths which are large compared to the characteristic lengths of the system. Specifically, we must assume that simultaneously

$$\lambda \gg a_0, \qquad \lambda \gg \langle d \rangle, \qquad \lambda \gg r_\rho. \tag{6.6}$$

Here, $\langle d \rangle$ is the average spacing of atoms in the medium (for dilute gases, the corresponding frequency range is far in the infrared). Furthermore, r_ρ is a length scale that characterizes density fluctuations inside the sample. For the frequency of the perturbation (light frequency), this means that simultaneously,

$$\nu \ll \frac{c}{a_0}, \qquad \nu \ll \frac{c}{\langle d \rangle}, \qquad \nu \ll \frac{c}{r_\rho}. \tag{6.7}$$

We can then choose a distance R, subject to the condition

$$\lambda \gg R \gg a_0, \langle d \rangle, r_\rho, \tag{6.8}$$

and average the fields over a volume of dimension R^3. This averaging operation (spatial average at constant time) leads to the macroscopic Maxwell equations formulated in terms of the magnetic field $\vec{H}(\vec{r}, t)$ and the dielectric displacement $\vec{D}(\vec{r}, t)$. These quantities include the response of the medium to the external perturbations, and an averaging of the fields over certain length scales.

6.2.2 *Macroscopic Fields from Microscopic Properties*

We consider the limit of low-frequency perturbations, characterized by the conditions (6.6) and (6.7),

$$\lambda \gg R \gg a_0, \langle d \rangle, r_\rho, \qquad \nu \ll \frac{c}{R} \ll \frac{c}{a_0}, \frac{c}{\langle d \rangle}, \frac{c}{r_\rho}, \tag{6.9}$$

where R^3 measures the dimension of our sampling (or averaging) volume. The dimension of the sampling volume is larger than the Bohr radius, larger than the average interatomic distance, and larger than the distance scale r_ρ over which density fluctuations in the sample average out. Our classical approach is inspired by the treatment described in Ref. [26] and provides for an intuitive interpretation of the macroscopic Maxwell equations.

We consider a material with a typical density of $\sim 10^{23}$ atoms (or molecules) per cubic centimeter, cm^3, i.e., one gram per mol. In a typical case, we can choose

$$100\,\text{Å} < R < 1000\,\text{Å}. \qquad (6.10)$$

In the volume R^3, there are approximately 10^6 atoms, and the charge constituents (electrons, nuclei) can be treated as point charges. There is a subtle point. We have previously classified the regime (6.9) as the regime of very-low-frequency light waves. Indeed, the condition

$$\lambda \gg a_0, \qquad \nu \ll \frac{c}{a_0}, \qquad (6.11)$$

means that the wavelength is much longer than the typical size of an atom. This condition is fulfilled by the frequencies corresponding to typical optical transitions of atoms. These wavelengths typically are much larger than the Bohr radius. Namely, we have for a typical atomic wavelength λ_0,

$$\lambda_0 \sim \frac{a_0}{\alpha} \sim 1000\,\text{Å}, \qquad \nu \sim \alpha\frac{c}{a_0} = \frac{\alpha^2 m_e c^2}{\hbar} \sim 3 \times 10^{15}\,\text{Hz}. \qquad (6.12)$$

This is just in the range of atomic transition frequencies, or, in a classical picture, equal to the typical frequency of electronic charge vibrations in a typical atom. In the optical frequency range, the formalism to be outlined below is applicable: One can define a meaningful, frequency-dependent dielectric permittivity $\epsilon(\omega)$ and magnetic permeability $\mu(\omega)$ for the optical frequency range. Of course, the formalism remains valid for even longer-than-optical wavelengths, or, low-frequency driving of the macroscopic sample.

In discussing the microscopic model to be outlined in the following, we might, in principle, distribute and classify the nuclear and electronic charges at our will. We denote the positions of the atomic nuclei by $\vec{r}_n$ and the relative positions of the j electrons bound to the nth nucleus as $\vec{r}_{jn}$, so that the absolute position of the jth electron of the nth atom is at $\vec{r}_n + \vec{r}_{jn}$. Because almost all the mass of the atom is in the nucleus, this approach is equivalent to defining $\vec{r}_n$ as the center of mass of the atom, up to reduced-mass corrections of order $m_e/m_p \approx 1/2000$, where m_e is the mass of the electron, and m_p is the mass of the proton.

We shall be interested in the averages of the microscopic electric field, $\vec{e}(\vec{r},t)$, and of the microscopic magnetic induction field $\vec{b}(\vec{r},t)$, at the point $\vec{r}$ and at time t. The vector $\vec{r}$ is understood to represent a position vector in a much larger region of space (of the order of cm^3) at which the macroscopic electric field $\vec{E}(\vec{r},t)$, and the macroscopic magnetic induction field $\vec{B}(\vec{r},t)$ are set equal to the average of the microscopic fields over the volume of dimension R^3. The fields shall be assumed to be redefined in the indicated sense, for the remainder of the derivation of the macroscopic Maxwell equations, i.e., within Sec. 6.2. With the definitions $\vec{r} - \vec{r}' \equiv \vec{\rho}$ and $\vec{r} - \vec{\rho} = \vec{r}'$, we have

$$\vec{E}(\vec{r},t) \equiv \left\langle \vec{e}(\vec{r},t) \right\rangle = \int_{\mathbb{R}^3} d^3\rho\, f(\vec{r} - \vec{\rho})\, \vec{e}(\vec{\rho},t) = \int_{\mathbb{R}^3} d^3 r'\, f(\vec{r}')\, \vec{e}(\vec{r} - \vec{r}',t), \qquad (6.13a)$$

$$\vec{B}(\vec{r},t) \equiv \left\langle \vec{b}(\vec{r},t) \right\rangle = \int_{\mathbb{R}^3} d^3\rho\, f(\vec{r} - \vec{\rho})\, \vec{b}(\vec{\rho},t) = \int_{\mathbb{R}^3} d^3 r'\, f(\vec{r}')\, \vec{b}(\vec{r} - \vec{r}',t). \qquad (6.13b)$$

Here $f(\vec{r}')$ is a normalized distribution function, which is used in order to average the field over a volume of radius R. Both forms of the averaging operation outlined in Eq. (6.13) will become useful in the following. The definite form of the normalized distribution function f is not crucial, and it is convenient to use a simple normalized Gaussian function

$$f(\vec{r}') = \frac{1}{\pi^{3/2} R^3} \exp\left(-\frac{r'^2}{R^2}\right), \qquad \int d^3 r' f(\vec{r}') = 1. \qquad (6.14)$$

A particular, important result is that the averaging of a Dirac-δ function reproduces the sampling function itself,

$$\langle \delta^{(3)}(\vec{r} - \vec{r}_0) \rangle = \int d^3 \rho\, f\left((\vec{r} - \vec{r}_0) - \vec{\rho}\right) \delta^{(3)}(\vec{\rho}) = f(\vec{r} - \vec{r}_0). \qquad (6.15)$$

This result is heavily used in the following. The function (6.14) has the advantage that the distribution falls off smoothly at the edge of the volume. It is clear that the gradient operator and the time differential operator commute with the averaging operation,

$$\vec{\nabla} \cdot \langle \vec{e}(\vec{r}, t) \rangle = \int f(\vec{r}')\, \vec{\nabla} \cdot \vec{e}(\vec{r} - \vec{r}', t)\, d^3 r' = \langle \vec{\nabla} \cdot \vec{e}(\vec{r}, t) \rangle, \qquad (6.16)$$

$$\frac{\partial}{\partial t} \langle \vec{e}(\vec{r}, t) \rangle = \int f(\vec{r}')\, \frac{\partial}{\partial t} \vec{e}(\vec{r} - \vec{r}', t)\, d^3 r' = \left\langle \frac{\partial}{\partial t} \vec{e}(\vec{r}, t) \right\rangle. \qquad (6.17)$$

The total charge density $\rho(\vec{r}, t)$ is the sum of a conduction band charge density $\rho_c(\vec{r}, t)$, with

$$\rho_c(\vec{r}, t) = \sum_c q_c\, \delta^{(3)}(\vec{r} - \vec{r}_c(t)), \qquad (6.18)$$

and an atomic/ionic charge density $\rho_a(\vec{r}, t)$, where

$$\rho_a(\vec{r}, t) = \sum_{nj} q_{jn}\, \delta^{(3)}(\vec{r} - \vec{r}_n(t) - \vec{r}_{jn}(t)), \qquad (6.19)$$

so that the total charge density $\rho(\vec{r}, t)$ is

$$\rho(\vec{r}, t) = \rho_c(\vec{r}, t) + \rho_a(\vec{r}, t). \qquad (6.20)$$

The charge and current densities mentioned above have the right physical dimension: The Dirac-$\delta^{(3)}$ has dimension of inverse length to the third power. The dimension of $\vec{J}$ is found to be equal to the charge per time per area, which is the correct dimension for a current density.

We reemphasize that we use a model in which all particles are described as classical point particles moving on classical trajectories. (This model, of course, has obvious limitations in describing a system of atoms, whose constituent particles are known to be described by quantum mechanics. Still, it is instructive to consider a classical model composed of point particles and to see how the averaging operation leads to the phenomenological Maxwell equations.)

In Eq. (6.19), $q_{1n} = -Z e$ is the charge of the nth nucleus, where e is the electron charge and Z is the nuclear charge number. The variable $j = 2, \ldots, j_{\max}(n)$

enumerates the charges $q_{jn} = e$ of the electrons in the nth atom (we still keep the subscript in order to identify the individual electrons). The current densities are

$$\vec{J_c}(\vec{r},t) = \sum_c q_c \frac{d\vec{r_c}}{dt} \delta^{(3)}(\vec{r} - \vec{r_c}(t)),\tag{6.21}$$

$$\vec{J_a}(\vec{r},t) = \sum_{nj} q_{jn} \left(\frac{d\vec{r_n}}{dt} + \frac{d\vec{r}_{jn}}{dt}\right)\delta^{(3)}(\vec{r} - \vec{r_n}(t) - \vec{r}_{jn}(t)),\tag{6.22}$$

$$\vec{J}(\vec{r},t) = \vec{J_c}(\vec{r},t) + \vec{J_a}(\vec{r},t).\tag{6.23}$$

The positions $\vec{r_n}$ and $\vec{r}_{jn}$ still depend on time.

The following definitions and relations are relevant to the discussion (the electron charge is denoted as e),

$$\rho(\vec{r},t) = \sum_c q_c \delta^{(3)}(\vec{r} - \vec{r_c}(t)) + \sum_{nj} q_{jn} \delta^{(3)}(\vec{r} - \vec{r_n}(t) - \vec{r}_{jn}(t)),\tag{6.24a}$$

$$\langle\rho(\vec{r},t)\rangle = \sum_c q_c\, f(\vec{r} - \vec{r_c}(t)) + \sum_{nj} q_{jn}\, f(\vec{r} - \vec{r_n}(t) - \vec{r}_{jn}(t)),\tag{6.24b}$$

$$q_{1n} = -Z\,e, \qquad \vec{r}_{1n} \approx \vec{0},\tag{6.24c}$$

$$\sum_{j=2}^{j_{\max}(n)} q_{jn} = Z'\,e, \qquad \sum_j q_{jn} = \sum_{j=1}^{j_{\max}(n)} q_{jn} = (Z' - Z)\,e = Q_n.\tag{6.24d}$$

Here, r_{1n} is the position of the nucleus of the nth neutral atom or nth ion, $Z'e$ is the charge of the electrons in the nth neutral atom or nth ion, $(Z'-Z)\,e$ is the total charge of the nth neutral atom or nth ion, and $\langle r^2 \rangle_n$ is the charge radius of the nth neutral atom or nth ion. For a neutral atom, we have $Q_n = 0$. Unless indicated otherwise, for the remainder of the derivation of the macroscopic Maxwell equations (see Sec. 6.2), the sum over j will be interpreted as

$$\sum_j \equiv \sum_{j=1}^{j_{\max}(n)},\tag{6.25}$$

where $j_{\max}(n)$ is the number of constituent charges in the nth atom or ion. The (averaged) microscopic Maxwell equations are as follows,

$$\langle\vec{\nabla}\cdot\vec{e}(\vec{r},t)\rangle = \frac{1}{\epsilon_0}\sum_c q_c\,\langle\delta^{(3)}(\vec{r} - \vec{r_c})\rangle$$

$$+ \frac{1}{\epsilon_0}\sum_{jn} q_{jn}\,\langle\delta^{(3)}(\vec{r} - \vec{r_n} - \vec{r}_{jn})\rangle,\tag{6.26a}$$

$$\langle\vec{\nabla}\cdot\vec{b}(\vec{r},t)\rangle = 0, \qquad \langle\vec{\nabla}\times\vec{e}(\vec{r},t)\rangle + \left\langle\frac{\partial}{\partial t}\vec{b}(\vec{r},t)\right\rangle = 0,\tag{6.26b}$$

$$\frac{\langle\vec{\nabla}\times\vec{b}(\vec{r},t)\rangle}{\mu_0} - \epsilon_0\left\langle\frac{\partial}{\partial t}\vec{e}(\vec{r},t)\right\rangle = \sum_c q_c \frac{d\vec{r_c}}{dt}\,\langle\delta^{(3)}(\vec{r} - \vec{r_c})\rangle$$

$$+ \sum_{jn} q_{jn}\left(\frac{d\vec{r_n}}{dt} + \frac{d\vec{r}_{jn}}{dt}\right)\langle\delta^{(3)}(\vec{r} - \vec{r_n} - \vec{r}_{jn})\rangle.\tag{6.26c}$$

Here, we have suppressed the explicit time dependence of the positions $\vec{r}_c(t)$ and $\vec{r}_{jn}(t)$. Written in terms of the averaged fields, we can alternatively express the inhomogeneous Maxwell equations as

$$\vec{\nabla} \cdot \vec{E}(\vec{r},t) = \frac{1}{\epsilon_0} \sum_c q_c \, f(\vec{r} - \vec{r}_c) + \frac{1}{\epsilon_0} \sum_{jn} q_{jn} f(\vec{r} - \vec{r}_n - \vec{r}_{jn}),$$

(6.27a)

$$\frac{1}{\mu_0} \vec{\nabla} \times \vec{B}(\vec{r},t) - \epsilon_0 \frac{\partial}{\partial t} \vec{E}(\vec{r},t) = \sum_c q_c \frac{d\vec{r}_c}{dt} f(\vec{r} - \vec{r}_c)$$

$$+ \sum_{jn} q_{jn} \left(\frac{d\vec{r}_n}{dt} + \frac{d\vec{r}_{jn}}{dt} \right) f(\vec{r} - \vec{r}_n - \vec{r}_{jn}). \quad (6.27b)$$

Let us verify that the charge and current densities fulfill the continuity equation, and start with ρ_a,

$$\frac{\partial}{\partial t} \langle \rho_a(\vec{r},t) \rangle = \sum_{jn} q_{jn} \frac{\partial}{\partial t} f(\vec{r} - \vec{r}_n(t) - \vec{r}_{jn}(t))$$

$$= -\vec{\nabla} \cdot \sum_n q_n \left(\frac{d\vec{r}_n}{dt} + \frac{d\vec{r}_{jn}}{dt} \right) f(\vec{r} - \vec{r}_n(t) - \vec{r}_{jn}(t)) = -\vec{\nabla} \cdot \langle \vec{J}_a(\vec{r},t) \rangle.$$

(6.28)

Here, the three-dimensional chain rule has been used, for the function $f(\vec{r} - \vec{r}_n(t) - \vec{r}_{jn}(t))$, where $f = f(\vec{x})$ is a function of a three-dimensional argument, while in turn $\vec{x} = \vec{r} - \vec{r}_n(t) - \vec{r}_{jn}(t)$ depends on the time t.

In order to understand the three-dimensional chain rule, let us first recall the basic formulas for a Taylor expansion to arbitrary order in one dimension,

$$g(x + \eta) = g(x) + \eta \frac{\partial}{\partial x} g(x) + \frac{1}{2!} \eta^2 \frac{\partial^2}{\partial x^2} g(x) + \ldots$$

$$= \sum_{n=0}^{\infty} \frac{\eta^n}{n!} \left[\frac{\partial^n}{\partial x^n} g(x) \right] = \exp\left(\eta \frac{\partial}{\partial x} \right) g(x). \quad (6.29)$$

The generalization to three dimensions is as follows, and we give it in various different formulations,

$$g(\vec{x} + \vec{\eta}) = g(\vec{x}) + \vec{\eta} \cdot \vec{\nabla} g(\vec{x}) + \frac{1}{2!} \eta_\alpha \eta_\beta \frac{\partial^2}{\partial x_\alpha \, \partial x_\beta} g(\vec{x})$$

$$+ \frac{1}{3!} \eta_\alpha \eta_\beta \eta_\gamma \frac{\partial^3}{\partial x_\alpha \, \partial x_\beta \, \partial x_\gamma} g(\vec{x}) + \ldots$$

$$= g(\vec{x}) + \vec{\eta} \cdot \vec{\nabla} g(\vec{x}) + \frac{1}{2!} \left(\vec{\eta} \cdot \vec{\nabla} \right) \left(\vec{\eta} \cdot \vec{\nabla} \right) g(\vec{x})$$

$$+ \frac{1}{3!} \left(\vec{\eta} \cdot \vec{\nabla} \right) \left(\vec{\eta} \cdot \vec{\nabla} \right) \left(\vec{\eta} \cdot \vec{\nabla} \right) g(\vec{x}) + \ldots$$

$$= \sum_{n=0}^{\infty} \frac{(\vec{\eta} \cdot \vec{\nabla})^n}{n!} g(\vec{x}) = \exp\left(\vec{\eta} \cdot \vec{\nabla} \right) g(\vec{x}). \quad (6.30)$$

Here, it is necessary to remember that the differential operators act only on the argument vector $\vec{x}$, not on the spatial displacement $\vec{\eta}$.

6.2.3 *Macroscopic Averaging and Charge Density*

In order to derive the phenomenological Maxwell equations, and relate them to the microscopic quantities, one expands the charge and current densities in powers of the displacements $\vec{r}_{jn}$ and its time derivative $\dot{\vec{r}}_{jn}$, assuming that these quantities are smaller than the position vector $\vec{r}_n$ that describe the centers of the individual atoms. The expansion of the charge density in powers of $\vec{r}_{jn}$ leads to

$$
\langle \rho(\vec{r},t) \rangle = \sum_c q_c\, f(\vec{r} - \vec{r}_c) + \sum_{nj} q_{jn}\, f(\vec{r} - \vec{r}_n - \vec{r}_{jn})
$$

$$
= \underbrace{\sum_c q_c\, f(\vec{r} - \vec{r}_c) + \sum_{nj} q_{jn}\, f(\vec{r} - \vec{r}_n) + \sum_{nj} q_{jn}\, (-\vec{r}_{jn} \cdot \vec{\nabla})\, f(\vec{r} - \vec{r}_n)}_{=\rho_0(\vec{r},t)}
$$

$$
+ \sum_{nj} \tfrac{1}{2} q_{jn}\, (-\vec{r}_{jn} \cdot \vec{\nabla})\,(-\vec{r}_{jn} \cdot \vec{\nabla})\, f(\vec{r} - \vec{r}_n)\,. \tag{6.31}
$$

Here, we identify the free charge density ρ_0 as the sum of the free charges in the conduction band q_c and the total ionic charges $Q_n = \sum_j q_{jn}$. Then,

$$
\langle \rho(\vec{r},t) \rangle = \rho_0(\vec{r},t) - \sum_n \left(\sum_j q_{jn} \vec{r}_{jn} \right) \cdot \vec{\nabla} f(\vec{r} - \vec{r}_n) + \sum_{nj} \tfrac{1}{2} q_{jn} \left(\vec{r}_{jn} \cdot \vec{\nabla} \right)\!\left(\vec{r}_{jn} \cdot \vec{\nabla} \right) f(\vec{r} - \vec{r}_n)
$$

$$
= \rho_0(\vec{r},t) - \vec{\nabla} \cdot \left(\sum_n \vec{p}_n f(\vec{r} - \vec{r}_n) - \sum_{nj} \tfrac{1}{2} q_{jn}\, \vec{r}_{jn} \left(\vec{r}_{jn} \cdot \vec{\nabla} \right) f(\vec{r} - \vec{r}_n) \right)
$$

$$
= \rho_0(\vec{r},t) - \vec{\nabla} \cdot \sum_n \vec{P}_n f(\vec{r} - \vec{r}_n)\,. \tag{6.32}
$$

Here, the polarization of the nth atom, including the quadrupole term, is given as

$$
\vec{P}_n = \vec{p}_n - \sum_j \tfrac{1}{2} q_{jn} \vec{r}_{jn} (\vec{r}_{jn} \cdot \vec{\nabla})\,, \tag{6.33}
$$

where the pure dipole contribution is $\vec{p}_n = \sum_n q_{jn}\, \vec{r}_{jn}$, and the second term includes the quadrupole correction. The averaged form (6.26a) of Gauss's law thus reads as follows,

$$
\epsilon_0 \left\langle \vec{\nabla} \cdot \vec{e}(\vec{r},t) \right\rangle = \rho_0(\vec{r},t) - \vec{\nabla} \cdot \sum_n \vec{P}_n f(\vec{r} - \vec{r}_n)\,, \tag{6.34}
$$

and it can be rewritten as

$$
\vec{\nabla} \cdot \vec{D}(\vec{r},t) = \rho_0(\vec{r},t)\,, \tag{6.35}
$$

provided we identify the dielectric displacement $\vec{D}(\vec{r},t)$ as

$$
\vec{D}(\vec{r},t) = \epsilon_0\, \vec{E}(\vec{r},t) + \vec{P}(\vec{r},t) = \epsilon_0 \left\langle \vec{e}(\vec{r},t) \right\rangle + \sum_n \vec{P}_n\, f(\vec{r} - \vec{r}_n)\,. \tag{6.36}
$$

This equation identifies the dielectric displacement as being due to the induced dipole moments of the constituent molecules, plus a correction proportional to the projection of the quadrupole moment onto the gradient of the sampling function.

The projection of the quadrupole term carries a certain dependence on the shape of the sampling function f. The pure dipole term $\vec{p}_n$ is averaged over the sampling function $f(\vec{r} - \vec{r}_n)$ and this average is not very dependent on the precise shape of the sampling function, but the integral of the gradient may exhibit a significant dependence. We can resolve this apparent contradiction in the following way: *(i)* In a quantum mechanical treatment, the dependence on the shape of the sampling function would be absorbed in a formulation of the scattering process under study. One has to analyze the multipole components of the incoming and outgoing waves and then write the multipole components of the atomic polarizabilities, in order to describe the multipole components of the scattering process. The functional form of the sampling function describes the structure of the atoms relevant for the scattering process under study. In other words, the appropriate form of the sampling function is determined by quantum mechanical considerations which are beyond our classical model. *(ii)* The shape-independent pure dipole term, as we shall see later, provides by far the dominant effect if the system is interrogated in the infrared. At optical wavelengths, the pure dipole term only receives corrections of relative order α, where α is the fine-structure constant. In most applications described in the literature, one may thus restrict the discussion to the pure dipole term.

6.2.4 *Macroscopic Averaging and Current Density*

The total current density is

$$\langle \vec{J}(\vec{r},t) \rangle = \sum_c q_c \frac{d\vec{r}_c}{dt} f(\vec{r} - \vec{r}_c) + \sum_{nj} q_{jn} \left(\frac{d\vec{r}_n}{dt} + \frac{d\vec{r}_{jn}}{dt} \right) f(\vec{r} - \vec{r}_n - \vec{r}_{jn}) . \qquad (6.37)$$

An expansion to second order in the $\vec{r}_{jn}$ leads to

$$\langle \vec{J}(\vec{r},t) \rangle = \sum_c q_c \frac{d\vec{r}_c}{dt} f(\vec{r} - \vec{r}_c) + \sum_{nj} q_{jn} \frac{d\vec{r}_n}{dt} f(\vec{r} - \vec{r}_n)$$

$$+ \sum_{nj} q_{jn} \frac{d\vec{r}_{jn}}{dt} f(\vec{r} - \vec{r}_n) + \sum_{nj} q_{jn} \left(\frac{d\vec{r}_n}{dt} + \frac{d\vec{r}_{jn}}{dt} \right) (-\vec{r}_{jn} \cdot \vec{\nabla}) f(\vec{r} - \vec{r}_n)$$

$$+ \sum_{nj} \frac{1}{2} q_{jn} \frac{d\vec{r}_n}{dt} (-\vec{r}_{jn} \cdot \vec{\nabla})(-\vec{r}_{jn} \cdot \vec{\nabla}) f(\vec{r} - \vec{r}_n) . \qquad (6.38)$$

The first two terms on the right-hand side can be identified in terms of the free current density $\vec{J}_0$,

$$\vec{J}_0(\vec{r},t) = \sum_c q_c \frac{d\vec{r}_c}{dt} f(\vec{r} - \vec{r}_c) + \sum_n Q_n \frac{d\vec{r}_n}{dt} f(\vec{r} - \vec{r}_n) , \qquad (6.39)$$

where $Q_n = \sum_j q_{jn}$. Furthermore, we identify $\vec{p}_n = \sum_j q_{jn} \vec{r}_n$ and define

$$\vec{v}_n = \frac{d\vec{r}_n}{dt} , \qquad \vec{v}_{jn} = \frac{d\vec{r}_{jn}}{dt} . \qquad (6.40)$$

We can thus write the current density as

$$\langle \vec{J}(\vec{r},t) \rangle = \vec{J}_0(\vec{r},t) + \sum_n \frac{d\vec{p}_n}{dt} f(\vec{r}-\vec{r}_n) - \sum_n \vec{v}_n \sum_j q_{jn}\, \vec{r}_{jn} \cdot \vec{\nabla} f(\vec{r}-\vec{r}_n)$$
$$- \sum_{nj} \vec{v}_{jn}\, q_{jn}\, \vec{r}_{jn} \cdot \vec{\nabla} f(\vec{r}-\vec{r}_n) + \sum_{nj} \tfrac{1}{2} q_{jn}\, \vec{v}_n\, (\vec{r}_{jn} \cdot \vec{\nabla})(\vec{r}_{jn} \cdot \vec{\nabla}) f(\vec{r}-\vec{r}_n).$$

(6.41)

We shall now transform the third term on the right-hand side,

$$\vec{T} = -\sum_n \vec{v}_n \sum_j q_{jn}\, \vec{r}_{jn} \cdot \vec{\nabla} f(\vec{r}-\vec{r}_n) = -\sum_n \vec{v}_n\, \vec{p}_n \cdot \vec{\nabla} f(\vec{r}-\vec{r}_n). \tag{6.42}$$

For any $\vec{a}$ and $\vec{b}$ which are independent of $\vec{r}$, we have $\vec{\nabla} \times (\vec{a} \times \vec{b}) = \vec{a}\,(\vec{b}\cdot\vec{\nabla}) - \vec{b}\,(\vec{a}\cdot\vec{\nabla})$. Thus,

$$\vec{\nabla} \times \sum_n (\vec{p}_n \times \vec{v}_n)\, f(\vec{r}-\vec{r}_n) = \sum_n \left[\vec{p}_n(\vec{v}_n \cdot \vec{\nabla}) - \vec{v}_n(\vec{p}_n \cdot \vec{\nabla}) \right] f(\vec{r}-\vec{r}_n), \tag{6.43}$$

and so

$$\vec{T} = -\sum_n \vec{p}_n(\vec{v}_n \cdot \vec{\nabla}) f(\vec{r}-\vec{r}_n) + \vec{\nabla} \times \sum_n (\vec{p}_n \times \vec{v}_n)\, f(\vec{r}-\vec{r}_n). \tag{6.44}$$

The total current can thus be expressed as

$$\langle \vec{J}(\vec{r},t) \rangle = \vec{J}_0(\vec{r},t) + \sum_n \frac{d\vec{p}_n}{dt} f(\vec{r}-\vec{r}_n) + \sum_n \vec{p}_n \left(-\frac{d\vec{r}_n}{dt} \cdot \vec{\nabla} \right) f(\vec{r}-\vec{r}_n)$$
$$+ \vec{\nabla} \times \sum_n (\vec{p}_n \times \vec{v}_n)\, f(\vec{r}-\vec{r}_n) - \sum_{nj} \vec{v}_{jn}\, q_{jn}\, \vec{r}_{jn} \cdot \vec{\nabla} f(\vec{r}-\vec{r}_n)$$
$$+ \sum_{nj} \tfrac{1}{2} q_{jn}\, \vec{v}_n\, (\vec{r}_{jn} \cdot \vec{\nabla})(\vec{r}_{jn} \cdot \vec{\nabla}) f(\vec{r}-\vec{r}_n). \tag{6.45}$$

With the help of the chain rule, the second and third terms on the right-hand side of Eq. (6.45) can conveniently be expressed in terms of the time derivative of the dipole contribution to the polarization density, resulting in the expression

$$\langle \vec{J}(\vec{r},t) \rangle = \vec{J}_0(\vec{r},t) + \frac{d}{dt}\left(\sum_n \vec{p}_n\, f(\vec{r}-\vec{r}_n) \right) + \vec{\nabla} \times \sum_n (\vec{p}_n \times \vec{v}_n)\, f(\vec{r}-\vec{r}_n)$$
$$- \sum_{nj} \vec{v}_{jn}\, q_{jn}\, \vec{r}_{jn} \cdot \vec{\nabla} f(\vec{r}-\vec{r}_n) + \sum_{nj} \tfrac{1}{2} q_{jn}\, \vec{v}_n\, (\vec{r}_{jn} \cdot \vec{\nabla})(\vec{r}_{jn} \cdot \vec{\nabla}) f(\vec{r}-\vec{r}_n). \tag{6.46}$$

The third term on the right-hand side reminds us of the curl of an induced density of magnetic moments, as will be further explored below. However, the fourth and fifth terms on the right-hand side of Eq. (6.46),

$$\vec{U} = -\sum_{nj} \vec{v}_{jn}\, q_{jn}\, \vec{r}_{jn} \cdot \vec{\nabla} f(\vec{r}-\vec{r}_n) + \sum_{nj} \tfrac{1}{2} q_{jn}\, \vec{v}_n\, (\vec{r}_{jn} \cdot \vec{\nabla})(\vec{r}_{jn} \cdot \vec{\nabla}) f(\vec{r}-\vec{r}_n), \tag{6.47}$$

remain to be treated. Key to the further analysis is the following identity, which allows us to rewrite $\vec{U}$ appropriately,

$$U = \vec{\nabla} \times \left(\sum_{nj} \tfrac{1}{2} q_{jn}\, (\vec{r}_{jn} \times \vec{v}_{jn}) \right) f(\vec{r}-\vec{r}_n) - \frac{d}{dt} \sum_{nj} \tfrac{1}{2} q_{jn}\, \vec{r}_{jn}(\vec{r}_{jn} \cdot \vec{\nabla}) f(\vec{r}-\vec{r}_n)$$
$$- \vec{\nabla} \times \sum_{nj} \tfrac{1}{2} q_{jn}(\vec{r}_{jn} \cdot \vec{\nabla})\, (\vec{r}_{jn} \times \vec{v}_n)\, f(\vec{r}-\vec{r}_n). \tag{6.48}$$

The three terms on the right-hand side of Eq. (6.48) will eventually be identified as the curl of the dipole magnetization, the time derivative of the quadrupole contribution to the polarization, and the curl of the quadrupole correction to the magnetization, respectively. We start with the right-hand side of Eq. (6.48) and rewrite the three terms with the help of the following identities. The first term is transformed according to

$$\vec{\nabla} \times \left(\sum_{nj} \tfrac{1}{2} q_{jn} \left(\vec{r}_{jn} \times \vec{v}_{jn} \right) \right) = \underbrace{\sum_{nj} \tfrac{1}{2} q_{jn} \, \vec{r}_{jn} \left(\vec{v}_{jn} \cdot \vec{\nabla} \right) f(\vec{r} - \vec{r}_n)}_{=T_1}$$

$$- \sum_{nj} \tfrac{1}{2} q_{jn} \, \vec{v}_{jn} \left(\vec{r}_{jn} \cdot \vec{\nabla} \right) f(\vec{r} - \vec{r}_n) , \qquad (6.49)$$

while the second term is transformed as follows,

$$- \frac{\mathrm{d}}{\mathrm{d}t} \sum_{nj} \tfrac{1}{2} q_{jn} \, \vec{r}_{jn} (\vec{r}_{jn} \cdot \vec{\nabla}) f(\vec{r} - \vec{r}_n) = - \sum_{nj} \tfrac{1}{2} q_{jn} \, \vec{v}_{jn} (\vec{r}_{jn} \cdot \vec{\nabla}) f(\vec{r} - \vec{r}_n)$$

$$\underbrace{- \sum_{nj} \tfrac{1}{2} q_{jn} \, \vec{r}_{jn} (\vec{v}_{jn} \cdot \vec{\nabla}) f(\vec{r} - \vec{r}_n)}_{=-T_1} \underbrace{- \sum_{nj} \tfrac{1}{2} q_{jn} \, \vec{r}_{jn} (\vec{r}_{jn} \cdot \vec{\nabla})(-\vec{v}_n \cdot \vec{\nabla}) f(\vec{r} - \vec{r}_n)}_{=T_2} . \qquad (6.50)$$

In the last term on the right-hand side of Eq. (6.50), we have two minus signs, one overall minus sign out front and a second minus sign in the term $(-\vec{v}_n \cdot \vec{\nabla})$. The third term on the right-hand side of Eq. (6.48) is transformed according to

$$-\vec{\nabla} \times \sum_{nj} \tfrac{1}{2} q_{jn} (\vec{r}_{jn} \cdot \vec{\nabla}) \left(\vec{r}_{jn} \times \vec{v}_n \right) f(\vec{r} - \vec{r}_n) = \underbrace{- \sum_{nj} \tfrac{1}{2} q_{jn} \vec{r}_{jn} \left(\vec{r}_{jn} \cdot \vec{\nabla} \right) \left(\vec{v}_n \cdot \vec{\nabla} \right) f(\vec{r} - \vec{r}_n)}_{=-T_2}$$

$$+ \sum_{nj} \tfrac{1}{2} q_{jn} \vec{v}_n \left(\vec{r}_{jn} \cdot \vec{\nabla} \right) \left(\vec{r}_{jn} \cdot \vec{\nabla} \right) f(\vec{r} - \vec{r}_n) , \qquad (6.51)$$

which proves Eq. (6.48). Adding Eqs. (6.49)–(6.51), one obtains Eq. (6.48).

Using the identity Eq. (6.48) in Eq. (6.46), one obtains the following decomposition for the current density,

$$\langle \vec{J}(\vec{r}, t) \rangle = \vec{J}_0(\vec{r}, t) + \frac{\mathrm{d}}{\mathrm{d}t} \left(\sum_n \left(\vec{p}_n - \sum_j \tfrac{1}{2} q_{jn} \, \vec{r}_{jn} (\vec{r}_{jn} \cdot \vec{\nabla}) \right) f(\vec{r} - \vec{r}_n) \right)$$

$$+ \vec{\nabla} \times \sum_n \left(\sum_j \tfrac{1}{2} q_{jn} \left(\vec{r}_{jn} \times \vec{v}_{jn} \right) \right) f(\vec{r} - \vec{r}_n)$$

$$+ \vec{\nabla} \times \left(\sum_n \left(\vec{p}_n - \sum_j \tfrac{1}{2} q_{jn} \vec{r}_{jn} (\vec{r}_{jn} \cdot \vec{\nabla}) \right) f(\vec{r} - \vec{r}_n) \times \vec{v}_n \right) , \qquad (6.52)$$

which can be rewritten as

$$\langle \vec{J}(\vec{r},t) \rangle = \vec{J}_0(\vec{r},t) + \frac{\mathrm{d}}{\mathrm{d}t}\left(\sum_n \vec{P}_n \, f(\vec{r}-\vec{r}_n) \right) + \vec{\nabla} \times \sum_n \vec{m}_n \, f(\vec{r}-\vec{r}_n)$$

$$+ \vec{\nabla} \times \left(\sum_n \vec{P}_n \, f(\vec{r}-\vec{r}_n) \times \vec{v}_n \right), \tag{6.53}$$

where we have encountered the polarization density operator $\vec{P}_n$ in Eq. (6.33) above,

$$\vec{P}_n = \vec{p}_n - \sum_j \tfrac{1}{2} q_{jn} \vec{r}_{jn} (\vec{r}_{jn} \cdot \vec{\nabla}). \tag{6.54}$$

The partial time derivative acting on a field means that one leaves the observation point $\vec{r}$ unchanged. We thus have the relation

$$\frac{\partial}{\partial t}\vec{P}(\vec{r},t) = \frac{\partial}{\partial t}\sum_n \vec{P}_n \, f(\vec{r}-\vec{r}_n) = \frac{\mathrm{d}}{\mathrm{d}t}\sum_n \vec{P}_n \, f(\vec{r}-\vec{r}_n(t)). \tag{6.55}$$

The magnetic moment of the nth atom is identified as follows,

$$\vec{m}_n = \frac{1}{2}\sum_j q_{jn} \left(\vec{r}_{jn} \times \vec{v}_{jn} \right). \tag{6.56}$$

This definition can be motivated in the following way. We first observe that the magnetic moment of a circular ring current I of time period T, covering a circle of radius R, is

$$\vec{M} = I \, A \, \hat{e}_z = \frac{q}{T} \pi R^2 \, \hat{e}_z. \tag{6.57}$$

The surface normal to the area covered by the area A is assumed to be directed along the positive z axis. The velocity of the atom is $2\pi R/T$, and so the magnetic moment can be expressed as

$$\vec{M} = \frac{q}{T} \pi R^2 \, \hat{e}_z = \frac{1}{2} qR \frac{2\pi R}{T} \, \hat{e}_z = \frac{1}{2} qRv \, \hat{e}_z = \frac{1}{2} q \, \vec{R} \times \vec{v}. \tag{6.58}$$

This justifies Eq. (6.56). We can finally identify the magnetization as

$$\vec{M}(\vec{r},t) = \sum_n \vec{m}_n \, f(\vec{r}-\vec{r}_n) + \sum_n \vec{P}_n \, f(\vec{r}-\vec{r}_n) \times \vec{v}_n, \tag{6.59}$$

where the second term is a "global" term due to "rotating" dipole moments.

6.2.5 *Phenomenological Maxwell Equations*

In view of Eqs. (6.52), (6.55), and (6.59), we have finally found a microscropic model for the separation

$$\langle \vec{J}(\vec{r},t) \rangle = \vec{J}_0(\vec{r},t) + \frac{\partial}{\partial t}\vec{P}(\vec{r},t) + \vec{\nabla} \times \vec{M}(\vec{r},t). \tag{6.60}$$

The averaged Ampere–Maxwell law (6.26c) thus reads as follows,

$$\frac{1}{\mu_0}\langle \vec{\nabla} \times \vec{b}(\vec{r},t) \rangle - \frac{1}{\mu_0 c^2}\frac{\partial}{\partial t}\langle \vec{e}(\vec{r},t) \rangle = \langle \vec{J}(\vec{r},t) \rangle$$

$$= \vec{J}_0(\vec{r},t) + \frac{\partial}{\partial t}\vec{P}(\vec{r},t) + \vec{\nabla} \times \vec{M}(\vec{r},t). \tag{6.61}$$

We may now define the dielectric displacement and the magnetic field,

$$\vec{H}(\vec{r},t) = \frac{1}{\mu_0} \left\langle \vec{b}(\vec{r},t) \right\rangle - \vec{M}(\vec{r},t), \qquad \vec{D}(\vec{r},t) = \epsilon_0 \left\langle \vec{e}(\vec{r},t) \right\rangle + \vec{P}(\vec{r},t), \qquad (6.62)$$

in which case the Ampere–Maxwell law reads as

$$\vec{\nabla} \times \vec{H}(\vec{r},t) - \frac{\partial}{\partial t} \vec{D}(\vec{r},t) = \vec{J}_0(\vec{r},t). \qquad (6.63)$$

Together with Eq. (6.35), this completes the derivation of the phenomenological Maxwell equations.

6.2.6   *Parameters of the Multipole Expansion*

A final word should be supplemented, regarding the multipole expansion. Let us recall the condition (6.8) under which our treatment is valid,

$$\lambda \gg R \gg a_0, \langle d \rangle, r_\rho, \qquad (6.64)$$

where a_0 is the Bohr radius, $\langle d \rangle$ is a typical atomic separation distance, and r_ρ is a length scale of density fluctuations. For our applications, we have

$$|\vec{r}_{jn}| \sim a_0, \qquad (6.65)$$

because the Bohr radius gives the dimension of the atom, and

$$\vec{\nabla} f \sim \frac{1}{R} f, \qquad (6.66)$$

in view of the fact that the gradient operator is distributed over a volume of dimension R. We recall that the sampling function f [defined in Eq. (6.14)] has a physical dimension of inverse volume. Indeed, the extent to which the quadrupole corrections and higher-order terms contribute to the averaged Maxwell equations depends on the extent of the sampling volume. Our expansion is formulated in terms of the expansion parameter, or merely expansion operator, $\vec{r}_{jn} \cdot \vec{\nabla}$. One can assign a parametric estimate to this operator by letting it act on a sampling function f,

$$\vec{r}_{jn} \cdot \vec{\nabla} f \sim \frac{a_0}{R} f \gg \frac{a_0}{\lambda} f, \qquad (6.67)$$

where we take into account the fact that $R \ll \lambda$. The parametric estimate thus is

$$\vec{r}_{jn} \cdot \vec{\nabla} \sim \frac{a_0}{R} \gg \frac{a_0}{\lambda}. \qquad (6.68)$$

This condition has to be taken with a grain of salt. For interactions ("interrogations of the material") with very long wavelengths, $\lambda \gg R$, the condition is certainly fulfilled. However, for interrogations at optical wavelengths, we can actually choose the averaging dimension R of the sampling volume to be larger, but we cannot choose it to be very much larger, than the (optical) wavelength. If we interrogate the sample with an optical wavelength, then, for typical wavelengths,

$$\frac{a_0}{\lambda} \sim \frac{a_0}{a_0/\alpha} \sim \alpha. \qquad (6.69)$$

The condition

$$\lambda \sim \frac{a_0}{\alpha} \gg R \gg a_0 , \qquad 137\, a_0 \gg R \gg a_0 , \tag{6.70}$$

severely restricts the range of permissible values of R. This shows, in particular, that the formulation in terms of the phenomenological Maxwell equations is justified if one interrogates condensed matter (where $\langle d \rangle$ is of the order of a_0) by optical radiation. Furthermore, in this regime, the corrections to the dipole approximation are of order α. The dipole approximation works even better at frequencies below the optical transitions, i.e., in the infrared. In the optical regime, the expansion in powers of the "expansion operator"

$$\vec{r}_{jn} \cdot \vec{\nabla} \sim \alpha , \tag{6.71}$$

is an expansion in powers of the fine-structure constant α for interrogations with optical wavelengths. Other examples for these relevant parametric estimates have recently been discussed in a related context, in Ref. [27].

6.3 Fourier Decomposition and Maxwell Equations in a Medium

We have just encountered a classical model for the description of the dielectric displacement $\vec{D}(\vec{r},t)$, and the magnetic field $\vec{H}(\vec{r},t)$, in a medium. In reality, even if the classical approach could be extended into the atomic (quantum) domain, it would be impossible to follow all the trajectories $\vec{r}_n(t)$ and $\vec{r}_{jn}(t)$ of the constituent particles in the sample.

One thus has to find a different way to describe the dielectric displacement in terms of a phenomenologically inspired model. It is customary to assume a functional relationship of the form

$$\vec{D}(\vec{r},t) = \epsilon_0 \,\vec{E}(\vec{r},t) + \vec{P}(\vec{r},t) = \int dt'\, \epsilon(\vec{r} - \vec{r}', t - t')\, \vec{E}(\vec{r},t') . \tag{6.72}$$

(All integrals extend of the maximum integration domain, namely, $\mathbb{R}$, if the range of integration is not explicitly specified.) Here, $\epsilon(\vec{r} - \vec{r}', t - t')$ is the dielectric permittivity of the material, which can be expressed in terms of the relative permittivity $\epsilon_r(\vec{r} - \vec{r}', t - t')$,

$$\epsilon(\vec{r} - \vec{r}', t - t') = \epsilon_0\, \epsilon_r(\vec{r} - \vec{r}', t - t') . \tag{6.73}$$

In many cases, one may assume that the polarization is generated locally, i.e., that

$$\epsilon_r(\vec{r} - \vec{r}', t - t') = \delta^{(3)}(\vec{r} - \vec{r}')\, \epsilon_r(t - t') , \tag{6.74}$$

in which case the dielectric displacement can be given as a convolution integral,

$$\vec{D}(\vec{r},t) = \epsilon_0 \int dt'\, \epsilon_r(t - t')\, \vec{E}(\vec{r},t') . \tag{6.75}$$

The same is true for the permeability

$$\mu(\vec{r} - \vec{r}', t - t') = \mu_0\, \mu_r(\vec{r} - \vec{r}', t - t') , \tag{6.76}$$

which also can be approximated by a local function in many cases,

$$\mu_r(\vec{r} - \vec{r}', t - t') = \delta^{(3)}(\vec{r} - \vec{r}') \, \mu_r(t - t') \,, \tag{6.77}$$

so that

$$\vec{B}(\vec{r}, t) = \mu_0 \int dt' \, \mu_r(t - t') \, \vec{H}(\vec{r}, t') \,. \tag{6.78}$$

Furthermore, it is quite common for typical materials that

$$\mu_r(t - t') \approx 1 \,, \tag{6.79}$$

while the relative permittivity $\epsilon_r(t - t')$ appreciably differs from unity on time scales $t - t' \sim 1/\nu$, where ν is a transition frequency of the constituent atoms in the medium.

Formulas (6.72)–(6.79) are valid for isotropic materials, in which case the relative permittivity and relative permeability are just scalar functions. For non-isotropic materials, these quantities acquire a tensor structure, $\epsilon_r \to \epsilon_{r,ij}$ and $\mu_r \to \mu_{r,ij}$, where $i, j = 1, 2, 3$. Off-diagonal elements in $\epsilon_{r,ij}$ with $i \neq j$ describe the possibility of a "sheared" induced polarization density, where the atomic dipoles align along directions which deviate from the direction of the applied external electric field. We shall contend ourselves, here, with the isotropic case and rather focus on the analytic properties of the relative permittivity ϵ_r in both the time and frequency domains. Indeed, it will turn out to be very convenient to transform the entire formalism into Fourier space. To this end, we shall briefly recall a few basic facts about Fourier transformations, extending the previous discussions in Sec. 2.2 and 4.1.1.

The Fourier transform of the electric field into the mixed position-frequency representation is

$$\vec{E}(\vec{r}, \omega) \equiv \int dt \, e^{i\omega t} \, \vec{E}(\vec{r}, t) \,, \tag{6.80}$$

with the Fourier backtransformation

$$\vec{E}(\vec{r}, t) = \int \frac{d\omega}{2\pi} \, e^{-i\omega t} \, \vec{E}(\vec{r}, \omega) \,. \tag{6.81}$$

In physics, as compared to some engineering disciplines, one usually distributes the factor 2π in the Fourier integrals asymmetrically, with the integral over the angular frequency having the factor $1/(2\pi)$. For a wave with oscillation period T, the frequency is just the reciprocal $1/T$, and the angular frequency is $\omega = 2\pi/T$. The integration measure $d\omega/(2\pi)$ therefore is nothing but $d\nu$, where $\nu = 1/T$ is the frequency. The phase conventions in Eqs. (6.80) and (6.81) are inspired by the fact that the application of the time derivative or "Hamilton" operator $i\partial_t$ to the function $e^{-i\omega t}$ yields just ω, which, multiplied by the unit of action $\hbar$ (the reduced Planck constant), yields the energy. As already emphasized in the text following Eq. (2.10), we recall that $\vec{E}(\vec{r}, t)$ and $\vec{E}(\vec{r}, \omega)$ are two different functions, distinguished by the physical dimension of the second argument.

Similarly, the space-time Fourier transform of the electric field is

$$\vec{E}(\vec{k}, \omega) \equiv \int d^3r \int dt \, e^{-i(\vec{k} \cdot \vec{r} - \omega t)} \, \vec{E}(\vec{r}, t) \,, \tag{6.82}$$

with the Fourier backtransformation

$$\vec{E}(\vec{r},t) = \int \frac{d^3k}{(2\pi)^3} \int \frac{d\omega}{2\pi} \, e^{i(\vec{k}\cdot\vec{r}-\omega t)} \, \vec{E}(\vec{k},\omega) \,. \tag{6.83}$$

Here, too, the factor 2π are unevenly distributed. For a wave with wavelength λ, the wave number is just the reciprocal $1/\lambda$, and the angular wave number is $k = 2\pi/\lambda$. The integration measure $dk/(2\pi)$ is equal to $d(1/\lambda)$, which is the number of full spatial oscillations of the wave packet per unit of length, measured in inverse meters in SI mksA units (a measure which also carries the name kayser). Formulas analogous to Eqs. (6.80)–(6.83) hold for the Fourier transforms of the magnetic induction field $\vec{B}$, as well as the dielectric displacement field $\vec{D}$ and the magnetic field $\vec{H}$.

Let us now reformulate Eq. (6.72) with the help of (6.73), under the assumption (6.74),

$$\vec{D}(\vec{r}) = \epsilon_0 \int_{-\infty}^{\infty} dt' \, \epsilon_r(t-t') \, \vec{E}(\vec{r},t-t') \,. \tag{6.84}$$

Then,

$$\begin{aligned}
\vec{D}(\vec{r},\omega) &= \epsilon_0 \int_{-\infty}^{\infty} dt \, e^{i\omega t} \int_{-\infty}^{\infty} dt' \, \epsilon_r(t-t') \, \vec{E}(\vec{r},t') \\
&= \epsilon_0 \int_{-\infty}^{\infty} dt \, e^{i\omega t} \int_{-\infty}^{\infty} dt' \int_{-\infty}^{\infty} \frac{d\omega'}{2\pi} \, \epsilon_r(\omega') \, e^{-i\omega'(t-t')} \, \vec{E}(\vec{r},t') \\
&= \epsilon_0 \int_{-\infty}^{\infty} dt \int_{-\infty}^{\infty} dt' \int_{-\infty}^{\infty} \frac{d\omega'}{2\pi} \, e^{i(\omega-\omega')t} \, \epsilon_r(\omega') \, e^{i\omega't'} \, \vec{E}(\vec{r},t') \\
&= \epsilon_0 \int_{-\infty}^{\infty} \frac{d\omega'}{2\pi} \, [2\pi\,\delta(\omega-\omega')] \, \epsilon_r(\omega') \, \vec{E}(\vec{r},\omega') \, = \epsilon_r(\omega) \, \vec{E}(\vec{r},\omega) \,. \tag{6.85}
\end{aligned}$$

This is a manifestation of the general wisdom that convolution in time simply means multiplication in frequency (Fourier) space. The relative permittivity $\epsilon_r(\omega)$ is dimensionless, while $\epsilon_r(t-t')$ has a dimension of inverse time.

From the representation (6.72), it is clear that the relative permittivity ϵ_r takes the role of a Green function, relating the electric field $\vec{E}$ to the induced polarization density $\vec{P}$ which is part of the dielectric displacement field $\vec{D}$. Causality dictates that we should interpret $\epsilon_r(t-t')$ as a retarded rather than advanced Green function. There is a great deal to learn from the analytic properties of $\epsilon_r(\omega)$, as we shall see in the following.

It is very useful to summarize the phenomenological Maxwell equations in both Fourier as well as coordinate space,

$$\vec{\nabla} \cdot \vec{D}(\vec{r},t) = \rho_0(\vec{r},t) \,, \tag{6.86a}$$

$$\vec{\nabla} \cdot \vec{B}(\vec{r},t) = 0 \,, \tag{6.86b}$$

$$\vec{\nabla} \times \vec{E}(\vec{r},t) = -\frac{\partial}{\partial t}\vec{B}(\vec{r},t) \,, \tag{6.86c}$$

$$\vec{\nabla} \times \vec{H}(\vec{r},t) = \vec{J}_0(\vec{r},t) + \frac{\partial}{\partial t}\vec{D}(\vec{r},t) \,. \tag{6.86d}$$

The second and third Maxwell equations have not changed. These are the homogeneous equations. The first and the fourth equation have been derived in Eqs. (6.35) and (6.63). It is interesting to note that these equations, in SI mksA units, carry no explicit prefactors like c, ϵ_0 and μ_0; the prefactors are all unity. In mixed coordinate-frequency space, one has

$$\vec{\nabla} \cdot \vec{D}(\vec{r}, \omega) = \rho_0(\vec{r}, \omega), \qquad \vec{\nabla} \cdot \vec{B}(\vec{r}, \omega) = 0, \qquad (6.87a)$$

$$\vec{\nabla} \times \vec{E}(\vec{r}, \omega) = i\omega \vec{B}(\vec{r}, \omega), \qquad \vec{\nabla} \times \vec{H}(\vec{r}, \omega) = \vec{J}_0(\vec{r}, \omega) - i\omega \vec{D}(\vec{r}, \omega), \qquad (6.87b)$$

while in wave-number-frequency space, after a Fourier transformation with respect to both space and time, one has

$$i\vec{k} \cdot \vec{D}(\vec{k}, \omega) = \rho_0(\vec{k}, \omega), \qquad \vec{k} \cdot \vec{B}(\vec{k}, \omega) = 0, \qquad (6.88a)$$

$$\vec{k} \times \vec{E}(\vec{k}, \omega) = \omega \, \vec{B}(\vec{k}, \omega), \qquad i\vec{k} \times \vec{H}(\vec{k}, \omega) = \vec{J}_0(\vec{k}, \omega) - i\omega \, \vec{D}(\vec{k}, \omega). \qquad (6.88b)$$

Central to the following analysis is the derivation of the wave equation in a dielectric medium, in mixed coordinate-frequency space. We start from Eq. (6.87) with $\vec{J}_0 = \vec{0}$,

$$\vec{\nabla} \times [\vec{\nabla} \times \vec{H}(\vec{r}, \omega)] = \frac{1}{\mu_0 \, \mu_r(\omega)} \vec{\nabla} \times [\vec{\nabla} \times \vec{B}(\vec{r}, \omega)]$$

$$= \frac{1}{\mu_0 \, \mu_r(\omega)} \left[-\vec{\nabla}^2 \vec{B}(\vec{r}, \omega) \right] = -i\omega \vec{\nabla} \times \vec{D}(\vec{r}, \omega). \qquad (6.89)$$

Furthermore,

$$\vec{\nabla} \times \vec{D}(\vec{r}, \omega) = \epsilon_0 \, \epsilon_r(\omega) \, \vec{\nabla} \times \vec{E}(\vec{r}, \omega) = \epsilon_0 \, \epsilon_r(\omega) \, i\omega \vec{B}(\vec{r}, \omega), \qquad (6.90)$$

The two above equations could be summarized as

$$\frac{1}{\mu_0 \, \mu_r(\omega)} \vec{\nabla}^2 \vec{B}(\vec{r}, \omega) = i\omega \vec{\nabla} \times \vec{D}(\vec{r}, \omega), \qquad (6.91a)$$

$$\vec{\nabla} \times \vec{D}(\vec{r}, \omega) = \epsilon_0 \, \epsilon_r(\omega) \, i\omega \vec{B}(\vec{r}, \omega). \qquad (6.91b)$$

One finally obtains the result

$$\left(\vec{\nabla}^2 + \epsilon_r(\omega) \, \mu_r(\omega) \, \frac{\omega^2}{c^2} \right) \vec{B}(\vec{r}, \omega) = \vec{0}. \qquad (6.92)$$

Similarly, for the electric field, one has

$$\left(\vec{\nabla}^2 + \epsilon_r(\omega) \, \mu_r(\omega) \, \frac{\omega^2}{c^2} \right) \vec{E}(\vec{r}, \omega) = \vec{0}. \qquad (6.93)$$

Equations (6.92) and (6.93) describe wave propagation in a medium. For $\mu_r(\omega) = 1$, one may set

$$\sqrt{\epsilon_r(\omega)} = n(\omega) + i\,\kappa(\omega), \qquad (6.94)$$

where the refractive index $n(\omega)$ describes diffraction, and $\kappa(\omega)$ is the absorption coefficient.

6.4 Dielectric Permittivity: Various Examples

6.4.1 *Sellmeier Equation*

The propagation of waves in a medium depends on the relative magnetic perme-
ability and relative electric permittivity functions $\mu_r(\omega)$ and $\epsilon_r(\omega)$ for the medium.
We will generally deal with systems in which $\mu_r(\omega)$ can be approximated as having
the value unity. A general form often used to approximate the electric permittivity
is the Sellmeier equation [28], or, a generalization thereof,

$$\epsilon(\omega) = \epsilon_r(\omega)\,\epsilon_0\,, \qquad \epsilon_r(\omega) = 1 + \sum_m \frac{\mathcal{A}_m}{\omega_m^2 - \omega^2 - i\gamma_m\omega}\,. \tag{6.95}$$

Here, $\mathcal{A}_m = \Omega_m^2$ has the dimension of the square of a frequency. In its original form,
the Sellmeier equation describes the dielectric response of a transparent medium
such as a glass and contains two or three terms of the form given in Eq. (6.95),
in the sum over m. Furthermore, the original Sellmeier equation is formulated
without the damping terms proportional to γ_m in Eq. (6.95) and thus is capable
of describing damped oscillations. Hence, we refer to Eq. (6.95) as a generalized
Sellmeier equation, which is based on the assumption that the system behaves as a
set of resonances (cf. Sec. 2.2).

The real and imaginary parts of the dielectric permittivity given in Eq. (6.95)
are as follows,

$$\mathrm{Re}\,\epsilon_r(\omega) = 1 + \sum_m \frac{(\omega_m^2 - \omega^2)\,\mathcal{A}_m}{(\omega^2 - \omega_m^2)^2 + \omega^2\,\gamma_m^2} \sim 1 - \sum_m \frac{\mathcal{A}_m}{\omega^2}\,, \quad \omega \to \infty\,, \tag{6.96}$$

and

$$\mathrm{Im}\,\epsilon_r(\omega) = \sum_m \frac{\omega\,\gamma_m\,\mathcal{A}_m}{(\omega^2 - \omega_m^2)^2 + \omega^2\,\gamma_m^2} \sim \sum_m \frac{\gamma_m\,\mathcal{A}_m}{\omega^3}\,, \quad \omega \to \infty\,. \tag{6.97}$$

The real part is an even function of ω, the imaginary part is an odd function of ω.

6.4.2 *Drude Model*

A specialization of the Sellmeier equation to a single term is capable of describing
electrical conductors, either metals or plasmas. One strives to model the dielectric
response of a plasma, where electrons move about freely. One considers only one
resonance, with $m = 0$ and $\omega_0 = 0$, but nonvanishing width $\gamma_0 \neq 0$. This theory is
relevant to the oscillations of an electron gas in a near-perfect conductor, where the
"resonant excitation frequency" of the electrons is zero. The relaxation time τ_0 is
defined as

$$\tau_0 = \frac{1}{\gamma_0}\,. \tag{6.98}$$

We thus have

$$\epsilon_r(\omega) = 1 - \frac{\sigma_0}{\epsilon_0}\,\frac{1}{\omega\,(\omega\tau_0 + i)}\,. \tag{6.99}$$

Here, σ_0 is a characteristic conductivity of the plasma, and τ_0 is the mean time between electron collisions in the plasma. The Drude model gives us an excellent opportunity to discuss the Green function G which determines the polarization density. One defines

$$G(\omega) = \epsilon_r(\omega) - 1, \qquad \epsilon_r(t - t') = \delta(t - t') + G(t - t'), \qquad (6.100a)$$

$$G(t - t') = \int \frac{d\omega}{2\pi} e^{-i\omega(t-t')} G(\omega). \qquad (6.100b)$$

For a causal behavior, $G(t - t')$ vanishes for $t - t' < 0$. The dielectric displacement is obtained from the electric field as

$$\vec{D}(\vec{r}, t) = \epsilon_0 \int_{-\infty}^{\infty} \delta(t - t') \vec{E}(\vec{r}, t') + \epsilon_0 \int_{-\infty}^{\infty} dt' \, G(t - t') \, \vec{E}(\vec{r}, t')$$

$$= \epsilon_0 \vec{E}(\vec{r}, t) + \epsilon_0 \int_0^{\infty} d\tau \, G(\tau) \, \vec{E}(\vec{r}, t - \tau). \qquad (6.101)$$

In frequency space, we have a simple multiplication,

$$\vec{D}(\vec{r}, \omega) = \epsilon_0 \left[1 + G(\vec{r}, \omega) \right] \vec{E}(\vec{r}, \omega). \qquad (6.102)$$

For small driving frequency, the polarization density $\vec{P}$ is obtained as

$$\vec{P}(\vec{r}, \omega) = \epsilon_0 G(\vec{r}, \omega) \vec{E}(\vec{r}, \omega) = -\sigma_0 \frac{1}{\omega \, (\omega \tau_0 + i)} \vec{E}(\vec{r}, \omega) \approx -\sigma_0 \frac{1}{i\omega} \vec{E}(\vec{r}, \omega). \qquad (6.103)$$

We now refer to Eq. (6.60) and identify the polarization current density $\vec{J}_p$ as the time derivative of the polarization density,

$$\vec{J}_p(\vec{r}, t) = \frac{\partial}{\partial t} \vec{P}(\vec{r}, t), \qquad \vec{J}_p(\vec{r}, \omega) = -i\omega \vec{P}(\vec{r}, \omega). \qquad (6.104)$$

The conductivity σ_0 relates the induced (polarization) current flowing in the material to the electric field,

$$\vec{J}_p(\vec{r}, \omega) = \sigma(\vec{r}, \omega) \vec{E}(\vec{r}, \omega), \qquad (6.105)$$

which allows us to identify

$$\sigma(\vec{r}, \omega) = -i\omega \epsilon_0 G(\vec{r}, \omega). \qquad (6.106)$$

Here, σ_0 has units of $(C/(m^2\, s))\, m/V = (C/(m\, s))\, (C/J) = C^2/(Jms)$. By contrast, ϵ_0 has units of $A\,s/(V\, m) = C^2/(J\, m)$. So, σ_0/ϵ_0 has the physical dimension of a frequency.

The Drude *ansatz* (6.99) for the relative permittivity has the right physical dimension (unity). Furthermore, the low-freqency limit of the conductivity is found as

$$\lim_{\omega \to 0} \sigma(\vec{r}, \omega) = \lim_{\omega \to 0} (-\epsilon_0 \, i\omega) \left(-\frac{1}{i\omega} \right) = \sigma_0, \qquad (6.107)$$

which is fully consistent with the usual definition of the current density and conductivity σ_0 provided the current flowing in the material is identified with the polarization current $\vec{J}_p(\vec{r}, \omega)$.

The Drude model also allows us to study the so-called plasma frequency of the electron gas. The plasma frequency ω_p is obtained from the relative permittivity,

$$\omega_p^2 \equiv \lim_{\omega \to \infty} \omega^2 \left[1 - \epsilon_r(\omega) \right] . \tag{6.108}$$

For the plasma model (6.99), one has

$$\omega_p^2 = \frac{\sigma_0}{\epsilon_0 \, \tau_0} . \tag{6.109}$$

We shall now try to use a microscopic theory of the plasma, and to find an alternative representation, as follows,

$$\omega_p^2 = \frac{N_V \, e^2}{\epsilon_0 \, m_e} , \qquad \sigma_0 = \frac{N_V \, e^2}{m_e} \, \tau_0 . \tag{6.110}$$

Here, N_V is the volume density of free electrons and m_e is the electron mass. A bulk material of electrons is assumed to have a volume density N_V, moving "freely" relative to a jellium of ionized cores, by a collective distance x_c. Let A be the (large) cross-sectional area of the box, and $V = A x_c$ its volume. The induced charges of the electrons, "sticking out" at either end of the box, are easily calculated as follows,

$$Q_1 = N_V \, e \, A \, x_c, \qquad Q_2 = -Q_1 . \tag{6.111}$$

We now apply Gauss's theorem in integrated form, to the layer in between the charged structures. The induced electric field obeys the equation $\oint \vec{E}_{\text{ind}} \cdot d\vec{A} = (\epsilon_0)^{-1} \int d^3 r \, \rho(\vec{r})$, with

$$E_{\text{ind}} \, A = -\frac{1}{\epsilon_0} \, N_V \, e \, A \, x_c, \qquad E_{\text{ind}} = -\frac{1}{\epsilon_0} \, N_V \, e \, x_c . \tag{6.112}$$

The sign follows from a geometric consideration, which is left as an exercise to the reader. We now express the dielectric constant as follows,

$$\epsilon_r(\omega) \approx \frac{E_{\text{ext}}(\omega) - E_{\text{ind}}(\omega)}{E_{\text{ext}}(\omega)} , \tag{6.113}$$

where $E_{\text{ext}}(\omega)$ is the external (driving) field. The sign in this equation can be justified by observing that the induced field is always directed against the driving field, and thus opposite to the induced polarization density. The polarization density, by contrast, has to be added to the expression $\epsilon_0 \, E_{\text{ext}}$ in order to obtain the dielectric displacement. Since the electrons in the bulk medium are free, we can approximate

$$m_e \frac{d^2 x_c}{dt^2} \approx e \, E_{\text{ext}}(t), \qquad -\omega^2 \frac{m_e}{e} \, x_c(\omega) \approx E_{\text{ext}}(\omega), \tag{6.114}$$

where in the latter step we go into Fourier space. Finally, one has

$$\epsilon_r(\omega) \approx \frac{-\omega^2 \dfrac{m_e}{e} \, x_c(\omega) + \dfrac{1}{\epsilon_0} \, N_V \, e \, x_c(\omega)}{-\omega^2 \dfrac{m_e}{e} \, x_c(\omega)} = 1 - \frac{N_V \, e^2}{\epsilon_0 \, m_e} \frac{1}{\omega^2} = 1 - \frac{\omega_p^2}{\omega^2} . \tag{6.115}$$

This "plasma" model

$$\epsilon_r(\omega) = 1 - \frac{\omega_p^2}{\omega^2} \tag{6.116}$$

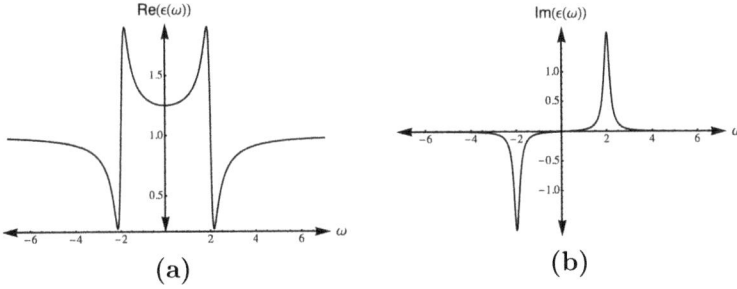

Fig. 6.1   Plot of $\text{Re}(\epsilon_r(\omega))$ and $\text{Im}(\epsilon_r(\omega))$ for $\omega_p = 1$, $\omega_0 = 2$, and $\gamma = 0.3$, according to Eq. (6.117). The functional shape of $\text{Re}(\epsilon_r(\omega))$ is consistent with in-phase driving below resonance, with a phase jump by π as one crosses the resonance. The deviation of $\text{Re}(\epsilon_r(\omega))$ from unity at high ω is proportional to $1/\omega^2$.

only has a real part and thus ignores the possibility of damping. As we later see, it does not fulfill the Kramers–Kronig relations, in view of the fact that $\epsilon_r(\omega)$ cannot be real everywhere along the real ω-axis. Our model (6.99) thus is physically more consistent than the "plasma" or "bulk" formula (6.116).

Infinitesimal damping can be studied by considering the limit $\tau_0 \rightarrow \infty$ in Eq. (6.99), while keeping the plasma frequency (6.109) constant. This leads to a well-defined result for any generalized model with a finite resonance frequency ω_0. We start from the Sellmeier equation (6.95), keep only the term with $m = 0$, while refraining from making the additional assumption that $\omega_0 = 0$, and write (see Fig. 6.1)

$$\epsilon_r(\omega) = 1 + \frac{\omega_p^2}{\omega_0^2 - \omega^2 - i\omega\gamma_0}, \qquad \gamma_0 \rightarrow 0. \qquad (6.117)$$

Using the prescription

$$\frac{1}{\omega_0 - \omega - i\gamma_0} = (\text{P.V.}) \frac{1}{\omega_0 - \omega} + i\pi\, \delta(\omega_0 - \omega), \qquad \gamma_0 \rightarrow 0, \qquad (6.118)$$

where (P.V.) denotes the principal value, one obtains

$$\frac{\omega_p^2}{\omega_0^2 - \omega^2 - i\omega\gamma_0} = (\text{P.V.}) \frac{1}{\omega_0^2 - \omega^2} + \frac{i\pi}{2\omega_0} \left[\delta(\omega_0 - \omega) + \delta(\omega_0 + \omega)\right]. \qquad (6.119)$$

For any finite ω_0, a model consistent with the Kramers–Kronig relations can thus be written down as

$$\epsilon_r(\omega) = 1 + (\text{P.V.}) \frac{1}{\omega_0^2 - \omega^2} + \frac{i\pi}{2\omega_0} \left[\delta(\omega - \omega_0) + \delta(\omega + \omega_0)\right]. \qquad (6.120)$$

The limit $\omega_0 \rightarrow 0$ cannot be taken, because it would lead to a divergence in the second term. A nonvanishing damping term is essential for the internal consistency of the model (6.99).

Finally, it is instructive to calculate the Fourier backtransform of Eq. (6.99),

$$G(\omega) = -\frac{\sigma_0}{\epsilon_0} \frac{1}{\omega (\omega \tau_0 + i)}, \qquad G(t - t') = -\frac{\sigma_0}{\epsilon_0} \int \frac{d\omega}{2\pi} e^{-i\omega (t-t')} \frac{1}{(\omega + i\eta) (\omega \tau_0 + i)},$$

(6.121)

where we introduce an infinitesimal imaginary part $i\eta$ in order to ensure causality. The two residues are

$$\operatorname*{Res}_{\omega = -i\eta} e^{-i\omega (t-t')} \frac{1}{(\omega + i\eta) (\omega \tau_0 + i)} = \frac{1}{i},$$

(6.122a)

$$\operatorname*{Res}_{\omega = -i/\tau_0} e^{-i\omega (t-t')} \frac{1}{(\omega + i\eta) (\omega \tau_0 + i)} = e^{-(t-t')/\tau_0} \left(-\frac{1}{i}\right).$$

(6.122b)

The prefactor multiplying the residues is $(-2\pi i)$, because the poles are encircled in the mathematically negative (clockwise) direction, for $t - t' > 0$. The end result is

$$G(t - t') = \Theta(t - t') \frac{\sigma_0}{\epsilon_0} \left[1 - \exp\left(-\frac{t - t'}{\tau_0}\right)\right].$$

(6.123)

This quantity has dimension of frequency, as it should, because it is obtained as the Fourier backtransform of a dimensionless quantity $\epsilon_r(\omega)$. For perfect electrical conductors, $\tau_0 \to \infty$, and

$$G(t - t') \to \frac{\sigma_0}{\epsilon_0}, \qquad \tau_0 \to \infty.$$

(6.124)

For a realistic conductor and finite τ_0, we have $G(t - t' = 0) = 0$, and the integral (6.101) converges.

6.4.3   *Dielectric Permittivity and Atomic Polarizability*

We have already seen that a very successful way of describing the relative permittivity of a typical material consists in the analogy with a collection of harmonic oscillators. One compares the functional form of the Green function of the harmonic oscillator (2.11) with the Sellmeier equation (6.95). For dilute gases, it is additionally possible to relate the given "collection of harmonic oscillators" with the atomic properties of the gas atoms themselves, which are summarized in the so-called dynamic atomic polarizability. This analogy will be explored in the following.

By definition, the Fourier component $\vec{p}_n(\omega)$ of a dipole moment of the nth atom is related to its polarizability $\alpha(\omega)$ and to the applied electric field $\vec{E}(\vec{r}, \omega)$ at the coordinate of the atom, as follows (we assume spatially uniform fields),

$$\vec{p}_n(\omega) = \alpha(\omega) \vec{E}(\vec{r}, \omega).$$

(6.125)

We may thus relate the polarization density $\vec{P}(\vec{r}, \omega)$ to the volume density of atoms, N/V, and the driving electric field $\vec{E}(\vec{r}, \omega)$, as follows,

$$\vec{P}(\vec{r}, \omega) = \frac{N}{V} \vec{p}_n = \frac{N}{V} \alpha(\omega) \vec{E}(\vec{r}, \omega).$$

(6.126)

This implies that

$$\epsilon_r(\omega) - 1 = \frac{N_V}{\epsilon_0}\,\alpha(\omega), \qquad N_V = \frac{N}{V}. \tag{6.127}$$

In atomic physics, one writes the corresponding expression for $\alpha(\omega)$ is

$$\alpha(\omega) = \sum_n \frac{f_{n0}}{E_{n0}^2 - i\Gamma_n \hbar\,\omega - (\hbar\omega)^2}, \tag{6.128}$$

where E_{n0} are the atomic transition energies from the quantum mechanical ground state (denoted by the subscript zero) to the nth excited state, and the Γ_n are the energy widths of the transitions. The angular transition frequency ω_{n0} and the damping factor γ_n of the transition are given as

$$\omega_{n0} = \frac{E_{n0}}{\hbar}, \qquad \gamma_{n0} = \frac{\Gamma_n}{\hbar}. \tag{6.129}$$

The factors f_{n0} in Eq. (6.128) are the so-called oscillator strengths of the transitions. An alternative representation of the atomic polarizability thus is

$$\alpha(\omega) = \sum_n \frac{f_{n0}}{\hbar^2} \frac{1}{\omega_{n0}^2 - \omega^2 - i\gamma_n \omega}. \tag{6.130}$$

Written in this form, the analogy to the Green function (2.11) of the damped harmonic oscillator is obvious,

$$g(\omega) = \frac{1}{\omega_0^2 - i\gamma\omega - \omega^2}. \tag{6.131}$$

Equation (6.130) represents the atomic polarizability as a sum over damped harmonic oscillators. Restoring prefactors, Eq. (6.127) can be written as

$$\epsilon_r(\omega) = 1 + \frac{N_V}{\epsilon_0}\,\alpha(\omega) = 1 + \sum_n \frac{f_{n0}}{\hbar^2} \frac{1}{\omega_{n0}^2 - \omega^2 - i\gamma_n \omega}. \tag{6.132}$$

A remark is in order. By convention, $\alpha(\omega) = \alpha_{\ell=1}(\omega)$ given in Eq. (6.128) is the dipole ($2^{\ell=1}$-pole) dynamic polarizability of the ground state $|\phi_0\rangle$ of an atom. It is the dominant term in the long-wavelength limit, and it describes the response of the atom to incident dipole radiation. Incident quadrupole radiation ($\ell = 2$) will dynamically ($\omega \neq 0$) induce a quadrupole moment in the atoms, while an incident octupole radiation ($\ell = 3$) will do the same for an atomic octupole moment. Quantum mechanically, it is useful to note at this stage that the 2^ℓ-multipole oscillator strength is

$$f_{n0}^{(\ell)} = 2\,\frac{4\pi e^2}{(2\ell+1)^2}\,E_{n0} \sum_m \sum_{m_n} \left| \left\langle \phi_n \left| \sum_i (r_i)^\ell\, Y_{\ell m}(\theta_i, \varphi_i) \right| \phi_0 \right\rangle \right|^2, \tag{6.133}$$

where we sum over the magnetic projections m of the spherical harmonic and over the magnetic projections m_n of the excited state $|\phi_n\rangle$. The ground state is denoted as $|\phi_0\rangle$. Furthermore, r_i is the radial coordinate of the ith electron, and $Y_{\ell m}(\theta_i, \varphi_i)$ is the spherical harmonic with the arguments being equal to the polar and azimuthal angles of the position of the ith electron. The sum is over all electrons within the atomic system; these are enumerated in the sum over i. The dipole oscillator strength is recovered as

$$f_{n0} = f_{n0}^{(\ell=1)}. \tag{6.134}$$

There is a sum rule,

$$\sum_n f_{n0} = Z\, e^2\, a_0^2\, E_h, \qquad a_0 = \frac{\hbar}{\alpha\, m_e\, c}, \qquad E_h = \alpha^2 m_e c^2. \tag{6.135}$$

Here, a_0 is the Bohr radius, E_h is the Hartree energy, α is the fine-structure constant, Z is the number of electrons in the atom, and e is the elementary charge. Let us check the consistency of physical units. The oscillator strength f_{n0} is measured in $(\mathrm{Cm})^2\,\mathrm{J}$, while ϵ_0 carries a unit of $\frac{\mathrm{A\,s}}{\mathrm{V\,m}}$ or $\frac{\mathrm{C}}{\mathrm{V\,m}}$. Hence,

$$\frac{N_V}{\epsilon_0}\, \alpha(\omega) \sim \frac{N_V}{\epsilon_0}\, \frac{f_{n0}}{E_{n0}^2} \sim \frac{1}{\mathrm{m}^3}\, \frac{\mathrm{m\,V}}{\mathrm{C}}\, \frac{(\mathrm{Cm})^2}{\mathrm{J}^2}\, \mathrm{J} = \frac{\mathrm{J\,C}}{\mathrm{V}} = 1, \tag{6.136}$$

as it should be.

6.4.4 Dielectric Permittivity for Dense Materials

Up to this point, we have ignored the backreaction of the induced electric field, inside the sample, onto the orientation of the dipole moments. This approximation is justified if the relative permittivity of the sample does not significantly differ from unity, or in other words, if sample is dilute. Indeed, for a dilute gas, the dielectric constant may be expressed as [Eq. (6.132)],

$$\epsilon_r(\omega) = 1 + \frac{N_V}{\epsilon_0}\, \alpha(\omega), \tag{6.137}$$

because the backreaction of the induced polarization field onto the atoms in the sample can be ignored. From Eqs. (6.33) and (6.36), we recall that, within the dipole approximation, the polarization is given as

$$\vec{P}(\vec{r}, t) = \sum_n \vec{p}_n\, f(\vec{r} - \vec{r}_n(t)), \tag{6.138}$$

where f constitutes the test function for the averaging volume V over which the macroscopic Maxwell equations are determined. The distribution function f is normalized according to Eq. (6.14), namely, $\int_V \mathrm{d}^3 r\, f(\vec{r}) = 1$. Thus, f has dimension of inverse volume. It is advantageous to choose the reference volume V, as a rectangular volume of dimensions Δx, Δy, and Δz. We choose Δz to be the distance of the two charges that represent the dipole (see Fig. 6.2). If the dipoles are distributed uniformly in the test volume, then

$$\vec{P} = \langle \vec{p} \rangle\, \frac{N}{V} = N_V\, \langle \vec{p} \rangle, \tag{6.139}$$

where $\langle \vec{p} \rangle$ is the average dipole moment. Let us suppose that the dipoles are oriented in the z direction. Then, a cut through the sample, in the xy plane, with N atoms in the volume, will result in a surface charge density (see also Fig. 6.2),

$$\sigma = N\, \frac{q}{\Delta x\, \Delta y} = \frac{N}{\Delta x\, \Delta y\, \Delta z}\, (q\Delta z) = N_V\, |\langle \vec{p} \rangle| = |\vec{P}|. \tag{6.140}$$

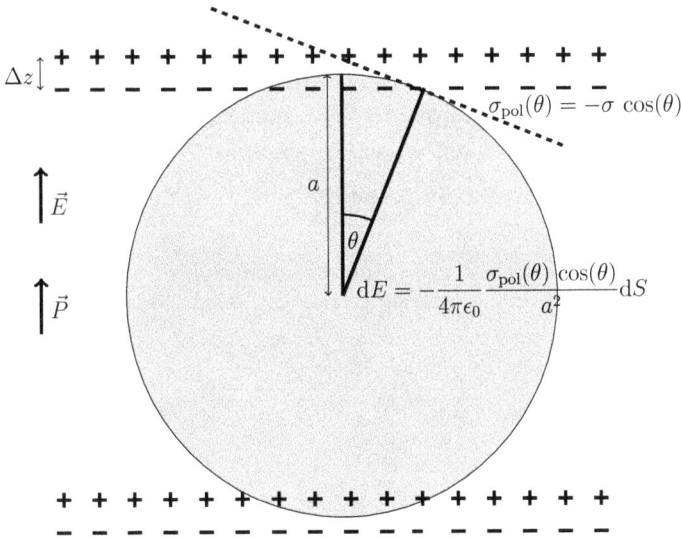

Fig. 6.2 Illustration of Eq. (6.143) and (6.144).

If the surface is tilted, then the appropriate modification is

$$\sigma = |\vec{P}| \cos\theta. \tag{6.141}$$

The sign of σ has to be determined based on additional considerations.

We recall that for a uniform medium, the polarization

$$\vec{P} = \epsilon_0 \left(\epsilon_r - 1\right) \vec{E} \tag{6.142}$$

is aligned and parallel to the applied electric field. Let us draw a sphere of radius a about the point $\vec{r}$ in the dense sample. Let us assume, furthermore, that the polarization points into the $+z$ direction, and the applied electric field also points along $+z$. We cut through the dipole layers in the middle of the dipoles, separating the dipoles, and ignore the outer portion of the sphere as well as the rest of the bulk medium in the following. Then, the surface charge density at an angle θ with respect to the symmetry axis of the polarization (the $+z$ axis is parallel, not anti-parallel, to $\vec{P}$) is

$$\sigma_{\text{pol}}(\theta) = -|\vec{P}| \cos\theta, \tag{6.143}$$

where θ is the polar angle with respect to the $+z$ axis (see also Fig. 6.2). Let us assume that the dipoles are oriented so that the positive charges are to the top and the negative ones to the bottom. The polarization vector points upward. We use the center of the sphere as an anchor point. This center is below the upper surface of the sphere. Cutting through the sample, positive charge is left on the upper side of the sphere, i.e., on the outer rim of the cut-out sphere. At the same time, $\theta = 0$.

What counts for us is the negative charge on the lower side of the cut. This justifies the negative sign in Eq. (6.143).

Now, imagine that we cut out a sphere of radius a about our reference point $\vec{r}$. The z component of the field due to the infinitesimal surface charge element of radius a, generated at the reference point $\vec{r}$, amounts to

$$\mathrm{d}E_z = -\frac{1}{4\pi\epsilon_0} \frac{\sigma_{\mathrm{pol}}(\theta)\,\cos\theta}{a^2}\,\mathrm{d}S = \frac{1}{4\pi\epsilon_0} \frac{+|\vec{P}|\,\cos^2\theta}{a^2}\,\mathrm{d}S, \tag{6.144}$$

because the positive charges at the top of the sphere lead to an electric field that points downward. The field generated by the polarized constituents at the point $\vec{r}$ is

$$E_z = \frac{1}{4\pi\epsilon_0} \int \frac{|\vec{P}|\,\cos^2\theta}{a^2}\,\mathrm{d}S = \frac{|\vec{P}|}{4\pi\epsilon_0} \frac{1}{a^2} \frac{4\pi a^2}{3} = \frac{|\vec{P}|}{3\epsilon_0}, \tag{6.145a}$$

$$\vec{E}_{\mathrm{pol}} = E_z\,\hat{e}_z = \frac{\vec{P}}{3\epsilon_0}. \tag{6.145b}$$

The total field of the remaining dipoles within the sphere of radius R is

$$\sum_i \vec{E}_i = \frac{1}{4\pi\epsilon_0} \sum_i \left(\frac{3(\vec{p}_i \cdot \vec{r}_i)\,\vec{r}_i}{r_i^5} - \frac{\vec{p}_i}{r_i^3} \right) \tag{6.146}$$

where i enumerates the molecules, and $\vec{r}_i$ is the direction vector from the location of the ith molecule to $\vec{r}$. Provided the entire sample is z polarized, we may average out the field according to the prescription

$$3(\vec{p}_i \cdot \vec{r}_i)\,\vec{r}_i \to 3\hat{e}_z\,p_i\,z_i^2 \to \frac{1}{3} \times 3\hat{e}_z\,p_i\,r_i^2, \tag{6.147}$$

where in the last step, we have performed the angular average $z_i^2 \to r_i^2/3$. In summary,

$$\sum_i \vec{E}_i = \vec{0}. \tag{6.148}$$

The polarization charge is exclusively given by the integral (6.145). Letting the radius of the sphere become large, we see that the result (6.145) holds universally.

The local field at the point $\vec{r}$ is thus found as the sum of the external field $\vec{E}$ and the field $\vec{E}_{\mathrm{pol}}$ due to the polarized constituents,

$$\vec{E}_{\mathrm{loc}} = \vec{E} + \vec{E}_{\mathrm{pol}} = \vec{E} + \frac{\vec{P}}{3\,\epsilon_0}. \tag{6.149}$$

Using Eq. (6.142), we obtain

$$\vec{E}_{\mathrm{loc}} = \vec{E} + \frac{1}{3\,\epsilon_0}\,\epsilon_0\,(\epsilon_r - 1)\,\vec{E} = \frac{\epsilon_r(\omega) + 2}{3}\,\vec{E}. \tag{6.150}$$

Hence, we have two relations,

$$\vec{P} = N_V\,\alpha(\omega)\,\vec{E}_{\mathrm{loc}} = N_V\,\alpha(\omega)\,\frac{\epsilon_r(\omega) + 2}{3}\,\vec{E}, \tag{6.151}$$

$$\vec{P} = \epsilon_0\,(\epsilon_r(\omega) - 1)\,\vec{E}, \tag{6.152}$$

the second of which is just the definition (6.142). Eliminating $\vec{P}$, then $\vec{E}$, one obtains

$$\frac{\epsilon_r(\omega) - 1}{\epsilon_r(\omega) + 2} = \frac{N_V \, \alpha(\omega)}{3 \, \epsilon_0} , \tag{6.153}$$

which is the Clausius-Mosotti equation. It is easy to show that

$$\frac{d\epsilon_r}{dN_V} = \frac{(\epsilon_r(\omega) - 1)(\epsilon_r(\omega) + 2)}{3 \, N_V} , \tag{6.154}$$

which is left as an exercise to the reader.

6.5 Propagation of Plane Waves in a Medium

6.5.1 *Refractive Index and Group Velocity*

The wave equation in a medium has been given in Eqs. (6.92) and (6.93). For a plane wave with angular frequency ω and wave vector $\vec{k}$, the dispersion is

$$k^2 = \epsilon_r(\omega) \frac{\omega^2}{c^2} , \tag{6.155}$$

where $k = |\vec{k}|$ and in general, $\epsilon_r(\omega)$ is complex rather than real. In consequence, the wave number $k - 2\pi/\lambda$ can be complex. The relation between the wave number and the angular frequency is given by $k = \mathrm{Re}\, k + i\,\mathrm{Im}\, k$, with

$$\mathrm{Re}\, k = n(\omega) \frac{\omega}{c} , \qquad \mathrm{Im}\, k = \kappa(\omega) \frac{\omega}{c} . \tag{6.156}$$

From the dispersion relation (6.155), one gets

$$k^2 = \epsilon_r(\omega) \frac{\omega^2}{c^2} = [\mathrm{Re}\, \epsilon_r(\omega) + i\,\mathrm{Im}\, \epsilon_r(\omega)] \frac{\omega^2}{c^2}$$
$$= [n(\omega) + i\,\kappa(\omega)]^2 \frac{\omega^2}{c^2} = [n^2(\omega) - \kappa^2(\omega) + 2\,i\,n(\omega)\,\kappa(\omega)] \frac{\omega^2}{c^2} . \tag{6.157}$$

This can be summarized as follows,

$$\mathrm{Re}\, \epsilon_r(\omega) = n^2(\omega) - \kappa^2(\omega) , \tag{6.158a}$$
$$\mathrm{Im}\, \epsilon_r(\omega) = 2\,n(\omega)\,\kappa(\omega) . \tag{6.158b}$$

Alternatively, one may write

$$\sqrt{\epsilon_r(\omega)} = n(\omega) + i\,\kappa(\omega) . \tag{6.159}$$

We note that in all our models and calculation, $\epsilon_r(\omega)$ has a positive imaginary and a positive real part. We identity $\sqrt{\epsilon_r(\omega)} = \sqrt{|\epsilon_r(\omega)|} \exp[\frac{i}{2} \arg(\epsilon_r(\omega))]$ with the "obvious" branch of the square root, consistent with the branch cut of the square root being along the negative real axis.

Just as for the relative permittivity $\epsilon_r(\omega)$, the real part of the index of refraction is an even function of ω and the imaginary part is an odd function of ω,

$$n(\omega) = n(-\omega), \qquad \kappa(\omega) = -\kappa(-\omega) . \tag{6.160}$$

Using these properties, a wave traveling in the $+x$ direction can be described by the expression

$$\vec{E}(x,t) = \int_{-\infty}^{\infty} \frac{d\omega}{2\pi} \, \vec{E}_0(\omega) \, \exp\left[i\left(\operatorname{Re} k(\omega) + i \operatorname{Im} k(\omega)\right) x - \omega t\right]$$

$$= \int_{-\infty}^{\infty} \frac{d\omega}{2\pi} \, \vec{E}_0(\omega) \, \exp\left[i\frac{\omega}{c}\left(n(\omega)x - ct\right)\right] \exp\left(-\frac{\omega}{c}\kappa(\omega)x\right). \tag{6.161}$$

This wave packet is the Fourier backtransformation of a wave packet composed of plane waves, each of which fulfills the wave equation

$$\left(\frac{\partial^2}{\partial x^2} + \epsilon_r(\omega)\frac{\omega^2}{c^2}\right) \vec{E}_0(\omega) \, \exp\left(i\frac{\omega}{c}n(\omega)x\right) \exp\left(-\frac{\omega}{c}\kappa(\omega)x\right) = 0, \tag{6.162}$$

which is a specialization of Eq. (6.93) to one spatial dimension, with $\mu_r(\omega) = 1$.

We restrict our analysis to the case in which $|\vec{E}_0(\omega)|$ has a single, narrow peak of width $\Delta\omega$ at ω_0. Here, $\Delta\omega$ is the characteristic width of the wave packet in angular frequency space. The peak will be considered narrow if

$$\frac{\Delta\omega}{n(\omega_0)} \frac{dn(\omega_0)}{d\omega_0} \ll 1. \tag{6.163}$$

For a wave composed of a narrow range of frequencies, the wave vector $\vec{k}(\omega)$ can be expanded about the central frequency, to obtain an approximate expression for the propagation of the wave. In our example, the central frequency is ω_0, and

$$\frac{\omega}{c}n(\omega) = \frac{\omega_0}{c}n(\omega_0) + \left[n(\omega_0) + \omega_0\frac{dn(\omega_0)}{d\omega_0}\right]\frac{\omega - \omega_0}{c} + \mathcal{O}(\omega - \omega_0)^2. \tag{6.164}$$

While a similar expression holds for $\omega\kappa(\omega)/c$, we here set $\kappa(\omega) = \kappa(\omega_0)$, assuming that the damping is not very strong, i.e., that $\kappa(\omega) \ll n(\omega)$. Working to first order in $\omega - \omega_0$, we find

$$\vec{E}(x,t) = \exp\left[i\frac{\omega_0}{c}\left(n(\omega_0)x - ct\right)\right] \exp\left(-\frac{\omega_0}{c}\kappa(\omega_0)x\right)$$

$$\times \int_{-\infty}^{\infty} \frac{d\omega}{2\pi} \vec{E}_0(\omega) \exp\left(i\frac{\omega - \omega_0}{c}\left\{\left[n(\omega_0) + \omega_0\frac{dn(\omega_0)}{d\omega_0}\right]x - ct\right\}\right). \tag{6.165}$$

The interpretation is as follows: The first factor identifies a term that oscillates at frequency ω_0 and propagates at the phase velocity $c/n(\omega_0)$. The second term is an integral (envelope function) that travels with the group velocity $v_g(\omega_0)$, where

$$\operatorname{Re} k = n(\omega)\frac{\omega}{c}, \qquad \frac{d\operatorname{Re} k}{d\omega}\bigg|_{\omega=\omega_0} = \frac{1}{c}\left(n(\omega_0) + \omega_0\frac{dn(\omega)}{d\omega}\bigg|_{\omega=\omega_0}\right) = \frac{1}{v_g(\omega_0)},$$

$$v_g(\omega_0) = c\left[n(\omega_0) + \omega_0\frac{dn(\omega_0)}{d\omega_0}\right]^{-1}. \tag{6.166}$$

One can write $\vec{E}(x,t)$ as

$$\vec{E}(x,t) = \exp\left[i\frac{\omega_0}{c}\left(n(\omega_0)x - ct\right)\right] \exp\left[-\frac{\omega_0}{c}\,\kappa(\omega_0)\,x\right] \vec{F}(x - v_g(\omega_0)\,t),$$

$$\vec{F}(x - v_g(\omega_0)\,t) = \int_{-\infty}^{\infty} \frac{d\omega}{2\pi}\,\vec{E}_0(\omega)\exp\left(i\frac{\omega - \omega_0}{v_g(\omega_0)}\,\{x - v_g(\omega_0)\,t\}\right), \qquad (6.167)$$

where the first factor describes a damped wave traveling at the phase velocity, while $\vec{F}(x - v_g(\omega_0)\,t)$ is an (undamped) wave traveling at the group velocity. The "broad" wave packet has an envelope function $\vec{F}$ traveling with speed $v_g(\omega_0)$, while the underlying wave has the phase velocity $c/n(\omega_0)$.

6.5.2 Wave Propagation and Method of Steepest Descent

The evaluation of Fourier backtransform integrals of the type (6.161) can be rather complicated and in many cases, not analytically feasible if the functional form of $n(\omega)$ or $\kappa(\omega)$ is sufficiently complex. However, in some cases, it can be helpful to resort to a method of integration otherwise known as the method of stationary phase, or method of steepest descent, which can be applied with good effect to integrals of the form (6.161).

We will ask the following question. Let $u(x,t)$ be a wave propagating in a medium with a permittivity $\epsilon_r(\omega)$ and let a signal $u(x,t)$ be written as the Fourier backtransform of a sum of waves, propagating in the $+x$ and/or $-x$ directions, as follows,

$$u(x,t) = \int_{-\infty}^{\infty} \frac{d\omega}{2\pi}\,e^{-i\omega t}\left[A(\omega)\exp\left(i\frac{\omega\,n(\omega)\,x}{c}\right) + B(\omega)\exp\left(-i\frac{\omega\,n(\omega)\,x}{c}\right)\right],$$
$$(6.168)$$

where for reference we temporarily set

$$n(\omega) \equiv \sqrt{\epsilon_r(\omega)}, \qquad (6.169)$$

unifying the refractive index $n(\omega)$ and the absorption coefficient $\kappa(\omega)$ into a single symbol.

Because all traveling waves have to pass any point of the x axis at some time, we can conjecture that knowledge of the signal $u(x,t)$ at a given point, but for all time, should enable us to calculate the signal for all (x,t). Our knowledge of the signal at the given point, say, the origin, but for all time, enables us to calculate the Fourier transform of the signal. Fourier backtransformation to space-time coordinates then enables us to calculate $u(x,t)$, as desired. Yet, in the Fourier backtransformation integrals, one may ask if contributions from times $t' > t$ might influence the signal at (x,t), arguing that Fourier integrals stretch from $-\infty$ to $+\infty$. This question and the eventual solution, as well as the evaluation of the resulting integral by the method of steepest descent, will be discussed in the following.

We consider a situation where $u(0,t)$ and $\partial u(x,t)/\partial x|_{x=0}$ are given for all t, and attempt to calculate the signal for all (x,t). First, from Eq. (6.168), we find that

$$u(0,t) = \int_{-\infty}^{\infty} e^{-i\omega t} \left[A(\omega) + B(\omega) \right] \frac{d\omega}{2\pi}. \tag{6.170}$$

Fourier transformation of $u(0,t)$ enables us to calculate the sum of the Fourier amplitudes,

$$\int_{-\infty}^{\infty} e^{i\omega' t} u(0,t) \, dt = \int_{-\infty}^{\infty} \int_{-\infty}^{\infty} e^{-i(\omega-\omega')t} \left[A(\omega) + B(\omega) \right] \frac{d\omega}{2\pi} \, dt$$

$$= \int_{-\infty}^{\infty} \left[A(\omega) + B(\omega) \right] \delta(\omega - \omega') \, d\omega = A(\omega') + B(\omega'). \tag{6.171}$$

Similarly, differentiating Eq. (6.168) with respect to time, one finds that

$$\left. \frac{\partial}{\partial x} u(x,t) \right|_{x=0} = \int_{-\infty}^{\infty} e^{-i\omega t} \left[i \frac{\omega n(\omega)}{c} A(\omega) - i \frac{\omega n(\omega)}{c} B(\omega) \right] \frac{d\omega}{2\pi}. \tag{6.172}$$

Fourier transformation of the spatial derivative of $u(x,t)$ at the origin thus gives us the difference of the Fourier amplitudes,

$$\int_{-\infty}^{\infty} dt\, e^{i\omega' t} \left. \frac{\partial}{\partial x} u(x,t) \right|_{x=0} = \int_{-\infty}^{\infty} \frac{d\omega}{2\pi} \left[A(\omega) - B(\omega) \right] i \frac{\omega n(\omega)}{c} \delta(\omega - \omega') \tag{6.173}$$

and so

$$-i \frac{c}{\omega' n(\omega')} \int_{-\infty}^{\infty} e^{i\omega' t} \left. \frac{\partial}{\partial x} u(x,t) \right|_{x=0} dt = A(\omega') - B(\omega'). \tag{6.174}$$

Solving Eqs. (6.171) and (6.174) for $A(\omega')$ and $B(\omega')$, and setting $\partial_x u(0,t) \equiv \left. \frac{\partial}{\partial x} u(x,t) \right|_{x=0}$, one obtains

$$A(\omega') = \frac{1}{2} \int_{-\infty}^{\infty} e^{i\omega' t} \left\{ u(0,t) - i \frac{c}{\omega' n(\omega')} \partial_x u(0,t) \right\} dt, \tag{6.175a}$$

$$B(\omega') = \frac{1}{2} \int_{-\infty}^{\infty} e^{i\omega' t} \left\{ u(0,t) + i \frac{c}{\omega' n(\omega')} \partial_x u(0,t) \right\} dt. \tag{6.175b}$$

As anticipated, the Fourier expansion coefficients $A(\omega')$ and $B(\omega')$ can be reconstructed from integrals over the signal and its derivative, evaluated at $x = 0$ but observed over all time. Under the interchange $\omega \leftrightarrow \omega'$ and $t \leftrightarrow t'$, these formulas can be used directly in Eq. (6.168),

$$u(x,t) = \frac{1}{2} \int_{-\infty}^{\infty} \frac{d\omega}{2\pi} \int_{-\infty}^{\infty} dt'\, e^{-i\omega(t-t')} \exp\left(i \frac{\omega n(\omega) x}{c} \right) \left\{ u(0,t') - \frac{i c\, \partial_x u(0,t')}{\omega n(\omega)} \right\}$$

$$+ \frac{1}{2} \int_{-\infty}^{\infty} \frac{d\omega}{2\pi} \int_{-\infty}^{\infty} dt'\, e^{-i\omega(t-t')} \exp\left(-i \frac{\omega n(\omega) x}{c} \right) \left\{ u(0,t') + \frac{i c\, \partial_x u(0,t')}{\omega n(\omega)} \right\}. \tag{6.176}$$

One might ask the following question: Under the integral signs on the right-hand side, we integrate over the entire range of t', i.e., $t' \in (-\infty, \infty)$. However, on the left-hand side, we have $u(x,t)$. The suspicion would be that the signal at $t' > t$ as it enters the integrand on the right-hand side might influence the left-hand side in a non-causal way. In order to ensure causality, we have to slightly deform the integration over ω into a contour C, which is conveniently chosen to lie infinitesimally above the real axis, just like for the retarded Green function (see Sec. 4.1.3). We thus write

$$
u(x,t) = \frac{1}{2} \int_{-\infty}^{\infty} dt' \int_{C} \frac{d\omega}{2\pi} \, e^{-i\omega(t-t')} \exp\left(i\frac{\omega n(\omega) x}{c}\right) \left\{ u(0,t') - \frac{ic\,\partial_x u(0,t')}{\omega n(\omega)} \right\}
$$
$$
+ \frac{1}{2} \int_{-\infty}^{\infty} dt' \int_{C} \frac{d\omega}{2\pi} \, e^{-i\omega(t-t')} \exp\left(-i\frac{\omega n(\omega) x}{c}\right) \left\{ u(0,t') + \frac{ic\,\partial_x u(0,t')}{\omega n(\omega)} \right\}.
$$
$$(6.177)$$

We now take the observed signal at $x = 0$ to be a single Dirac-δ peak at $t = 0$,

$$
u(0,t) = u_0\, \delta(t), \qquad \left. \frac{\partial u(x,t)}{\partial x} \right|_{x=0} = 0. \tag{6.178}
$$

These boundary conditions will provide identical waves traveling in the $+x$ and $-x$ directions. Since the Fourier transform of a Dirac-δ function is a constant in frequency space, this acts as a "white light" source. The generated signal is, by virtue of Eq. (6.177),

$$
u(x,t) = \frac{1}{2} \int_{-\infty}^{\infty} dt' \int_{C} \frac{d\omega}{2\pi} \, e^{-i\omega(t-t')} \left[\exp\left(i\frac{\omega n(\omega) x}{c}\right) + \exp\left(-i\frac{\omega n(\omega) x}{c}\right) \right] u(0,t')
$$
$$
= u_0 \int_{C} \frac{d\omega}{2\pi} \, e^{-i\omega t} \cos\left(\frac{\omega n(\omega) x}{c}\right). \tag{6.179}
$$

One can split the signal into waves traveling in either the $+x$ or $-x$ direction,

$$
u(x,t) = \frac{u_0}{2} \int_{C} \frac{d\omega}{2\pi} \exp\left[-i\omega\left(t - \frac{n(\omega) x}{c}\right)\right] + \frac{u_0}{2} \int_{C} \frac{d\omega}{2\pi} \exp\left[-i\omega\left(t + \frac{n(\omega) x}{c}\right)\right]
$$
$$
= u_+(x,t) + u_-(x,t). \tag{6.180}
$$

We investigate the wave traveling in the $+x$ direction,

$$
u_+(x,t) = \frac{1}{2} u_0 \int_{C} \frac{d\omega}{2\pi} \exp\left[-i\omega\left(t - \frac{n(\omega) x}{c}\right)\right], \tag{6.181a}
$$

$$
u_+(x,t) = \frac{1}{2} u_0 \int_{0}^{\infty} \frac{d\omega}{2\pi} \exp\left[-i\omega\left(t - \frac{n(\omega) x}{c}\right)\right]
$$
$$
+ \frac{1}{2} u_0 \int_{-\infty}^{0} \frac{d\omega}{2\pi} \exp\left[-i\omega\left(t - \frac{n(\omega) x}{c}\right)\right]
$$
$$
= \frac{1}{2} u_0 \int_{0}^{\infty} \frac{d\omega}{2\pi} \exp\left[-i\omega\left(t - \frac{n(\omega) x}{c}\right)\right] + \text{c.c.}. \tag{6.181b}
$$

We have used the fact that $n(\omega) = n^*(-\omega)$, and divided the integration along the contour C into one which extends from $-\infty$ to 0 and another one from 0 to $+\infty$ (ignoring the infinitesimal displacement off the real axis). Thus,

$$u_+(x,t) = u_0 \, \mathrm{Re} \int_0^\infty \frac{d\omega}{2\pi} \exp\left[-i\omega\left(t - \frac{n(\omega)x}{c}\right)\right]. \tag{6.182}$$

The integrand contains an exponential with a complicated structure. Typically, integrals of this type are very well suited for the application of the method of steepest descent. One looks for the points in the complex w plane where the "phase", i.e., the argument of the exponential, is stationary. In our case, this would imply that one defines

$$f(\omega) = \omega t - \omega\frac{n(\omega)x}{c}, \qquad \frac{df(\omega)}{\omega} = t - \left(\omega\frac{dn(\omega)}{d\omega} + n(\omega)\right)\frac{x}{c}. \tag{6.183}$$

This equation determines the saddle point in the integral over ω, via the condition

$$\left.\frac{df(\omega)}{d\omega}\right|_{\omega=\omega_c} = 0. \tag{6.184}$$

Referring to Eq. (6.166), we establish that the saddle point occurs at

$$x = v_g(\omega_c)t, \tag{6.185}$$

which is the point where the wave signal is expected to travel, according to the interpretation of the group velocity.

One therefore approximates

$$u_+(x,t) = u_0 \, \mathrm{Re} \int_0^\infty \frac{d\omega}{2\pi} \exp\left[-i f(\omega)\right]$$

$$\approx u_0 \, \mathrm{Re} \, \exp\left[-i f(\omega_c)\right] \int_0^\infty \frac{d\omega}{2\pi} \exp\left(-i \left.\frac{d^2 f(\omega)}{d\omega^2}\right|_{\omega=\omega_c} (\omega-\omega_c)^2\right). \tag{6.186}$$

The spirit of the method of steepest descent dictates to choose in the complex plane, a contour which passes through the saddle point, and leads to the most rapid decrease in the magnitude of the integrand. One then extends this contour to ∞, in both directions, to evaluate the Gaussian integral, under the assumption that the integrand decreases sufficiently rapidly to make this extension possible. In other words, one heads down to the valley of the integrand as soon as possible. This may imply that one has to distort the path of integration in the complex plane, in order to "hit" the saddle point at the "right" angle. For example, if

$$\left.\frac{d^2 f(\omega)}{d\omega^2}\right|_{\omega=\omega_c} > 0, \tag{6.187}$$

then one would set

$$\omega - \omega_c = \omega(w) = \exp\left(-i\frac{\pi}{4}\right)w = \left(\frac{1}{\sqrt{2}} - \frac{i}{\sqrt{2}}\right)w. \tag{6.188}$$

Extending the integration contour to the interval $w \in (-\infty, \infty)$, one obtains

$$J = \exp\left(-i\frac{\pi}{4}\right) \int_0^\infty \exp\left(-\frac{1}{2}f''(\omega_c)\,w^2\right) dw$$

$$\approx \exp\left(-i\frac{\pi}{4}\right) \int_{-\infty}^\infty dw \exp\left(-\frac{1}{2}f''(\omega_c)\,w^2\right) = e^{-i\pi/4}\sqrt{\frac{2\pi}{f''(\omega_c)}}, \qquad (6.189)$$

and thus

$$u_+(x,t) \approx u_0 \,\mathrm{Re}\, \exp\left[-i f(\omega_c)\right] e^{-i\pi/4}\sqrt{\frac{2\pi}{f''(\omega_c)}}, \qquad (6.190)$$

where the primary phase factor is

$$\exp\left[-i f(\omega_c)\right] = \exp\left[-i\omega_c\left(t - \frac{x}{v_g(\omega_c)}\right)\right]. \qquad (6.191)$$

One of the simplest possible applications concerns the plasma model without damping term,

$$n(\omega) \approx \left(1 - \frac{\omega_p^2}{\omega^2}\right)^{1/2}, \qquad \omega\,n(\omega) = \begin{cases} \sqrt{\omega^2 - \omega_p^2}, & (\omega > \omega_p) \\ i\sqrt{\omega_p^2 - \omega^2}, & (\omega_p > \omega) \end{cases}. \qquad (6.192)$$

With these observations, the function can be written as

$$u_+(x,t) = \frac{u_0}{2\pi}\,\mathrm{Re}\,\int_{\omega_p}^\infty dw \exp\left(-i\omega\left[t - \sqrt{\frac{\omega^2 - \omega_p^2}{\omega^2}}\,\frac{x}{c}\right]\right)$$

$$+ \frac{u_0}{2\pi}\,\mathrm{Re}\,\int_0^{\omega_p} dw \exp\left(-\omega\left[it + \sqrt{\frac{\omega^2 - \omega_p^2}{\omega^2}}\,\frac{x}{c}\right]\right). \qquad (6.193)$$

We thus estimate

$$u_+(x,t) \approx \frac{u_0}{2\pi}\,\mathrm{Re}\,\int_{\omega_p}^\infty \exp\left(-i\omega\left[t - \sqrt{\frac{\omega^2 - \omega_p^2}{\omega^2}}\,\frac{x}{c}\right]\right) dw = \frac{u_0}{2\pi}\,\mathrm{Re}\,\int_{\omega_p}^\infty e^{-i f(\omega)}\,dw$$

$$\qquad (6.194)$$

by the method of steepest descent. We have defined the complex phase of the integrand as

$$f(\omega) = \omega\left(t - \sqrt{\frac{\omega^2 - \omega_p^2}{\omega^2}}\,\frac{x}{c}\right). \qquad (6.195)$$

The critical value $\omega = \omega_c$ is determined as follows,

$$\left.\frac{\partial f}{\partial \omega}\right|_{\omega=\omega_c} = 0, \qquad \omega = \omega_c = \frac{\omega_p\,ct}{\sqrt{(ct)^2 - x^2}}, \qquad f(\omega_c) = \sqrt{(ct)^2 - x^2}\,\frac{\omega_p}{c}, \qquad (6.196)$$

$$f''(\omega_c) = \left.\frac{\partial^2 f}{\partial \omega^2}\right|_{\omega=\omega_c} = \frac{\left((ct)^2 - x^2\right)^{3/2}}{c\,x^2\,\omega_p}, \qquad f(\omega) \approx f(\omega_c) + \frac{1}{2}f''(\omega_c)(\omega - \omega_c)^2.$$

$$\qquad (6.197)$$

We have assumed $ct > x$ in writing these expressions, which implies, in particular, that $f''(\omega_c) > 0$. Hence, we may use Eqs. (6.188) and (6.189) with the result

$$J = \int_{\omega_p}^{\infty} d\omega \, e^{-i\frac{1}{2} f''(\omega_c)(\omega-\omega_c)^2} \approx e^{-i\pi/4} \int_{-\infty}^{\infty} d\omega \, e^{-\frac{1}{2} f''(\omega_c) \, \omega^2} = e^{-i\pi/4} \sqrt{\frac{2\pi}{f''(\omega_c)}} \,.$$

(6.198)

The final result for the signal is

$$u_+(x,t) \approx \frac{u_0}{2\pi} \operatorname{Re}\left\{ \exp\left[-i\left(f(\omega_c) + \frac{\pi}{4}\right)\right] \sqrt{\frac{2\pi}{f''(\omega_c)}} \right\} = \frac{u_0}{2\pi} \sqrt{\frac{2\pi}{f''(\omega_c)}} \cos\left(f(\omega_c) + \frac{\pi}{4}\right)$$

$$= \frac{u_0}{2\pi} \sqrt{2\pi} \left(\frac{\left((ct)^2 - x^2\right)^{3/2}}{c \, x^2 \, \omega_p}\right)^{-1/2} \cos\left(\sqrt{(ct)^2 - x^2} \, \frac{\omega_p}{c} + \frac{\pi}{4}\right)$$

$$= \frac{u_0}{\sqrt{2\pi}} \frac{\sqrt{c}\sqrt{\omega_p} \, x}{\left((ct)^2 - x^2\right)^{3/4}} \cos\left(\frac{\sqrt{(ct)^2 - x^2}}{c} \omega_p + \frac{\pi}{4}\right).$$

(6.199)

This result fulfills the condition that $u(x = 0, t) = u_0 \, \delta(t) = 0$ for $t > 0$, in that it vanishes at $x = 0$. In the "acausal" region $x > ct$, the value of the critical phase $f(\omega_c)$ becomes imaginary, and the contribution of the saddle point is exponentially suppressed. Hence, the result given in Eq. (6.199) for the "causal" region $ct > x$ is the complete result obtained by the steepest descent method.

6.6 Kramers–Kronig Relationships

6.6.1 *Analyticity and the Kramers–Kronig Relationships*

We recall once more the basic relations relevant for the dielectric constant and the electric field, summarized in Eqs. (6.100)–(6.102),

$$\epsilon_r(\omega) = 1 + G(\omega), \qquad \epsilon_r(t - t') = \delta(t - t') + G(t - t'),$$

(6.200a)

$$\vec{D}(\vec{r}, \omega) = \epsilon_0 \left[1 + G(\omega)\right] \vec{E}(\vec{r}, \omega),$$

(6.200b)

$$\vec{D}(\vec{r}, t) = \epsilon_0 \, \vec{E}(\vec{r}, t) + \epsilon_0 \int_0^{\infty} d\tau \, G(\tau) \, \vec{E}(\vec{r}, t - \tau).$$

(6.200c)

Here, $G(t - t')$ is a retarded Green function, which implies that it vanishes for $t - t' < 0$. Furthermore, it means that $G(\omega)$ cannot have poles in the upper half of the complex plane. We can thus formulate the Fourier transform as

$$\epsilon_r(\omega) - 1 = \int_0^{\infty} d\tau \, G(\tau) \, \exp(i\omega\tau).$$

(6.201)

Since $\vec{D}$ and $\vec{E}$ are real, $G(\tau)$ must be real, and we have

$$\epsilon_r(\omega) = \epsilon_r(-\omega)^*.$$

(6.202)

Because $\epsilon_r(\omega) - 1$ is an analytic function in the upper half complex plane, we have by Cauchy's residue theorem (see Sec. 3.3.5),

$$\epsilon_r(\omega) = 1 + \frac{1}{2\pi i} \oint_C d\omega' \frac{\epsilon_r(\omega') - 1}{\omega' - \omega}, \qquad \text{Im } \omega > 0, \tag{6.203}$$

with the imaginary part of all points on the curve C restricted to be greater than or equal to zero. The path of integration is closed at infinite radius, in the upper complex half plane, which ensures that the only pole of the integrand in Eq. (6.203), inside the contour of integration, occurs at $\omega' = \omega$, with Im $\omega > 0$. Because $\epsilon_r(\omega)$ can be interpreted as a retarded Green function, it does not contribute any further pole terms.

For initially real ω, we can enforce the position of the pole to be in the upper half plane, by adding a small imaginary part to ω, replacing ω in Eq. (6.203) by $\omega + i\delta$. Then,

$$\epsilon_r(\omega) = 1 + \frac{1}{2\pi i} \lim_{\delta \to 0} \int_{-\infty}^{\infty} d\omega' \frac{\epsilon_r(\omega') - 1}{\omega' - (\omega + i\delta)}, \qquad \text{Im } \omega = 0, \tag{6.204}$$

where we assume that both the real as well as the imaginary parts of $\epsilon_r(\omega)$ vanish sufficiently rapidly for $|\omega| \to \infty$ so that the contribution from the closing half-circle can be neglected. When drawing the integration contour, it is immediately obvious that the integral taken below the pole equals the principal part of the integral plus πi times the residue at $\omega' = \omega$. This is encompassed in the formula,

$$\frac{1}{\omega' - \omega - i\delta} = (\text{P.V.}) \frac{1}{\omega' - \omega} + i\pi \, \delta(\omega' - \omega). \tag{6.205}$$

Therefore,

$$\epsilon_r(\omega) = 1 + \frac{1}{2\pi i} \left((\text{P.V.}) \int_{-\infty}^{\infty} \frac{\epsilon_r(\omega') - 1}{\omega' - \omega} \, d\omega' + i\pi \left(\epsilon_r(\omega) - 1 \right) \right)$$

$$= 1 + \frac{1}{2\pi i} (\text{P.V.}) \int_{-\infty}^{\infty} \frac{\epsilon_r(\omega') - 1}{\omega' - \omega} \, d\omega' + \frac{1}{2} \left(\epsilon_r(\omega) - 1 \right)$$

$$= \frac{1}{2} \epsilon_r(\omega) + \frac{1}{2} + \frac{1}{2\pi i} (\text{P.V.}) \int_{-\infty}^{\infty} \frac{\epsilon_r(\omega') - 1}{\omega' - \omega}. \tag{6.206}$$

Taking the real and imaginary parts of this equation, one obtains

$$\text{Re } \epsilon_r(\omega) = 1 + \frac{1}{\pi} (\text{P.V.}) \int_{-\infty}^{\infty} \frac{\text{Im } \epsilon_r(\omega')}{\omega' - \omega} \, d\omega', \tag{6.207a}$$

$$\text{Im } \epsilon_r(\omega) = -\frac{1}{\pi} (\text{P.V.}) \int_{-\infty}^{\infty} \frac{\text{Re } \epsilon_r(\omega') - 1}{\omega' - \omega} \, d\omega'. \tag{6.207b}$$

Using $\operatorname{Re} \epsilon_r(\omega) = \operatorname{Re} \epsilon_r(-\omega)$ and $\operatorname{Im} \epsilon_r(\omega) = -\operatorname{Im} \epsilon_r(-\omega)$, this can be written as

$$\operatorname{Re} \epsilon_r(\omega) = 1 + \frac{1}{\pi} (\text{P.V.}) \left[\int_{-\infty}^{0} \frac{\operatorname{Im} \epsilon_r(\omega')}{\omega' - \omega} \, d\omega' + \int_{0}^{\infty} \frac{\operatorname{Im} \epsilon_r(\omega')}{\omega' - \omega} \, d\omega' \right]$$

$$= 1 + \frac{1}{\pi} (\text{P.V.}) \left[\int_{0}^{\infty} \frac{\operatorname{Im} \epsilon_r(-\omega')}{-\omega' - \omega} \, d\omega' + \int_{0}^{\infty} \frac{\operatorname{Im} \epsilon_r(\omega')}{\omega' - \omega} \, d\omega' \right]$$

$$= 1 + \frac{1}{\pi} (\text{P.V.}) \left[\int_{0}^{\infty} \frac{\operatorname{Im} \epsilon_r(\omega')}{\omega' + \omega} \, d\omega' + \int_{0}^{\infty} \frac{\operatorname{Im} \epsilon_r(\omega')}{\omega' - \omega} \, d\omega' \right]$$

$$= 1 + \frac{2}{\pi} (\text{P.V.}) \int_{0}^{\infty} \frac{\omega' \operatorname{Im} \epsilon_r(\omega')}{\omega'^2 - \omega^2} \, d\omega'. \tag{6.208a}$$

For the imaginary part, we have

$$\operatorname{Im} \epsilon_r(\omega) = -\frac{1}{\pi} (\text{P.V.}) \left[\int_{-\infty}^{0} \frac{\operatorname{Re} \epsilon_r(\omega') - 1}{\omega' - \omega} \, d\omega' + \int_{0}^{\infty} \frac{\operatorname{Re} \epsilon_r(\omega') - 1}{\omega' - \omega} \, d\omega' \right]$$

$$= -\frac{1}{\pi} (\text{P.V.}) \left[\int_{0}^{\infty} \frac{\operatorname{Re} \epsilon_r(-\omega') - 1}{-\omega' - \omega} \, d\omega' + \int_{0}^{\infty} \frac{\operatorname{Re} \epsilon_r(\omega') - 1}{\omega' - \omega} \, d\omega' \right]$$

$$= -\frac{1}{\pi} (\text{P.V.}) \left[-\int_{0}^{\infty} \frac{\operatorname{Re} \epsilon_r(\omega') - 1}{\omega' + \omega} \, d\omega' + \int_{0}^{\infty} \frac{\operatorname{Re} \epsilon_r(\omega') - 1}{\omega' - \omega} \, d\omega' \right]$$

$$= -\frac{2\omega}{\pi} (\text{P.V.}) \int_{0}^{\infty} \frac{\operatorname{Re} \epsilon_r(\omega') - 1}{\omega'^2 - \omega^2} \, d\omega'. \tag{6.208b}$$

Equations (6.208a) and (6.208b) are the classic Kramers–Kronig relations between the real and imaginary parts of the electric permittivity. Similar relations hold for the frequency response function for any causal system. They are very useful in quantum mechanics (particularly in scattering theory), circuit analysis in electrical engineering, etc.

6.6.2 *Applications of the Kramers–Kronig Relationships*

It would be impossible to give a complete overview of the many applications of the Kramers–Kronig relationships in physics. We need to restrict ourselves to a few examples. For example, it is possible to use known values of $\epsilon_r(\omega)$, in support of the calculation of unknown values. Suppose we have measured or calculated the imaginary part of the index of refraction. This may happen by a measurement of the absorption of the system at given angular frequency. Also, suppose that the real part of $\epsilon_r(\omega)$ is known, but only for a single value $\omega = \omega_1$. We are seeking the real part $\operatorname{Re} \epsilon_r(\omega)$ for given ω. In view of Eq. (6.208), $\epsilon_r(\omega_1)$ satisfies

$$\operatorname{Re} \epsilon_r(\omega_1) = 1 + \frac{2}{\pi} (\text{P.V.}) \int_{0}^{\infty} \frac{\omega' \operatorname{Im} \epsilon_r(\omega')}{\omega'^2 - \omega_1^2} \, d\omega'. \tag{6.209}$$

It follows that

$$\text{Re}\,\epsilon_r\,(\omega) - \text{Re}\,\epsilon_r\,(\omega_1) = \frac{2}{\pi}\,(\text{P.V.})\int_0^\infty \omega'\,\text{Im}\,\epsilon_r(\omega')\left(\frac{1}{\omega'^2 - \omega^2} - \frac{1}{\omega'^2 - \omega_1^2}\right)d\omega'$$

$$= \frac{2}{\pi}\left(\omega^2 - \omega_1^2\right)(\text{P.V.})\int_0^\infty \frac{\omega'\,\text{Im}\,\epsilon_r\,(\omega')}{(\omega'^2 - \omega^2)\,(\omega'^2 - \omega_1^2)}\,d\omega'.$$

$$(6.210)$$

This form for $\epsilon_r\,(\omega)$ reduces the integral's dependence on the values of $\text{Im}\,\epsilon_r\,(\omega')$ at large ω', because of the ω'^{-4} dependence for large ω'. This is known as a "subtracted dispersion relation". The process of subtraction can be repeated for each known value of $\epsilon_r(\omega)$.

Other applications concern so-called sum rules. We noted that, at high frequency, the expression $\epsilon_r(\omega) - 1$ behaves as $\omega^{-2} + \mathcal{O}\left(\omega^{-3}\right)$ [see Eq. (6.96)]. Physically, at high frequencies, the binding forces for the electrons become negligible and the system responds as a plasma. With this interpretation, we define the plasma frequency for the system as [see also Eq. (6.108)]

$$\omega_p^2 \equiv \lim_{\omega \to \infty}\left[\omega^2\,(1 - \epsilon_r\,(\omega))\right]. \qquad (6.211)$$

This definition of the plasma frequency includes the contributions of the responses of all charged particles in the system. In the sense of Eq. (6.95), the sources of the resonances include optically active phonons, plasmons, electronic excitations, electronic ionizations, and others.

For a functional form of the kind introduced in Eq. (6.95),

$$\epsilon_r\,(\omega) = 1 + \sum_m \frac{\mathcal{A}_m}{\omega_m^2 - \omega^2 - i\gamma_m\omega}, \qquad (6.212)$$

one has

$$\omega_p^2 = \sum_m \mathcal{A}_m \qquad (6.213)$$

by direct inspection of the limit. Using the Kramers–Kronig relation, it is possible to derive a more universal integral representation of ω_p^2 as defined in Eq. (6.211). Namely, if $\text{Im}\,\epsilon_r\,(\omega)$ varies as $\omega^{-3} + \mathcal{O}\left(\omega^{-4}\right)$ for high frequencies [see Eq. (6.97)], then Eq. (6.208) gives

$$\omega_p^2 = \lim_{\omega \to \infty}\left[\omega^2\,(1 - \text{Re}\,\epsilon_r\,(\omega)) - \omega^2 i\,\text{Im}\,\epsilon_r\,(\omega)\right] = \lim_{\omega \to \infty}\left[\omega^2\,(1 - \text{Re}\,\epsilon_r\,(\omega))\right]$$

$$= \lim_{\omega \to \infty}\left[-\omega^2\,\frac{2}{\pi}\,(\text{P.V.})\int_0^\infty d\omega'\frac{\omega'\,\text{Im}\,\epsilon_r\,(\omega')}{\omega'^2 - \omega^2}\right]$$

$$= \lim_{\omega \to \infty}\left[-\frac{2}{\pi}\,(\text{P.V.})\int_0^\infty d\omega'\,\underbrace{\frac{1}{[\omega'^2/\omega^2 - 1]}}_{\to -1}\,\omega'\,\text{Im}\,\epsilon_r\,(\omega')\right]$$

$$= \frac{2}{\pi}\int_0^\infty d\omega\,\omega\,\text{Im}\,\epsilon_r\,(\omega). \qquad (6.214)$$

In typical cases, the imaginary part itself has no singularities along the integration contour, and we can dispose of the principal value prescription, as done in the last step.

One can derive other sum rules. Let us assume, again, that the real part of the dielectric function goes as $\mathrm{Re}\,\epsilon_r\,(\omega) - 1 \sim -\omega_p^2/\omega^2 + \mathcal{O}\left(\omega^{-4}\right)$ for large ω whereas the imaginary part goes as $\mathrm{Im}\,\epsilon_r(\omega) \sim \mathcal{O}(\omega^{-3})$ for large ω [see Eq. (6.97)]. The asymptotic relationship between the real and imaginary parts of $\epsilon_r\,(\omega)$ can be used to take advantage of the asymptotic properties in the range of large ω, say, $\omega \gg \omega_L > \omega_p$. We write the Kramers–Kronig relation as

$$\mathrm{Im}\,\epsilon_r\,(\omega) = \frac{2}{\pi\omega}\left[(\mathrm{P.V.})\int_0^{\omega_L} d\omega'\,\frac{\mathrm{Re}\,\left[\epsilon_r\,(\omega') - 1\right]}{1 - (\omega'/\omega)^2} + \int_{\omega_L}^\infty d\omega'\,\frac{\mathrm{Re}\,\left[\epsilon_r\,(\omega') - 1\right]}{1 - (\omega'/\omega)^2}\right].$$

(6.215)

Again, we assume that ω_L is sufficiently large that the asymptotic form for the real part of $\epsilon_r\,(\omega)$, namely, $\mathrm{Re}\,\epsilon_r\,(\omega) \approx 1 - \omega_p^2/\omega^2$, provides a valid approximation. Hence,

$$\mathrm{Im}\,\epsilon_r\,(\omega) \approx \frac{2}{\pi\omega}(\mathrm{P.V.})\int_0^{\omega_L}\frac{\mathrm{Re}\,\left[\epsilon_r\,(\omega') - 1\right]}{1 - (\omega'/\omega)^2}d\omega'$$

$$+ \frac{2}{\pi\omega}(\mathrm{P.V.})\int_{\omega_L}^\infty\frac{1}{1 - (\omega'/\omega)^2}\left[-\frac{\omega_p^2}{\omega'^2} + \mathcal{O}\left(\omega'^{-4}\right)\right]d\omega'$$

$$\approx \frac{2}{\pi\omega}\left[(\mathrm{P.V.})\int_0^{\omega_L}\mathrm{Re}\,\left[\epsilon_r\,(\omega') - 1\right]d\omega' - \frac{\omega_p^2}{\omega_L} + \mathcal{O}\left(\omega^{-2}\right)\right]. \qquad (6.216)$$

One may now reason as follows. On the one hand, we have derived an approximation for the coefficient of order ω^{-1} of $\mathrm{Im}\,\epsilon_r\,(\omega)$, for large ω. On the other hand, we have by assumption $\mathrm{Im}\,\epsilon_r\,(\omega) \sim \mathcal{O}\left(\omega^{-3}\right)$ for large ω. It follows that the term in square brackets in the last line of Eq. (6.216) must vanish, asymptotically, as

$$(\mathrm{P.V.})\int_0^{\omega_L}\mathrm{Re}\,[\epsilon_r\,(\omega') - 1]\,d\omega' = \frac{\omega_p^2}{\omega_L} + \mathcal{O}\left(\omega^{-2}\right). \qquad (6.217)$$

For large ω_L, we must therefore have the property

$$\frac{1}{\omega_L}(\mathrm{P.V.})\int_0^{\omega_L}\mathrm{Re}\,\epsilon_r\,(\omega')\,d\omega' = 1 + \frac{\omega_p^2}{\omega_L^2} + \mathcal{O}(\omega_L^{-3}). \qquad (6.218)$$

This is sometimes called a superconvergence relation.

6.7   Exercises

• **Exercise 6.1**: Show that the charge density (6.20) and the current density (6.21) fulfill the continuity equation.

• **Exercise 6.2**: Generalize the derivation (6.28) to the conduction band electrons.

- **Exercise 6.3**: Consider a field configuration with a single Fourier mode,

$$\vec{E}(\vec{r},t) = \vec{E}(\vec{r}) \exp(-i\omega t), \qquad \vec{E}(\vec{r},\omega') = \vec{E}(\vec{r}) \, 2\pi \, \delta(\omega' - \omega), \qquad (6.219a)$$

$$\vec{B}(\vec{r},t) = \vec{B}(\vec{r}) \exp(-i\omega t), \qquad \vec{B}(\vec{r},\omega') = \vec{B}(\vec{r}) \, 2\pi \, \delta(\omega' - \omega). \qquad (6.219b)$$

Show that the dielectric displacement, in direct space (coordinates and time) is related to $\epsilon_r(\omega)$ as a proportionality factor in

$$\vec{D}(\vec{r},t) = \epsilon_0 \, \epsilon_r(\omega) \, \vec{E}(\vec{r}) \exp(-i\omega t), \qquad (6.220)$$

where ω is the carrier frequency. Show that for waves composed of a single Fourier mode, the wave equations (7.5a) and (7.5b) are obtained.

- **Exercise 6.4**: Derive Eq. (6.93) from the phenomenological Maxwell equations in coordinate-frequency space, given in Eq. (6.87).

- **Exercise 6.5**: Consider a plane wave traveling in free space with $\epsilon(\omega) = \epsilon_0$ and $\mu(\omega) = \mu_0$,

$$\vec{E}(\vec{r},t) = \vec{E}_0 \, e^{i\vec{k}\cdot r - i\omega t}, \qquad \vec{B}(\vec{r},t) = \vec{B}_0 \, e^{i\vec{k}\cdot r - i\omega t}, \qquad (6.221a)$$

$$\vec{E}(\vec{r}) = \vec{E}_0 \, e^{i\vec{k}\cdot r}, \qquad \vec{B}(\vec{r}) = \vec{B}_0 \, e^{i\vec{k}\cdot r}, \qquad \rho(\vec{r},t) = 0, \qquad \vec{J}(\vec{r},t) = \vec{0}. \qquad (6.221b)$$

Show that the Maxwell equations become

$$\vec{k} \cdot \vec{E}_0 = 0, \qquad\qquad \vec{k} \cdot \vec{B}_0 = 0, \qquad (6.222a)$$

$$i\vec{k} \times \vec{E}_0 - i\omega\vec{B}_0 = 0, \qquad i\vec{k} \times \vec{B}_0 + i\frac{\omega}{c^2}\vec{E}_0 = \vec{0}. \qquad (6.222b)$$

Assume that the magnetic field amplitude is given as

$$\vec{B}_0 = \frac{\vec{k}}{\omega} \times \vec{E}_0. \qquad (6.223)$$

Show that the Ampere-Maxwell law becomes

$$\vec{k} \times \vec{B}_0 = -\frac{\omega}{c^2} \, \vec{E}_0, \qquad (6.224)$$

where we have used the relation $\vec{k}^2 = \omega^2/c^2$. With reference to the considerations preceding Eq. (5.106), calculate the time-averaged Poynting vector as

$$\vec{S} = \frac{1}{2\mu_0} \vec{E}_0 \times \vec{B}_0^* = \frac{\vec{k}}{2\omega\mu_0} \left|\vec{E}_0\right|^2. \qquad (6.225)$$

- **Exercise 6.6**: Derive (6.112) based on a geometric consideration.
- **Exercise 6.7**: Derive (6.116) from Eq. (6.114), observing that you can obtain the polarization density as $N_V e x_c(\omega)$, and expressing the dielectric displacement appropriately.
- **Exercise 6.8**: Derive (6.143) based on a geometric consideration. This task is a little subtle, because one must carefully consider the limit $\Delta z \to 0$, i.e., the limit of small dipoles, which corresponds to the continuum limit of the dipole density. First, carefully consider what happens when you displace the separating plane from the mid-dipole position in Fig. 6.2.

• **Exercise 6.9**: Derive Eq. (6.154).

• **Exercise 6.10**: Use the steepest descent method for the Gamma function, to obtain Stirling's approximation to the factorial. Then, study the literature on the utility of the steepest descent method for the evaluation of Bessel functions.

• **Exercise 6.11**: Consider a model with a simple resonance,

$$\epsilon_r(\omega) = \frac{\alpha}{\omega_0 - \omega - i\frac{1}{2}\gamma} + \frac{\alpha}{\omega_0 + \omega + i\frac{1}{2}\gamma} \approx 1 + \frac{2\alpha\omega_0}{\omega_0^2 - \omega^2 - i\gamma\omega}. \tag{6.226}$$

Calculate the limit of the imaginary part as $\gamma \to 0$, express it in terms of Dirac-δ functions, and use the Kramers–Kronig relationships to reproduce the real part.

• **Exercise 6.12**: Practice the calculation of principal-value integrals. Remember that a principal value integration can be done in many ways, in particular, using the following alternative possibilities. *(i)* One does the integral analytically, then uses the upper and lower boundaries of the integration interval, after "throwing away" the imaginary part. *(ii)* If x is the integration variable and $x = a$ is the location of the singularity, then one integrates up to $x = a - \epsilon$, and starts again from $x = a + \epsilon$, observing that the the divergent term in $1/\epsilon$ cancels. *(iii)* One uses a complex contour above and immediately below the singularities on the real axis, and takes the mean. The different sense of revolving around the singularity then means that its pole contribution cancels.

Employing all three prescriptions, show that

$$(\text{P.V.}) \int_0^3 dx \, \frac{1}{x - 1} = \ln(2), \tag{6.227a}$$

$$(\text{P.V.}) \int_0^3 dx \, \frac{1}{x^2 - 1} = \frac{1}{2} \ln(2), \tag{6.227b}$$

$$(\text{P.V.}) \int_{-\infty}^{\infty} dx \, \frac{1}{x^2 - 1} = 0. \tag{6.227c}$$

Using your knowledge, show that the Drude model (6.99) satisfies the Kramers–Kronig relationships (6.208a) and (6.208b).

• **Exercise 6.13**: Use the result derived in Eq. (6.214),

$$\omega_p^2 = \frac{2}{\pi} \int_0^{\infty} d\omega \, \omega \, \text{Im} \, \epsilon_r(\omega), \tag{6.228}$$

and consider the imaginary part (6.97),

$$\text{Im} \, \epsilon_r(\omega) = \sum_m \frac{\omega \, \gamma_m \, \mathcal{A}_m}{(\omega^2 - \omega_m^2)^2 + \omega^2 \, \gamma_m^2}. \tag{6.229}$$

By contour integration, show that

$$\omega_p^2 = \frac{2}{\pi} \sum_m \int_0^{\infty} d\omega \, \frac{\omega^2 \, \gamma_m \, \mathcal{A}_m}{(\omega_m^2 - \omega^2)^2 + (\gamma_m \omega)^2} = \sum_m \mathcal{A}_m. \tag{6.230}$$

Finally, simplify the result for a dilute gas of atoms, expressing the result with the help of the sum rule (6.135) for the oscillator strengths.

• **Exercise 6.14**: Derive a relation similar to Eq. (6.210) for the imaginary part of the relative permittivity.

• **Exercise 6.15**: Explain why is it a good idea to model the relative permittivity of a solid (dense material) according to the *ansatz* [see Eq. (6.95)]

$$\frac{\epsilon_r(\omega) - 1}{\epsilon_r(\omega) + 2} = \sum_m \frac{\mathcal{A}_m}{\omega_m^2 - \omega^2 - i\gamma_m\omega} \,. \tag{6.231}$$

Study the application of this *ansatz* to α-quartz, according to Ref. [27], and try to find other suitable approximations for different materials, according to data published in Ref. [29].

Chapter 7

Waveguides and Cavities

7.1 Overview

When we are talking about a laser diode, we often say that it is "excited in the $TE_{1,0}$ mode", which is the laser mode with the lowest possible frequency of the laser cavity—hence we can be sure that no other laser modes are excited, and the laser light is monochromatic. An illustration of the precise meaning of this statement is the subject of the current chapter. We shall try to understand why the spectrum of electromagnetic waves in cavities and waveguides is discrete and, in typical cases, classified according to transverse electric (TE) and transverse magnetic (TM) modes. When we say "discrete" we mean that we can assign "reference numbers" to the waves, counting the number of nodes of the electric or magnetic fields in different directions of space.

Along the same lines, when one analyzes the spectrum of atoms, say, the hydrogen spectrum, one may classify the "matter waves" of the bound electron according to the number of nodes in the (radial) wave functions [17, 18]. Quantum mechanics entails the assumption that the trajectories of point particles are in fact smeared out and constitute matter waves. The transition from classical to wave mechanics is analogous to the transition from ray to wave optics (were it not for the additional postulate regarding the collapse of the wave function upon measurement, see Refs. [17, 18]). Yet, the analysis of electromagnetic waves in cavities and waveguides teaches us that the spectrum of waves can be discrete, and is a good preparatory exercise for quantum mechanics.

A few more explanatory words are in order; these may not be fully discernible for the novice but are still included here for better reference. You may have heard that the electromagnetic field itself needs to be quantized at some point, leading to the theory of quantum electrodynamics [20]. This latter quantization works differently as compared to the quantum matter waves; namely, one assumes that the energy stored in the wave-like excitations is discrete and comes in "packages" of size $\hbar\omega$, where $\hbar$ is the reduced quantum unit of action (Planck's constant divided by 2π), and ω is the angular frequency. This amounts to a quantization of the electric and magnetic fields themselves; the fields "wiggle" at every given point in space-time.

The wave character of the electromagnetic excitations is taken for granted in this context; in fact, if we could determine the value of the electric and magnetic fields simultaneously with absolute precision at any given space-time point, then we would not need to quantize the field. Even the classical electromagnetic theory allows for the presence of waves. The quantization of the electromagnetic field therefore is called "second quantization"; quantum electrodynamics is a "second-quantized" theory.

From this point of view, we should add that, what we do in the current chapter is assume that the fields in the cavities and waveguides are sufficiently strong so that we do not need to quantize them; the "wiggling" is a small effect and negligibly affects strong macroscopic fields. We thus treat the problems on the level of "first quantization" which in the case of electrodynamics, is just the classical wave theory of light.

Before we set forth in this endeavor, let us indicate a few more paradigmatic applications of waveguides and cavities, to illustrate their practical importance. Indeed, in the generation of radar waves, the so-called klystron has been instrumental. So-called X band radar has a typical frequency of 7175 MHz, corresponding to a wavelength of roughly 4.2 cm; it has been used in Doppler tracking of the Cassini spacecraft in superior conjunction on its way to Saturn [30]. Cavities also find applications in accelerators; the basic idea here is that the electric field in the fundamental mode in a cylindrical cavity has oscillations in the direction parallel to the symmetry axis of the cavity; this direction in turn can be aligned with the beam direction [31].

Last, but certainly not least, the study of the field modes inside a cavity is a necessary preparatory exercise for a quantum electrodynamic (yes!) calculation of the attraction between perfectly conducting plates (see Sec. 8.4). Two large parallel plates a small distance apart form an approximate "cavity" whose modes are distinct from what would be expected in free space. In the fully quantized theory, one associates a so-called "zero-point" energy with every possible excitation of the electromagnetic field. The difference of the so-called "zero-point" energy of the modes inside the cavity, compared to the modes in free space, results in a distance-dependent energy shift. This shift leads to a force between the plates known as the Casimir effect, as discussed in more detail in Sec. 8.4.

7.2 Waveguides

7.2.1 *General Formalism*

Let a waveguide be oriented along the z axis, with the (hollow) structure being cylindrical or rectangular. For a wave propagating along the positive z direction, we shall assume that

$$\vec{E}(\vec{r}) = \vec{E}(x,y)\, e^{ikz}, \qquad \vec{B}(\vec{r}) = \vec{B}(x,y)\, e^{ikz}. \tag{7.1}$$

Let us see if we can express the E_x and E_y components as functions of E_z only (for the electric field), and likewise for the magnetic field. There are no sources inside the conductor, but we shall assume the material inside the waveguide is isotropic with electric permittivity

$$\epsilon(\omega) = \epsilon_r(\omega)\epsilon_0 \equiv \epsilon_r\,\epsilon_0\,, \tag{7.2}$$

and magnetic permeability,

$$\mu(\omega) = \mu_r(\omega)\,\mu_0 \equiv \mu_r\,\mu_0\,. \tag{7.3}$$

Without boundary conditions, the speed of the propagating wave in the medium is $c/\sqrt{\epsilon_r\mu_r}$. We work in mixed position-frequency space, with fields $\vec{E} = \vec{E}(\vec{r},\omega)$ and $\vec{B} = \vec{B}(\vec{r},\omega)$. The Maxwell equations yield, with $\epsilon = \epsilon_r\,\epsilon_0$ and $\mu = \mu_r\mu_0$, in a source-free region,

$$\vec{\nabla} \times \vec{E}(\vec{r},t) = -\frac{\partial}{\partial t}\vec{B}(\vec{r},t) \Rightarrow \vec{\nabla} \times \vec{E}(\vec{r},\omega) = i\omega\vec{B}(\vec{r},\omega)\,, \tag{7.4a}$$

$$\vec{\nabla} \times \vec{H}(\vec{r},t) = \frac{\partial}{\partial t}\vec{D}(\vec{r},t) \Rightarrow \vec{\nabla} \times \vec{B}(\vec{r},\omega) = -i\frac{\omega}{c^2}\epsilon_r\mu_r\vec{E}(\vec{r},\omega)\,. \tag{7.4b}$$

Inserting the curl of the first equation into the second, or vice versa, one obtains the following wave equations,

$$\left(\vec{\nabla}^2 + \epsilon_r\,\mu_r\,\frac{\omega^2}{c^2}\right)\vec{B}(\vec{r},\omega) = \vec{0}\,, \tag{7.5a}$$

$$\left(\vec{\nabla}^2 + \epsilon_r\,\mu_r\,\frac{\omega^2}{c^2}\right)\vec{E}(\vec{r},\omega) = \vec{0}\,. \tag{7.5b}$$

From now on, we restrict the discussion to a single Fourier mode,

$$\vec{E}(\vec{r},t) = \vec{E}(\vec{r})\exp(-i\omega t)\,, \qquad \vec{E}(\vec{r},\omega') = \vec{E}(\vec{r})\,2\pi\,\delta(\omega'-\omega)\,, \tag{7.6a}$$

$$\vec{B}(\vec{r},t) = \vec{B}(\vec{r})\exp(-i\omega t)\,, \qquad \vec{B}(\vec{r},\omega') = \vec{B}(\vec{r})\,2\pi\,\delta(\omega'-\omega)\,. \tag{7.6b}$$

With Eq. (7.1), the wave equations become

$$\left(\vec{\nabla}_\parallel^2 + \epsilon_r\,\mu_r\,\frac{\omega^2}{c^2} - k^2\right)\vec{E}(x,y) = \vec{0}\,, \tag{7.7a}$$

$$\left(\vec{\nabla}_\parallel^2 + \epsilon_r\,\mu_r\,\frac{\omega^2}{c^2} - k^2\right)\vec{B}(x,y) = \vec{0}\,, \tag{7.7b}$$

where

$$\vec{\nabla}_\parallel^2 = \frac{\partial^2}{\partial x^2} + \frac{\partial^2}{\partial y^2}\,, \qquad \vec{\nabla}_\parallel = \hat{e}_x\frac{\partial}{\partial x} + \hat{e}_y\frac{\partial}{\partial y}\,. \tag{7.8}$$

We shall denote the "in-plane" component of a vector (i.e., the component parallel to, or "inside", the xy plane) by the subscript $\parallel$. We can decompose the fields into z components, and in-plane components,

$$\vec{E}_\parallel = \vec{E} - \hat{e}_z\,E_z = (\hat{e}_z \times \vec{E}) \times \hat{e}_z = -\hat{e}_z \times (\hat{e}_z \times \vec{E}_\parallel)\,, \tag{7.9a}$$

$$\vec{B}_\parallel = \vec{B} - \hat{e}_z\,B_z = (\hat{e}_z \times B) \times \hat{e}_z = -\hat{e}_z \times (\hat{e}_z \times \vec{B}_\parallel)\,. \tag{7.9b}$$

The $\hat{e}_z \times (\hat{e}_z \times \ldots)$-operation thus amounts to a simple minus sign for in-plane vectors. Let us try to take this a little further and decompose Faraday's law into parallel and transverse components. We start from the relation

$$\vec{\nabla} \times \vec{E} = \left(\hat{e}_z \nabla_z + \vec{\nabla}_{\|} \right) \times \left(\hat{e}_z \, E_z + \vec{E}_{\|} \right) = i \omega \, \vec{B}(\vec{r}) \,. \tag{7.10}$$

Then, in view of $\hat{e}_z \times \hat{e}_z = \vec{0}$, we have

$$\vec{\nabla} \times \vec{E} = \underbrace{\hat{e}_z \nabla_z \times \vec{E}_{\|} - \hat{e}_z \times \vec{\nabla}_{\|} E_z}_{\text{in-plane}} + \underbrace{\vec{\nabla}_{\|} \times \vec{E}_{\|}}_{z\text{-oriented}} = \underbrace{i \omega \, \vec{B}_{\|}}_{\text{in-plane}} + \underbrace{i \omega \, \hat{e}_z \, B_z}_{z\text{-oriented}} \,, \tag{7.11}$$

$$\vec{\nabla} \times \vec{B} = \underbrace{\hat{e}_z \, \nabla_z \times \vec{B}_{\|} - \hat{e}_z \times \vec{\nabla}_{\|} B_z}_{\text{in-plane}} + \underbrace{\vec{\nabla}_{\|} \times \vec{B}_{\|}}_{z\text{-oriented}} = \underbrace{-i \frac{\omega}{c^2} \epsilon_r \, \mu_r \, \vec{E}_{\|}}_{\text{in-plane}} - \underbrace{i \frac{\omega}{c^2} \epsilon_r \, \mu_r \, \hat{e}_z \, E_z}_{z\text{-oriented}} \,. \tag{7.12}$$

The in-plane component of Eq. (7.11) is

$$\hat{e}_z \times \nabla_z \, \vec{E}_{\|} - \hat{e}_z \times \vec{\nabla}_{\|} E_z = i \, \omega \, \vec{B}_{\|} \,. \tag{7.13}$$

We now apply the $(\hat{e}_z \times \ldots)$-operation to both sides and in view of Eq. (7.9), we obtain

$$\nabla_z \vec{E}_{\|} + i \omega \left(\hat{e}_z \times \vec{B}_{\|} \right) = \vec{\nabla}_{\|} E_z \,. \tag{7.14}$$

Finally, in view of Eq. (7.1), we may replace $\nabla_z \to i k$ and write

$$i k \vec{E}_{\|} = \vec{\nabla}_{\|} \, E_z - i \omega (\hat{e}_z \times \vec{B}_{\|}) \,. \tag{7.15}$$

Analogously, the in-plane component of Eq. (7.12) reads

$$\hat{e}_z \times \nabla_z \vec{B}_{\|} - \hat{e}_z \times \vec{\nabla}_{\|} B_z = -i \frac{\omega}{c^2} \, \epsilon_r \, \mu_r \, \vec{E}_{\|} \,. \tag{7.16}$$

Replacing ∇_z by $i k$, we have

$$\hat{e}_z \times \vec{B}_{\|} = \frac{1}{i k} (\hat{e}_z \times \vec{\nabla}_{\|} B_z) - \frac{\omega \, \epsilon_r \, \mu_r}{c^2 \, k} \, \vec{E}_{\|} \,. \tag{7.17}$$

Applying the $(\hat{e}_z \times \ldots)$-operation to both sides of Eq. (7.16),

$$\nabla_z \vec{B}_{\|} - i \frac{\omega}{c^2} \, \epsilon_r \, \mu_r \left(\hat{e}_z \times \vec{E}_{\|} \right) = \vec{\nabla}_{\|} B_z \,. \tag{7.18}$$

Using Eq. (7.1) and replacing $\nabla_z \to i k$, one obtains

$$i k \, \vec{B}_{\|} - i \frac{\omega}{c^2} \, \epsilon_r \, \mu_r \left(\hat{e}_z \times \vec{E}_{\|} \right) = \vec{\nabla}_{\|} B_z \,. \tag{7.19}$$

Also, since the divergence of both electric and magnetic fields vanishes,

$$\vec{\nabla}_{\|} \cdot \vec{E}_{\|} + \nabla_z \, E_z = 0 \,, \qquad \vec{\nabla}_{\|} \cdot \vec{B}_{\|} + \nabla_z \, B_z = 0 \,. \tag{7.20}$$

Finally, we solve for $\vec{E}_{\|}$ and $\vec{B}_{\|}$ if E_z and B_z are known (and not both are zero). Combining Eqs. (7.15) and (7.17), we may eliminate $\vec{B}_{\|}$ and write

$$i k \vec{E}_{\|} = \vec{\nabla}_{\|} E_z - \frac{\omega}{k} \left(\hat{e}_z \times \vec{\nabla}_{\|} B_z \right) + i \frac{\omega^2 \, \epsilon_r \, \mu_r}{c^2 k} \, \vec{E}_{\|} \,. \tag{7.21}$$

From the last equation, multiplying by ik, we find that

$$\left(\frac{\omega^2}{c^2}\epsilon_r\mu_r - k^2\right)\vec{E}_\parallel = ik\,\vec{\nabla}_\parallel E_z - i\omega\,(\hat{e}_z \times \vec{\nabla}_\parallel B_z)\,. \tag{7.22}$$

We can finally solve for $\vec{E}_\parallel$,

$$\vec{E}_\parallel = i\,\frac{1}{\omega^2\,\epsilon_r\,\mu_r/c^2 - k^2}\,\left(k\,\vec{\nabla}_\parallel E_z - \omega\,(\hat{e}_z \times \vec{\nabla}_\parallel)B_z\right)\,. \tag{7.23}$$

In Eq. (7.19), we can replace $\vec{E}_\parallel$ with the result just obtained,

$$\vec{B}_\parallel = i\,\frac{1}{\omega^2\,\epsilon_r\,\mu_r/c^2 - k^2}\,\left(k\,\vec{\nabla}_\parallel B_z + \frac{\omega\,\epsilon_r\,\mu_r}{c^2}\,(\hat{e}_z \times \vec{\nabla}_\parallel)E_z\right)\,. \tag{7.24}$$

We have discussed waves which manifestly travel in the positive z direction. For waves traveling in the opposite direction, one just changes k to $-k$. We should also point out that in vacuum, we have $\epsilon_r \to 1$ and $\mu_r \to 1$, so that the denominator $\omega^2\,\epsilon_r\,\mu_r/c^2 - k^2$ vanishes. However, in this case, the z components of the electric and magnetic fields also vanish; we have assumed that the wave propagates in the z direction, and the magnetic and electric fields are strictly transverse in vacuum. Indeed, we shall soon realize that the dispersion relation for electromagnetic waves in a waveguide is different from the dispersion relation in vacuum, i.e., $\omega \neq c|\vec{k}|$, even if $\epsilon_r = \mu_r = 1$.

Equations (7.23) and (7.24) are valid for the general propagation of waves in a medium with permittivity ϵ_r and permeability μ_r, not necessarily for a waveguide. The waveguide aspect comes in through the boundary conditions, which will be discussed next.

7.2.2 *Boundary Conditions at the Surface*

Let us investigate special boundary conditions for the case of a perfect electric conductor; the walls of a waveguide are typically idealized as conducting walls. Inside the conductor, the electric field is zero, because any conceivably entering field is immediately compensated by a rearrangement of conducting electrons. Hence, we have the boundary condition

$$\hat{n} \times \vec{E}\big|_S = 0\,, \qquad \text{or} \qquad E_{\parallel,S} = 0\,, \tag{7.25}$$

where $\hat{n}$ is the surface normal and the subscript S denotes the evaluation on a point (x,y,z) that satisfies the equation $F(x,y,z) = 0$ defining the surface of the conductor (see Sec. 3.2.7). We use the notation $E_{\parallel,S}$ in order to denote the component of $\vec{E}$ parallel and inside the surface.

There is a further boundary condition, which concerns the magnetic field. First, we observe that in principle, there is a no obstacle against a static magnetic field entering a perfect conductor. However, let us suppose that a time varying, oscillating field enters the conductor. Then, in view of Faraday's law,

$$\vec{\nabla} \times \vec{E} + \frac{\partial}{\partial t}\vec{B} = 0\,, \tag{7.26}$$

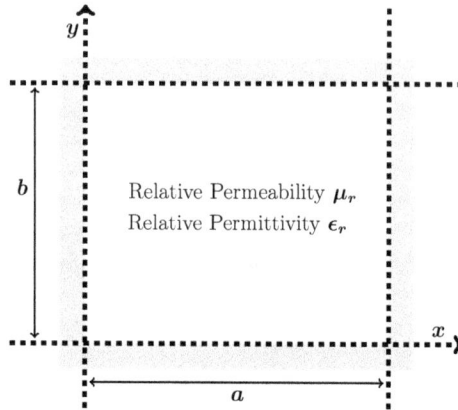

Fig. 7.1   Picture of the rectangular waveguide.

one would generate a time-varying electric field in the conductor, which in turn cannot exist. Specifically, let us suppose that $\hat{n}$ is the surface normal of the conductor. The magnetic field is an oscillating field by assumption. If the projection of the magnetic field on the surface normal $\hat{n} \cdot \vec{B}\big|_S$ were nonzero, we would generate a time-varying electric field inside the plane defining the surface. In order to see this more clearly, let us assume that the surface normal is the z axis. If $\partial_t \vec{B} \| \vec{B} \propto \hat{e}_z \, B_z$, then $\vec{\nabla}_\| \times \vec{E}_\|$ also is directed along the z axis; the latter quantity, however, has to vanish. Hence, we need another boundary condition,

$$\hat{n} \cdot \vec{B}\big|_S = 0, \qquad \text{or} \qquad B_{\perp, S} = 0. \tag{7.27}$$

However, a nonvanishing and oscillating component of the magnetic field, in the boundary plane, may exist in the vicinity of the conductor. By Faraday's law, this oscillating magnetic field gives rise to an oscillating electric field, perpendicular to the surface, and this is not in contradiction to the boundary condition for the electric field.

If we assume the waveguide to be aligned along the z axis, then for a rectangular waveguide (see Fig. 7.1), the boundaries are the xy, xz and yz planes. There are two kinds of propagating electromagnetic modes for which the boundary conditions (7.25) and (7.27) are fulfilled. These are called the transverse magnetic (TM) and transverse electric (TE) modes and we shall discuss them in detail. The basis of our considerations lies in Eqs. (7.23) and (7.24), which imply that any fields inside the waveguide can be decomposed into those generated for a vanishing field $E_z = 0$ (in which case $\vec{E} = \vec{E}_\|$, and the electric field is transverse), and those generated for a vanishing field $B_z = 0$ (in which case $\vec{B} = \vec{B}_\|$, and the magnetic field is transverse).

We shall discuss boundary conditions for both TE and TM modes in the following. The first and most straightforward case concerns the TM mode,

$$[\text{TM defining property:}] \qquad B_z = 0 \qquad \text{everywhere}, \qquad \vec{B} = \vec{B}_\| . \tag{7.28}$$

Here, "transverse" is to be understood as perpendicular with respect to the beam direction, i.e., perpendicular to the z axis. For TM modes, we also have to require that according to Eq. (7.25), the component of the electric field tangent to the interface (E_z) must vanish at the surface, i.e., $E_{z,S} = 0$, giving us

$$[\text{TM conditions:}] \qquad B_z = 0 \quad \text{everywhere}, \qquad E_z\big|_S = 0. \qquad (7.29)$$

Furthermore, for TM modes, we have according to Eq. (7.24),

$$\vec{B}_\parallel \propto \hat{e}_z \times \vec{\nabla}_\parallel E_z. \qquad (7.30)$$

At the boundary, since $E_{z,S} = 0$ by assumption, the vector $\vec{\nabla}_\parallel E_z$ (just like the entire gradient $\vec{\nabla} E_z$) must be normal to the surface. Forming the vector product of $\vec{\nabla}_\parallel E_z$ with $\hat{e}_z$ at the surface, we thus get a vector lying inside the surface, i.e., we have shown that $\vec{B}_\parallel$ must necessarily be parallel to the surface in its immediate vicinity. We shall verify this observation below, for the explicit formulas we shall derive for the TM modes. For now, in view of Eqs. (7.23), (7.24) and (7.29), we summarize the relations for the transverse magnetic (TM) modes,

$$[\text{TM wave:}] \qquad B_z = 0 \quad \text{everywhere and} \quad E_z\big|_S = 0, \qquad (7.31a)$$

$$\vec{E}_\parallel = \mathrm{i}\, \frac{1}{\epsilon_r\, \mu_r\, \omega^2/c^2 - k^2}\, k\, \vec{\nabla}_\parallel E_z, \qquad (7.31b)$$

$$\vec{B}_\parallel = \mathrm{i}\, \frac{1}{\epsilon_r\, \mu_r\, \omega^2/c^2 - k^2}\, \omega\, \epsilon_r\, \mu_r \left(\hat{e}_z \times \vec{\nabla}_\parallel E_z \right), \qquad (7.31c)$$

$$0 = \left(\vec{\nabla}_\parallel^2 + \epsilon_r\, \mu_r\, \frac{\omega^2}{c^2} - k^2 \right) E_z(x,y). \qquad (7.31d)$$

For TE modes, we have different relations,

$$[\text{TE defining property:}] \qquad E_z = 0 \quad \text{everywhere}, \qquad \vec{E} = \vec{E}_\parallel, \qquad (7.32)$$

and the boundary condition (7.25) for the electric field is automatically fulfilled. In view of Eq. (7.24), we must have $\vec{B}_\parallel \propto \vec{\nabla}_\parallel B_z$. Thus, the normal component $\hat{n} \cdot \vec{B}_\parallel = \hat{n} \cdot \vec{B}$ can alternatively be obtained by projecting $\vec{\nabla}_\parallel B_z$ onto the surface normal. The corresponding condition on B_z is

$$0 = \hat{n} \cdot \vec{B}\big|_S \propto \hat{n} \cdot \vec{B}_\parallel\big|_S \propto \left(\hat{n} \cdot \vec{\nabla}_\parallel \right) B_z\big|_S = \left(\hat{n} \cdot \vec{\nabla} \right) B_z\big|_S \equiv \frac{\partial B_z}{\partial n}\bigg|_S. \qquad (7.33)$$

The latter expression merely is a definition motivated by the relation

$$\frac{\partial B_z(\vec{r} + \xi\, \hat{n})}{\partial \xi}\bigg|_{\xi=0} = \hat{n} \cdot \vec{\nabla} B_z = \hat{n} \cdot \vec{\nabla}_\parallel B_z. \qquad (7.34)$$

We summarize that for a TM mode,

$$[\text{TE conditions:}] \qquad E_z = 0 \quad \text{everywhere}, \qquad \frac{\partial B_z}{\partial n}\bigg|_S = 0. \qquad (7.35)$$

In view of Eqs. (7.23), (7.24) and (7.35), we find for the transverse electric (TE) modes,

$$[\text{TE wave:}] \qquad E_z = 0 \quad \text{everywhere and} \quad \left.\frac{\partial B_z}{\partial n}\right|_S = 0, \tag{7.36a}$$

$$\vec{B}_\parallel = i\frac{1}{\epsilon_r\,\mu_r\,\omega^2/c^2 - k^2}\,k\,\vec{\nabla}_\parallel B_z, \tag{7.36b}$$

$$\vec{E}_\parallel = i\frac{1}{\epsilon_r\,\mu_r\,\omega^2/c^2 - k^2}\left[-\omega\left(\hat{e}_z \times \vec{\nabla}_\parallel B_z\right)\right], \tag{7.36c}$$

$$0 = \left(\vec{\nabla}_\parallel^2 + \epsilon_r\mu_r\,\frac{\omega^2}{c^2} - k^2\right)B_z(x,y). \tag{7.36d}$$

The boundary conditions give rise to eigenvalues of k (dependent on ω) for which the propagation is allowed. These eigenvalues may be continuous. Since the boundary conditions for E_z and B_z are different, the eigenvalues are also different for TE as opposed to TM modes. The allowed TE and TM waves (and the TEM wave, if it exists) provide a complete set of waves from which one can construct an arbitrary electromagnetic disturbance in the waveguide or cavity.

7.2.3 *Modes in a Rectangular Waveguide*

We aim to determine the TE modes in a rectangular waveguide with edge length a in the x direction and b in the y direction (with $a > b$), so that $0 < x < a$ and $0 < y < b$ (see Fig. 7.1). So, $\vec{E}_\parallel$ will be found from B_z alone, according to the relation (7.36c),

$$\begin{aligned}
\vec{E}_\parallel &= i\,\frac{1}{\epsilon_r\,\mu_r\,\omega^2/c^2 - k^2}\left[-\omega(\hat{e}_z \times \vec{\nabla}_\parallel)B_z\right] \\
&= -\frac{\omega}{k}\,\hat{e}_z \times \left(i\,\frac{k}{\epsilon_r\,\mu_r\,\omega^2/c^2 - k^2}\,\vec{\nabla}_\parallel B_z\right) = -\frac{\omega}{k}\,\hat{e}_z \times \vec{B}_\parallel.
\end{aligned} \tag{7.37}$$

This, in particular, means that $\vec{E}_\parallel$ and $\vec{B}_\parallel$ are still perpendicular to each other. The general solution of the wave equation (7.7b) for B_z is

$$B_z(x,y) = C_1\,e^{+i\vec{k}\cdot\vec{r}_\parallel} + C_2\,e^{-i\vec{k}\cdot\vec{r}_\parallel}, \qquad \vec{r}_\parallel = \hat{e}_x\,x + \hat{e}_y\,y. \tag{7.38}$$

The form for $B_z(x,y)$ which is non-zero when $x = y = 0$ is

$$B_z(x,y) = B_0\,\cos(k_x\,x)\,\cos(k_y\,y). \tag{7.39}$$

We recall the boundary condition (7.35), namely, $\left(\hat{n}\cdot\vec{\nabla}_\parallel\right)B_z|_S = 0$, for transverse magnetic (TM) modes in component form,

$$\left.\frac{\partial}{\partial x}B_0\,\cos(k_x\,x)\,\cos(k_y\,y)\right|_{x=0,a} = 0, \tag{7.40a}$$

$$\left.\frac{\partial}{\partial y}B_0\,\cos(k_x\,x)\,\cos(k_y\,y)\right|_{y=0,b} = 0. \tag{7.40b}$$

Any presence of a sine function in these two equations would lead to a nonvanishing derivative unless B_0 vanishes and thus the whole term. This justifies, a posteriori, our *ansatz* with a cosine. The boundary condition $(\hat{n} \cdot \vec{\nabla}_\parallel) B_z|_S = 0$ leads to the following condition on the components of the wave vector,

$$\sin(k_x\, a) = \sin(k_x\, 0) = 0\,, \qquad k_x = \frac{m\pi}{a}\,, \qquad m = 0, 1, 2, \ldots\,, \qquad (7.41a)$$

$$\sin(k_y\, b) = \sin(k_y\, 0) = 0\,, \qquad k_y = \frac{n\pi}{b}\,, \qquad n = 0, 1, 2, \ldots\,. \qquad (7.41b)$$

This means that for given m and n, a dispersion relation can be established between ω and k,

$$B_z(x, y) = B_{0,mn}\, \cos\left(\frac{m\pi x}{a}\right) \cos\left(\frac{n\pi y}{b}\right)\,, \qquad (7.42)$$

$$\left(\frac{m\pi}{a}\right)^2 + \left(\frac{n\pi}{b}\right)^2 = \epsilon_r\, \mu_r\, \frac{\omega^2}{c^2} - k^2\,, \qquad m, n = 0, 1, 2, 3, \ldots\,, \qquad (7.43)$$

$$k = \sqrt{\epsilon_r\, \mu_r\, \frac{\omega^2}{c^2} - \left(\frac{m\pi}{a}\right)^2 - \left(\frac{n\pi}{b}\right)^2}\,. \qquad (7.44)$$

Propagation in the (m, n) mode cannot proceed unless the solution for k is real,

$$\omega > \omega_{mn} \equiv \frac{c}{\sqrt{\epsilon_r \mu_r}} \left[\left(\frac{m\pi}{a}\right)^2 + \left(\frac{n\pi}{b}\right)^2\right]^{1/2} = \frac{\pi c}{\sqrt{\epsilon_r \mu_r}} \left[\left(\frac{m}{a}\right)^2 + \left(\frac{n}{b}\right)^2\right]^{1/2}\,. \qquad (7.45)$$

For $\omega < \omega_{\min}$, the wave number k becomes purely imaginary, $k = i|k|$, and the wave, propagating in the z direction, being proportional to $\exp(-|k|\, z)$, becomes evanescent. Moreover, ω_{mn} is the m and n-dependent waveguide angular frequency. The minimum frequency is obtained for

$$\omega_{\min} = \omega_{\min;\mathrm{TE}} = \frac{\pi}{\sqrt{\epsilon_r \mu_r}} \frac{c}{a}\,, \qquad m = 1, \qquad n = 0, \qquad a > b\,. \qquad (7.46)$$

In view of the occurrence of the factor c/a, the minimum frequency for propagation in the waveguide is commensurate with the inverse time it takes light to travel the spatial dimension of the waveguide. For a non-trivial solution, m and n cannot both be zero. When the condition (7.45) is fulfilled, real rather than imaginary solutions can be found for the wave vector.

Let us summarize the field configuration for a TE_{mn} mode, and start with the z component,

$$B_z(x, y) = B_{z,mn}(x, y) = B_{0,mn}\, \cos\left(\frac{m\pi x}{a}\right) \cos\left(\frac{n\pi y}{b}\right)\,, \qquad E_z(x, y) = 0\,, \quad (7.47a)$$

The in-plane component of the $\vec{B}$ field reads as follows,

$$\vec{B}_\parallel(m, n) = \frac{i\, k}{\omega^2\, \epsilon_r\, \mu_r / c^2 - k^2} \vec{\nabla}_\parallel B_{z,mn}\, e^{i\,(k\, z - \omega\, t)} = -i\, k \left[\left(\frac{m\pi}{a}\right)^2 + \left(\frac{n\pi}{b}\right)^2\right]^{-1} B_{0,mn}$$

$$\times \left[\hat{e}_x \frac{m\pi}{a} \sin\left(\frac{m\pi x}{a}\right) \cos\left(\frac{n\pi y}{b}\right) + \hat{e}_y \frac{n\pi}{b} \cos\left(\frac{m\pi x}{a}\right) \sin\left(\frac{n\pi y}{b}\right)\right] e^{i\,(k\, z - \omega\, t)}\,,$$

$$(7.47b)$$

while the electric field is given by

$$
\vec{E}_\parallel(m,n) = -\frac{\omega}{k} \left[\hat{e}_z \times \vec{B}_\parallel(m,n) \right] = -i\omega \left[\left(\frac{m\pi}{a} \right)^2 + \left(\frac{n\pi}{b} \right)^2 \right]^{-1} B_{0,mn}
$$

$$
\times \left[\hat{e}_x \frac{n\pi}{b} \cos\left(\frac{m\pi x}{a} \right) \sin\left(\frac{n\pi y}{b} \right) - \hat{e}_y \frac{m\pi}{a} \sin\left(\frac{m\pi x}{a} \right) \cos\left(\frac{n\pi y}{b} \right) \right] e^{i(kz-\omega t)} .
$$

$$(7.47c)$$

The modulus of the wave vector k is given by Eq. (7.44). It is instructive to indicate the solution for the fundamental $TE_{1,0}$ mode with $m = 1$ and $n = 0$ separately,

$$
B_z(x,y) = B_0 \cos\left(\frac{\pi x}{a} \right) e^{i(kz-\omega t)} ,
$$

$$(7.48a)$$

$$
\vec{B}_\parallel(m=1, n=0) = -i \frac{ka}{\pi} B_0 \, \hat{e}_x \sin\left(\frac{\pi x}{a} \right) e^{i(kz-\omega t)} ,
$$

$$(7.48b)$$

$$
\vec{E}_\parallel(m=1, n=0) = i \frac{\omega a}{\pi} B_0 \, \hat{e}_y \sin\left(\frac{\pi x}{a} \right) e^{i(kz-\omega t)} .
$$

$$(7.48c)$$

Transverse Electric Mode $TE_{1,0}$

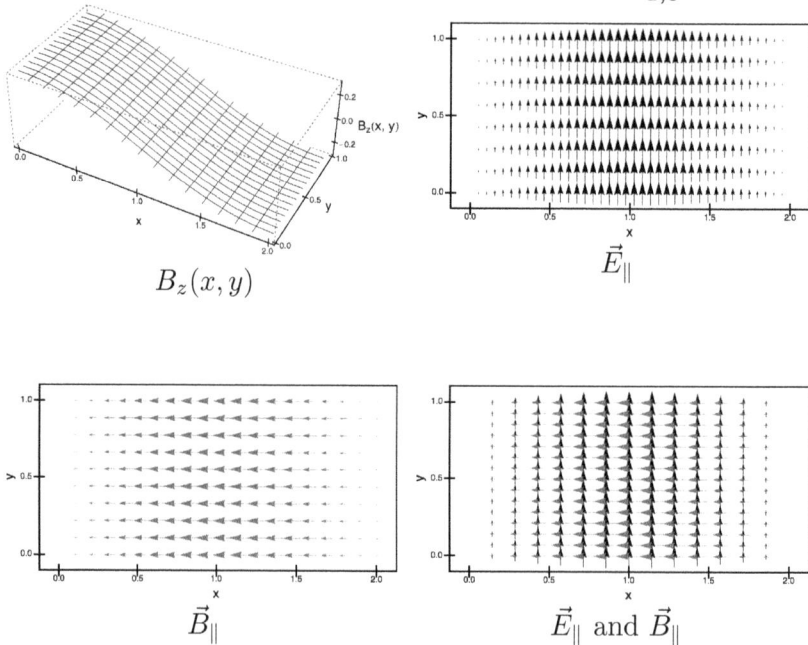

$B_z(x,y)$

$\vec{E}_\parallel$

$\vec{B}_\parallel$

$\vec{E}_\parallel$ and $\vec{B}_\parallel$

Fig. 7.2 Four plots illustrating the transverse electric mode $TE_{1,0}$. The leftmost upper plot shows $B_z(x,y)$ as a function of x and y. As the mode is transverse, $E_z(x,y)$ vanishes. The rightmost upper plot shows, as a vector plot, $\vec{E}_\parallel(x,y)$ in a plane of constant z, at a particular time of the sinusoidal oscillation. All vector arrows oscillate sinusoidally. The left lower plot shows $\vec{B}_\parallel(x,y)$ in a plane of constant z, at a particular time of the sinusoidal oscillation. It is clearly visible that $(\hat{n} \cdot \vec{\nabla}_\parallel)B_z$ vanishes at the surface, as is required for all TE modes. In the right lower plot, we illustrate the perpendicular character of the two fields.

Transverse Electric Mode TE$_{1,1}$

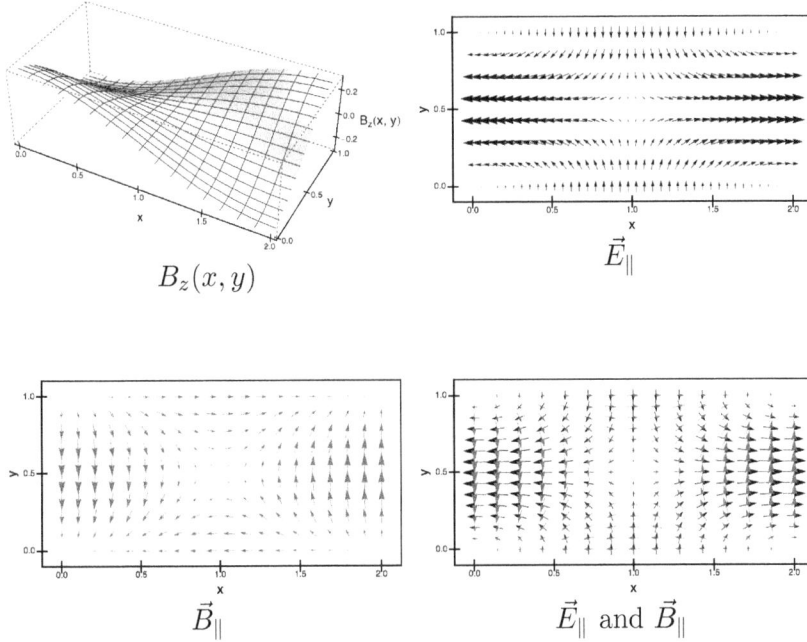

$B_z(x, y)$

$\vec{E}_{\parallel}$

$\vec{B}_{\parallel}$

$\vec{E}_{\parallel}$ and $\vec{B}_{\parallel}$

Fig. 7.3 Same as Fig. 7.2, for the transverse electric mode TE$_{1,1}$.

The dispersion relation for the fundamental mode is

$$k = k_{10} = \sqrt{\epsilon_r \mu_r \left(\frac{\omega}{c}\right)^2 - \left(\frac{\pi}{a}\right)^2} = \sqrt{\epsilon_r \mu_r} \frac{\omega}{c} \sqrt{1 - \frac{1}{\epsilon_r \mu_r} \left(\frac{\pi c}{\omega a}\right)^2}. \tag{7.48d}$$

We recall that propagation happens only for

$$\omega^2 > \omega_{\min;\text{TE}} = \frac{\pi}{\sqrt{\epsilon_r \mu_r}} \frac{c}{a}. \tag{7.48e}$$

Note the 90° phase difference between B_x, B_y and B_z arising from the $-i = e^{-i\pi/2}$ factor. The in-plane components $\vec{B}_{\parallel}$ and $\vec{E}_{\parallel}$ are 180° out of phase. Graphical representations of the first TE modes can be found in Figs. 7.2, 7.3, 7.4 and 7.5.

The dispersion relation (7.44) can trivially be rewritten as

$$k = \frac{\sqrt{\epsilon_r \mu_r}}{c} \sqrt{\omega^2 - \omega_{mn}^2}, \qquad \omega_{mn} = \frac{\pi c}{\sqrt{\epsilon_r \mu_r}} \left[\left(\frac{m}{a}\right)^2 + \left(\frac{n}{b}\right)^2\right]^{1/2}, \tag{7.49}$$

where ω_{mn} is the cutoff frequency for the mode. The dispersion relations in Figs. 7.6, 7.7 and 7.8 represent the functional relationships $\omega = \omega(k)$ for the first few modes of a typical waveguide.

It is often convenient to choose the dimensions of the waveguide so that at the operating frequency, only the lowest mode TE$_{1,0}$ can occur. Since the wave number,

Transverse Electric Mode TE$_{2,1}$

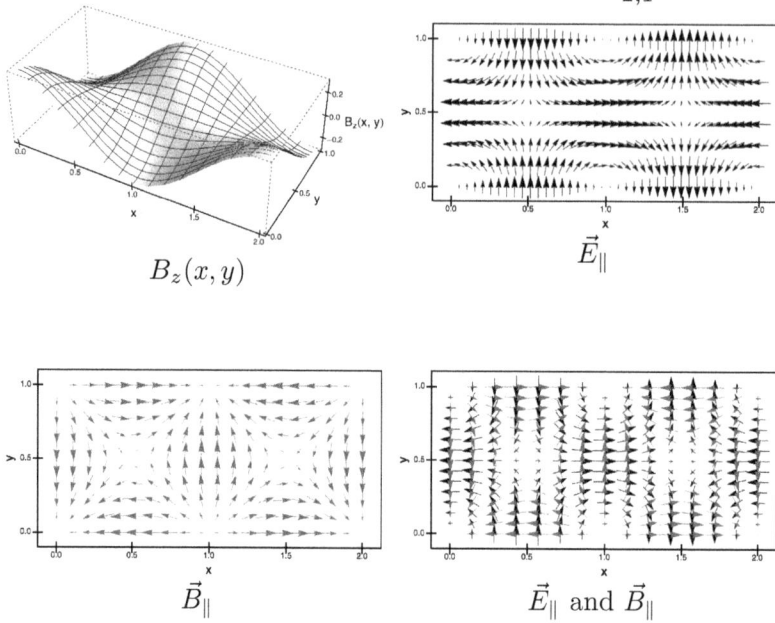

$B_z(x,y)$

$\vec{E}_\parallel$

$\vec{B}_\parallel$

$\vec{E}_\parallel$ and $\vec{B}_\parallel$

Fig. 7.4 Same as Figs. 7.2 and 7.3, for the transverse electric mode TE$_{2,1}$.

Transverse Electric Mode TE$_{2,2}$

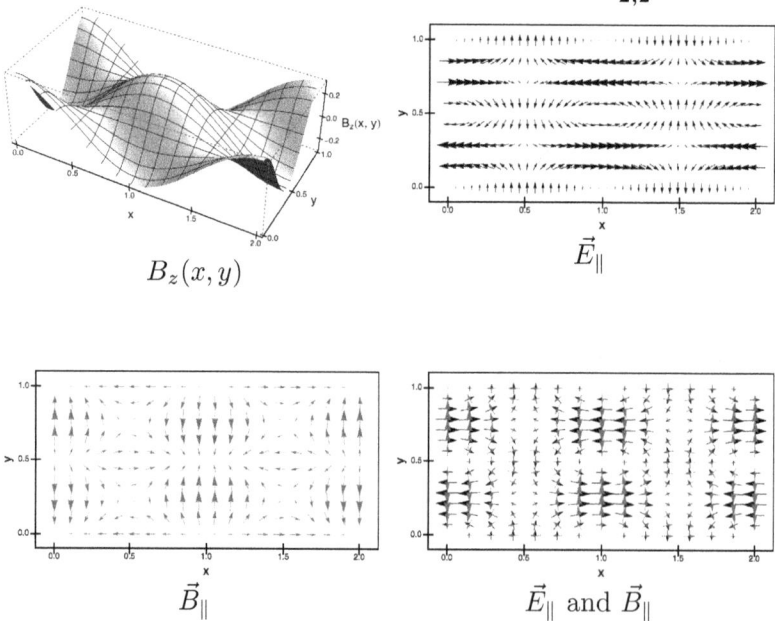

$B_z(x,y)$

$\vec{E}_\parallel$

$\vec{B}_\parallel$

$\vec{E}_\parallel$ and $\vec{B}_\parallel$

Fig. 7.5 Same as Figs. 7.2, 7.4, and 7.5, but for the transverse electric mode TE$_{2,2}$.

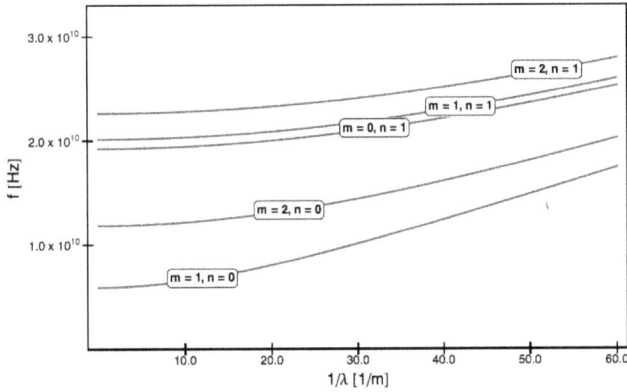

Fig. 7.6   Dispersion relations for the first few modes of a rectangular waveguide, for all modes with $m \leq 2$ and $n \leq 2$. The dimensions of the waveguide are $a = 2.3\,\text{cm}$ and $b = 0.71\,\text{cm}$. The relative permittivity of the waveguide medium is $\epsilon_r = 1.2$, and the permeability is $\mu_r = 1.0$, independent of the angular frequency ω. The abscissa gives the inverse wavelength $1/\lambda$ in units of inverse meter (m), where $1/\lambda = k/(2\pi)$, and the ordinate axis gives the light frequency $f = \omega/(2\pi)$ in units of cycles per second, which is equal to Hertz (Hz). We note that the SI mksA unit of the angular frequency ω is radians per second. A full oscillation period per second corresponds to one cycle per second, or 2π radians per second (rad/s).

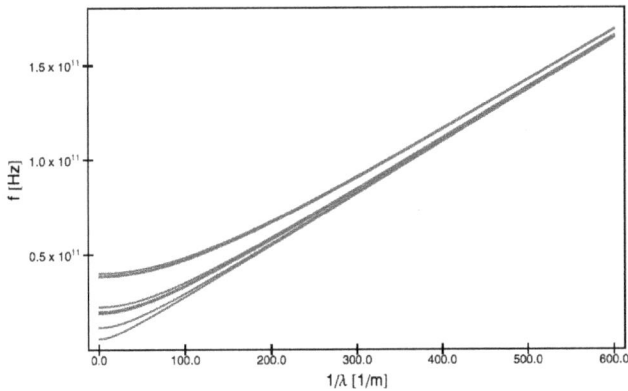

Fig. 7.7   Same as Fig. 7.6, but for a region of medium wave number, where the modes of the waveguide approach the free dispersion relation $\omega = ck/\sqrt{\epsilon_r\,\mu_r}$.

k_{mn}, is always less than the "free space" value, the wavelength in the waveguide is always larger than the free space wavelength. Surprisingly, this means that the phase velocity in free space is always greater than the free-space value, and equal to ω/k_{mn}.

To supplement Eq. (7.47), let us also give the formulas for the TM modes,

$$E_z(x,y) = E_{0,mn} \sin\left(\frac{m\pi x}{a}\right) \sin\left(\frac{n\pi y}{b}\right) e^{i\,(k\,z-\omega t)}, \qquad B_z(x,y) = 0. \qquad (7.50a)$$

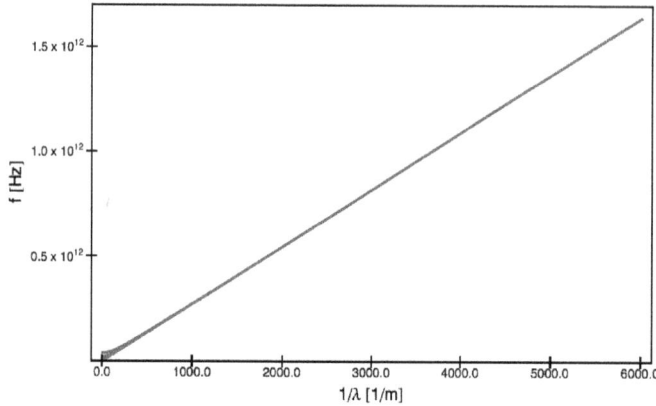

Fig. 7.8   Same as Figs. 7.6 and 7.7, but for even larger wave numbers. The deviation from the dispersion relation without waveguide, $\omega = ck/\sqrt{\epsilon_r \mu_r}$ is hardly discernible.

The in-plane component of the electric field is

$$\vec{E}_\parallel(m,n) = i\frac{k}{\epsilon_r \mu_r \omega^2/c^2 - k^2}\vec{\nabla}_\parallel E_z = ik\left[\left(\frac{m\pi}{a}\right)^2 + \left(\frac{n\pi}{b}\right)^2\right]^{-1} E_{0,mn}$$

$$\times\left[\hat{e}_x\frac{m\pi}{a}\cos\left(\frac{m\pi x}{a}\right)\sin\left(\frac{n\pi y}{b}\right) + \hat{e}_y\frac{n\pi}{b}\sin\left(\frac{m\pi x}{a}\right)\cos\left(\frac{n\pi y}{b}\right)\right]e^{i(kz-\omega t)},$$

(7.50b)

while the same component for the $\vec{B}$ field reads as follows,

$$\vec{B}_\parallel(m,n) = \frac{\omega\,\epsilon_r\,\mu_r}{k\,c^2}\left(\hat{e}_z \times \vec{E}_\parallel\right) = i\frac{\epsilon_r\,\mu_r\,\omega}{c^2}\left[\left(\frac{m\pi}{a}\right)^2 + \left(\frac{n\pi}{b}\right)^2\right]^{-1} E_{0,mn}$$

$$\times\left[-\hat{e}_x\frac{n\pi}{b}\sin\left(\frac{m\pi x}{a}\right)\cos\left(\frac{n\pi y}{b}\right) + \hat{e}_y\frac{m\pi}{a}\cos\left(\frac{m\pi x}{a}\right)\sin\left(\frac{n\pi y}{b}\right)\right]e^{i(kz-\omega t)}.$$

(7.50c)

Illustrations are provided in Figs. 7.9, 7.10 and 7.11. The dispersion relation for TM modes is just the same as for TE modes, but the available values of m and n are different,

$$\left(\frac{m\pi}{a}\right)^2 + \left(\frac{n\pi}{b}\right)^2 = \epsilon_r\,\mu_r\,\frac{\omega^2}{c^2} - k^2, \qquad m,n = 1,2,3,\ldots. \qquad (7.51)$$

The z component of the electric field is $E_z(x,y) = 0$ at $x = 0$ and $x = a$, and at $y = 0$ and $y = b$. In the TM modes, if $n = 0$ or if $m = 0$, then we have $E_z = 0$, as a consequence of the properties of the sine versus the cosine function. So, the $m = 1$, $n = 0$ mode is not available for TM waves because the fields simply vanish. The lowest

Transverse Magnetic Mode TM$_{1,1}$

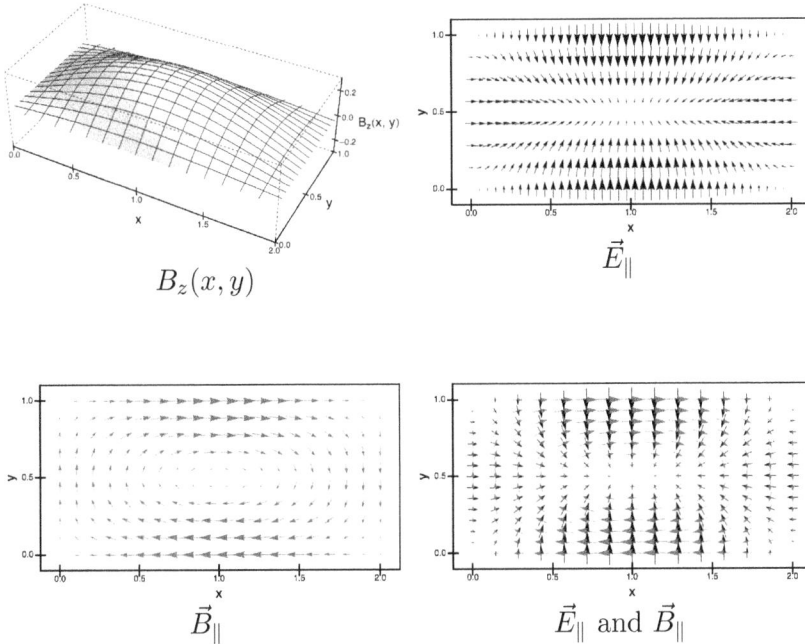

$B_z(x, y)$

$\vec{E}_{\parallel}$

$\vec{B}_{\parallel}$

$\vec{E}_{\parallel}$ and $\vec{B}_{\parallel}$

Fig. 7.9 Four plots illustrating the transverse magnetic mode TM$_{1,1}$. The leftmost upper plot shows $E_z(x,y)$ as a function of x and y. As the mode is transverse magnetic, $B_z(x,y)$ vanishes. The rightmost upper plot shows, as a vector plot, $\vec{E}_{\parallel}(x,y)$ in a plane of constant z, at a particular time of the sinusoidal oscillation. All vector arrows oscillate sinusoidally. The left lower plot shows $\vec{B}_{\parallel}(x,y)$ in a plane of constant z, at a particular time of the sinusoidal oscillation. The magnetic field lines form "closed loops" in the xy plane, as they do for all TM modes. In the right lower plot, the perpendicular character of the two fields is illustrated.

possible mode is $n = m = 1$ with

$$\omega_{\text{min;TM}} = \omega_{11} = \frac{\pi c}{\sqrt{\epsilon_r \mu_r}} \left[\left(\frac{1}{a} \right)^2 + \left(\frac{1}{b} \right)^2 \right]^{1/2} > \omega_{\text{min;TE}} = \omega_{10} = \frac{\pi}{\sqrt{\epsilon_r \mu_r}} \frac{c}{a}. \qquad (7.52)$$

Thus, the transverse electric mode TE$_{10}$ provides the smallest available frequency for propagation in the waveguide. For $a > b$, for frequencies

$$\omega_{\text{min;TE}} < \omega < \omega_{\text{min;TM}}, \qquad (7.53)$$

only one mode is available for wave propagation. The phase velocity is given by

$$v_p = \frac{\omega}{k} = \frac{c}{\sqrt{\epsilon_r \mu_r}} \frac{\sqrt{k^2 + (\omega_{mn}/c)^2}}{k} = \frac{c}{\sqrt{\epsilon_r \mu_r}} \frac{\omega}{\sqrt{\omega^2 - \omega_{mn}^2}}$$

$$= \frac{c}{\sqrt{\epsilon_r \mu_r}} \left(1 - \frac{\omega_{mn}^2}{\omega^2} \right)^{-1/2} > c, \qquad (7.54)$$

Transverse Magnetic Mode TM$_{2,1}$

$B_z(x,y)$

$\vec{E}_\|$

$\vec{B}_\|$

$\vec{E}_\|$ and $\vec{B}_\|$

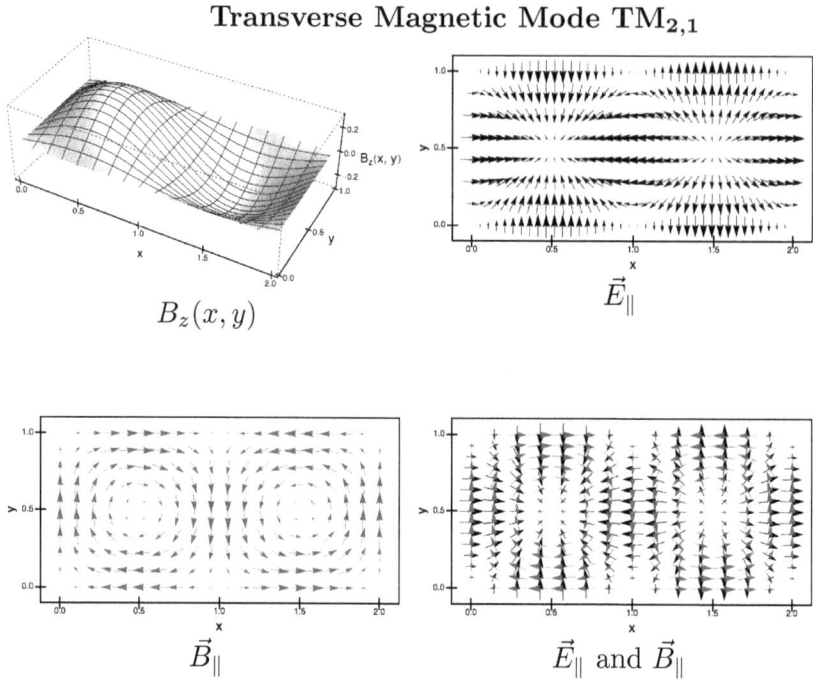

Fig. 7.10   Four plots illustrating the transverse magnetic mode TM$_{2,1}$, otherwise the same as Fig. 7.9.

where the last inequality holds for $\epsilon_r = \mu_r = 1$ and $\omega_{mn} \neq 0$. The group velocity is given by

$$v_g = \frac{\mathrm{d}\omega}{\mathrm{d}k} = \frac{c}{\sqrt{\epsilon_r \mu_r}} \frac{\mathrm{d}}{\mathrm{d}k} [(ck)^2 + \omega_{mn}^2]^{1/2} = \frac{c^2 k}{\sqrt{\epsilon_r \mu_r}} \frac{1}{\sqrt{(ck)^2 + \omega_{mn}^2}}$$

$$= \frac{c}{\sqrt{\epsilon_r \mu_r}} \frac{\sqrt{\omega^2 - \omega_{mn}^2}}{\omega} = \frac{c}{\sqrt{\epsilon_r \mu_r}} \left(1 - \frac{\omega_{mn}^2}{\omega^2}\right)^{1/2} \to 0, \qquad \omega \to \omega_{mn}, \qquad (7.55)$$

where we indicate a few possible, alternative forms. The latter limit indicates the possibility of "slow light" because the group velocity is the speed at which a light pulse (or, "light pulse train") travels. Furthermore, the functional form $\sqrt{\omega^2 - \omega_{mn}^2}/\omega$ is consistent with the group velocity always remaining smaller than the speed of light. The index $n(\omega)$ of refraction can be determined from

$$k = n(\omega) \frac{\omega}{c} = \frac{\sqrt{\omega^2 - \omega_{mn}^2}}{c}, \qquad (7.56a)$$

$$n(\omega) = \sqrt{1 - \frac{\omega_{mn}^2}{\omega^2}} = \frac{1}{\sqrt{\epsilon_r \mu_r}} \frac{c}{v_p} < 1, \qquad (7.56b)$$

where the latter inequality holds for $v_p > c$.

Transverse Magnetic Mode TM$_{2,2}$

$B_z(x,y)$

$\vec{E}_{\parallel}$

$\vec{B}_{\parallel}$

$\vec{E}_{\parallel}$ and $\vec{B}_{\parallel}$

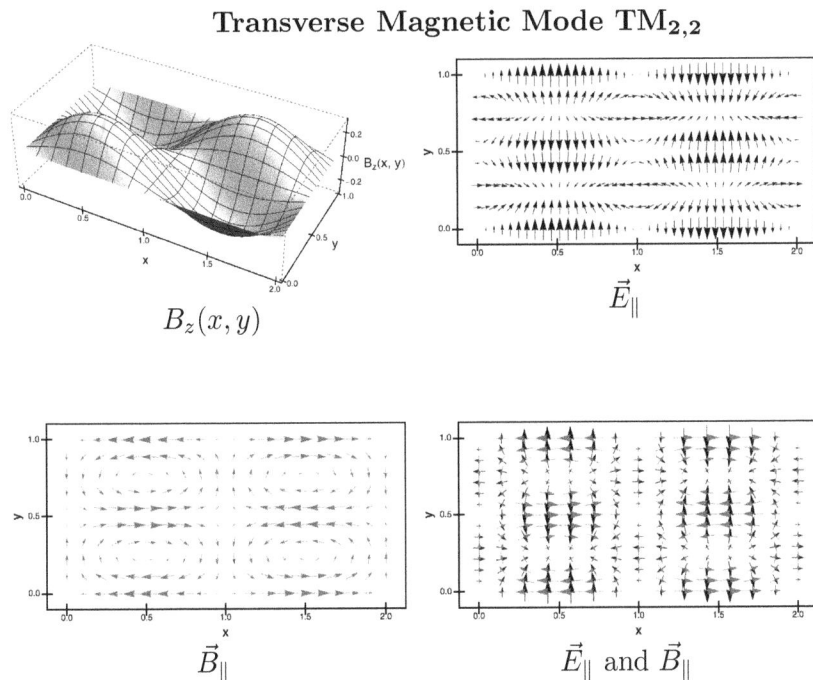

Fig. 7.11   Four plots illustrating the transverse magnetic mode TM$_{2,2}$, otherwise the same as Figs. 7.10 and 7.11.

7.3   Resonant Cavities

7.3.1   *Resonant Cylindrical Cavities*

In a waveguide, modes are characterized by two discrete quantum numbers m and n, and a continuous quantum number k (or ω). The oscillations in the xy plane are quantized. In a cylindrical cavity (as opposed to a waveguide), the oscillations in the third direction (z direction) are also quantized, and one ends up with a fully discrete set of modes. In some sense, the waves are "bound" into the cavities, and the additional quantization conditions induced by the cavity correspond to the "bound states" in atoms, which are also discrete. The resonant modes are standing waves, not traveling waves as in the z-oriented waveguide, and resemble the "resonant modes" of, say, a drum or the string of a violin. In a waveguide, the allowed frequencies are of the form $\omega = \omega_{mn}(k)$, where m and n are discrete numbers (integers) but k is a continuous variable. The cavity adds one more boundary condition, in the z direction, because both the lower as well as the upper end are covered by a perfect conductor. This implies a further quantization condition, and we anticipate that the allowed cavity mode frequencies are of the form ω_{mnp}, with three integer subscripts.

We start the discussion with the TM modes and use cylindrical coordinates ρ, φ, and z. Let the cavity extend from $z = 0$ to $z = d$ and from $\rho = 0$ (symmetry axis) to $\rho = R$. The first step is to separate the spatial and time dependence of the fields, just as we did for the waveguide,

$$\vec{E}(\vec{r}, t) = \vec{E}(\vec{r})\, e^{-i\omega t}, \quad \vec{B}(\vec{r}, t) = \vec{B}(\vec{r})\, e^{-i\omega t}. \tag{7.57}$$

Let us assume first that the magnetic field is transverse, i.e., that only the electric field has a z component,

$$E_z = f(\rho)\, g(\varphi)\, h(z) = f(\rho)\, e^{i m \varphi} \left[C \sin\left(\frac{p \pi z}{d}\right) + D \cos\left(\frac{p \pi z}{d}\right) \right]. \tag{7.58}$$

The Helmholtz equation requires that

$$\left(\vec{\nabla}^2 + \epsilon_r \mu_r \frac{\omega^2}{c^2} \right) E_z = \left(\frac{1}{\rho} \frac{\partial}{\partial \rho} \rho \frac{\partial}{\partial \rho} + \frac{1}{\rho^2} \frac{\partial^2}{\partial \varphi^2} + \frac{\partial^2}{\partial z^2} + \epsilon_r \mu_r \frac{\omega^2}{c^2} \right) E_z$$

$$= \left[\frac{1}{\rho} \frac{\partial}{\partial \rho} \rho \frac{\partial}{\partial \rho} + \underbrace{\frac{1}{\rho^2} \frac{\partial^2}{\partial \varphi^2}}_{=-m^2/\rho^2} + \underbrace{\left(-\left(\frac{p \pi}{d}\right)^2 + \epsilon_r \mu_r \frac{\omega^2}{c^2} \right)}_{=\gamma_p^2} \right] E_z$$

$$= \left(\frac{\partial^2}{\partial \rho^2} + \frac{1}{\rho} \frac{\partial}{\partial \rho} - \frac{m^2}{\rho^2} + \gamma_p^2 \right) E_z = 0. \tag{7.59}$$

The parameter γ_p is defined as

$$\gamma_p^2 = \epsilon_r \mu_r \frac{\omega^2}{c^2} - \left(\frac{p \pi}{d}\right)^2. \tag{7.60}$$

We recall the defining differential equation for the ordinary Bessel function Eq. (2.144),

$$\left(\frac{\partial^2}{\partial \rho^2} + \frac{1}{\rho} \frac{\partial}{\partial \rho} - \frac{n^2}{\rho^2} + 1 \right) J_n(\rho) = 0. \tag{7.61}$$

In the *ansatz* (7.58) for the TM modes, we can thus set $f(\rho) = J_m(\gamma_p \rho)$,

$$[\text{TM:}] \quad E_z = E_0\, J_m(\gamma_p \rho)\, e^{i m \varphi} \left[C \sin\left(\frac{p \pi z}{d}\right) + D \cos\left(\frac{p \pi z}{d}\right) \right], \tag{7.62}$$

while, of course, $B_z = 0$ for the TM mode. The constants C and D have to be determined from the boundary conditions. The gradient operator, in spherical coordinates, has the representation

$$\vec{\nabla} = \hat{e}_\rho \frac{\partial}{\partial \rho} + \hat{e}_\varphi \frac{1}{\rho} \frac{\partial}{\partial \varphi} + \hat{e}_z \frac{\partial}{\partial z} = \vec{\nabla}_\parallel + \hat{e}_z \nabla_z. \tag{7.63}$$

In general, because the z axis is still a valid symmetry axis of the problem, we can apply Eq. (7.14) to read as follows,

$$\nabla_z \vec{E}_\parallel + i\omega \left(\hat{e}_z \times \vec{B}_\parallel \right) = \vec{\nabla}_\parallel E_z. \tag{7.64}$$

Furthermore, we recall Eq. (7.18), which for $B_z = 0$ (TM mode) reads as

$$\nabla_z \vec{B}_\parallel - i\frac{\omega}{c^2}\,\epsilon_r\,\mu_r\left(\hat{e}_z \times \vec{E}_\parallel\right) = 0\,. \tag{7.65}$$

In a waveguide, we could replace $\nabla_z\vec{E}_\parallel = ik\vec{E}_\parallel$ and $\nabla_z\vec{B}_\parallel = ik\vec{B}_\parallel$ without altering the z dependence, which was exclusively manifest in the e^{ikz} prefactor. We could thus solve Eq. (7.65) for $\vec{B}_\parallel$, after replacing $\nabla_z\vec{B}_\parallel \to ik\,\vec{B}_\parallel$, and insert the result in Eq. (7.64), finally obtaining $\vec{E}_\parallel$ as a function of E_z.

For the cavity, $\vec{E}_\parallel$ and E_z will have different $\sin(p\pi z/d)$ and $\cos(p\pi z/d)$ dependences. and we have to integrate Eq. (7.65) with respect to z before using the relation in Eq. (7.64). For an electric field of the form (7.62), whose z dependence is sinusoidal or cosinusoidal, we have the relations

$$\nabla_z\vec{E}_\parallel = -\left(\frac{\pi p}{d}\right)^2\int \vec{E}_\parallel\,\mathrm{d}z\,, \qquad \vec{E}_\parallel = -\left(\frac{\pi p}{d}\right)^2\iint \vec{E}_\parallel\,\mathrm{d}z\,\mathrm{d}z\,. \tag{7.66}$$

A double integration with respect to $\mathrm{d}z$ thus is equivalent to a multiplication by $-(d/(\pi p))^2$.

We can integrate Eq. (7.65) with respect to z, and solve for $\vec{B}_\parallel$, with the result

$$\vec{B}_\parallel = i\frac{\omega}{c^2}\,\epsilon_r\,\mu_r\left(\hat{e}_z \times \int \vec{E}_\parallel\mathrm{d}z\right)\,, \tag{7.67}$$

and insert the result in the z-integrated form of Eq. (7.64),

$$\vec{E}_\parallel + i\omega\left(\hat{e}_z \times \int \vec{B}_\parallel\mathrm{d}z\right) = \int \vec{\nabla}_\parallel E_z\mathrm{d}z\,. \tag{7.68}$$

So, now, inserting (7.67) into (7.68), we have

$$\vec{E}_\parallel + i\omega\left(\hat{e}_z \times i\frac{\omega}{c^2}\,\epsilon_r\,\mu_r\left(\hat{e}_z \times \iint \vec{E}_\parallel\,\mathrm{d}z\mathrm{d}z\right)\right) = \int \vec{\nabla}_\parallel E_z\mathrm{d}z\,. \tag{7.69}$$

We now use Eq. (7.9) for the in-plane component $\vec{E}_\parallel$,

$$\vec{E}_\parallel - i^2\frac{\omega^2}{c^2}\,\epsilon_r\,\mu_r\iint \vec{E}_\parallel\,\mathrm{d}z\,\mathrm{d}z = \int \vec{\nabla}_\parallel E_z\mathrm{d}z\,. \tag{7.70}$$

Using Eq. (7.66), we can transform the double integration with respect to z into a multiplication by $-(d/(p\pi))^2$, and so

$$\vec{E}_\parallel\left[1 - \frac{\omega^2}{c^2}\,\epsilon_r\,\mu_r\left(\frac{d}{p\pi}\right)^2\right] = \int \vec{\nabla}_\parallel E_z\,\mathrm{d}z\,. \tag{7.71}$$

We now pull out a prefactor,

$$\left(\frac{d}{p\pi}\right)^2\vec{E}_\parallel\left[\left(\frac{p\pi}{d}\right)^2 - \frac{\omega^2}{c^2}\,\epsilon_r\,\mu_r\right] = \int \vec{\nabla}_\parallel E_z\,\mathrm{d}z\,, \tag{7.72}$$

so that with Eq. (7.60),

$$-\left(\frac{p\pi}{d}\right)^{-2}\gamma_p^2\,\vec{E}_\parallel = \int \vec{\nabla}_\parallel E_z\,\mathrm{d}z\,. \tag{7.73}$$

It now becomes clear why in the *ansatz* given by Eq. (7.62), we need to choose the term with the cosine as opposed to the sine. Namely, the in-plane field $\vec{E}_\parallel$ needs to vanish inside the endcap surfaces at $z = 0$ and $z = d$. According to Eq. (7.73), $\vec{E}_\parallel$ is given by an integral of E_z with respect to z; the in-plane gradient operator $\vec{\nabla}_\parallel$ [Eq. (7.63)] does not play a role in this consideration. The sine term in Eq. (7.62) would lead to a term proportional to $\cos(p\pi z/d)$ in $\vec{E}_\parallel$, which would lie inside the endcap surfaces, leading to a contradiction with our assumption of a perfect conductor. Hence, we choose the cosine term in Eq. (7.62), which upon integrating with respect to z leads to following result for $\vec{E}_\parallel$,

$$\vec{E}_\parallel = -\left(\frac{p\pi}{d}\right)^2 \left(\frac{d}{p\pi}\right) \frac{1}{\gamma_p^2} E_0 \vec{\nabla}_\parallel J_m(\gamma_p \rho) \sin\left(\frac{p\pi z}{d}\right) e^{im\varphi} e^{-i\omega t}. \tag{7.74}$$

Here, E_0 is an overall multiplicative constant, and we recall Eq. (7.63) for the definition of $\vec{\nabla}_\parallel$. In view of Eq. (7.67), which we recall for convenience, $\vec{B}_\parallel = i\frac{\omega}{c^2}\epsilon_r \mu_r \left(\hat{e}_z \times \int \vec{E}_\parallel dz\right)$, the final result for $\vec{B}_\parallel$ is

$$\vec{B}_\parallel = E_0 \left(i\frac{\omega}{c^2}\epsilon_r \mu_r\right) \frac{1}{\gamma_p^2} \left(\hat{e}_z \times \vec{\nabla}_\parallel J_m(\gamma_p \rho)\right) \cos\left(\frac{p\pi z}{d}\right) e^{im\varphi} e^{-i\omega t}. \tag{7.75}$$

We have used the result [see Eq. (7.63)] Finally, we summarize the result for TM modes in a cylindrical cavity,

$$E_z = E_0 J_m(\gamma_p \rho) \cos\left(\frac{p\pi z}{d}\right) e^{im\varphi} e^{-i\omega t}, \qquad B_z = 0, \tag{7.76a}$$

$$\vec{E}_\parallel = -\frac{p\pi}{d} \frac{1}{\gamma_p^2} E_0 \vec{\nabla}_\parallel J_m(\gamma_p \rho) \sin\left(\frac{p\pi z}{d}\right) e^{im\varphi} e^{-i\omega t}, \tag{7.76b}$$

$$\vec{B}_\parallel = E_0 \left(i\frac{\omega}{c^2}\epsilon_r \mu_r\right) \frac{1}{\gamma_p^2} \left(\hat{e}_z \times \vec{\nabla}_\parallel J_m(\gamma_p \rho)\right) \cos\left(\frac{p\pi z}{d}\right) e^{im\varphi} e^{-i\omega t}. \tag{7.76c}$$

Here, we suppress the arguments of the electric field in our notation, i.e., we understand that $E_z = E_z(\rho, \varphi, z, t)$, $\vec{E}_\parallel = \vec{E}_\parallel(\rho, \varphi, z, t)$, and $\vec{B}_\parallel = \vec{B}_\parallel(\rho, \varphi, z, t)$.

There is one more point to clarify. Namely, E_z becomes parallel to the outer surfaces of the cylinder near $\rho = R$, where R is the cylinder radius. Hence, we need to require that

$$J_m(\gamma_p R) = 0, \quad \gamma_p R = x_{mn}, \quad J_m(x_{mn}) = 0, \quad m = 0, 1, 2, \ldots, \quad n = 1, 2, 3, \ldots, \tag{7.76d}$$

where x_{mn} is the nth zero of the Bessel function of order m [see Eq. (2.158)]. So, because of the condition that E_z must vanish on the outer rim of the cylinder, we must postulate that

$$\gamma_p \overset{!}{=} \gamma_{mn} = \gamma_{mnp} = \frac{x_{mn}}{R}, \qquad n = 1, 2, 3, \ldots. \tag{7.76e}$$

This condition gives us an equation which needs to be fulfilled by the "quantum numbers" m, n, and p,

$$\gamma_p^2 = \gamma_{mnp}^2 = \frac{\omega^2}{c^2} \epsilon_r \mu_r - \left(\frac{p\pi}{d}\right)^2 = \left(\frac{x_{mn}}{R}\right)^2 , \tag{7.76f}$$

$$\omega = \omega_{mnp} = \frac{c}{\sqrt{\epsilon_r \mu_r}} \sqrt{\left(\frac{p\pi}{d}\right)^2 + \left(\frac{x_{mn}}{R}\right)^2} , \tag{7.76g}$$

$$m = 0, 1, 2, \ldots , \quad n = 1, 2, 3, \ldots , \quad p = 0, 1, 2, \ldots . \tag{7.76h}$$

The quantum numbers are m (azimuthal), n (radial) and p (longitudinal). Indeed, the quantum number n start from $n = 1$ because we must integrate at least up to the first node of the Bessel function, m starts from zero because the magnetic quantum number may vanish, and p also can be zero because E_z is proportional to $\cos(p\pi z/d)$, not $\sin(p\pi z/d)$, so the fields do not all vanish if we set $p = 0$. The quantum number n counts the nodes of the radial wave function and acts, in some sense, as the principal quantum number (the number of nodes actually is $n - 1$ for given n). The quantum number m acts as a projection quantum number characterizing the "angular momentum" about the quantization axis, $L_z = -i\partial/\partial\varphi$ [see Eq. (2.30c)]. The quantum number p characterizes the number of oscillations in the z direction.

The fundamental TM mode in a cylindrical cavity is the 010 mode ($m = 0$, $n = 1$, and $p = 0$). The numerical value of the first zero of the Bessel function of order $m = 0$ is $x_{01} = 2.404\,825\,557\ldots$ (see the discussion in Sec. 2.4.3 and the numerical values in Table 2.1). It has the lowest frequency available,

$$\omega_{010} = \frac{c}{\sqrt{\epsilon_r \mu_r}} \frac{x_{01}}{R} \approx \frac{c}{\sqrt{\epsilon_r \mu_r}} \frac{2.404\,825}{R} . \tag{7.77}$$

As will be shown in the following, this frequency typically is less than the corresponding minimum frequency for the lowest TE mode, and so the fundamental mode for a cylindrical cavity is TM rather than TE. The ground state is a radially symmetric state (no φ dependence) with quantum numbers $m = 0$, $n = 1$, and $p = 0$.

The TM states with $m = 0$ and $p = 0$ are characterized by the field configurations ("wave functions"),

$$E_z = E_0 \, J_0(\gamma_{0n}\, \rho)\, e^{-i\omega_{0n0}\, t} , \qquad \vec{E}_\parallel = 0 ,$$

$$\vec{B}_\parallel = E_0 \frac{i\omega\epsilon_r\mu_r}{\gamma_{0n}^2 c^2} \left(\hat{e}_z \times \vec{\nabla}_\parallel J_0(\gamma_{0n}\, \rho)\right) e^{-i\omega_{0n0}\, t} , \qquad B_z = 0 . \tag{7.78}$$

Of course, $J_0(\gamma_{0n} R) = 0$. The wave function with $n = 1$ has one node, with $n = 2$, two nodes, etc., much like in quantum mechanics. This is illustrated in Fig. 7.12 for the ground state. The electric and magnetic fields of the ground state with $m = p = 0$ and $n = 1$ are shown in Fig. 7.13. In view of Eq. (7.78), the oscillations of the magnetic and electric fields are 90° out of phase for the standing wave. Because of the longitudinal electric field, TM modes are useful for accelerator cavities in storage rings.

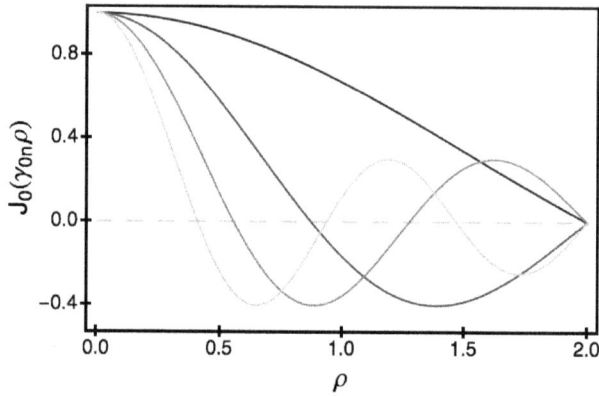

Fig. 7.12   Illustration of the first few radial wave functions for the cylindrical cavity.

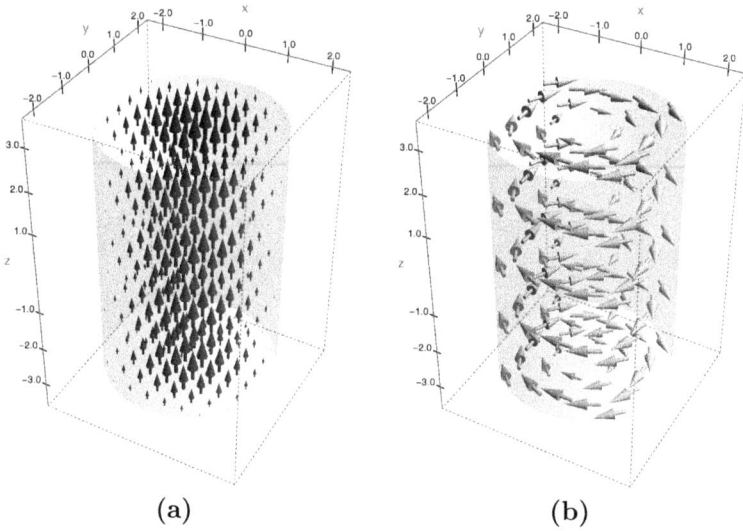

(a)   (b)

Fig. 7.13   Illustration of the electric field for the TM ground state of the cylindrical cavity (left plot), and of the magnetic field (right plot). The electric field has a z component (and no transverse component), while the magnetic field is transverse and circulating around the symmetry axis.

We now switch to the TE modes. Our *ansatz* for the z components of the electric and magnetic fields of a TE mode is

$$[\text{TE:}] \qquad B_z = B_0 \, J_m(\overline{\gamma}_p \rho) \sin\left(\frac{p\pi z}{d}\right) \mathrm{e}^{\mathrm{i}m\varphi} \, \mathrm{e}^{-\mathrm{i}\omega t} \, , \qquad E_z = 0 \, , \tag{7.79}$$

The $\overline{\gamma}_p$ are assigned differently as compared to the γ_p for the TM modes. The normal component of the $\vec{B}$ field has to vanish on the outer rim of the cylinder, in view of the boundary condition (7.27). Now, Eq. (7.63) mandates that the gradient operator, in cylindrical coordinates, has a component in the radial direction which is just given by the partial derivative with respect to ρ. We thus have to require that

$$\overline{\gamma}_p = \overline{\gamma}_{mnp} = \frac{\overline{x}_{mn}}{R}, \qquad J'_m(\overline{x}_{mn}) = 0, \tag{7.80}$$

where we appeal to Eq. (2.159). For TE modes, in view of Eqs. (7.14) and (7.18), we have in general,

$$\nabla_z \vec{E}_\parallel + i\omega \left(\hat{e}_z \times \vec{B}_\parallel\right) = \vec{\nabla}_\parallel E_z = \vec{0}, \tag{7.81a}$$

$$\nabla_z \vec{B}_\parallel - i\frac{\omega}{c^2} \epsilon_r \mu_r \left(\hat{e}_z \times \vec{E}_\parallel\right) = \vec{\nabla}_\parallel B_z. \tag{7.81b}$$

We integrate the latter equation with respect to z and obtain

$$\vec{B}_\parallel - i\frac{\omega}{c^2} \epsilon_r \mu_r \left(\hat{e}_z \times \int \vec{E}_\parallel dz\right) = \int \vec{\nabla}_\parallel B_z dz. \tag{7.82}$$

From Eq. (7.81a), we have upon integration with respect to z,

$$\vec{E}_\parallel = -i\omega \left(\hat{e}_z \times \int \vec{B}_\parallel dz\right). \tag{7.83}$$

Inserting Eq. (7.83) into (7.82), we have the following relation,

$$\vec{B}_\parallel - \frac{\omega^2}{c^2} \epsilon_r \mu_r \left(\hat{e}_z \times \left(\hat{e}_z \times \int \vec{B}_\parallel dz\right)\right) = \int \vec{\nabla}_\parallel B_z dz. \tag{7.84}$$

Using Eq. (7.9), this becomes

$$\vec{B}_\parallel + \frac{\omega^2}{c^2} \epsilon_r \mu_r \iint \vec{B}_\parallel dz dz = \int \vec{\nabla}_\parallel B_z dz. \tag{7.85}$$

Within the *ansatz* (7.79), a double integration with respect to z is equivalent to a multiplication by $-(d/(p\pi))^2$ [see also Eq. (7.66)], and so

$$\vec{B}_\parallel \left[1 - \frac{\omega^2}{c^2} \epsilon_r \mu_r \left(\frac{d}{p\pi}\right)^2\right] = \int \vec{\nabla}_\parallel B_z dz. \tag{7.86}$$

We can rewrite this as

$$\left(\frac{d}{p\pi}\right)^2 \vec{B}_\parallel \left[\left(\frac{p\pi}{d}\right)^2 - \frac{\omega^2}{c^2} \epsilon_r \mu_r\right] = \int \vec{\nabla}_\parallel B_z dz. \tag{7.87}$$

Because the z component of the magnetic field given in (7.79) has to fulfill the Helmholtz equation (7.59) separately, we require that

$$\overline{\gamma}_p^2 = \frac{\omega^2}{c^2} \epsilon_r \mu_r - \left(\frac{p\pi}{d}\right)^2. \tag{7.88}$$

Hence,

$$-\left(\frac{p\pi}{d}\right)^{-2} \overline{\gamma}_p^2 \vec{B}_\parallel = \int \vec{\nabla}_\parallel B_z dz. \tag{7.89}$$

We can then trivially integrate the *ansatz* (7.79) with respect to z, to simplify the right-hand side,

$$-\left(\frac{p\pi}{d}\right)^{-2} \overline{\gamma}_p^2 \vec{B}_\parallel = -\left(\frac{p\pi}{d}\right) B_0 \vec{\nabla}_\parallel J_m(\overline{\gamma}_p \rho) \cos\left(\frac{p\pi z}{d}\right) e^{im\varphi} e^{-i\omega t}. \tag{7.90}$$

Solving for $\vec{B}_\parallel$, we obtain

$$\vec{B}_\parallel = B_0 \left(\frac{p\pi}{d}\right) \frac{1}{\overline{\gamma}_p^2} \vec{\nabla}_\parallel J_m(\overline{\gamma}_p \rho) \cos\left(\frac{p\pi z}{d}\right) e^{im\varphi} e^{-i\omega t}. \tag{7.91}$$

The boundary condition is fulfilled if

$$\overline{\gamma}_p^2 = \frac{\omega^2}{c^2} \epsilon_r \mu_r - \left(\frac{p\pi}{d}\right)^2 = \left(\frac{\overline{x}_{mn}}{R}\right)^2. \tag{7.92}$$

Solving for ω and setting $\omega = \omega_{mnp}$, we have

$$\omega_{mnp} = \frac{c}{\sqrt{\epsilon_r \mu_r}} \sqrt{\left(\frac{p\pi}{d}\right)^2 + \left(\frac{\overline{x}_{mn}}{R}\right)^2}. \tag{7.93}$$

Finally, we can determine $\vec{E}_\parallel$ from Eq. (7.81a) after z integration,

$$\vec{E}_\parallel = -i\omega \left(\hat{e}_z \times \int \vec{B}_\parallel dz\right),$$

$$= -i\omega B_0 \frac{1}{\overline{\gamma}_p^2} \left(\hat{e}_z \times \vec{\nabla}_\parallel J_m(\overline{\gamma}_p \rho)\right) \sin\left(\frac{p\pi z}{d}\right) e^{im\varphi} e^{-i\omega t}. \tag{7.94}$$

It now becomes clear why in the *ansatz* (7.79), we need to choose the term with the sine as opposed to the cosine. Just as for the TM modes, the requirement is that the in-plane field $\vec{E}_\parallel$ vanish inside the endcap surfaces at $z = 0$ and $z = d$. A cosine term in Eq. (7.79) would lead to a cosine term in Eq. (7.94), leading to an inconsistency inside the endcap surfaces at $z = 0$ and $z = d$.

On the other hand, the sine term in Eq. (7.79) [as opposed to the cosine term in Eq. (7.62)] may lead to a problem in the case $p = 0$, because in this case, the "generating" z component of the magnetic field (7.79) and the in-plane component of the electric field (7.94) vanish altogether. Specifically,

$$B_z = B_0 J_m(\overline{\gamma}_p \rho) \sin\left(\frac{p\pi z}{d}\right) e^{im\varphi} e^{-i\omega t} = 0, \qquad (p = 0), \tag{7.95}$$

and

$$\vec{E}_\parallel = -i\omega B_0 \frac{1}{\overline{\gamma}_p^2} \hat{e}_z \times \vec{\nabla}_\parallel J_m(\overline{\gamma}_p \rho) \sin\left(\frac{p\pi z}{d}\right) e^{im\varphi} e^{-i\omega t} = \vec{0}, \qquad (p = 0). \tag{7.96}$$

However, one might counter argue that the in-plane component (7.91) of the magnetic field does not vanish for $p = 0$,

$$\vec{B}_\parallel = B_0 \left(\frac{p\pi}{d}\right) \frac{1}{\overline{\gamma}_p^2} \vec{\nabla}_\parallel J_m(\overline{\gamma}_p \rho) \cos\left(\frac{p\pi z}{d}\right) e^{im\varphi} e^{-i\omega t} \neq 0, \qquad (p = 0), \tag{7.97}$$

and so the mode with $p = 0$ might still exist as a nontrivial field configuration. The final blow to the idea of a $p = 0$ comes when we consider that we cannot fulfill Eq. (7.18),

$$\nabla_z \vec{B}_\parallel - i \frac{\omega}{c^2} \epsilon_r \mu_r \left(\hat{e}_z \times \vec{E}_\parallel\right) = \vec{\nabla}_\parallel B_z, \tag{7.98}$$

because $B_z = 0$, and $\vec{E}_\parallel = \vec{0}$, but $\nabla_z \vec{B}_\parallel \neq 0$. This means that the mode with $p = 0$ is excluded for TE modes, in contrast to TM modes.

We summarize the result for TE modes in a cylindrical cavity,

$$B_z = B_0 J_m(\overline{\gamma}_p \rho) \sin\left(\frac{p\pi z}{d}\right) e^{im\varphi} e^{-i\omega t}, \qquad E_z = 0, \tag{7.99a}$$

$$\vec{E}_\parallel = -i\omega B_0 \frac{1}{\overline{\gamma}_p^2} \hat{e}_z \times \vec{\nabla}_\parallel J_m(\overline{\gamma}_p \rho) \sin\left(\frac{p\pi z}{d}\right) e^{im\varphi} e^{-i\omega t}, \tag{7.99b}$$

$$\vec{B}_\parallel = B_0 \left(\frac{p\pi}{d}\right) \frac{1}{\overline{\gamma}_p^2} \vec{\nabla}_\parallel J_m(\overline{\gamma}_p \rho) \cos\left(\frac{p\pi z}{d}\right) e^{im\varphi} e^{-i\omega t}, \tag{7.99c}$$

where the radial quantization is taken into account by the formulas,

$$\overline{\gamma}_p^2 = \overline{\gamma}_{mnp}^2 = \frac{\omega^2}{c^2} \epsilon_r \mu_r - \left(\frac{p\pi}{d}\right)^2 = \left(\frac{\overline{x}_{mn}}{R}\right)^2, \tag{7.99d}$$

$$\omega = \overline{\omega}_{mnp} = \frac{c}{\sqrt{\epsilon_r \mu_r}} \sqrt{\left(\frac{p\pi}{d}\right)^2 + \left(\frac{\overline{x}_{mn}}{R}\right)^2}, \qquad J_m'(\overline{x}_{mn}) = 0. \tag{7.99e}$$

According to Eq. (7.80), $\overline{x}_{mn}$ is the nth zero of the derivative of the Bessel function of order m. Just like for TM modes, the quantum numbers are m (azimuthal), n (radial) and p (longitudinal). One might think that the fundamental TM mode in a cylindrical cavity is the 011 mode ($m = 0$, $n = 1$, and $p = 1$). The reasoning is as follows: The lowest-order Bessel function has $m = 0$, the first zero of its derivative is at $n = 1$, and we have to have at least $p = 1$ because the magnetic field B_z is otherwise vanishing. The numerical value is easily found as $\overline{x}_{01} = 3.831\,705\,970\ldots$. However, according to Table 2.2, the first zero of the Bessel function of order $m = 1$ actually occurs before the first zero of $J_0(x)$, namely, at $\overline{x}_{11} = 1.841\,183\,781\ldots$. Hence, the corresponding angular frequency for the fundamental TE mode is

$$\overline{\omega}_{111} = \frac{c}{\sqrt{\epsilon_r \mu_r}} \sqrt{\left(\frac{\pi}{d}\right)^2 + \left(\frac{\overline{x}_{11}}{R}\right)^2} \approx \frac{c}{\sqrt{\epsilon_r \mu_r}} \sqrt{\left(\frac{\pi}{d}\right)^2 + \left(\frac{1.841\,183}{R}\right)^2}. \tag{7.100}$$

TE modes are useful for giving a transverse deflection to a beam in an accelerator, but are not much use for providing acceleration.

7.3.2 *Resonant Rectangular Cavities*

In order to describe the resonant TE and TM eigenmodes of a rectangular cavity, it is advantageous to start from the vector potential, i.e., from an *ansatz* where

$$\vec{A}(\vec{r}, t) = \vec{A}(\vec{r}) e^{-i\omega t}, \qquad \vec{\nabla} \cdot \vec{A}(\vec{r}) = 0, \qquad \Phi(\vec{r}, t) = 0, \tag{7.101}$$

so that

$$\vec{B}(\vec{r}, t) = \vec{\nabla} \times \vec{A}(\vec{r}, t), \qquad \vec{E}(\vec{r}, t) = -\frac{\partial}{\partial t} \vec{A}(\vec{r}, t). \tag{7.102}$$

We shall first derive a wave equation for the vector potential, and start from the Ampere–Maxwell law in the absence of source terms,

$$\vec{\nabla} \times \vec{H}(\vec{r}, \omega) = -i\omega \vec{D}(\vec{r}, \omega). \tag{7.103}$$

In a material with relative permittivity ϵ_r and relative permeability μ_r, one has

$$\vec{H}(\vec{r},\omega) = \frac{1}{\mu_r(\omega)\,\mu_0}\,\vec{B}(\vec{r},\omega) = \frac{1}{\mu_r(\omega)\,\mu_0}\,\vec{\nabla}\times\vec{A}(\vec{r},\omega) \tag{7.104}$$

and

$$\vec{D}(\vec{r},\omega) = \epsilon_r(\omega)\,\epsilon_0\,\vec{E}(\vec{r},\omega) = i\,\epsilon_r\,\epsilon_0\,\omega\vec{A}(\vec{r},\omega)\,. \tag{7.105}$$

Inserting Eqs. (7.104) and (7.105) into (7.103), one obtains

$$\frac{1}{\mu_r(\omega)\mu_0}\,\vec{\nabla}\times\left(\vec{\nabla}\times\vec{A}(\vec{r},\omega)\right) = \epsilon_r(\omega)\,\epsilon_0\,\omega^2\,\vec{A}(\vec{r},\omega)\,, \tag{7.106}$$

which for $\vec{\nabla}\cdot\vec{A} = 0$ [see Eq. (7.101)] results in

$$\vec{\nabla}\times\left(\vec{\nabla}\times\vec{A}(\vec{r},\omega)\right) = \vec{\nabla}\,\left(\vec{\nabla}\cdot\vec{A}(\vec{r},\omega)\right) - \vec{\nabla}^2\vec{A}(\vec{r},\omega)$$

$$= -\vec{\nabla}^2\vec{A}(\vec{r},\omega) = \epsilon_r(\omega)\mu_r(\omega)\frac{\omega^2}{c^2}\,\vec{A}(\vec{r},\omega)\,. \tag{7.107}$$

We thus obtain the wave equation for the vector potential,

$$\left(\vec{\nabla}^2 + \epsilon_r(\omega)\mu_r(\omega)\frac{\omega^2}{c^2}\right)\vec{A}(\vec{r},\omega) = 0\,. \tag{7.108}$$

We write the vector $\vec{k}$ as follows,

$$k_x = k\,\sin\theta\,\cos\varphi\,, \qquad k_y = k\,\sin\theta\,\sin\varphi\,, \qquad k_z = k\,\cos\theta\,, \tag{7.109a}$$

$$\vec{k} = k_x\,\hat{e}_x + k_y\,\hat{e}_y + k_z\,\hat{e}_z\,. \tag{7.109b}$$

For the later discussion, it is instructive to remember that the angles θ and φ belong to $\vec{k}$, not to the coordinate vector $\vec{r}$. We now define the two polarization vectors for TE and TM modes,

$$\hat{\epsilon}_{\vec{k},\mathrm{TE}} = \sin\varphi\,\hat{e}_x - \cos\varphi\,\hat{e}_y\,, \tag{7.110a}$$

$$\hat{\epsilon}_{\vec{k},\mathrm{TM}} = -\cos\theta\,\cos\varphi\,\hat{e}_x - \cos\theta\,\sin\varphi\,\hat{e}_y + \sin\theta\,\hat{e}_z\,. \tag{7.110b}$$

$$\vec{k}\cdot\hat{\epsilon}_{\vec{k},\mathrm{TE}} = \vec{k}\cdot\hat{\epsilon}_{\vec{k},\mathrm{TM}} = 0 \qquad \hat{\epsilon}_{\vec{k},\mathrm{TM}}\times\hat{\epsilon}_{\vec{k},\mathrm{TE}} = \hat{k} = \vec{k}/|\vec{k}|\,. \tag{7.110c}$$

We write the vector potential for the mode functions as

$$\vec{A}_{\vec{k},\lambda}(\vec{r},t) = A_{\vec{k},\lambda,x}(\vec{r},t)\,\hat{e}_x + A_{\vec{k},\lambda,y}(\vec{r},t)\,\hat{e}_y + A_{\vec{k},\lambda,z}(\vec{r},t)\,\hat{e}_z\,, \tag{7.111}$$

where λ can be TE or TM, and

$$A_{\vec{k},\lambda,x} = \mathcal{A}_0\sqrt{\frac{8}{V}}\,\hat{\epsilon}_{\vec{k},\lambda,x}\,\cos(k_x x)\,\sin(k_y y)\,\sin(k_z z)\,\mathrm{e}^{-i\omega t}\,, \tag{7.112a}$$

$$A_{\vec{k},\lambda,y} = \mathcal{A}_0\sqrt{\frac{8}{V}}\,\hat{\epsilon}_{\vec{k},\lambda,y}\,\sin(k_x x)\,\cos(k_y y)\,\sin(k_z z)\,\mathrm{e}^{-i\omega t}\,, \tag{7.112b}$$

$$A_{\vec{k},\lambda,z} = \mathcal{A}_0\sqrt{\frac{8}{V}}\,\hat{\epsilon}_{\vec{k},\lambda,z}\,\sin(k_x x)\,\sin(k_y y)\,\cos(k_z z)\,\mathrm{e}^{-i\omega t}\,. \tag{7.112c}$$

Here, $V = L_x L_y L_z$, and $\mathcal{A}_0$ is a global amplitude normalized so that

$$\int_V d^3r \left| \vec{A}_{\vec{k},\lambda}(\vec{r},t) \right|^2 = \mathcal{A}_0^2. \tag{7.113}$$

The polarization vector for the mode with wave vector $\vec{k}$ and polarization λ is

$$\hat{\epsilon}_{\vec{k},\lambda} = \epsilon_{\vec{k},\lambda,x}\, \hat{e}_x + \epsilon_{\vec{k},\lambda,y}\, \hat{e}_y + \epsilon_{\vec{k},\lambda,z}\, \hat{e}_z\,; \tag{7.114}$$

it can describe a TE or TM mode, according to Eq. (7.110). The components of the wave vectors for the discrete modes are given as follows,

$$k_x = \frac{\ell\pi}{L_x}, \qquad k_y = \frac{m\pi}{L_y}, \qquad k_z = \frac{n\pi}{L_z}, \qquad \vec{k}_{\ell mn} = \frac{\ell\pi}{L_x}\hat{e}_x + \frac{m\pi}{L_y}\hat{e}_y + \frac{n\pi}{L_z}\hat{e}_z\,. \tag{7.115}$$

where ℓ, m, n are integers. The electric field corresponding to the polarization $\lambda = \text{TE}$ is transverse, i.e., its z component vanishes, while the magnetic induction field corresponding to the polarization $\lambda = \text{TM}$ is transverse, i.e., its z component vanishes. The boundary conditions for the electric field and the boundaries of the rectangular cavity are fulfilled if Eq. (7.115) is fulfilled.

The transversality condition (7.101) for the vector potential is verified as follows,

$$\vec{\nabla} \cdot \vec{A}_{\vec{k},\lambda}(\vec{r}) = - \mathcal{A}_0 \sqrt{\frac{8}{V}} \left(\vec{k} \cdot \hat{\epsilon}_{\vec{k},\lambda} \right) \sin(k_x x)\, \sin(k_y y)\, \sin(k_z z) = 0\,, \tag{7.116}$$

which is fulfilled if $\vec{k} \cdot \hat{\epsilon}_{\vec{k},\lambda} = 0$.

The wave equation (7.108), applied to the vector potential (7.112), results in the condition

$$\left(-\vec{k}^2 + \epsilon_r(\omega)\mu_r(\omega)\frac{\omega^2}{c^2} \right) \vec{A}(\vec{r},\omega) = 0\,, \tag{7.117}$$

and thus in component form,

$$k_x^2 + k_y^2 + k_z^2 = \frac{\epsilon_r\,\mu_r\,\omega^2}{c^2}\,, \tag{7.118}$$

where we suppress the ω argument of ϵ_r and μ_r. Together with Eq. (7.115), this leads to the "quantization condition"

$$\omega = \omega_{\ell mn} = \frac{c}{\sqrt{\epsilon_r\,\mu_r}} \sqrt{ \left(\frac{\ell\pi}{L_x} \right)^2 + \left(\frac{m\pi}{L_y} \right)^2 + \left(\frac{n\pi}{L_z} \right)^2 }\,,$$

$$\ell = 0,1,2,\ldots, \qquad m = 0,1,2,\ldots, \qquad n = 0,1,2,\ldots\,. \tag{7.119}$$

In the rectangular waveguide, we have no continuous symmetry axis. Still, three quantum numbers and the polarization λ suffice to characterize all possible modes.

Let us try to identify the fundamental mode. In order for the vector potential in Eq. (7.112) to be nonvanishing, two of the "quantum numbers" ℓ, m, and n must be nonvanishing. Otherwise, the entire vector potential in Eq. (7.112) vanishes. First, let L_y and L_z be the longest edges of the rectangular cavity. Then, the angular frequency of the fundamental mode is

$$
\omega_{011} = \frac{c}{\sqrt{\epsilon_r \mu_r}} \sqrt{\left(\frac{\pi}{L_y}\right)^2 + \left(\frac{\pi}{L_z}\right)^2}. \tag{7.120}
$$

Expressed differently, the fundamental mode carries the quantum numbers $\ell = 0$, $m = 1$ and $n = 1$, with frequency ω_{011} given in Eq. (7.120). For the fundamental mode, we thus have $k_x = \ell\,\pi/L_x = 0$ and hence the azimuth angle of the $\vec{k}$ vector is $\varphi = 90° = \pi/2$. According to Eq. (7.110), the two polarization vectors are given by

$$
\hat{\epsilon}_{\vec{k}_{011},\mathrm{TE}} = \hat{e}_x, \qquad \hat{\epsilon}_{\vec{k}_{011},\mathrm{TM}} = -\cos\theta\,\hat{e}_y + \sin\theta\,\hat{e}_z, \tag{7.121a}
$$

$$
\cos\theta = \frac{k_z}{\sqrt{k_y^2 + k_z^2}} = \frac{1/L_z}{\sqrt{(1/L_y)^2 + (1/L_z)^2}}. \tag{7.121b}
$$

For $\ell = 0$, $m = 1$ and $n = 1$, the only nonvanishing component of $\vec{A}$ is the x component $A_{\vec{k}_{011},\mathrm{TE},x} \neq 0$ [see Eq. (7.112)]. One may combine Eqs. (7.115) with Eq. (7.112) and observe that $\sin(k_x x) = 0$. Furthermore, because only $\hat{\epsilon}_{\vec{k}_{011},\mathrm{TE}}$ has a nonvanishing x component, the fundamental mode is TE if the longest edges are L_y and L_z.

Second, let L_x and L_y be the longest edges of the rectangular cavity. In this case, the angular frequency of the fundamental mode is

$$
\omega_{110} = \frac{c}{\sqrt{\epsilon_r \mu_r}} \sqrt{\left(\frac{\pi}{L_x}\right)^2 + \left(\frac{\pi}{L_y}\right)^2}. \tag{7.122}
$$

In this case, the polar angle of the $\vec{k}$ vector is $\theta = 90°$. According to Eq. (7.110), the two relevant polarization vectors are

$$
\hat{\epsilon}_{\vec{k}_{110},\mathrm{TE}} = \sin\varphi\,\hat{e}_x - \cos\varphi\,\hat{e}_y, \qquad \hat{e}_{\vec{k}_{110},\mathrm{TM}} = \hat{e}_z, \tag{7.123a}
$$

$$
\cos\varphi = \frac{k_x}{\sqrt{k_x^2 + k_y^2}} = \frac{1/L_x}{\sqrt{(1/L_x)^2 + (1/L_y)^2}}. \tag{7.123b}
$$

For $\ell = 1$, $m = 1$ and $n = 0$, the only nonvanishing component of $\vec{A}$ is the z component $A_{\vec{k}_{110},\mathrm{TE},z} \neq 0$. Furthermore, because only $\hat{\epsilon}_{\vec{k}_{110},\mathrm{TM}}$ has a nonvanishing z component, the fundamental mode is a TM mode if the shortest edge is L_z.

Let us try to verify once more, the explicit fulfillment of the boundary conditions by the electric and magnetic fields generated by the vector potentials indicated in Eq. (7.112). We recall that, in view of Eq. (7.102), we have $\vec{E} = -\partial\vec{A}/\partial t = i\omega\vec{A}$. The transversal (normal) component of the electric field does not need to vanish at the boundaries. Indeed we have, e.g., for the planes with $z = 0$ or at $z = L_z$,

$$
E_{\vec{k},\lambda,z}(x, y, L = 0) = E_{\vec{k},\lambda,z}(x, y, L = L_z) = -i\omega\,\mathcal{A}_0 \sqrt{\frac{8}{V}}\,\epsilon_{\vec{k},\lambda,z}\,\sin(k_x x)\,\sin(k_y y). \tag{7.124}
$$

However, the mode functions (7.112) are such that there is a cosine function if the Cartesian component of the wave vector in the argument of the trigonometric function is the same Cartesian component as the vector potential component itself. The presence of the sine functions implies that the tangential component of the vector potential (the component lying inside the boundary planes) vanishes for all three boundaries,

$$A_{\vec{k},\lambda,x}(x,y,z=0)=A_{\vec{k},\lambda,x}(x,y,z=L_z)=A_{\vec{k},\lambda,x}(x,y=0,z)=A_{\vec{k},\lambda,x}(x,y=L_y,z)=0,$$
$$(7.125a)$$

$$A_{\vec{k},\lambda,y}(x,y,z=0)=A_{\vec{k},\lambda,y}(x,y,z=L_z)=A_{\vec{k},\lambda,y}(x=0,y,z)=A_{\vec{k},\lambda,y}(x=L_x,y,z)=0,$$
$$(7.125b)$$

$$A_{\vec{k},\lambda,z}(x=0,y,z)=A_{\vec{k},\lambda,z}(x=L,y,z)=A_{\vec{k},\lambda,z}(x,y=0,z)=A_{\vec{k},\lambda,z}(x,y=L_y,z)=0.$$
$$(7.125c)$$

In view of $\vec{E} = -\partial\vec{A}/\partial t = i\omega\vec{A}$, the conditions (7.125) are exactly equivalent to the boundary condition for the electric field, given in Eq. (7.25), namely, $\vec{E}_{\parallel,S} = 0$.

The next step is to look at the $\vec{B}$ field. An explicit calculation of the curl of the vector potential given in Eq. (7.112) leads to the following result for $\vec{B} = \vec{\nabla} \times \vec{A}$,

$$B_{\vec{k},\lambda,x} = \mathcal{A}_0 \sqrt{\frac{8}{V}} \left(\epsilon_{\vec{k},\lambda,z}\, k_y - \epsilon_{\vec{k},\lambda,y}\, k_z\right) \sin(k_x x)\, \cos(k_y y)\, \cos(k_z z)\, e^{-i\omega t}, \quad (7.126a)$$

$$B_{\vec{k},\lambda,y} = \mathcal{A}_0 \sqrt{\frac{8}{V}} \left(\epsilon_{\vec{k},\lambda,x}\, k_z - \epsilon_{\vec{k},\lambda,z}\, k_x\right) \cos(k_x x)\, \sin(k_y y)\, \cos(k_z z)\, e^{-i\omega t}, \quad (7.126b)$$

$$B_{\vec{k},\lambda,z} = \mathcal{A}_0 \sqrt{\frac{8}{V}} \left(\epsilon_{\vec{k},\lambda,y}\, k_x - \epsilon_{\vec{k},\lambda,x}\, k_y\right) \cos(k_x x)\, \cos(k_y y)\, \sin(k_z z)\, e^{-i\omega t}. \quad (7.126c)$$

Here, the sine term depends on the same coordinate as the component of the $\vec{B}$ field; it means that the component in the direction of the normal to the boundary surface vanishes on the boundaries, fulfilling Eq. (7.27).

In our considerations leading up to Eqs. (7.121) and (7.123), we considered the fundamental mode for the cases of the shortest edge length being in the x and z directions. We found that we have a special situation where one of the mode functions given in Eq. (7.112) vanishes, and we only have one polarization available. In this case, one of the quantum numbers ℓ, m, n is zero. Let us try to generalize the consideration somewhat. If either ℓ, m, n is zero, then according to Eqs. (7.115) and (7.112), only one of the components of the vector potential $\vec{A}$ "survives"; the others vanish. The only valid polarization vector points into the same Cartesian direction as the vanishing component of $\vec{k}$. If $\ell = 0$, then we need to have $\hat{\epsilon}_{\vec{k},\lambda} = \hat{e}_x$, and if $n = 0$, then $\hat{\epsilon}_{\vec{k},\lambda} = \hat{e}_z$. This can be verified on the examples given in Eqs. (7.121) and (7.123) and generalizes to any mode with one vanishing quantum number.

7.4 Exercises

• **Exercise 7.1**: Reproduce the derivation from Eq. (7.1) to (7.24) in your own words, conceivably letting $\epsilon_r = \mu_r = 1$, but being aware of the fact that $\omega \neq ck$ for waves in the waveguide.

• **Exercise 7.2**: Start from the dispersion relation (7.44) for electromagnetic waves in a rectangular waveguide,

$$k = \frac{\sqrt{\epsilon_r \mu_r}}{c} \sqrt{\omega^2 - \omega_{mn}^2}, \tag{7.127}$$

$$\omega_{mn} = \frac{\pi c}{\sqrt{\epsilon_r \mu_r}} \left[\left(\frac{m}{a}\right)^2 + \left(\frac{n}{b}\right)^2 \right]^{1/2}. \tag{7.128}$$

For each mode, the k_{mn} varies with frequency $\omega > \omega_{mn}$. The ω_{mn} is the cutoff frequency for the mode. Question: How can you generate "slow light" (as far as the group velocity is concerned) in a waveguide? For which frequencies ω does the group velocity become zero?

• **Exercise 7.3**: Tabulate the minimum frequencies ω_{mn} for the TE and TM modes of a rectangular waveguide of dimension $a = 2.3\,\text{cm}$ and $b = 1.7\,\text{cm}$. Perform the same task for the TE and TM modes with $m \le 2$ and $n \le 2$. Then, plot the dispersion relation for a rectangular waveguide and try to reproduce the findings of Figs. 7.6–7.8 regarding the small-frequency and large-frequency asymptotics. Give the results to an accuracy of at least 6 decimals.

• **Exercise 7.4**: Tabulate the first few energy eigenvalues $E_{mnp} = \hbar \omega_{mnp}$ in the spectrum of a perfectly conducting cylindrical cavity with $R = 2.3\,\text{cm}$ and height $d = 1.9\,\text{cm}$. Consider the cases $m = 0, 1$, $n = 1, 2, 3$, and $p = 0, 1, 2$. Give the results to an accuracy of at least 6 decimals.

• **Exercise 7.5**: Try to devise dimensions of a cylindrical cavity adapted to generate radiation whose frequency is equal to that of X band radar at 7175 MHz.

• **Exercise 7.6**: We have determined the waves of a rectangular waveguide, as well as the modes of cylindrical and rectangular cavities. Try to determine the waves of a cylindrical waveguide. Remember that you only need to devise an *ansatz* for the z component of the fields; the rest follows.

    Hint: You might use the following *ansatz* for the z component of the electric field, for TM modes,

$$E_z = E_0 \, J_m \left(\frac{x_{mn}}{R} \rho \right) e^{im\varphi} \, e^{-i\omega t}, \tag{7.129}$$

where x_{mn} is defined in Eq. (2.158). Note that E_z must vanish on the boundaries of the cylindrical waveguide. For TE modes, you might choose

$$B_z = B_0 \, J_m \left(\frac{\overline{x}_{mn}}{R} \rho \right) e^{im\varphi} \, e^{-i\omega t}, \tag{7.130}$$

where $\overline{x}_{mn}$ is defined in Eq. (2.159). Note that B_z must vanish on the boundaries of the cylindrical waveguide.

• **Exercise 7.7**: Show that (7.101) fulfills both Coulomb gauge as well as Lorenz gauge conditions.

• **Exercise 7.8**: Convince yourself that you can write

$$\vec{\nabla}_{\parallel} J_m(\gamma_p \rho) = \hat{e}_\rho \, J'_m(\gamma_p \rho). \tag{7.131}$$

Also, show that

$$\hat{e}_z \times \vec{\nabla}_{\parallel} J_m(\gamma_p \rho) = (\hat{e}_z \times \hat{e}_\rho) J'_m(\gamma_p \rho) = \hat{e}_\varphi J'_m(\gamma_p \rho) \tag{7.132}$$

and same for $\gamma_p \to \overline{\gamma}_p$. With these results at hand, try to decompose the formulas given in Eqs. (7.76) and (7.99) into components parallel to $\hat{e}_\rho$ and $\hat{e}_\varphi$.

• **Exercise 7.9**: Adjust the constant $\mathcal{A}_0$ in Eq. (7.113) so that the field energy stored in the fundamental mode is equal to the energy of a single photon, $\hbar\omega$, in contrast to the normalization condition (7.113).

Chapter 8

Advanced Topics

8.1 Overview

In the current chapter, we shall attempt to cover a few advanced topics, such as Lorentz transformations of the electromagnetic fields and the transition to quantum field theory. It is instructive to put this endeavor into context.

When the Maxwell equations were discovered, and the formalism of (special) relativity was developed, it was observed that the Maxwell theory was in fact compatible with the formalism of special relativity. Thus, it was not necessary to add any "relativistic corrections" to these equations, as opposed to the equations of classical mechanics, which require numerous relativistic correction terms. How exactly do the electromagnetic fields transform under the action of the Lorentz group? How do we transform, e.g., the Ampere–Maxwell law into a moving coordinate system?

We shall try to approach this problem carefully, by first recalling a few basic facts about Lorentz transformations, then developing some basic aspects of the representation theory of the Lorentz group, before applying this formalism to the Maxwell theory. In the latter step, we shall assign a specific behavior to the potentials and fields under Lorentz transformations, and understand the concept of a four-vector, which groups certain entities into a vector with four components, all of which transform jointly under Lorentz transformations. An example is the four-vector $(\rho, c^{-1}\vec{J})$ composed of charge and current density, which transforms as a joint four-vector under a Lorentz transformation, i.e., the charge density transforms into the current density and vice versa. This endeavor will comprise Chaps. 8.2 and 8.3.

Up to this current chapter, all subjects treated in this book required only the classical theory as input. In Sec. 8.4, we shall attempt to "build a bridge" toward the quantum theory of the electromagnetic field, known as quantum electrodynamics (QED). In principle, quantum electrodynamics is a relativistic quantum field theory. Here, the word "relativistic" means that the field operators, and all the couplings, are written down such as to be compatible with Lorentz invariance [20]. In Sec. 8.2, we shall see that Lorentz and spinor indices qualify themselves as indices which transform under specific representations of the Lorentz group. QED describes processes in which virtual quanta are created and annihilated, and particles (and antiparticles) are annihilated and created. Furthermore, the important entities (scalar and vector potentials, and fields) are functions of space and time. By contrast, classical electrodynamics is a field theory because all potentials (scalar and vector) as well as the fields are functions of space-time, but classical electrodynamics does not constitute a quantum theory, where quanta of the electromagnetic radiation (photons) are created and annihilated.

In Sec. 8.4, we attempt to approach the concept of a quantum field theory by treating the field quantization. We shall see that it is possible to calculate the Casimir attractive force between plates in a semi-quantized formalism where we simply assign a nonvanishing zero-point energy to all harmonic modes of the electromagnetic field. This formalism avoids a full quantization of the electromagnetic field as well as any mention of relativity while allowing us to study a truly quantum effect, namely, the attractive force between perfectly conducting plates due to the shift in the zero-point energy of the modes in the space in between them. Specifically, some otherwise available electromagnetic oscillation modes are missing from the spectrum of available oscillator modes under the influence of the boundary conditions; these missing modes generate a "lack of zero-point energy" which leads to an attractive force. The closer the two plates are, the greater is the influence of the quantization, and the more "zero-point energy is missing." Hence, the system can lower its energy by moving the plates closer, and this generates the Casimir force.

Two of the most important concepts in quantum field theory are the regularization and renormalization. We shall discuss these in Sec. 8.4, in a "nutshell'. So-called loop integrals in quantum field theory require regularization and renormalization in higher orders. An aura of mystery and paradox still overshines these concepts. It is less well known that regularization and renormalization can also be applied with advantage, to the calculation of classical, electrostatic potentials (see Sec. 8.5). Divergences typically result when the expression $1/|\vec{r} - \vec{r}'|$ is integrated over source configurations [charge densities $\rho(\vec{r}')$] with an infinite extent. In this case, the naturally occurring divergences can be regularized, and the potential can be renormalized, using concepts borrowed from field theory. The calculation gives us an opportunity to introduce dimensional regularization, which is an important concept in modern field theory, and the concept of a renormalization condition. After these preparations, a reader of this book should be well prepared to study further literature on quantum electrodynamics, or indeed, quantum field theory [20].

Finally, in Sec. 8.6, we study open problems in the field, such as the complete understanding of the radiative reaction problem, which is still a subject of active research, and the connection of relativistic electrodynamics and general relativity. In the latter endeavor, we need to generalize the gradient vector, and the Laplacian, to their respective forms relevant to curved space-times.

8.2 Lorentz Transformations, Generators and Matrices

8.2.1 *Lorentz Boosts*

We shall begin the endeavor by first recalling a few basic facts about Lorentz transformations. A Lorentz boost is a Lorentz transformation which describes the transformation of time and space into moving frames of reference. When combined with rotations of space, Lorentz boosts describe all possible coordinate transformations of space and time compatible with special relativity. In SI mksA units, the most basic Lorentz transformation for the transformation from the unprimed frame (laboratory frame) into the moving (primed) frame reads as

$$t' = \gamma t - \frac{v}{c^2}\gamma x, \qquad x' = \gamma x - v\gamma t, \qquad y' = y, \qquad z' = z. \tag{8.1}$$

Here, we assume that the moving frame has a uniform velocity v in the positive x direction (so-called Lorentz boost in the x direction). The Lorentz factor is

$$\gamma = \frac{1}{\sqrt{1 - v^2/c^2}}. \tag{8.2}$$

The transformation goes from the laboratory frame to the moving frame; the primed quantities are those measured in the moving frame. It is customary to define the four-vector (ct, x, y, z) of space-time coordinates so that all entries have the same physical dimension (corresponding to a spatial coordinate), i.e.,

$$ct' = \gamma ct - \left(\gamma \frac{v}{c}\right) x, \qquad x' = \gamma x - \left(\gamma \frac{v}{c}\right)(ct), \qquad y' = y, \qquad z' = z. \tag{8.3}$$

It is interesting to verify that the backtransformation from the moving to the rest frame is achieved by replacing $v \to -v$. That means that

$$ct = \gamma ct' + \left(\gamma \frac{v}{c}\right) x', \qquad x = \gamma x' + \left(\gamma \frac{v}{c}\right) ct', \qquad y = y', \qquad z = z'. \tag{8.4}$$

We shall now generalize the formalism to a Lorentz boost into a frame that moves with velocity $\vec{v}$,

$$ct' = \gamma ct - \left(\gamma \frac{v}{c}\right)(\hat{v} \cdot \vec{r}) = \gamma \left(ct - \frac{\vec{v}}{c} \cdot \vec{r}\right), \tag{8.5a}$$

$$\vec{r}' = \vec{r} + (\gamma - 1)\frac{\vec{v} \cdot \vec{r}}{\vec{v}^2}\vec{v} - \left(\gamma \frac{v}{c}\right)\hat{v}(ct), \tag{8.5b}$$

where $\hat{v} = \vec{v}/v$. The quantity $(ct')^2 - \vec{r}'^2 = (ct)^2 - \vec{r}^2$ is invariant under all Lorentz boosts.

The transformation property (8.5) describes how time and space transform into each other when we change the frame of reference. It is based on the fact that time and space are not independent physical quantities under Lorentz transformations, but in fact, transform into each other as we change the frame of reference. Indeed, $(ct, \vec{r})$ constitutes a so-called four-vector. Just as much as the quantities $(ct, \vec{r})$ constitute a four-vector, there is also a four-vector composed of energy and momentum

$(E, c\vec{p})$. The transformation properties of all Lorentz four-vectors have to be the same; otherwise we could not call them Lorentz *vectors*. We thus conclude that

$$E' = \gamma E - \left(\gamma \frac{v}{c}\right) \hat{v} \cdot (c\vec{p}) = \gamma \left(E - \vec{v} \cdot \vec{p}\right), \tag{8.6a}$$

$$c\vec{p}' = c\vec{p} + (\gamma - 1) \frac{\vec{v} \cdot (c\vec{p})}{\vec{v}^2} \vec{v} - \left(\gamma \frac{v}{c}\right) \hat{v} E. \tag{8.6b}$$

For a massless quantum particle, we can relate the energy to the angular frequency $(E = \hbar\omega)$ and the momentum to the wave vector $(p = \hbar k)$. Quite generally, the entities $(\omega, c\vec{k})$ constitute a four-vector for any electromagnetic or massless matter wave, and so we have

$$\omega' = \gamma \omega - \left(\gamma \frac{v}{c}\right) \hat{v} \cdot (c\vec{k}) = \gamma \left(\omega - \vec{v} \cdot \vec{k}\right), \tag{8.7a}$$

$$c\vec{k}' = c\vec{k} + (\gamma - 1) \frac{\vec{v} \cdot (c\vec{k})}{\vec{v}^2} \vec{v} - \left(\gamma \frac{v}{c}\right) \hat{v} \omega. \tag{8.7b}$$

We are now in the position to investigate the consequences of Lorentz transformations regarding basic phenomenology of special relativity.

8.2.2   *Time Dilation and Lorentz Contraction*

Time dilation and Lorentz contraction are basic relativistic phenomena which need to be discussed. The derivation is somewhat nontrivial because one has to carefully distinguish between the two observers, the moving observer and the one at rest. The question is who sees which event at which time.

Let us start with time dilation. We investigate one event which is at the origin in both the moving as well as the rest frame,

$$x_1 = 0, \qquad t_1 = 0, \qquad x_1' = 0, \qquad t_1' = 0. \tag{8.8}$$

Now, let us assume that the particle moves at the same speed as the moving frame, namely v, and therefore remains at rest in the moving frame. So, if we look some time later (time t_2'), the position in the moving frame is $x_2' = x_1' = 0$. How much time τ has elapsed for the moving observer? Well, in the rest frame, the particle is at position $x_2 = vt$ at time t, and the second event has the following space-time coordinates,

$$x_2 = vt, \qquad t_2 = t, \qquad x_2' = 0, \qquad t_2' = \tau. \tag{8.9}$$

For consistency, we verify that the relation

$$x_2' = \gamma x_2 - v\gamma t_2 = \gamma vt - v\gamma t = 0 \tag{8.10}$$

is trivially fulfilled. The time t_2', which is the proper time elapsed in the moving coordinate system, can be expressed in terms of the unprimed quantities t_2 and x_2, by virtue of the Lorentz transformation,

$$t_2' = \tau = \gamma t_2 - \frac{v}{c^2} \gamma x_2 = \gamma t - \frac{v}{c^2} \gamma vt = \gamma \left(1 - \frac{v^2}{c^2}\right) t = \sqrt{1 - \frac{v^2}{c^2}} \, t = \frac{1}{\gamma} t. \tag{8.11}$$

Note that the equation

$$\tau = \frac{1}{\gamma} t \le t \tag{8.12}$$

implies that the time that has passed in the frame of the moving observer is less than or equal to the time that has elapsed in the rest frame ($\tau \le t$). This is called time dilation and experimentally confirmed, e.g., by the enhancement of the muon lifetime at accelerator storage rings. The time τ is also called the proper time that elapses in the rest frame of the moving particle.

Length contraction works as follows. We first have to carefully consider what a length measurement actually means. It means that we have two events, one, where we measure the position of the left end of a moving slab, and another one, where the position of the right end of a moving slab is measured. We assume that the first event (left end of moving slab) happens at the following space-time coordinates,

$$x_1 = x_<, \qquad t_1 = 0, \qquad x_1' = x_<', \qquad t_1' = 0. \tag{8.13}$$

The measurement of the right end of the slab happens at the same time $t_2 = 0$ in the rest frame. It results in a position measurement at $x_2' = x_>'$ which is not necesarily happening at $t_2' = 0$. We have

$$x_2 = x_>, \qquad t_2 = 0, \qquad x_2' = x_>', \qquad t_2' = \gamma t_2 - \frac{v}{c^2} \gamma x_2 = -\frac{v}{c^2} \gamma x_>. \tag{8.14}$$

Then,

$$x_>' - x_<' = x_2' - x_1' = \gamma(x_2 - x_1) - v\gamma(t_2 - t_1) = \gamma(x_> - x_<), \tag{8.15}$$

or

$$\Delta x = x_> - x_< = \frac{1}{\gamma}(x_>' - x_<') = \frac{1}{\gamma}\Delta x' \le \Delta x'. \tag{8.16}$$

The length Δx seen in the rest frame is smaller or equal to the length $\Delta x'$ in the moving frame; this is the result of Lorentz contraction.

8.2.3 Addition Theorem

From nonrelativistic intuition, we would say that if a particle moves with velocity u in the positive x direction, then the velocity will be measured as $u' = u - v$ from the point of view of an observer who moves with uniform velocity v in the positive x direction. One simply subtracts the relative velocity v. However, this is not the case anymore if we consider special relativity. The particle velocities are $u = dx/dt$ and $u' = dx'/dt'$, respectively, and the Lorentz transformation tells us how the infinitesimal changes in the positions of the particles transform into each other. We have, according to Eq. (8.3),

$$t' = \gamma t - \frac{v}{c^2} \gamma t, \qquad x' = \gamma x - v\gamma t, \qquad y' = y, \qquad z' = z. \tag{8.17}$$

The differential form of the Lorentz transformation is

$$dx' = \gamma\, dx - v\gamma\, dt, \qquad dt' = \gamma\, dt - \frac{v}{c^2}\gamma\, dx. \tag{8.18}$$

The velocity addition theorem follows quite immediately, dividing dx' by dt',

$$u' = \frac{dx'}{dt'} = \frac{\gamma\,dx - v\,\gamma\,dt}{\gamma\,dt - \dfrac{v}{c^2}\,\gamma\,dx} = \frac{\gamma\,u - \gamma\,v}{\gamma - \dfrac{v}{c^2}\,\gamma\,u} = \frac{u - v}{1 - \dfrac{u\,v}{c^2}}\,. \tag{8.19}$$

The reverse transformation is obtained by inverting $v \to -v$, as with the ordinary Lorentz transformation (8.4),

$$u = \frac{u' + v}{1 + \dfrac{u'\,v}{c^2}}\,. \tag{8.20}$$

From this representation, we see that the magnitude of u can never exceed that of c if the velocity in the moving frame fulfills $-c < u' < c$. This is because u grows monotonically with u' and v, and it reaches its terminal velocity $u = c$ when either u' or v (or both) reach a magnitude of c. Lorentz transformations single out the velocity of light as a limiting velocity.

When the particle velocity $\vec{u}$ in the rest frame and the velocity of the moving frame $\vec{v}$ are not collinear, then the vector-valued velocity $\vec{u}'$ can be calculated as

$$\vec{u}' = \frac{d\vec{x}'}{dt} = \frac{d\vec{x} + (\gamma - 1)\dfrac{\vec{v}\cdot d\vec{x}}{\vec{v}^2}\,\vec{v} - \left(\gamma\dfrac{v}{c}\right)\vec{v}\,(c\,dt)}{\gamma\,c\,dt - \left(\gamma\dfrac{v}{c}\right)(\hat{v}\cdot d\vec{x})} \tag{8.21}$$

$$= \frac{\vec{u} + (\gamma - 1)\dfrac{\vec{v}\cdot\vec{u}}{\vec{v}^2}\,\vec{v} - \gamma\,\vec{v}}{\gamma - \gamma\dfrac{\vec{u}\cdot\vec{v}}{c^2}} = \frac{\vec{u} + (\gamma - 1)\dfrac{\vec{v}\cdot\vec{u}}{v^2}\,\vec{v} - \gamma\,\vec{v}}{\gamma\left(1 - \dfrac{\vec{v}\cdot\vec{u}}{c^2}\right)}\,. \tag{8.22}$$

If $\vec{u}$ is parallel to $\vec{v}$, then we can replace $\vec{v}\cdot\vec{u}/c^2 \to 1$ and have

$$\vec{u}' = \frac{\vec{u} + (\gamma - 1)\dfrac{\vec{v}\cdot\vec{u}}{v^2}\,\vec{v} - \gamma\,\vec{v}}{\gamma\left(1 - \dfrac{\vec{v}\cdot\vec{u}}{c^2}\right)} \xrightarrow{\vec{u}\,\|\,\vec{v}} \frac{\gamma\,(\vec{u} - \vec{v})}{\gamma\left(1 - \dfrac{\vec{v}\cdot\vec{u}}{c^2}\right)} = \frac{\vec{u} - \vec{v}}{1 - \dfrac{v\,u}{c^2}}\,, \tag{8.23}$$

which is the same as the vector-valued version of the addition theorem (8.19).

8.2.4 Generators of the Lorentz Group

Up to now, we have written the Lorentz transformations of various quantities in terms of separate transformations for the "time-like" and "space-like" components. The transformations are linear, which calls for a matrix representation. It is customary to summarize the Lorentz vectors (four-vectors) which we have encountered so far, into an explicit vector form, with components being labeled as $\mu = 0, 1, 2, 3$, and $\mu = 0$ being reserved for the "time-like" component. This will eventually lead to a unification of the transformation properties of time and space under a common matrix representation. For the space-time position vector, we would write

$$x^\mu = (c\,t, \vec{r})\,, \qquad x^0 = c\,t\,, \qquad x^1 = x\,, \qquad x^2 = y\,, \qquad x^3 = z\,. \tag{8.24}$$

In the physics literature, one has almost universally adopted a notation with $\mu = 0, 1, 2, 3$, which corresponds to "C conventions" instead of "Fortran conventions" for the labeling of the vector or array structure (we recall that array indices canonically start from zero for programs written in C, while they start from one in the case of a Fortran program). In the natural unit system, one would set $\hbar = c = \epsilon_0 = 1$ and then the components x^μ of the relativistically covariant four-vector vector would be the same as those of the spatial vector x^i. Here, because we use SI mksA units, we need to supply an extra factor c in x^0. The convention is that x^μ is understood as a four-component object which, if μ (Greek index) attains the value $\mu = 0$, is equal to ct, while for $\mu = i$ (Latin index), we have $x^i = (\vec{r})^i$, equal to the ith component of the position vector $\vec{r}$. In the following, we adopt the general convention that Greek superscripts and subscripts denote a Lorentz index that assumes the values $\mu = 0, 1, 2, 3$, whereas Latin superscripts and subscripts denotes spatial indices $i = 1, 2, 3$. The distinction between contravariant (upper index) and covariant (lower index) components in relativity becomes necessary because of the presence of a nontrivial metric, to be discussed below.

The four-vectors we have encountered so far are

$$x^\mu = (ct, \vec{r}), \qquad \mathcal{P}^\mu = (E, c\vec{p}), \qquad \mathcal{K}^\mu = (\omega, c\vec{k}),$$

$$A^\mu = (\Phi, c\vec{A}), \qquad \mathcal{J}^\mu = (\rho, c^{-1}\vec{J}). \tag{8.25}$$

We use calligraphic symbols in order to denote those four-vectors where we introduce a factor c in the spatial components; for the vector potential, the spatial components are thus $\mathcal{A}^i = c\,A^i$. In a unit system with $c = 1$, such as the natural unit system, one may universally denote the four-vectors and its spatial components by the same symbol, but in the SI mksA unit system, this would lead to an ambiguity for the spatial components $A^i = \mathcal{A}^i/c$.

Under Lorentz transformations (with matrix representation $\Lambda^\mu{}_\nu$), the four-vectors transform as follows,

$$x'^\mu = \Lambda^\mu{}_\nu\, x^\nu, \qquad p'^\mu = \Lambda^\mu{}_\nu\, p^\nu, \qquad k'^\mu = \Lambda^\mu{}_\nu\, k^\nu, \tag{8.26}$$

etc., where the sum over $\nu = 0, 1, 2, 3$ is implicitly assumed by the Einstein summation convention, and Λ is the Lorentz transformation. We now have to find the explicit matrix representation of Λ. The Lorentz transformation (8.3) can be written in matrix form,

$$\begin{pmatrix} ct' \\ x' \\ y' \\ z' \end{pmatrix} = \begin{pmatrix} \gamma & -\gamma\frac{v}{c} & 0 & 0 \\ -\gamma\frac{v}{c} & \gamma & 0 & 0 \\ 0 & 0 & 1 & 0 \\ 0 & 0 & 0 & 1 \end{pmatrix} \begin{pmatrix} ct \\ x \\ y \\ z \end{pmatrix}. \tag{8.27}$$

The Lorentz boost can be interpreted as a hyperbolic rotation,

$$\gamma = \cosh\rho, \qquad \gamma\frac{v}{c} = \sinh\rho, \qquad \cosh^2\rho - \sinh^2\rho = 1, \tag{8.28}$$

where ρ is the rapidity. We can convince ourselves by explicit calculation that

$$(ct')^2 - x'^2 - y'^2 - z'^2 = [(ct)\cosh(\rho) - x\sinh(\rho)]^2$$
$$- [x\cosh(\rho) - (ct)\sinh(\rho)]^2 - y'^2 - z'^2 = (ct)^2 - x^2 - y^2 - z^2.$$
$$(8.29)$$

This amounts to the conservation of a scalar product with a negative weight being assigned to the spatial components, i.e., with regard to a metric with the signature $(+,-,-,-)$. If we define the matrix representation of the elementary Lorentz boost $\Lambda^\mu{}_\nu$ and of the metric $g_{\mu\nu}$, as follows,

$$(\Lambda^\mu{}_\nu) = \begin{pmatrix} \cosh\rho & -\sinh\rho & 0 & 0 \\ -\sinh\rho & \cosh\rho & 0 & 0 \\ 0 & 0 & 1 & 0 \\ 0 & 0 & 0 & 1 \end{pmatrix}, \quad (g_{\mu\nu}) = (g^{\mu\nu}) = \begin{pmatrix} 1 & 0 & 0 & 0 \\ 0 & -1 & 0 & 0 \\ 0 & 0 & -1 & 0 \\ 0 & 0 & 0 & -1 \end{pmatrix}, \quad (8.30)$$

then the following quantity ("scalar product") remains invariant under Lorentz boosts,

$$(ct')^2 - \vec{r}'^2 = g_{\mu\nu}x'^\mu x'^\nu = g_{\mu\nu}\Lambda^\mu{}_\rho x^\rho \Lambda^\nu{}_\delta x^\delta = g_{\rho\delta}x^\rho x^\delta = (ct)^2 - \vec{r}^2. \quad (8.31)$$

Because x^ρ and x^δ are arbitrary, we can establish that

$$g_{\mu\nu}\Lambda^\mu{}_\rho \Lambda^\nu{}_\delta = (\Lambda^T)_\rho{}^\mu g_{\mu\nu}\Lambda^\nu{}_\delta = g_{\rho\delta}, \quad g = \Lambda^T g \Lambda. \quad (8.32)$$

The latter equation expresses the defining property of a Lorentz transformation in matrix form. One can show that any matrix Λ that fulfills Eq. (8.32) actually is a valid Lorentz transformation, composed of a rotation of space followed by a Lorentz boost. Because the signature of the conserved metric is $(+,-,-,-)$, we speak of the group $SO(1,3)$. Furthermore, the connected component of the Lorentz group (the one with determinant equal to $+1$ instead of -1) is all composed of Lorentz boosts and spatial rotations.

If a^μ is a Lorentz vector of contravariant components, then the vector a_μ of covariant components transforms as follows under a Lorentz transformation,

$$a_\mu = g_{\mu\nu}a^\nu, \quad a'_\mu = g_{\mu\rho}a'^\rho = g_{\mu\rho}\Lambda^\rho{}_\delta a^\delta = g_{\mu\rho}\Lambda^\rho{}_\delta g^{\delta\nu}a_\nu = ((\Lambda^{-1})^T)_\mu{}^\nu a_\nu, \quad (8.33)$$

where we use Eq. (8.32). So, the vector a_μ is transformed with the transpose of the inverse Lorentz matrix. For pure rotation matrices, the transpose is equal to the inverse, so that the distinction between covariant and contravariant components becomes unnecessary. Under Lorentz transformations, the scalar product of any of the quantities listed in Eq. (8.25) is a scalar, i.e., if a^μ and b^μ are two vectors, then

$$a' \cdot b' = a'^\mu b'_\mu = g_{\mu\nu}a'^\mu b'^\nu = g_{\mu\nu}\Lambda^\mu{}_\rho \Lambda^\nu{}_\delta a^\rho b^\delta = g_{\rho\delta}a^\rho b^\delta = a^\mu b_\mu = a \cdot b. \quad (8.34)$$

So, the scalar product $a \cdot b$ transforms under the elementary scalar representation of the group $SO(1,3)$; it is a Lorentz scalar and a conserved quantity under Lorentz transformations.

8.2.5 Representations of the Lorentz Group

We have seen that the scalar product $a^\mu b_\mu$ of two space-time vectors a^μ and b^μ is conserved under Lorentz boosts, which describe the transformation from a rest frame into a moving frame, without any spatial rotation. We may generalize this concept and define the Lorentz transformations as the group of linear space-time transformation that leave the scalar product invariant, i.e., as the group $SO(1,3)$. From the above discussion, it is evident that the identity transformation $\Lambda^\mu{}_\nu = \delta^\mu{}_\nu$ is a member of the Lorentz group, with unit determinant. The parity transformation changes the sign of the spatial coordinates and has the matrix representation $\mathrm{diag}(1,-1,-1,-1)$. Its determinant is equal to -1. It does not belong to the connected component of the unit matrix. The connected component of the identity transformation is called the proper Lorentz group. Furthermore, we would like the direction of time to be preserved under the Lorentz transformation, which defines the orthochronous Lorentz group. In physics, we are thus primarily interested in the proper orthochronous subgroup of the Lorentz group, although the important discrete transformations of time inversion [time inversion carries the matrix representation $\mathrm{diag}(-1,1,1,1)$] and parity also play an important role.

The space-time representation of the Lorentz group defined above plays an egregious role in physics. Indeed, proper orthochronous Lorentz transformations are products of spatial rotations and Lorentz boosts. Let Λ_1 and Λ_2 be two Lorentz transformations. We find a suitable representation D of the Lorentz group if

$$D(\Lambda_1 \Lambda_2) = D(\Lambda_1) D(\Lambda_2). \tag{8.35}$$

A trivial representation of the Lorentz group is obtained when we simply put $D(\Lambda_1) = \mathbb{1}$. In the following, starting from spatial rotations, we shall develop the concept of a generator of spatial rotations and Lorentz transformations, as well as the defining structure equations of the Lie group which generates the Lorentz transformations upon exponentiation. Having obtained the structure constants of the Lie group, we shall indicate a second representation of the Lorentz group, based on Dirac matrices, which describes spin-1/2 particles. However, before we do so, we need to go back once more and investigate the orbital and spin angular momenta, previously introduced in Secs. 2.3.2 and 5.4.2, in a different light.

Let us start with the matrix representations of a rotation of two-dimensional space by a rotation angle θ. In the following, we shall denote a matrix by a double-lined symbol such as $\mathbb{R}$ for a rotation matrix or $\mathbb{L}$ for the matrix representation of a Lorentz transformation. The two-dimensional rotation matrix reads as

$$\mathbb{R} = \begin{pmatrix} \cos\varphi & \sin\varphi \\ -\sin\varphi & \cos\varphi \end{pmatrix} \approx \begin{pmatrix} 1 & 0 \\ 0 & 1 \end{pmatrix} + \begin{pmatrix} 0 & \varphi \\ -\varphi & 0 \end{pmatrix}. \tag{8.36}$$

For a small rotation angle φ, we have

$$\begin{pmatrix} x' \\ y' \end{pmatrix} \approx \begin{pmatrix} x \\ y \end{pmatrix} + \varphi \times \underbrace{\begin{pmatrix} 0 & 1 \\ -1 & 0 \end{pmatrix}}_{=\mathbb{M}} \cdot \begin{pmatrix} x \\ y \end{pmatrix} = \begin{pmatrix} x + \varphi\,\delta y \\ y - \varphi\,\delta x \end{pmatrix}, \qquad \delta x^i = \varphi\,\epsilon^{ij}\,x^j, \tag{8.37}$$

where $\epsilon^{12} = -\epsilon^{21} = 1$ is an antisymmetric tensor, and the Einstein summation convention has been used. Geometrically, Eq. (8.37) finds a natural interpretation as a rotation by an angle $-\varphi$, which corresponds to the "passive" interpretation of the rotation, where the coordinates are rotated by the angle $-\varphi$, so that the transformed and original coordinates, under a transformation of the coordinate system by an angle $+\varphi$, represent the same vector in real space.

Throughout this derivation, we denote the Cartesian components of the coordinate vector $\vec{r}$ as x^i, not as r^i, so that the Cartesian representation reads as

$$\vec{r} = \hat{e}_1 \, x^1 + \hat{e}_2 \, x^2 + \hat{e}_3 \, x^3 = \hat{e}_1 \, x + \hat{e}_2 \, y + \hat{e}_3 \, z \, . \tag{8.38}$$

The notation is not completely free of small idiosyncrasies; the problem with the notation r^i for the Cartesian components is that the symbol $r = |\vec{r}|$ also is used for the modulus of the coordinate vector. Thus, r^i could very well be misunderstood as the ith power of $|\vec{r}|$, whereas the risk of a misunderstanding is somewhat reduced for the Cartesian component p^i, which is hard to misunderstand as $|\vec{p}|^i$, for reasons that have to do with general human intuition and cannot easily be quantified. In any case, Eq. (8.38) defines our notation uniquely.

A scalar function transforms as $f'(\vec{r}\,') = f(\vec{r})$ where $\vec{r}\,' = \mathbb{R} \cdot \vec{r}$. Here,

$$\mathbb{R} = \exp(\varphi \, \mathbb{M}) \approx 1 + \varphi \, \mathbb{M}, \qquad \mathbb{R}^{-1} = \exp(-\varphi \, \mathbb{M}) \approx 1 - \varphi \, \mathbb{M}, \tag{8.39}$$

for infinitesimal φ. Then, at the same coordinate $\vec{r}$ as for the original argument, but with the transformed function f', we have

$$f'(\vec{r}) = f(\mathbb{R}^{-1} \cdot \vec{r}) = f(\vec{r} - \varphi \, \mathbb{M} \cdot \vec{r}) = f(r^i - \varphi \, \epsilon^{ij} \, r^j) \tag{8.40}$$

$$= f(\vec{r}) + \varphi \, \epsilon^{ij} \, x^i \, \frac{\partial}{\partial x^j} f(\vec{r}) = f(\vec{r}) + \mathrm{i} \, \varphi \left(-\mathrm{i} \, \epsilon^{ij} \, x^i \, \frac{\partial}{\partial x^j} \right) f(\vec{r}) = f(\vec{r}) + \mathrm{i} \, \varphi \, \hat{L} \, f(\vec{r}) \, .$$

We have used the antisymmetry of the ϵ tensor and have defined the operator $\hat{L}$,

$$\hat{L} = -\mathrm{i} \, \epsilon^{ij} \, x^i \, \frac{\partial}{\partial x^j} \, . \tag{8.41}$$

The angular momentum operator $\hat{L}$ only has one component in two dimensions.

Let us define, as a generalization of the matrix $\mathbb{M}$ given above, three matrices $\mathbb{S}^k$ with $k = 1, 2, 3$, which are given as follows (in component form) [see Eq. (5.138)]

$$(\mathbb{S}^k)^{ij} = -\mathrm{i}\epsilon^{kij} \, , \qquad \epsilon^{123} = 1 \, , \tag{8.42}$$

where ϵ^{ijk} is antisymmetric upon interchange of any two arguments. These matrices are identical to the ones indicated in Eq. (5.139), the only difference being the different convention for the Cartesian coordinate indices which are indicated as superscripts rather than subscripts. The matrix $\mathbb{M}$ given above is identified as the upper right 2×2 submatrix of $\mathbb{S}^3$. The matrices and the corresponding components of the angular momentum vector fulfill the algebraic relations

$$\left[\mathbb{S}^i, \mathbb{S}^j \right] = \mathrm{i}\epsilon^{ijk} \, \mathbb{S}^k \, , \qquad \left[L^i, L^j \right] = \mathrm{i}\epsilon^{ijk} \, L^k \, , \tag{8.43}$$

the first of which was already mentioned in Eq. (5.140). The ϵ tensor both enters the explicit form of the $\mathbb{S}$ matrices but also the structure constants of the Lie algebra. The rotation matrix is given as

$$\mathbb{R} = \exp\left(\mathrm{i} \sum_{k=1}^{3} \varphi^k \, \mathbb{S}^k\right) = \exp\left(\mathrm{i}\vec{\varphi}\cdot\vec{\mathbb{S}}\right), \qquad \vec{r}\,' = \mathbb{R}\cdot\vec{r}, \tag{8.44}$$

where $\vec{\mathbb{S}}$ is the vector of $\mathbb{S}$ matrices, and $\vec{\varphi}$ is a vector of rotation angles about the three axes. Then,

$$\left[\exp\left(\mathrm{i}\vec{\varphi}\cdot\vec{\mathbb{S}}\right)\right]^{ij} \approx \left[\mathbb{1} + \mathrm{i}\vec{\varphi}\cdot\vec{\mathbb{S}}\right]^{ij} \approx \delta^{ij} + \varphi^k\,\epsilon^{kij}. \tag{8.45}$$

For small rotation angles $\vec{\varphi} = \hat{n}\,\delta\varphi$, the components of a vector $\vec{v}$ thus transform as follows,

$$\left[\exp\left(\mathrm{i}\vec{\varphi}\cdot\vec{\mathbb{S}}\right)\right]^{ij} v^j \approx \left(\delta^{ij} + \varphi^k\,\epsilon^{kij}\right) v^j = v^i - \left(\vec{\varphi}\times\vec{v}\right)^i, \tag{8.46}$$

which is just the representation of an infinitesimal rotation by an angle $\vec{\varphi}$, in the case of a passive interpretation of the rotation. That means that the coordinate system turns by an angle $\vec{\varphi}$, and the coordinate vector $\vec{r} \to \vec{r}\,'$ needs to turn by an angle $-\vec{\varphi}$, with the net result that $\vec{r}\,'$ and $\vec{r}$ represent the same vector in real space.

Let us briefly investigate if the transformation $\vec{v} \to \vec{v}\,' = \vec{v} - \left(\vec{\varphi}\times\vec{v}\right)^i$ under infinitesimal rotations is compatible with the transformation of a vector product. Indeed, the vector product $\vec{w} = \vec{u}\times\vec{v}$ transforms into $\vec{w}\,' = \vec{u}\,'\times\vec{v}\,'$ with

$$\begin{aligned}
\vec{w}\,' = \vec{u}\,'\times\vec{v}\,' &= \left(\vec{u} - \vec{\varphi}\times\vec{u}\right)\times\left(\vec{v} - \vec{\varphi}\times\vec{v}\right)\\
&\approx \vec{u}\times\vec{v} - \left(\vec{\varphi}\times\vec{u}\right)\times\vec{v} - \vec{u}\times\left(\vec{\varphi}\times\vec{v}\right) = \vec{u}\times\vec{v} - \vec{\varphi}\times\left(\vec{u}\times\vec{v}\right) = \vec{w} - \vec{\varphi}\times\vec{w},
\end{aligned} \tag{8.47}$$

where we expand to first order in φ, confirming the consistency. We have used the identity

$$\left(\vec{\varphi}\times\vec{u}\right)\times\vec{v} + \vec{u}\times\left(\vec{\varphi}\times\vec{v}\right) = \vec{\varphi}\times\left(\vec{u}\times\vec{v}\right). \tag{8.48}$$

The generalization of (8.40) to three-dimensional rotations is

$$\begin{aligned}
f'(\vec{r}) \approx f(\mathbb{R}^{-1}\cdot\vec{r}) &= f\left(x^i - \varphi^k\,\epsilon^{kij}\,x^j\right) \approx f(\vec{r}) - \varphi^k\,\epsilon^{kij}\,x^j\,\frac{\partial}{\partial x^i}\,f(\vec{r})\\
&= f(\vec{r}) + \mathrm{i}\,\varphi^k\left(-\mathrm{i}\,\epsilon^{kij}\,x^i\,\frac{\partial}{\partial x^j}\right)f(\vec{r}) = f(\vec{r}) + \mathrm{i}\vec{\varphi}\cdot\vec{L}\,f(\vec{r}),
\end{aligned} \tag{8.49}$$

where we take notice of the interchange $i \leftrightarrow j$ in the third step of the transformation. The angular momentum operator $\vec{L}$ is defined in Eq. (2.29). For finite rotations, we summarize the transformation law of a vector $\vec{r}$ and of a scalar function as

$$\vec{r}\,' = \exp(\mathrm{i}\vec{\varphi}\cdot\vec{S})\cdot\vec{r}, \qquad f'(\vec{r}) = \exp(\mathrm{i}\vec{\varphi}\cdot\vec{L})\,f(\vec{r}). \tag{8.50}$$

The question now is how to generalize the above formalism to Lorentz transformations. The answer can be given as follows. As mentioned above, all connected or orthochronous Lorentz transformations can be generated by rotations and boosts. In analogy to rotations in quantum mechanics, we want to investigate the form of infinitesimal Lorentz transformations. According to Eq. (8.50), the generators of spatial rotations are the components of the angular momentum operator [see Eqs. (2.29) and (8.41)]. If we are to generalize the spatial rotations to rotations of space-time, then it becomes clear that we need to find a suitable Lorentz-covariant generalization of the angular momentum operator. This generalization is given by generators $L^{\mu\nu}$ with $\mu, \nu = 0, 1, 2, 3$, where

$$L^{\mu\nu} = \mathrm{i}\left(x^\mu\,\partial^\nu - x^\nu\,\partial^\mu\right) = \mathrm{i}\left(x^\mu\,\frac{\partial}{\partial x_\nu} - x^\nu\,\frac{\partial}{\partial x_\mu}\right). \tag{8.51}$$

On the occasion, we define the derivatives with respect to covariant and contravariant components of space-time as

$$x^\mu = (ct, \vec{r}), \qquad x_\mu = (ct, -\vec{r}), \qquad \partial^\mu \equiv \frac{\partial}{\partial x_\mu}, \qquad \partial_\mu \equiv \frac{\partial}{\partial x^\mu}. \tag{8.52}$$

For spatial indices $i, j, k = 1, 2, 3$, the generators are connected to the angular momentum by the relation

$$L^i = \frac{1}{2}\,\epsilon^{ik\ell}\,L^{k\ell}, \qquad L^{k\ell} = -\mathrm{i}\left(x^k\,\frac{\partial}{\partial x^\ell} - x^\ell\,\frac{\partial}{\partial x^k}\right), \tag{8.53}$$

because X^i is the Cartesian component, not X_i. The generalization of the $\mathbb{S}$ matrices defined in Eq. (5.138) is given by

$$(\mathbb{M}_{\alpha\beta})^{\mu\nu} - g^\mu{}_\alpha\,g^\nu{}_\beta \quad g^\mu{}_\beta\,g^\nu{}_\alpha. \tag{8.54}$$

Here, we interpret $(\mathbb{M}_{\alpha\beta})^{\mu\nu}$ to be the element with Lorentz indices μ, ν of the generator matrix $\mathbb{M}_{\alpha\beta}$. That is to say, a priori we have a total of $16 = 4 \times 4$ matrices $\mathbb{M}_{\alpha\beta}$, each of which have 16 components $(\mathbb{M}_{\alpha\beta})^{\mu\nu}$. However, because of the antisymmetry $\mathbb{M}_{\alpha\beta} = -\mathbb{M}_{\beta\alpha}$, only 6 matrices are actually nonvanishing and different from each other. Both the Lorentz indices of the matrices as well as the component indices can be lowered and raised with the metric, i.e.,

$$(\mathbb{M}_{\alpha\beta})^\mu{}_\nu = g^\mu{}_\alpha\,g_{\nu\beta} - g^\mu{}_\beta\,g_{\nu\alpha}. \tag{8.55}$$

With generalized rotation angles $\omega^{\alpha\beta} = -\omega^{\beta\alpha}$, infinitesimal Lorentz transformations read as

$$\Lambda^\mu{}_\nu = g^\mu{}_\nu + \frac{1}{2}\,\omega^{\alpha\beta}\,(\mathbb{M}_{\alpha\beta})^\mu{}_\nu = g^\mu{}_\nu + \frac{1}{2}\left(\omega^\mu{}_\nu - \omega^\nu{}_\mu\right) = g^\mu{}_\nu + \omega^\mu{}_\nu. \tag{8.56}$$

For antisymmetric rotation angles $\omega^{\alpha\beta} = -\omega^{\beta\alpha}$, we have, in view of Eq. (8.54),

$$\frac{1}{2}\,\omega^{\alpha\beta}(M^{\mu\nu})_{\alpha\beta} = \omega^{\mu\nu}. \tag{8.57}$$

The matrix with elements $\omega^{\mu\nu}$ ($\mu, \nu = 0, 1, 2, 3$) has the structure of a general, antisymmetric matrix with zeros on the diagonal. It has six independent nonvanishing elements corresponding to the six generator matrices $\mathbb{M}^{\alpha\beta}$. They fulfill the algebraic relations:

$$[\mathbb{M}^{\mu\nu}, \mathbb{M}^{\kappa\lambda}] = g^{\mu\lambda}\,\mathbb{M}^{\nu\kappa} + g^{\nu\kappa}\,\mathbb{M}^{\mu\lambda} - g^{\mu\kappa}\,\mathbb{M}^{\nu\lambda} - g^{\nu\lambda}\,\mathbb{M}^{\mu\kappa}, \tag{8.58}$$

$$[L^{\mu\nu}, L^{\kappa\lambda}] = \mathrm{i}\left(g^{\mu\lambda}L^{\nu\kappa} + g^{\nu\kappa}L^{\mu\lambda} - g^{\mu\kappa}L^{\nu\lambda} - g^{\nu\lambda}L^{\mu\kappa}\right). \tag{8.59}$$

These commutation relations define the Lie algebra of the group and are fundamental. Exponentiating a matrix representation of the Lie group yields a matrix
representation of the Lorentz group. The generalization of Eq. (8.49) to finite
Lorentz transformations is

$$f'(x) = f(\Lambda^{-1} x) = f\left(x^\mu - \frac{1}{2}\omega^{\alpha\beta} (\mathrm{M}_{\alpha\beta})^\mu{}_\nu x^\nu\right) = f(x) - \frac{1}{2}\omega^{\alpha\beta} (\mathrm{M}_{\alpha\beta})^\mu{}_\nu x^\nu \partial_\mu f(X)$$

$$= f(x) - \omega^\alpha{}_\beta x^\beta \partial_\alpha f(x) = f(x) + \omega^{\alpha\beta} x_\alpha \partial_\beta f(x)$$

$$= f(x) - \frac{\mathrm{i}}{2}\omega^{\alpha\beta} (\mathrm{i}(x_\alpha \partial_\beta - x_\alpha \partial_\beta)) f(x) = f(x) - \frac{\mathrm{i}}{2}\omega^{\alpha\beta} L_{\alpha\beta} f(x). \tag{8.60}$$

We have the following Lorentz transformations for the four-vector X and the transformed scalar function f', expressed as a function of the old coordinates,

$$x'^\mu = \exp\left(\frac{1}{2}\omega^{\alpha\beta} (\mathrm{M}_{\alpha\beta})^\mu{}_\nu\right) x^\nu, \qquad f'(x) = \exp\left(-\frac{\mathrm{i}}{2}\omega^{\alpha\beta} L_{\alpha\beta}\right) f(x). \tag{8.61}$$

Here, the expression $\exp\left[\frac{1}{2}\omega^{\alpha\beta} (\mathrm{M}_{\alpha\beta})^\mu{}_\nu\right]$ needs to be interpreted as follows: One
first multiplies the generalized rotation angles $\omega^{\alpha\beta}$ with the generator matrices
$\mathrm{M}_{\alpha\beta}$, with the components of $\mathrm{M}_{\alpha\beta}$ taken to be the Lorentz components of upper
index μ and lower index ν. One then exponentiates the resultant matrix with
components $\frac{1}{2}\omega^{\alpha\beta} (M_{\alpha\beta})^\mu{}_\nu$ and obtains a Lorentz transformation matrix which
can be multiplied with a Lorentz vector component x^ν, to obtain the transformed
vector component x'^μ.

The phase conventions in Eq. (8.61) differ from those used in Eq. (8.50), and
are not uniform. (One can bring them into a more uniform appearance under the
replacement $L^{\mu\nu} \to \mathrm{i}M^{\mu\nu}$.) In order to understand this problem deeper, let us
consider the specialization of a Lorentz transformation to the case of a rotation
about the z axis, i.e.,

$$\omega^{12} = -\omega^{21} = -\varphi, \qquad \omega^{13} = \omega^{23} = \omega^{0i} = 0, \tag{8.62}$$

for $i = 1, 2, 3$. In this case,

$$X = \frac{1}{2}\omega^{\alpha\beta} (\mathrm{M}_{\alpha\beta})^\mu{}_\nu = \left[\begin{pmatrix} 0 & 0 & 0 & 0 \\ 0 & 0 & \varphi & 0 \\ 0 & -\varphi & 0 & 0 \\ 0 & 0 & 0 & 0 \end{pmatrix}\right]^\mu{}_\nu, \qquad [\exp(X)]^\mu{}_\nu = \left[\begin{pmatrix} 0 & 0 & 0 & 0 \\ 0 & \cos\varphi & \sin\varphi & 0 \\ 0 & -\sin\varphi & \cos\varphi & 0 \\ 0 & 0 & 0 & 0 \end{pmatrix}\right]^\mu{}_\nu.$$

$$\tag{8.63}$$

We consider the spatial part of the matrix X, in the xy plane. This is the submatrix
of X with $\mu, \nu = 1, 2$, corresponding to the second and third rows and columns of
the matrix X defined in Eq. (8.63), which is equal to the matrix M defined in
Eq. (8.37). We conclude that the specialization of the transformation law (8.61), to
spatial generators with $\alpha, \beta = 1, 2, 3$, leads to the same transformation law for the
spatial components of a four-vector as the transformation law (8.44), but one has
to be careful with the sign in the identification of the rotation angle.

Let us also investigate the expression

$$Y = -\frac{i}{2}\omega^{\alpha\beta} L_{\alpha\beta} = -\frac{i}{2}\omega^{12} L_{12} - \frac{i}{2}\omega^{21} L_{21} = -i\omega^{12} L_{12}$$

$$= i\varphi L_{12} = i\varphi \left(ix_1\,\partial_2 - ix_2\,\partial_1\right) = i\varphi \left(-ix^1\frac{\partial}{\partial x^2} + ix^2\frac{\partial}{\partial x^1}\right) = i\varphi\,L^z . \qquad (8.64)$$

The correspondence to the formula (8.50) is thus established; the expression $-\frac{i}{2}\omega^{\alpha\beta} L_{\alpha\beta}$ in Eq. (8.61) replaces $i\vec{\varphi}\cdot\vec{L}$ in Eq. (8.50).

In order to complement the discussion, let us now try to find at least one further nontrivial matrix representation of the Lorentz group. To this end, we define the Dirac γ^μ matrices as follows,

$$\gamma^0 = \beta = \begin{pmatrix} \mathbb{1}_{2\times2} & 0 \\ 0 & -\mathbb{1}_{2\times2} \end{pmatrix}, \qquad \vec{\gamma} = \begin{pmatrix} 0 & \vec{\sigma} \\ -\vec{\sigma} & 0 \end{pmatrix}. \qquad (8.65)$$

For $i = 1, 2, 3$, γ^i has a structure with two 2×2 Pauli matrices on the off-diagonal. These Dirac matrices are 4×4 matrices, and the Pauli σ^k matrices with $k = 1, 2, 3$ are given as

$$\sigma^1 = \begin{pmatrix} 0 & 1 \\ 1 & 0 \end{pmatrix}, \qquad \sigma^2 = \begin{pmatrix} 0 & -i \\ i & 0 \end{pmatrix}, \qquad \sigma^3 = \begin{pmatrix} 1 & 0 \\ 0 & -1 \end{pmatrix}. \qquad (8.66)$$

They fulfill the commutator relations

$$\left[\tfrac{1}{2}\sigma^i, \tfrac{1}{2}\sigma^j\right] = i\,\epsilon^{ijk}\,\tfrac{1}{2}\sigma^k . \qquad (8.67)$$

These commutators look similar to those for the Cartesian components of the angular momentum given in Eq. (8.43). These relations suggest that the matrices $\frac{1}{2}\vec{\sigma}$ might have something to do with angular momentum, or intrinsic angular momentum (spin). Indeed, we can easily convince ourselves that the $\Sigma^{\mu\nu}$ matrices, defined as

$$\Sigma^{\mu\nu} = \frac{i}{2}[\gamma^\mu, \gamma^\nu], \qquad (8.68)$$

fulfill

$$\Sigma^i = \epsilon^{ijk}\,\Sigma^{jk} , \qquad \Sigma^k = \begin{pmatrix} \sigma^k & 0 \\ 0 & \sigma^k \end{pmatrix}. \qquad (8.69)$$

The relation in Eq. (8.69) is basically the same as Eq. (8.53) only with the spin instead of the angular momentum operators. The spin matrices $\frac{1}{2}\Sigma^{\mu\nu}$ fulfill the same algebraic relations for generators of the Lorentz group, in full analogy to the angular momenta $L^{\mu\nu}$ [see Eq. (8.59)],

$$\left[\tfrac{1}{2}\Sigma^{\mu\nu}, \tfrac{1}{2}\Sigma^{\kappa\lambda}\right] = i\left(g^{\mu\lambda}\tfrac{1}{2}\Sigma^{\nu\kappa} + g^{\nu\kappa}\tfrac{1}{2}\Sigma^{\mu\lambda} - g^{\mu\kappa}\tfrac{1}{2}\Sigma^{\nu\lambda} - g^{\nu\lambda}\tfrac{1}{2}\Sigma^{\mu\kappa}\right) . \qquad (8.70)$$

Given generators $\omega_{\mu\nu}$, a matrix representation of the Lorentz group is thus given by the second term in Eq. (8.61), with $L^{\mu\nu}$ replaced by $\frac{1}{2}\Sigma^{\mu\nu}$. So, we obtain a spinor representation of the Lorentz group element Λ as follows,

$$S(\Lambda) = \exp\left(-\frac{i}{4}\omega^{\alpha\beta}\,\Sigma_{\alpha\beta}\right). \qquad (8.71)$$

This is a 4×4 matrix, which "lives" in some internal, spin space of the particle, not in space-time. The element ab of $S(\Lambda)$, for infinitesimal Lorentz transformations, reads as

$$S(\Lambda)_{ab} \approx \delta_{ab} - \frac{i}{4} \omega^{\alpha\beta} (\Sigma_{\alpha\beta})_{ab} . \tag{8.72}$$

Here, $a, b = 1, 2, 3, 4$ are spinor indices; they live in the "internal" world of the particle.

8.3 Relativistic Classical Field Theory

8.3.1 Maxwell Tensor and Lorentz Transformations

One may have noticed that the list of four-vectors given in Eq. (8.25) contains the electromagnetic scalar and vector potentials, but not the field strengths. The reason for this curious phenomenon is to be explained. Namely, the field strengths are actually obtained as derivatives of the potentials with respect to the space-time coordinates and therefore transform as a second-rank tensor under Lorentz transformations, and not as a four-vector.

Let us define the antisymmetric electromagnetic field-strength tensor (obviously of second rank) as

$$F^{\mu\nu} = \partial^\mu \mathcal{A}^\nu - \partial^\nu \mathcal{A}^\mu , \qquad \mathcal{A}^\mu = (\Phi, c \vec{A}) , \tag{8.73}$$

where we recall (8.25). In the current derivation, we suppress the space-time arguments $(\vec{r}, t)$ of the fields. Under Lorentz transformations, a second-rank tensor transforms as follows

$$F'^{\mu\nu} = \Lambda^\mu{}_\rho \Lambda^\nu{}_\delta F^{\rho\delta} . \tag{8.74}$$

This transformation property is different from that of the scalar and vector potential. We take notice of the explicit form of the components F^{i0} and write

$$F^{i0} = \partial^i \mathcal{A}^0 - \partial^0 \mathcal{A}^i = -\partial_i \Phi - c \partial^0 A^i = -\frac{\partial}{\partial x^i} \Phi - \frac{1}{c} \frac{\partial}{\partial t} \left(c A^i \right) = E^i , \tag{8.75}$$

while $F^{0i} = -F^{i0} = -E^i$, where $i = 1, 2, 3$ (spatial index). The component F^{23} reads

$$F^{23} = \partial^2 \mathcal{A}^3 - \partial^3 \mathcal{A}^2 = c \left(-\frac{\partial}{\partial y} A^z + \frac{\partial}{\partial z} A^y \right) = -c (\vec{\nabla} \times \vec{A})^x = -c B^x . \tag{8.76}$$

A more general form is

$$F^{i0} = E^i , \qquad F^{0i} = -E^i , \qquad F^{ij} = -c \epsilon^{ijk} B^k . \tag{8.77}$$

In SI mksA units, the electric field strength $\vec{E}$ has the same physical dimension as $c\vec{B}$. We can show the second equation in (8.77) as follows,

$$c \epsilon^{\ell m i} B^i = -\frac{1}{2} \epsilon^{\ell m i} \epsilon^{ijk} F^{jk} = -c \epsilon^{\ell m i} \epsilon^{ijk} \partial^j A^k = -\frac{c}{2} \left(\delta^{\ell j} \delta^{mk} - \delta^{\ell k} \delta^{mj} \right) \partial^j A^k$$

$$= c \left(\partial^m A^\ell - \partial^\ell A^m \right) = F^{m\ell} = -F^{\ell m} . \tag{8.78}$$

We have used the relation $\epsilon^{i\ell m} \epsilon^{ijk} = \delta^{\ell j} \delta^{mk} - \delta^{\ell k} \delta^{mj}$.

The field strengths can be reconstructed from the components of the field-strength tensor as

$$E^i = F^{i0}, \qquad cB^i = -\frac{1}{2}\epsilon^{ijk}F^{jk} = \frac{1}{2}\epsilon_{ijk}F^{jk}, \tag{8.79}$$

where $\epsilon^{123} = 1$ and $\epsilon_{123} = -1$. The second equation in (8.79) can be shown as follows,

$$B^i = \epsilon^{ijk}\partial_j A^k = \frac{1}{2}\epsilon^{ijk}\left(\partial_j A^k - \partial_k A^j\right) = -\frac{1}{2}\epsilon^{ijk}\left(\partial^j A^k - \partial^k A^j\right) = -\frac{1}{2c}\epsilon^{ijk}F^{jk}, \tag{8.80}$$

The structure of the field-strength tensor in all detail, in a component-wise representation, thus reads as follows,

$$F^{\mu\nu} = \left[\begin{pmatrix} 0 & -E^x & -E^y & -E^z \\ E^x & 0 & -cB^z & cB^y \\ E^y & cB^z & 0 & -cB^x \\ E^z & -cB^y & cB^x & 0 \end{pmatrix}\right]^{\mu\nu}. \tag{8.81}$$

With the metric $g_{\mu\nu}$, we can lower the indices to obtain $F_{\mu\nu} = g_{\mu\rho}g_{\nu\delta}F^{\rho\delta}$,

$$F_{\mu\nu} = \left[\begin{pmatrix} 0 & E^x & E^y & E^z \\ -E^x & 0 & -cB^z & cB^y \\ -E^y & cB^z & 0 & -cB^x \\ -E^z & -cB^y & cB^x & 0 \end{pmatrix}\right]_{\mu\nu}. \tag{8.82}$$

The Lagrangian density (in short: the Lagrangian) of classical electrodynamics is a Lorentz scalar and reads as

$$\mathcal{L} = \frac{\epsilon_0}{2}\left(\vec{E}^2 - c^2\vec{B}^2\right) = -\frac{\epsilon_0}{4}F^{\mu\nu}F_{\mu\nu}. \tag{8.83}$$

Together with currents, we have

$$\mathcal{L} = -\frac{\epsilon_0}{4}F^{\mu\nu}F_{\mu\nu} - J^\mu A_\mu, \qquad J^\mu = (\rho, c^{-1}\vec{J}). \tag{8.84}$$

Appealing to the formalism introduced in Sec. 3.2.2, the variation of the four-vector potential then leads to the equation

$$\begin{aligned} \delta S &= \int d^4x \left[\mathcal{L}(A + \delta A) - \mathcal{L}(A)\right] \\ &= -\frac{\epsilon_0}{4}\int d^4x \left[-\frac{1}{4}\left(-2\partial_\mu F^{\mu\nu}\,\delta A_\nu + 2\partial_\nu F^{\mu\nu}\,\delta A_\mu\right) - \frac{1}{\epsilon_0}J^\nu\,\delta A_\nu\right] \\ &= -\frac{\epsilon_0}{4}\int d^4x \left[\partial_\mu F^{\mu\nu}\,\delta A_\nu - \frac{1}{\epsilon_0}J^\nu\,\delta A_\nu\right] \\ &= -\frac{\epsilon_0}{4}\int d^4x \left(\partial_\mu F^{\mu\nu} - \frac{1}{\epsilon_0}J^\nu\right)\delta A_\nu \overset{!}{=} 0. \end{aligned} \tag{8.85}$$

By virtue of the arbitrariness of the variation of the four-vector potential, the Euler–Lagrange equations are obtained,

$$\frac{\partial \mathcal{L}}{\partial A_\nu} = \frac{\partial}{\partial x^\mu}\frac{\partial \mathcal{L}}{\partial(\partial_\mu A_\nu)}, \qquad \partial_\mu F^{\mu\nu} = \frac{1}{\epsilon_0}J^\nu. \tag{8.86}$$

The Cartesian components of the equation $\partial_\mu F^{\mu\nu} = \mathcal{J}^\nu/\epsilon_0$ read as follows,

$$\nu = 0: \qquad \frac{\partial}{\partial x}E^x + \frac{\partial}{\partial y}E^y + \frac{\partial}{\partial z}E^z = \frac{1}{\epsilon_0}\rho, \tag{8.87a}$$

$$\nu = 1: \qquad -\frac{1}{c}\frac{\partial}{\partial t}E^x + c\frac{\partial}{\partial y}B^z - c\frac{\partial}{\partial z}B^y = \frac{1}{\epsilon_0 c}J^x, \tag{8.87b}$$

$$\nu = 2: \qquad -\frac{1}{c}\frac{\partial}{\partial t}E^y + c\frac{\partial}{\partial z}B^x - c\frac{\partial}{\partial x}B^z = \frac{1}{\epsilon_0 c}J^y, \tag{8.87c}$$

$$\nu = 3: \qquad -\frac{1}{c}\frac{\partial}{\partial t}E^z + c\frac{\partial}{\partial x}B^y - c\frac{\partial}{\partial y}B^x = \frac{1}{\epsilon_0 c}J^z. \tag{8.87d}$$

With $\mu_0 = 1/(\epsilon_0 c^2)$, these equations can be summarized, in vector form, as the inhomogeneous Maxwell equations,

$$\vec{\nabla} \cdot \vec{E} = \frac{1}{\epsilon_0}\rho, \qquad \vec{\nabla} \times \vec{B} - \frac{1}{c^2}\frac{\partial}{\partial t}\vec{E} = \mu_0 \vec{J}. \tag{8.88}$$

The question is how to get the homogeneous Maxwell equations. First of all, we define the dual field-strength tensor $\widetilde{F}^{\mu\nu}$ as

$$\widetilde{F}^{\mu\nu} = \frac{1}{2}\epsilon^{\mu\nu\rho\delta} F_{\rho\delta}. \tag{8.89}$$

It reads, in components,

$$\widetilde{F}^{\mu\nu} = \left[\begin{pmatrix} 0 & -cB^x & -cB^y & -cB^z \\ cB^x & 0 & E^z & -E^y \\ cB^y & -E^z & 0 & E^x \\ cB^z & E^y & -E^x & 0 \end{pmatrix}\right]^{\mu\nu}. \tag{8.90}$$

The homogeneous Maxwell equations, which cannot be derived from the variation of the action $\mathcal{S} = \int \mathrm{d}^4x\, \mathcal{L}$, are given by

$$\partial_\mu \widetilde{F}^{\mu\nu} = \epsilon^{\mu\nu\rho\delta} \partial_\nu F_{\rho\delta} = 0. \tag{8.91}$$

In Cartesian components, these equations read

$$\nu = 0: \qquad \frac{\partial}{\partial x}(cB^x) + \frac{\partial}{\partial y}(cB^y) + \frac{\partial}{\partial z}(cB^z) = 0, \tag{8.92a}$$

$$\nu = 1: \qquad -\frac{1}{c}\frac{\partial}{\partial t}(cB^x) + \frac{\partial}{\partial z}E^y - \frac{\partial}{\partial y}E^z = 0, \tag{8.92b}$$

$$\nu = 2: \qquad -\frac{1}{c}\frac{\partial}{\partial t}(cB^y) + \frac{\partial}{\partial x}E^z - \frac{\partial}{\partial z}E^x = 0, \tag{8.92c}$$

$$\nu = 3: \qquad -\frac{1}{c}\frac{\partial}{\partial t}(cB^z) + \frac{\partial}{\partial y}E^x - \frac{\partial}{\partial x}E^y = 0. \tag{8.92d}$$

In vector notation, one can summarize them into the homogeneous Maxwell equations,

$$\vec{\nabla} \cdot \vec{B} = 0, \qquad \vec{\nabla} \times \vec{E} + \frac{\partial}{\partial t}\vec{B} = 0. \tag{8.93}$$

The Lorentz transformation of the elements of the field-strength tensor has already been indicated in Eq. (8.74), in symbolic form, as $F'^{\mu\nu} = \Lambda^\mu_{\ \rho} \Lambda^\nu_{\ \delta} F^{\rho\delta}$. We now need to clarify how this translates into the corresponding transformations of the field strengths $\vec{E}$ and $\vec{B}$.

A little algebra reveals that the transformations intertwine the electric and magnetic fields under a Lorentz boost into an inertial frame moving with velocity $\vec{v}$, and are given as follows,

$$\vec{E}' = \gamma \left(\vec{E} + \vec{v} \times \vec{B} \right) + (1 - \gamma) \frac{\vec{E} \cdot \vec{v}}{v^2} \vec{v}, \tag{8.94a}$$

$$\vec{B}' = \gamma \left(\vec{B} - \frac{1}{c^2} \vec{v} \times \vec{E} \right) + (1 - \gamma) \frac{\vec{B} \cdot \vec{v}}{v^2} \vec{v}. \tag{8.94b}$$

This transformation generalizes the transformation laws given in Eqs. (8.5), (8.6) and (8.7), to the case of a second-rank tensor. The transformation from the moving to the rest frame of the electric and magnetic induction fields is given by Eq. (8.94) under the replacement $\vec{v} \to -\vec{v}$,

$$\vec{E} = \gamma \left(\vec{E}' - \vec{v} \times \vec{B}' \right) + (1 - \gamma) \frac{\vec{E}' \cdot \vec{v}}{v^2} \vec{v}, \tag{8.95a}$$

$$\vec{B} = \gamma \left(\vec{B}' + \frac{1}{c^2} \vec{v} \times \vec{E}' \right) + (1 - \gamma) \frac{\vec{B}' \cdot \vec{v}}{v^2} \vec{v}. \tag{8.95b}$$

It is easy to check by explicit calculation that the following two quantities are Lorentz invariant,

$$F'^{\mu\nu} F'_{\mu\nu} = 2 \left(c^2 \vec{B}'^2 - \vec{E}'^2 \right) = 2 \left(c^2 \vec{B}^2 - \vec{E}^2 \right) = F^{\mu\nu} F_{\mu\nu}, \tag{8.96a}$$

$$\widetilde{F}'^{\mu\nu} F'_{\mu\nu} = - 4c \, \vec{E}' \cdot \vec{B}' = -4c \, \vec{E} \cdot \vec{B} = \widetilde{F}^{\mu\nu} F_{\mu\nu}. \tag{8.96b}$$

Current and charge density transform into the moving frame as follows,

$$\vec{J}' = \vec{J} - \gamma \rho \vec{v} + (\gamma - 1) \frac{\vec{J} \cdot \vec{v}}{v^2} \vec{v}, \tag{8.97a}$$

$$\rho' = \gamma \left(\rho - \frac{1}{c^2} \vec{J} \cdot \vec{v} \right), \tag{8.97b}$$

and back,

$$\vec{J} = \vec{J}' + \gamma \rho' \vec{v} + (\gamma - 1) \frac{\vec{J}' \cdot \vec{v}}{v^2} \vec{v}, \tag{8.98a}$$

$$\rho = \gamma \left(\rho' + \frac{1}{c^2} \vec{J}' \cdot \vec{v} \right). \tag{8.98b}$$

Finally, let us briefly comment on a somewhat subtle point. According to Eq. (8.81), the field-strength tensor $F^{\mu\nu}$ is composed of elements which constitute vectors (electric field) and pseudo-vectors (magnetic field). Under parity, a pseudo-vector does not change sign while a vector does. Still, the Lorentz transformation of the tensor is given by Eq. (8.74). How can we intuitively understand why $F^{\mu\nu}$ is composed of entries which transform differently under parity? In order to answer this question, let us step back and consider the velocity four-vector $\beta^\mu = (\gamma, \gamma \vec{v})$, the relativistic momentum $m\gamma\vec{v}$, and the proper time τ. The Lorentz force (1.110) reads, relativistically,

$$F^\mu = \frac{\mathrm{d}}{\mathrm{d}\tau} p^\mu = \frac{\mathrm{d}}{\mathrm{d}\tau} (m\,\beta^\mu) = q\,F^{\mu\nu}\,\beta_\mu\,. \tag{8.99}$$

The spatial components of this equation read as follows,

$$\vec{F} = \frac{\mathrm{d}}{\mathrm{d}\tau} (m\gamma\vec{v}) = \gamma\, q\,\left(\vec{E} + \vec{v} \times \vec{B}\right)\,. \tag{8.100}$$

Identifying the infinitesimal laboratory frame time interval as $\mathrm{d}t = \gamma\,\mathrm{d}\tau$, one finds

$$\frac{\mathrm{d}}{\mathrm{d}t} (m\gamma\vec{v}) = q\,\left(\vec{E} + \vec{v} \times \vec{B}\right)\,. \tag{8.101}$$

The Lorentz force F^μ transforms as a vector. In order for the expression $F^{\mu\nu}\,\beta_\mu$ to also transform as a vector, we consider the case of spatial index $\mu = i = 1, 2, 3$ and we must investigate the cases $\nu = 0$ and $\nu = j = 1, 2, 3$. For $\nu = 0$, we have $\beta^0 \to \beta^0$ under parity (no sign change), and so we conclude that F^{i0} must be the component of a vector. For $\nu = j$, we have $\beta^j \to -\beta^j$ under parity (sign change), and so, for F^i to transform as a vector, we conclude that F^{ij} must be the component of a pseudo-vector. Indeed, an inspection of the field-strength tensor reveals that the entries F^{ij} constitute a pseudo-vector: namely, the components of the magnetic field.

8.3.2 *Maxwell Stress–Energy Tensor*

We have discussed the relativistically covariant field-strength tensor, which allowed us to identify the transformation properties of the electric and magnetic fields under Lorentz transformations. One ensuing question then is how to generalize the Maxwell stress tensor $\mathbb{T}$, discussed in Sec. 1.3.2, to the Maxwell stress–energy tensor, i.e., into covariant form, in which case the tensor also includes a reference to the field energy (hence the addition to its name, where the "stress" becomes "stress–energy"). Indeed, the tensor

$$T^{\mu\nu} = -\frac{1}{\mu_0}\left(F^{\mu\alpha}\,F^\nu{}_\alpha - \frac{1}{4}\,g^{\mu\nu}\,F_{\alpha\beta}\,F^{\alpha\beta}\right) \tag{8.102}$$

has the structure

$$T^{\mu\nu} = \begin{pmatrix} \frac{1}{2}\left(\epsilon_0\,\vec{E}^2 + \vec{B}^2/\mu_0\right) & S_x/c & S_y/c & S_z/c \\ S_x/c & -\mathbb{T}_{xx} & -\mathbb{T}_{xy} & -\mathbb{T}_{xz} \\ S_y/c & -\mathbb{T}_{yx} & -\mathbb{T}_{yy} & -\mathbb{T}_{yz} \\ S_z/c & -\mathbb{T}_{zx} & -\mathbb{T}_{zy} & -\mathbb{T}_{zz} \end{pmatrix}, \tag{8.103}$$

where $\vec{S}$ is the Poynting vector, defined in Eq. (1.105), i.e., $\vec{S} = \left(\vec{E} \times \vec{B}\right)/\mu_0$, and $\mathbb{T}$ is the Maxwell stress tensor, defined in Eq. (1.120),

$$\mathbb{T} = \epsilon_0\left(\vec{E} \otimes \vec{E} - \frac{\vec{E}^2}{2}\,\mathbb{1}_{3\times3}\right) + \frac{1}{\mu_0}\left(\vec{B} \otimes \vec{B} - \frac{\vec{B}^2}{2}\,\mathbb{1}_{3\times3}\right) = \mathbb{T}_{ij}\,\hat{e}_i \otimes \hat{e}_j\,. \tag{8.104}$$

We had previously written the equation (1.118) for the force density,

$$\vec{f} = \vec{\nabla} \cdot \mathbb{T} - \frac{1}{c^2} \frac{\partial}{\partial t} \vec{S}. \tag{8.105}$$

This equation finds a natural expression in terms of the components $T^{\mu\nu}$ of the stress–energy tensor, given in Eq. (8.103),

$$f^i = \frac{\partial}{\partial x^j} \mathbb{T}^{ji} - \frac{\partial}{\partial x^0} \frac{S^i}{c} = -\frac{\partial}{\partial x^j} T^{ji} - \frac{\partial}{\partial x^0} T^{0i} = -\frac{\partial}{\partial x^\mu} T^{\mu i}, \tag{8.106}$$

where $i, j = 1, 2, 3$ (Latin indices are spatial), whereas Greek indices are space-time indices ($\mu, \nu = 0, 1, 2, 3$).

Relativistically, one should interpret the force density as

$$\vec{f} = \frac{\partial}{\partial t} \vec{\mathcal{P}} = \frac{\partial}{\partial t} \vec{\mathcal{P}}(\vec{r}, t), \tag{8.107}$$

where $\vec{\mathcal{P}}$ is the relativistic momentum density [here, the differentiation is with regard to t, not τ, as suggested by Eq. (8.101)], and it takes place at constant $\vec{r}$. Promoting the momentum density to a four vector, one identifies the energy density $\mathcal{P}^0 = w/c$, where w is the electromagnetic energy density.

In the sense of special relativity, the force density $\vec{f}$ should be generalized to a four-vector f^μ, which implies that we should find a physical interpretation for f^0. We write for the zeroth component,

$$\begin{aligned}
f^0 &= -\frac{\partial}{\partial x^\mu} T^{\mu 0} = -\frac{1}{c} \frac{\partial}{\partial t} T^{00} - \frac{\partial}{\partial x^i} T^{i0} \\
&= -\frac{1}{c} \frac{\partial}{\partial t} \left(\frac{\epsilon_0}{2} \vec{E}^2 + \frac{\vec{B}^2}{2\mu_0} \right) - \frac{\partial}{\partial x^i} \frac{S^i}{c} = -\frac{1}{c} \frac{\partial}{\partial t} u - \frac{1}{c} \vec{\nabla} \cdot \vec{S},
\end{aligned} \tag{8.108}$$

where

$$u = \frac{\epsilon_0}{2} \vec{E}^2 + \frac{\vec{B}^2}{2\mu_0} \tag{8.109}$$

is the field-energy density. We should thus identify

$$f^0 = \frac{\partial}{\partial t} \frac{w}{c}, \qquad w = \frac{\Delta W}{\Delta V}, \qquad \frac{\partial w}{\partial t} = \frac{\Delta P}{\Delta V} = \vec{E} \cdot \vec{J}, \tag{8.110}$$

as the work done per unit volume on the charges, by the electric field (here, ΔW denotes the work dissipated into the mechanical motion of the particles, and ΔP the power). The equation $f^0 = -\partial T^{\mu 0}/\partial x^\mu$ thus is equivalent to

$$\vec{\nabla} \cdot \vec{S} + \frac{\partial}{\partial t} u + \frac{\partial}{\partial t} w = 0, \tag{8.111}$$

which is just the Poynting theorem (1.106). Finally, the relativistic generalization of Eq. (1.122) is found as

$$f^\nu = -\frac{\partial}{\partial x^\mu} T^{\mu\nu}. \tag{8.112}$$

In the absence of material forces $(f^\mu = 0)$, one can interpret this equation as follows. The explicit separation into time and spatial derivatives

$$\frac{\partial}{\partial t}\frac{T^{0\nu}}{c} + \frac{\partial}{\partial x^i}T^{i\nu} = 0, \tag{8.113}$$

is reminiscent of the continuity equation,

$$\frac{\partial}{\partial t}\rho + \frac{\partial}{\partial x^i}J^i = 0. \tag{8.114}$$

So, for given ν, the vector with components $T^{i\nu}$ can be interpreted as the current for the "charge density" $T^{0\nu}$. In some sense, as much as $\vec{J}$ measures the flow of the charge, $T^{i\nu}$ measures the flow of the field momentum density $T^{0\nu}$.

If $\nu = 0$, then the "charge density" T^{00} is equal to the energy density of the electromagnetic field, while the "current density" with components T^{i0} is the Poynting vector or, the momentum density of the field times c in the ith Cartesian direction. Indeed, from Eq. (8.105), we learn that the momentum density of the field is $\vec{S}/c^2$. If $\nu = j$ with $j = 1, 2, 3$, then the "charge density" T^{0j} is the momentum density in the jth Cartesian direction, while the "current density" with components T^{ij} is the vector $\hat{e}_i\, T^{ij} = -\hat{e}_i\, \mathbb{T}^{ij}$, i.e., the (negative of the) jth column vector of the Maxwell stress tensor. Thus, we can say that the jth column vector of the Maxwell stress tensor measures the flow of the jth component of the field momentum.

8.3.3 *Lorentz Transformation and Biot–Savart Law*

Magnetic and electric fields transform into each other by a Lorentz transformation. Arguably, magnetic-field effects are the only relativistic effects that can easily be seen in a laboratory and remind us of the enormous strength of the electromagnetic interaction, which over long distances is suppressed by the neutrality of macroscopic objects. We start from Eq. (8.95b),

$$\vec{B} = \gamma\left(\vec{B}' + \frac{1}{c^2}\vec{v}\times\vec{E}'\right) + (1-\gamma)\frac{\vec{B}'\cdot\vec{v}}{v^2}\vec{v}. \tag{8.115}$$

Let us consider a moving charged object and aim to calculate its magnetic field. We might use the Liénard–Wiechert potentials and calculate the curl of the vector potential (5.261b). However, in a conductor (stationary current distribution), the electrons move slowly, and it is otherwise possible to find a suitable approximation to the magnetic field generated by the moving charge, as follows. We observe that in a comoving frame, by definition, the moving object is at rest, and therefore it does not generate a magnetic field, i.e., $\vec{B}' = \vec{0}$. Then, in view of Eq. (8.115),

$$\vec{B} = \gamma\frac{\vec{v}}{c^2}\times\vec{E}', \tag{8.116}$$

where $\vec{E}'$ is the electric field strength in the moving frame. We investigate the infinitesimal magnetic field $\mathrm{d}\vec{B}$ corresponding to the infinitesimal electric field $\mathrm{d}\vec{E}'$, where $\mathrm{d}\vec{E}'$ is the electric field generated by an infinitesimal charge element,

$$\mathrm{d}\vec{B} = \gamma\frac{\vec{v}}{c^2}\times\mathrm{d}\vec{E}', \qquad \mathrm{d}\vec{E}' = \frac{\mathrm{d}q}{4\pi\epsilon_0}\frac{\vec{r}-\vec{r}'}{|\vec{r}-\vec{r}'|^3} = \frac{1}{4\pi\epsilon_0}\frac{\vec{r}-\vec{r}'}{|\vec{r}-\vec{r}'|^3}\rho'(\vec{r}')\,\mathrm{d}^3r'. \tag{8.117}$$

The coordinate $\vec{r}'$ is given in the moving frame; the element of charge $dq = \rho(\vec{r}')\,d^3r'$ is located at $\vec{r}'$. However, we shall ignore the transformation of the coordinates, assuming that $\vec{r}$ and $\vec{r}'$ are equal in the rest frame of the particle and in the laboratory frame, effectively performing our calculations in the limit $\gamma \to 1$. Within this approximation, the moving charge generates a magnetic field of the form

$$d\vec{B} \approx \frac{1}{4\pi\epsilon_0 c^2}\,\gamma\,\rho'(\vec{r}')\left(\vec{v}\times\frac{\vec{r}-\vec{r}'}{|\vec{r}-\vec{r}'|^3}\right)d^3r' = \frac{\mu_0}{4\pi}\left(\vec{J}(\vec{r}')\times\frac{\vec{r}-\vec{r}'}{|\vec{r}-\vec{r}'|^3}\right)d^3r'. \quad (8.118)$$

This is the Biot–Savart law for a point charge. Here, we have identified the current density corresponding to a charge density $\rho(\vec{r}')$ as

$$\vec{J}(\vec{r}') = \gamma\,\rho'(\vec{r}')\,\vec{v}. \quad (8.119)$$

While the factor γ is beyond the order of approximation considered in our derivation, in view of approximations carried out previously, it is perhaps interesting to point out that it can easily be understood. We consider the Lorentz contraction of the charge density seen in the moving frame, $\rho'(\vec{r}') = dq/(d\ell_0\,dA)$, into the laboratory frame, where ℓ_0 is supposed to be directed parallel to the velocity $\vec{v}$, and dA is an infinitesimal cross-sectional area element. Then, ℓ_0 is Lorentz-contracted, acquires a factor $1/\gamma$, and we have Eq. (8.119). Alternatively, this follows from Eq. (8.98a), setting $\vec{J}' = \vec{0}$. Integrating, we find the Biot–Savart law,

$$\vec{B}(\vec{r}) = \frac{\mu_0}{4\pi}\int d^3r'\left(\vec{J}(\vec{r}')\times\frac{\vec{r}-\vec{r}'}{|\vec{r}-\vec{r}'|^3}\right). \quad (8.120)$$

This is the Biot–Savart law for extended, stationary current distributions. Again, magnetic fields are the only relativistic effects we are going to see in our everyday life, probably. They exist because the electromagnetic force is that much stronger than the gravitational force; the former is being compensated on macroscopic scales by the presence of unlike and compensating charges.

8.3.4 *Relativity and Magnetic Force*

As we have just seen, it is possible to derive the Biot–Savart law, and thus, the magnetic fields generated by moving charges, on the basis of the Lorentz transformation. This concerns the magnetic field, not the magnetic (Lorentz) force. Likewise, it is instructive to investigate whether or not the magnetic term in the Lorentz force can be obtained as a relativistic effect. To this end, we investigate the Lorentz contraction of moving charges in a conductor, where the test charge moving alongside the conductor would see a net electric field because the positive and negative charges in the conductor move at different speeds and experience a different degree of Lorentz contraction. The net result is that the conductor is electrically charged from the point of view of the moving charge, and a net electric force results. It turns out to be a little nontrivial to formalize this general idea, and this is why we present the

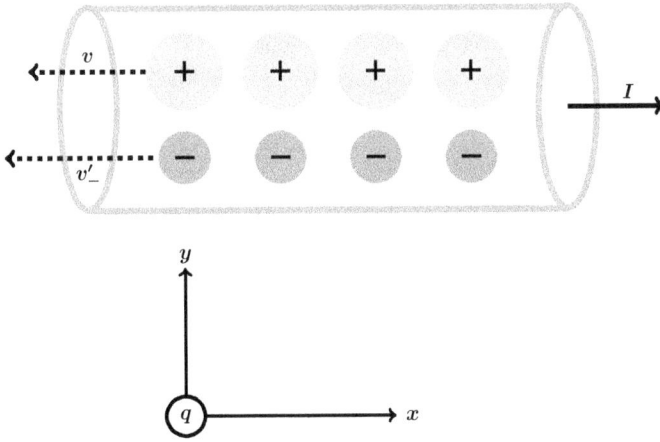

Fig. 8.1   We investigate the geometry of the wire as seen from the frame of reference of the moving charge q. A current current I' flows to the right. The electrons in the wire flow to the left at velocity v'_-. The positively charged ions are stationary and move to the left at the same velocity as the reference frame, namely, at velocity v. By the addition theorem, the drift velocity of the electrons in the moving frame is equal to $v'_- = (v + v_d)/(1 + v\,v_d/c^2)$, where v_d is the drift velocity of the electrons in the rest frame of the wire.

derivation in some detail. We also recall that the Lorentz force cannot be derived from the Maxwell equations alone.

Let us assume that a particle is moving alongside and parallel to a conductor, with its trajectory parallel to the conductor. The conductor is an infinite wire along the x axis. Inside the wire, negative charges move with a drift velocity v_d to the left. A charged particle moves at velocity v to the right, i.e., in the positive x direction, at a distance R from the conductor, and parallel to the conductor. The particle in its own rest frame is described by primed quantities like I', which is the current seen in the rest frame of the moving particle. The positively charged particles are at rest with respect to the wire (see Fig. 8.1). From the rest frame of the particle, the positive charges inside the wire move to the left with velocity v, i.e., with the same velocity as the entire conducting wire moves to the left. The negatively charged particles move to the left, with velocity v'_- (as seen in the frame of the particle), where v'_- is the sum of the velocity v of the conducting wire and the drift velocity v_d of the electrons, the latter being determined in the laboratory frame.

The velocity of the negative charges moving to the left, seen in the reference frame of the moving particle at distance R, is given by the addition theorem (8.19) as

$$v'_- = \frac{v + v_d}{1 + \dfrac{v\,v_d}{c^2}}, \qquad \beta'_- = \frac{\beta + \beta_d}{1 + \beta\,\beta_d}, \qquad (8.121)$$

where we define the scaled velocity $\beta = v/c$ (and analogously for all other velocities). Let the linear charge density inside the conductor be denoted as λ_+ and λ_- for the positive and negative charges, respectively. The charge densities seen in the reference frame of the moving particles are denoted by a prime. The total charge density λ seen in the laboratory frame and λ' seen by the moving particle are given as

$$\lambda = \lambda_+ + \lambda_- = 0, \qquad \lambda' = \lambda'_+ + \lambda'_- \neq 0, \tag{8.122}$$

where the first equality holds because the conductor remains electrically neutral in its own rest frame (laboratory frame). Due to Lorentz contraction, the moving observer (at velocity v) sees the positively charged particles as being contracted by

$$\lambda'_+ = \frac{1}{\sqrt{1 - v^2/c^2}} \lambda_+, \tag{8.123}$$

where we assume that v is the velocity of the particle moving alongside the wire. We assume that the positively charged atomic nuclei are stationary with respect to the wire. Now, the moving observer sees the negative linear charge density enhanced by a factor

$$\lambda'_- = \frac{1}{\sqrt{1 - v'^2_-/c^2}} (\lambda_-)_0, \tag{8.124}$$

where v'_- is the drift velocity of the negative moving charges with respect to the particle flying by [see Eq. (8.121)]. In Eq. (8.124), the quantity $(\lambda_-)_0$ is the charge density of the electrons at rest. The charge density λ_- seen in the laboratory frame, which is identical to the rest frame of the conductor, is different from the one in the rest frame of the moving particle and also different from the one in the rest frame of the positive charges in the conducting wire. Namely, λ_- reads as

$$\lambda_- = \frac{1}{\sqrt{1 - v_d^2/c^2}} (\lambda_-)_0, \qquad (\lambda_-)_0 = \sqrt{1 - \beta_d^2}\, \lambda_-, \tag{8.125}$$

where $\beta_d = v_d/c$. Hence,

$$\lambda'_- = \frac{\sqrt{1 - \beta_d^2}}{\sqrt{1 - \beta'^2_-}} \lambda_- = \frac{\sqrt{1 - \beta_d^2}}{\sqrt{1 - \left(\dfrac{\beta + \beta_d}{1 + \beta\beta_d}\right)^2}} \lambda_- = (1 + \beta\beta_d) \frac{\sqrt{1 - \beta_d^2}}{\sqrt{(1 + \beta\beta_d)^2 - (\beta + \beta_d)^2}} \lambda_-$$

$$= (1 + \beta\beta_d) \frac{\sqrt{1 - \beta_d^2}}{\sqrt{(1 - \beta^2)(1 - \beta_d^2)}} \lambda_- = \frac{1 + \beta\beta_d}{\sqrt{1 - \beta^2}} \lambda_-. \tag{8.126}$$

So, relating the charge densities λ'_+ and λ'_- in the rest frame of the moving particle to the charge densities λ_+ and λ_- in the rest frame of the conductor, we have

$$\lambda' = \lambda'_+ + \lambda'_- = \frac{1}{\sqrt{1 - \beta^2}} \lambda_+ + \frac{1 + \beta\beta_d}{\sqrt{1 - \beta^2}} \lambda_- = \frac{\beta\beta_d \lambda_-}{\sqrt{1 - \beta^2}}$$

$$= -\frac{1}{\sqrt{1 - \beta^2}} \frac{v}{c^2} (-v_d \lambda_-) = -\frac{1}{\sqrt{1 - \beta^2}} \frac{v\, I}{c^2} = -\gamma \frac{v\, I}{c^2}, \tag{8.127}$$

where $\gamma = (1 - \beta^2)^{-1/2}$ is the relativistic Lorentz factor. We have used Eq. (8.122). For the moving particle at velocity v, the conductor behaves as if it were a long, electrically charged wire with linear charge density $\lambda' = \lambda'_+ + \lambda'_-$. Note that the current $I = -\lambda_- v_d$ is calculated to be moving to the right, and it is calculated in the laboratory frame (we would assign a negative value to v_d as the electrons move to the left). We must now relate the current I seen in the rest frame of the conductor, to the current I' in the rest frame of the moving particle. We calculate the magnitude of I' by relating the current I' seen by the moving particle, to the current I in the laboratory frame. We define I' to be the current moving to the right. If v'_- is the modulus of the velocity and λ'_- is the charge per unit length, defined to be negative, then

$$-I' = \lambda'_+ v + \lambda'_- v'_- = \left(\frac{1}{\sqrt{1 - \beta^2}} \lambda_+ \right) v + \left(\frac{1 + \beta \beta_d}{\sqrt{1 - \beta^2}} \lambda_- \right) v'_-$$

$$= -\frac{1}{\sqrt{1 - \beta^2}} \lambda_- v + \left(\frac{1 + \beta \beta_d}{\sqrt{1 - \beta^2}} \right) \left(\frac{v + v_d}{1 + \beta \beta_d} \right) \lambda_-$$

$$= \frac{\lambda_- v_d}{\sqrt{1 - \beta^2}} = -\frac{I}{\sqrt{1 - \beta^2}}. \tag{8.128}$$

Hence, based on Eq. (8.127), we have

$$\lambda' = -\frac{v I'}{c^2} \tag{8.129}$$

in the rest frame of the particle; the result is exact to all orders in the relativistic corrections. The infinitesimal potential generated by a line element dx on the conductor reads

$$d\Phi' = \frac{1}{4\pi\epsilon_0} \frac{\lambda'}{\sqrt{x^2 + R^2}} \, dx, \tag{8.130}$$

where R is the transverse distance from the particle's trajectory to the wire. The electric field acting in the rest frame of the particle reads

$$E'_z = \frac{1}{4\pi\epsilon_0} \int_{-\infty}^{\infty} dx \left(-\frac{\partial}{\partial R} \right) \frac{\lambda'}{\sqrt{x^2 + R^2}} = \frac{\lambda'}{2\pi\epsilon_0 R}. \tag{8.131}$$

Here, R denotes the distance from the wire, and we implicitly assume that the test particle is located above the wire; a positive value of λ implies that the particle is pushed upward. If the test particle were below the wire, then the direction of E'_z is reversed. So, the electric force on the test particle is

$$F'_z = q E'_z = \frac{d}{d\tau} p'_z = \gamma \frac{d}{dt} p_z = q \frac{\lambda'}{2\pi\epsilon_0 R} = -\frac{q}{2\pi\epsilon_0 R} \frac{v}{c^2} I' = -\frac{q \mu_0 v I'}{2\pi R} = -\gamma \frac{q \mu_0 v I}{2\pi R},$$
$$\tag{8.132}$$

which is directed in the negative z direction, i.e., toward the conducting wire. Here, $d\tau = dt/\gamma$ is the proper time interval. We note that $p'_z = p_z$ for the momentum component because the relativistic Lorentz boost is in the x direction. Our result is in agreement with intuition, because the negative particles move to the right faster than the positively charged particles. The drift velocity is added to the relative velocity of the wire; the negative charges experience a more pronounced Lorentz contraction, and the test charge q of the particle (which we assume to be positive) is attracted to the conducting wire because of the negative net charge of the latter.

In the rest frame, thus, we have

$$F_z = \frac{d}{dt} p_z = -\frac{q \, \mu_0 \, v \, I}{2\pi R} \, . \tag{8.133}$$

Let us briefly show that this result is compatible with the usual formulation of the Lorentz force. The magnetic field can be calculated from Ampere's law as follows (in the rest frame of the particle). We reduce all vector-valued quantities to their moduli and consider a straightforward application of the Ampere–Maxwell law in steady state, where its integral form is $\oint \vec{B} \cdot d\vec{A} = \mu_0 \, I$. This leads to the formulas

$$2\pi R B = \mu_0 I \, , \qquad B = \frac{\mu_0 I}{2\pi R} \, , \qquad F = q v B = \frac{q \, \mu_0 \, v \, I}{2\pi R} \, . \tag{8.134}$$

The direction of the force (toward the conducting wire) is in line with the right-hand rule applied to the magnetic force as obtained for the current configuration, as implied by the formula $\vec{F} = q \vec{v} \times \vec{B}$. The current to be used in Ampere's law $\oint \vec{B} \cdot d\vec{A} = \mu_0 \, I$ points into the positive x direction because the negatively charged electrons inside the conductor move to the left, and we are investigating a positively charged test particle above the conductor. Then, the magnetic induction field above the conductor is directed toward us, and we recall that the particle is moving to the right. So, the right-hand rule for the vector product dictates that the magnetic force is pointing in the downward direction, toward the conductor.

8.3.5 *Covariant Form of the Liénard–Wiechert Potentials*

We have discussed the Liénard–Wiechert potentials in Sec. 5.6.1, in a manifestly noncovariant form. For relativistic moving objects, it is instructive to express the potentials in covariant form, implying that we shall encounter the proper time of the particle. Due to time dilation, the proper time interval $d\tau$ for a particle moving at velocity v is given as [see Eq. (8.12)]

$$d\tau = \frac{1}{\gamma} dt \, , \qquad \gamma = \left(1 - \frac{v^2}{c^2}\right)^{-1/2} \, , \qquad d\tau < dt \, , \tag{8.135}$$

where $d\tau$ is the time interval measured in the rest frame of the particle, and dt is the time interval in the laboratory frame. The velocity four-vector is

$$\beta^\mu = \left(\gamma, \gamma \frac{\vec{v}}{c}\right) \, , \qquad \beta^0 = \gamma \, , \qquad \vec{\beta} = \gamma \frac{\vec{v}}{c} \, , \qquad \beta^\mu \beta_\mu = \gamma^2 \left(1 - \left(\frac{\vec{v}}{c}\right)^2\right) = 1 \, . \tag{8.136}$$

We recall the manifestly non-covariant form of the Liénard–Wiechert potentials as given in Eqs. (5.261a) and (5.261b),

$$\Phi\left(\vec{r},t\right) = \frac{q}{4\pi\epsilon_0} \frac{1}{|\vec{r} - \vec{R}(t_{\text{ret}})|} \left(1 - \frac{\dot{\vec{R}}(t_{\text{ret}})}{c} \cdot \frac{\vec{r} - \vec{R}(t_{\text{ret}})}{|\vec{r} - \vec{R}(t_{\text{ret}})|}\right)^{-1}, \tag{8.137a}$$

$$\vec{A}\left(\vec{r},t\right) = \frac{q}{4\pi\epsilon_0 c^2} \frac{\dot{\vec{R}}(t_{\text{ret}})}{|\vec{r} - \vec{R}(t_{\text{ret}})|} \left(1 - \frac{\dot{\vec{R}}(t_{\text{ret}})}{c} \cdot \frac{\vec{r} - \vec{R}(t_{\text{ret}})}{|\vec{r} - \vec{R}(t_{\text{ret}})|}\right)^{-1}. \tag{8.137b}$$

These potentials are still given in a not completely covariant form, in the sense that the quantities in them carry spatial, not Lorentz, indices. We therefore explore the question how the Liénard–Wiechert potentials qualify themselves as manifestly covariant formulas. In this endeavor, we start with the scalar potential, which we rewrite as follows,

$$\Phi\left(\vec{r},t\right) = \frac{\gamma\, q}{4\pi\epsilon_0} \frac{1}{\gamma\,|\vec{r} - \vec{R}(t_{\text{ret}})| - \left(\gamma\,\dfrac{\dot{\vec{R}}(t_{\text{ret}})}{c}\right) \cdot (\vec{r} - \vec{R}(t_{\text{ret}}))}. \tag{8.138}$$

The retarded time fulfills the relations,

$$t_{\text{ret}} = t - \frac{|\vec{r} - \vec{R}(t_{\text{ret}})|}{c} < t, \qquad c(t - t_{\text{ret}}) = |\vec{r} - \vec{R}(t_{\text{ret}})|. \tag{8.139}$$

We are now in the position to order the space-time coordinate of the particle at the retarded time, when the radiation is emitted, into a four-vector form,

$$R^\mu(t_{\text{ret}}) = (c\,t_{\text{ret}}, \vec{R}(t_{\text{ret}})), \qquad \beta^\mu(t_{\text{ret}}) = \left(\gamma, \gamma\,\frac{\dot{\vec{R}}(t_{\text{ret}})}{c}\right). \tag{8.140}$$

The space-time coordinate where the potentials are observed is written as $x^\mu = (ct, \vec{r})$. With $\beta^0 = \gamma$, we can thus write $\Phi\left(\vec{r},t\right)$ as

$$\Phi\left(\vec{r},t\right) = \frac{\gamma\, q}{4\pi\epsilon_0} \frac{1}{\gamma\,c\,(t - t_{\text{ret}}) - \left(\gamma\,\dfrac{\dot{\vec{R}}(t_{\text{ret}})}{c}\right) \cdot (\vec{r} - \vec{R}(t_{\text{ret}}))}$$

$$= \frac{q}{4\pi\epsilon_0} \frac{\beta^0}{\beta^\mu(t_{\text{ret}})\,(x_\mu - R_\mu(t_{\text{ret}}))}. \tag{8.141}$$

Repeating the same steps for the vector potential, we obtain the formulas

$$\Phi\left(\vec{r},t\right) = \frac{q}{4\pi\epsilon_0} \frac{\beta^0(t_{\text{ret}})}{\beta^\mu(t_{\text{ret}})\,(x_\mu - R_\mu(t_{\text{ret}}))}, \tag{8.142a}$$

$$c\,\vec{A}\left(\vec{r},t\right) = \frac{q}{4\pi\epsilon_0} \frac{\vec{\beta}(t_{\text{ret}})}{\beta^\mu(t_{\text{ret}})\,(x_\mu - R_\mu(t_{\text{ret}}))}, \tag{8.142b}$$

or, using the · notation for the four-vector scalar product and ordering $\mathcal{A}^\mu = (\Phi, c\,\vec{A})$ into a four-vector, we obtain

$$\mathcal{A}^\mu(X) = \frac{q}{4\pi\epsilon_0} \left. \frac{\beta^\mu(t)}{\beta(t) \cdot (x - R(t))} \right|_{t=t_{\text{ret}}}. \tag{8.143}$$

This result is given in SI mksA units.

8.4   Towards Quantum Field Theory

8.4.1   *Casimir Effect and Quantum Electrodynamics*

In the preceding sections, we have approached the relativistic quantum field theory regime from the relativistic side; now, we approach it from the quantum side. To this end, it is instructive to realize that fluctuations in both the macroscopic as well as the quantum world can have important observable consequences. Seamen know this effect; ships in the ocean traveling at constant speed, and close to each other, are known to attract each other because the waves are broken in between the ships. In other words, the wave motion is attenuated between the ships. By contrast, no such attenuation occurs on the respective "outward" sides of the ships, i.e., on the sides not opposing each other. Therefore, the waves exert forces on the ships; these are attenuated in between but not attenuated from outside, and the two ships attract each other. Therefore ships on the open sea have to keep a minimum distance from each other; and that includes ships traveling in convoys. The Casimir attraction of perfectly conducting plates is of the same physical origin.

Our brief overview of the effect gives us an opportunity to look at field quantization, which is necessary in order to describe the effect. The most important paradigm in the derivation is the "fundamental conceptual equation" of quantum electrodynamics:

$$\text{Physical\_Observable} = \text{Bare\_Observable}\big|_{\text{reg.}} - \text{Counter\_Term}\big|_{\text{reg.}}$$

$$= \lim_{\text{regulator}\to 0} \int \left(\text{Bare\_Observable}\big|_{\text{reg.-int.}} - \text{Counter\_Term}\big|_{\text{reg.-int.}} \right) \tag{8.144}$$

The meaning of these expressions is as follows: *(i)* Bare_Observable is an an expression describing the physical observable, as derived from first principles, invoking the formalism of quantized fields. In quantum field theory, one would use the "normal Feynman rules of QED" (see Ref. [20]). *(ii)* The Bare_Observable$\big|_{\text{reg.}}$ is a regularized version of the expression for the bare observable, where the regulator respects basic symmetries of the physical problem, as discussed in greater detail in the following. *(iii)* A Counter_Term is is a quantity which needs to be absorbed into a parameter of the theory, i.e. interpreted as a physically unobservable contribution to an energy, or to a scattering amplitude, which needs to be subtracted in order to obtain a physically sensible answer. *(iv)* In the regularized version Counter_Term$\big|_{\text{reg.}}$ of the counter term, one needs to choose a regularization which is compatible with the one used for the bare observable. *(v)* The Bare_Observable$\big|_{\text{reg.-int.}}$ is the integrated form of the regularized bare observable. Without specifying the variables which need to be integrated over, we denote the regularized integrand by the symbol "reg.-int.". *(vi)* Of course, the Counter_Term$\big|_{\text{reg.-int.}}$: is the regularized integrand for the counter-term. *(vii)* Finally, as we let the regulator go to zero, $\lim_{\text{regulator}\to 0}$, the regularization is removed after all other operations (including integrations) have been carried out. Only this sequence ensures that no spurious finite contributions remain in the final expressions, which could otherwise lead to inconsistent results.

The basic, fundamental Eq. (8.144) is illustrated in the current section. The physical situation is illustrated in Fig. 8.2. Two perfectly conducting plates are a distance R apart, and the space in between them is empty. However, there may still be a nonvanishing interaction, i.e. a nonvanishing (attractive) force between the plates, because the vacuum (the "nothing") is influenced by the presence of the plates. In other words, the discretization of the electromagnetic field modes, as implied by the presence of the plates, generates a small residual effect, which leads to an attractive interaction of the plates. All terms listed in the fundamental Eq. (8.144) are thus important.

We need a counter term. The reason is that the plates are only felt as residual effects, as a change of the zero-point energy in view of the presence of the plates, as opposed to a situation where the plates are absent. However, the zero-point energy in the absence of the plates cannot be physically observable and therefore, must be subtracted. The approach consists in calculating the sum of the zero-point energies of the discretized modes in the presence of the plates, minus the sum of the zero-point energies of all the modes in the absence of the plates. The total zero-point energy in the presence of the plates should be interpreted as the unrenormalized ("bare") quantity, whereas the total zero-point energy in the absence of the plates is the counter term.

We need a regulator. The reason is that both the sum of the zero-point energies of the electromagnetic modes in the presence as well as in the absence of the plates are highly divergent quantities. From physical considerations, only the infrared region of long wavelengths is influenced by the quantization. The ultraviolet region does not receive any corrections. The divergence of both the unrenormalized quantity as well as of the counter term can be remedied in a physically sensible way by the introduction of an ultraviolet regulator, i.e. by the introduction of a quantity which suppresses the divergence induced by the high-frequency electromagnetic modes. This ultraviolet regulator should respect basic symmetries of the problem. In the current case, it means that this ultraviolet regulator should respect, e.g., all geometric symmetry of the total arrangement in the limit of large plates, where the edge length L in Fig. 8.2 tends to infinity.

We should be careful about the order of integrations. The regularization parameter which induces an exponential damping of the contribution from the ultraviolet modes, should be kept up until the very end of the calculation. Denoting the

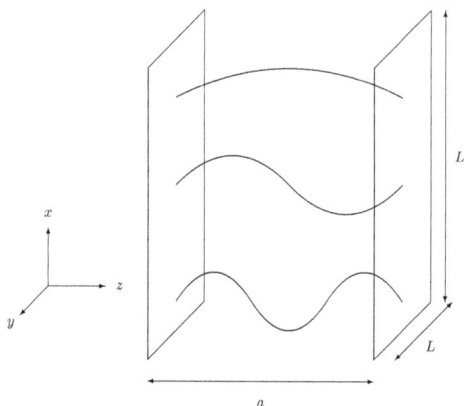

Fig. 8.2 The panel shows a Casimir box composed of two large plates of side length L, a distance R apart, with $L \gg R$. Some of the discrete modes of the electromagnetic field (standing waves with vanishing electric-field amplitude on the plates) are sketched.

regularization parameter as η, we should be careful not to perform the limit $\eta \to 0$ before all other integrations are done.

Our consideration starts with the calculation of a formal expression for the sum of zero-point energies of the electromagnetic modes between the plates. We shall interpret every mode of the electromagnetic field as a harmonic oscillator, which is n-fold excited when there are n photons present in the mode, and we simply take this concept for granted in the current derivation. The derivation relies in an absolutely essential way on the quantization of the electromagnetic field.

8.4.2 *Zero-Point Energy*

In order to describe the vacuum fluctuations of the electromagnetic field of angular frequency mode ω, we draw an analogy to the zero-point energy of a harmonic oscillator of frequency ω. The Hamiltonian in this case is

$$ H = \frac{p^2}{2m} + \frac{1}{2} m \omega^2 q^2 , \tag{8.145} $$

where p is the momentum operator, and m is the mass. Annihilation and creation operators may be defined as follows,

$$ a = \sqrt{\frac{1}{2}} \left(\frac{x}{x_0} + i \frac{x_0 \, p}{\hbar} \right) , \qquad a^+ = \sqrt{\frac{1}{2}} \left(\frac{x}{x_0} - i \frac{x_0 \, p}{\hbar} \right) . \tag{8.146} $$

The length scale x_0 is given by

$$ x_0 = \sqrt{\frac{\hbar}{m \omega}} . \tag{8.147} $$

The Hamiltonian may thus be rewritten as

$$H = \hbar\omega \left(a^+ a + \frac{1}{2} \right), \tag{8.148}$$

where $a^+ a$ is a number operator that counts the number oscillation quanta. Its spectrum consists of nonnegative integers. The zero-point energy (of the fundamental mode) can thus be inferred as

$$e_0 = \tfrac{1}{2} \hbar\omega. \tag{8.149}$$

Normally, this zero-point energy has no physically observable effect. However, as shown by Casimir and Polder [32], a small change in the total zero-point energies of a collection of oscillators (in this case, modes of the electromagnetic field), between two parallel perfect mirrors, has a small effect which results in an attractive interaction between the plates.

We consider two square perfectly conducting metal plates (of dimension $L \times L$, see Fig. 8.2), which define planes parallel to the xy plane, a distance $\Delta z = R$ apart. The plates are supposed to act as perfect mirrors over the entire frequency range of possible photon modes, which is, of course, an idealization. Furthermore, we assume that the plate dimension L is much larger than the plate distance, $L \gg R$.

We can now explore the connection to the discussion of the fundamental modes of the rectangular cavity, with $L = L_x = L_y$ being the "long" sides of the rectangle, and $L_z = R$ being the separation of the plates. In this case, we have seen in Eq. (7.123) that the fundamental mode is a TM mode, i.e., there is only one mode with $n = n_z = 0$, but two modes for all other nonzero integer n.

According to Eq. (7.115), we need to have

$$k_z = n_z \frac{\pi}{R}, \qquad n_z = 0, 1, 2, \dots. \tag{8.150}$$

For the x and y components of the photon momentum vector, no such discretization is indicated, and we have

$$k_x = n_x \frac{\pi}{L}, \qquad k_y = n_y \frac{\pi}{L}, \qquad n_x, n_y \in \mathbb{N}. \tag{8.151}$$

In the limit of large L, summations over n_x and n_y can therefore be replaced by integrations,

$$\sum_{n_x=0}^{\infty} \sum_{n_y=0}^{\infty} \to \left(\frac{L}{\pi}\right)^2 \int_0^{\infty} dk_x \int_0^{\infty} dk_y \to \left(\frac{L}{2\pi}\right)^2 \int_{-\infty}^{\infty} dk_x \int_{-\infty}^{\infty} dk_y, \tag{8.152}$$

where we assume that the integral operator is applied to a function symmetric under $k_x \leftrightarrow -k_x$ and $k_y \leftrightarrow -k_y$.

According to Eq. (8.149), the zero-point energy of a harmonic oscillator of frequency ω is $e_0 = \frac{1}{2}\hbar\omega$. We may thus calculate the sum of the zero-point energies of the photon modes in the space confined by the two plates as follows, keeping track of the summation limits and the photon polarizations λ. Denoting by $\lambda_{\max}$

the maximum number of polarization modes available for a given electromagnetic mode with wave vector $\vec{k}$, one has

$$
E_{\text{bare}} = \sum_{n_x=-\infty}^{\infty} \sum_{n_y=-\infty}^{\infty} \sum_{n_z=0}^{\infty} \sum_{\lambda=1}^{\lambda_{\max}} e_0 = \sum_{n_x=-\infty}^{\infty} \sum_{n_y=-\infty}^{\infty} \sum_{n_z=0}^{\infty} \sum_{\lambda=1}^{\lambda_{\max}} \frac{1}{2} \hbar c |\vec{k}_{n_x,n_y,n_z}|
$$

$$
= \frac{\hbar c}{2} L^2 \int_{-\infty}^{\infty} \frac{dk_x}{2\pi} \int_{-\infty}^{\infty} \frac{dk_y}{2\pi} \sum_{n_z=0}^{\infty} \sum_{\lambda=1}^{\lambda_{\max}} \sqrt{(k_x)^2 + (k_y)^2 + \frac{n_z^2 \pi^2}{R^2}}
$$

$$
= \frac{\hbar c L^2}{8\pi^2} \int_{\mathbb{R}^2} d^2 k_{\|} \sum_{n_z=0}^{\infty} \sum_{\lambda=1}^{\lambda_{\max}} \sqrt{\vec{k}_{\|}^2 + \frac{n_z^2 \pi^2}{R^2}} \, . \tag{8.153}
$$

The specification "bare" means that it refers to an unrenormalized, unregularized quantity which may be divergent (and actually is divergent in our case). The notation $d^2 k_{\|}$ is chosen for the integral $dk_x \, dk_y$. Our considerations from Sec. 7.3.2 imply that $\lambda_{\max} = 1$ for $n_z = 0$, while $\lambda_{\max} = 2$ for $n_z \in \mathbb{N}$ (in the latter case, we have both TE as well as TM modes available). It is interesting to observe that the different number of allowed polarizations for $n_z = 0$ as opposed to $n_z \geq 1$ finds a natural mathematical complement in terms of the Euler–Maclaurin summation formula, as discussed in the following. For now, we obtain

$$
E_{\text{bare}} = \frac{\hbar c L^2}{8\pi^2} \int_{\mathbb{R}^2} d^2 k_{\|} \left(|\vec{k}_{\|}| + 2 \sum_{n_z=1}^{\infty} \sqrt{\vec{k}_{\|}^2 + \frac{n_z^2 \pi^2}{R^2}} \right) . \tag{8.154}
$$

The zero-point energy of any particular harmonic oscillator mode is finite, but there is an infinite number of such modes. So, one might have expected that the total zero-point energy of the entire system should be a highly divergent quantity. This is confirmed by Eq. (8.154). Since the quantity on the right-hand side is highly divergent, the equality only holds formally.

8.4.3   *Regularization and Renormalization*

We must now transform $E_{\text{bare}} \to E_{\text{bare}}\big|_{\text{reg.}}$ in the sense of Eq. (8.144), by the introduction of a suitable regularization. Specifically, we introduce a regularization by replacing the moduli $k_{\|} = |\vec{k}_{\|}|$ of the photon wave vectors,

$$
k_{\|} \to h_{\eta}(k_{\|}) \equiv k_{\|} \, f_{\eta}(k_{\|}) , \tag{8.155a}
$$

where the regulator $f_{\eta}(k_{\|})$ fulfills the following conditions, $f_{\eta}(k_{\|}) \approx 1$ for $k_{\|} \sim R^{-1}$, and $f_{\eta}(k_{\|}) \to 0$ for $k_{\|} \gg R^{-1}$, i.e. the regulator suppresses the contribution of those modes for which the wave length is much smaller than the plate distance. Also, the function should preserve basic symmetries of problem. One possibility is

$$
f_{\eta}(k_{\|}) = \exp\left(-\eta \, k_{\|}\right) , \tag{8.155b}
$$

where $\eta \ll R$ is a free parameter; the limit $\eta \to 0$ is taken after all other operations are carried out. We then have

$$E_{\text{bare}}\big|_{\text{reg.}} = \frac{\hbar c L^2}{8\pi^2} \int_{\mathbb{R}^2} d^2 k_{\parallel} \left[h_{\eta}(k_{\parallel}) + 2 \sum_{n=1}^{\infty} h_{\eta}\left(\sqrt{k_{\parallel}^2 + \frac{n_z^2 \pi^2}{R^2}} \right) \right]. \tag{8.156}$$

This quantity is $\text{Bare\_Observable}\big|_{\text{reg.}}$ in the sense of Eq. (8.144), and it becomes obvious how to interpret the "regularized integrand" in our case.

However, the introduction of a regularization is not sufficient; we also need to perform a subtraction (counter term). Namely, we know that the zero-point energy of the photon modes without the plates present is physically unobservable. Therefore, we have to subtract the energy of these modes, duly regularized in the same manner as the energy of the modes in the modified vacuum. The difference of these two terms then gives the physically observable energy shift, i.e., the Casimir energy, in the sense of Eq. (8.144). The counter-term is given by the vacuum energy, with the boundary conditions removed, i.e., transforming the discrete sum over the allowed modes in Eq. (8.154) into an integral,

$$C = \frac{\hbar c L^2}{4\pi^2} \int_{\mathbb{R}^2} d^2 k_{\parallel} \int_0^{\infty} dn_z \sqrt{k_{\parallel}^2 + \frac{n_z^2 \pi^2}{R^2}},$$

$$C\big|_{\text{reg.}} = \frac{\hbar c L^2}{4\pi^2} \int_{\mathbb{R}^2} d^2 k_{\parallel} \int_0^{\infty} dn_z \, h_{\eta}\left(\sqrt{k_{\parallel}^2 + \frac{n_z^2 \pi^2}{R^2}} \right). \tag{8.157}$$

We, therefore, obtain the physically observable, renormalized zero-point energy shift $E_{\text{ren}}(R)$ as a function of the distance R of the plates as $E_{\text{ren}}(R) = E_{\text{bare}}\big|_{\text{reg.}} - C\big|_{\text{reg.}}$,

$$E_{\text{ren}}(R) = \frac{\hbar c L^2}{2\pi} \int_0^{\infty} dk_{\parallel} \, k_{\parallel} \left[\frac{1}{2} h_{\eta}(k_{\parallel}) + \sum_{n_z=1}^{\infty} h_{\eta}\left(\sqrt{k_{\parallel}^2 + \frac{n_z^2 \pi^2}{R^2}} \right) \right.$$

$$\left. - \int_0^{\infty} dn_z \, h_{\eta}\left(\sqrt{k_{\parallel}^2 + \frac{n_z^2 \pi^2}{R^2}} \right) \right], \tag{8.158}$$

where the integral over the azimuth angle has been carried out. We now substitute

$$k_{\parallel} = \frac{\pi}{R} \sqrt{u}, \qquad u = \frac{R^2 k_{\parallel}^2}{\pi^2}. \tag{8.159}$$

This implies the relations

$$\frac{dk_{\parallel}}{du} = \frac{\pi}{2R\sqrt{u}}, \qquad k_{\parallel} \, dk_{\parallel} = \frac{\pi^2}{2R^2} du, \qquad h_{\eta}(k_{\parallel}) = \frac{\pi}{R} \sqrt{u} \, f_{\eta}\left(\frac{\pi}{R} \sqrt{u} \right), \tag{8.160}$$

so that

$$h_{\eta}\left(\sqrt{k_{\parallel}^2 + \frac{n^2 \pi^2}{R^2}} \right) = \frac{\pi}{R} \sqrt{u + n^2} \, f_{\eta}\left(\frac{\pi}{R} \sqrt{u + n^2} \right). \tag{8.161}$$

The substitution $k_\parallel \to u$, applied to Eq. (8.158), leads to

$$\frac{E_{\text{ren}}(R)}{L^2} = \lim_{\eta \to 0} \frac{\hbar c \pi^2}{4 R^3} \int_0^\infty du \left[\frac{1}{2} \sqrt{u} \, f_\eta \left(\frac{\pi}{R} \sqrt{u} \right) + \sum_{n_z=1}^\infty \sqrt{u + n_z^2} \, f_\eta \left(\frac{\pi}{R} \sqrt{u + n_z^2} \right) \right.$$

$$\left. - \int_0^\infty dn_z \sqrt{u + n_z^2} \, f_\eta \left(\frac{\pi}{R} \sqrt{u + n_z^2} \right) \right]$$

$$= \lim_{\eta \to 0} \frac{\hbar c \pi^2}{4 R^3} \left[\frac{1}{2} F_\eta(0) + \sum_{n_z=1}^\infty F_\eta(n_z) - \int_0^\infty dn_z \, F_\eta(n_z) \right], \qquad (8.162)$$

where we exchange the order of the u- and n_z-integrations for the last term, and define the function $F_\eta(n_z)$ by

$$F_\eta(n_z) = \int_0^\infty du \sqrt{u + n_z^2} \, f_\eta \left(\frac{\pi}{R} \sqrt{u + n_z^2} \right) = \int_{n_z^2}^\infty dv \sqrt{v} \, f_\eta \left(\frac{\pi}{R} \sqrt{v} \right), \qquad (8.163)$$

where the latter form is given by the trivial substitution $v = u + n_z^2$. The exchange of the order of integration is possible because the regularization has led to integrals which are absolutely convergent. This aspect is very important as it illustrates the necessity of keeping all regulators finite up until the very last steps of the calculation.

The expression in square brackets in the last line of Eq. (8.162) has the structure of a discrete sum minus a corresponding integral. The prefactor $\frac{1}{2}$ in front of the first term finds a natural, geometrical interpretation. Quite fortunately, the Euler–Maclaurin formula comes to the rescue in our quest of evaluating the tiny difference between the sum and the integral in the last line of Eq. (8.162). The Euler–Maclaurin summation formula is given in many textbooks [see, e.g., Eqs. (2.01) and (2.02) on p. 285 of [9]]. In its full form, the Euler–Maclaurin formula reads (see Fig. 8.3)

$$\frac{1}{2}f(N) + \frac{1}{2}f(M) + \sum_{n=N+1}^{M-1} f(n) - \int_N^M dn \, f(n)$$

$$= \sum_{j=1}^q \frac{B_{2j}}{(2j)!} \left[f^{(2j-1)}(M) - f^{(2j-1)}(N) \right] + R_q, \qquad (8.164a)$$

where the remainder term R_q is

$$R_q = -\frac{1}{(2q)!} \int_N^M dx \, B_{2q}\big(x - [[x]]\big) f^{(2q)}(x). \qquad (8.164b)$$

Here, $[[x]]$ is the integral part of x, i.e., the largest integer m satisfying $m \leq x$, and $B_k(x)$ is a Bernoulli polynomial defined by the generating function [see Eq. (1.06) on p. 281 of Ref. [9]]

$$\frac{t \exp(xt)}{\exp(t) - 1} = \sum_{m=0}^\infty B_m(x) \frac{t^m}{m!}, \qquad |t| < 2\pi. \qquad (8.165)$$

The $B_m = B_m(0)$ are the Bernoulli numbers, for which we can indicate some example cases,

$$B_1 = -\frac{1}{2}, \quad B_2 = \frac{1}{6}, \quad B_3 = 0, \quad B_4 = -\frac{1}{30}, \quad B_5 = 0, \quad B_6 = \frac{1}{42}, \dots. \qquad (8.166)$$

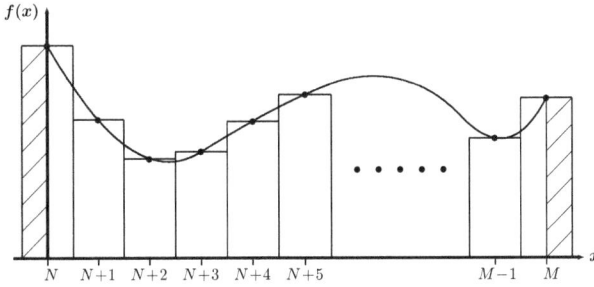

Fig. 8.3 Illustration of Eq. (8.164).

Geometrically, it is clear that the expression $\frac{1}{2}f(N) + \frac{1}{2}f(M) + \sum_{n=N+1}^{M-1} f(n)$ approximates the integral $\int_N^M \mathrm{d}x\, f(x)$ according to the trapezoid rule. The remainder term obtained by forming the difference is then expressed, by the Euler–Maclaurin formula, as a sum over expressions containing higher-order derivatives of the function at the boundaries of the domain of integration. In practical applications, the assumption is that the remainder term R_q on the right-hand side of Eq. (8.164b) tends to zero as $q \to \infty$, and that the first few terms on the right-hand side of Eq. (8.164a) yield increasingly better approximations to the remainder term given by the difference of the sum and the integral over n.

For our case [see Eq. (8.162)], we identify $N \to 0$, $M \to \infty$, $f \to F_\eta$ and $n \to n_z$. The convergence of the integral $\int_0^\infty \mathrm{d}n_z\, F_\eta(n_z)$ is ensured by the exponential damping at large n_z given by the regulator. The same exponential suppression of the contribution from large n_z is responsible for the fact that all derivatives of the function $F_\eta(n_z)$ vanish in the limit $n_z \to \infty$. We can thus rewrite the Eq. (8.162) considering only the derivatives of the function $F_\eta(n_z)$ at $n_z = 0$,

$$\frac{1}{2}F_\eta(0) + \sum_{n_z=1}^\infty F_\eta(n_z) - \int_0^\infty \mathrm{d}n_z\, F_\eta(n_z) = -\sum_{j=1}^\infty \frac{1}{(2j)!} B_{2j} F_\eta^{(2j-1)}(0)$$

$$= -\frac{1}{2!} B_2 F_\eta'(0) - \frac{1}{4!} B_4 F_\eta'''(0) - \dots.$$
$$(8.167)$$

We have expressed the renormalized vacuum energy shift as

$$\frac{E_{\mathrm{ren}}(R)}{L^2} = \lim_{\eta \to 0} \frac{\hbar c \pi^2}{4 R^3} \left[-\frac{1}{2!} B_2 F_\eta'(0) - \frac{1}{4!} B_4 F_\eta'''(0) - \dots \right]. \qquad (8.168)$$

The evaluation of the derivatives now proceeds using the representation for $F_\eta(n_z)$ given in Eq. (8.163), and we recall here for convenience this definition as well as the definition of $f_\eta(k)$ given in Eq. (8.155b),

$$F_\eta(n_z) = \int_{n_z^2}^\infty \mathrm{d}v\, \sqrt{v}\, f_\eta\!\left(\frac{\pi}{R}\sqrt{v}\right), \qquad f_\eta(k) = \exp(-\eta k). \qquad (8.169)$$

The differentiation with respect to n_z^2 can easily be reformulated as a differentiation with respect to n_z,

$$\frac{\partial F_\eta(n_z)}{\partial n_z^2} = -n_z \, f_\eta\left(\frac{\pi}{R}\, n_z\right) = \frac{1}{2n_z}\frac{\partial F_\eta(n_z)}{\partial n_z}, \tag{8.170}$$

and so

$$\frac{\partial F_\eta(n_z)}{\partial n_z} = -2n_z^2 \, f_\eta\left(\frac{\pi}{R}\, n_z\right) = -2n_z^2 \, \exp\left(-\frac{\pi\eta}{R}\, n_z\right). \tag{8.171}$$

This means that

$$F_\eta'(n_z) = -2n_z^2 \, \exp\left(-\eta\,\frac{\pi}{R}\, n_z\right), \tag{8.172a}$$

$$F_\eta''(n_z) = \left(-4n_z + \frac{2n_z^2\pi\eta}{R}\right)\exp\left(-\frac{\pi\eta}{R}\, n_z\right), \tag{8.172b}$$

$$F_\eta'''(n_z) = \left(-4 + \frac{8n_z\pi\eta}{R} - \frac{2n_z^2\pi^2\eta^2}{R^2}\right)\exp\left(-\frac{\pi\eta}{R}\, n_z\right), \tag{8.172c}$$

$$F_\eta''''(n_z) = \left(\frac{12\pi\eta}{R} - \frac{12n_z\pi^2\eta^2}{R^2} + \frac{2n_z^2\pi^3\eta^3}{R^3}\right)\exp\left(-\frac{\pi\eta}{R}\, n_z\right), \tag{8.172d}$$

$$F_\eta'''''(n_z) = \left(-\frac{24\pi^2\eta^2}{R^2} + \frac{16n_z\pi^3\eta^3}{R^3} - \frac{2n_z^2\pi^4\eta^4}{R^4}\right)\exp\left(-\frac{\pi\eta}{R}\, n_z\right). \tag{8.172e}$$

We have performed all integrations and differentiations with a finite cutoff η. The actual zero-point energy as well as the counter term have been regularized, and the regularization parameter η can now be removed at the very end of the calculation. Namely, we can easily read off the results

$$\lim_{\eta\to 0} F_\eta'(0) = \qquad \lim_{\eta\to 0} F_\eta''(0) = 0, \qquad \lim_{\eta\to 0} F_\eta'''(0) = -4, \qquad \lim_{\eta\to 0} F_\eta^{(j\geq 4)}(0) = 0. \tag{8.173}$$

All higher derivatives vanish because further powers of η are generated by the differentiation of the exponential function. Finally, after removing the regularization, we obtain the Casimir energy shift as

$$\frac{E_{\rm ren}(R)}{L^2} = -\frac{\hbar c\pi^2}{4\,R^3}\frac{1}{4!}\, B_4 \lim_{\eta\to 0} F_\eta'''(0) = -\frac{\hbar c\pi^2}{4\,R^3}\frac{1}{24}\left(-\frac{1}{30}\right)(-4) = -\frac{\hbar c\pi^2}{720\,R^3}. \tag{8.174}$$

The energy shift is negative, corresponding to an attractive interaction, which confirms the intuitive picture regarding the attractive force between ships on the open sea. The Casimir force per area (Casimir pressure) on the plate at $z = R$ is consequently negative (attractive) and amounts to

$$\frac{F_{\rm ren}(R)}{L^2} = -\frac{\partial}{\partial R}\frac{E_{\rm ren}(R)}{L^2} = -\frac{\hbar c\pi^2}{240\,R^4}. \tag{8.175}$$

Our brief investigation of the Casimir effect familiarizes us with the ideas of regularization and renormalization, which are extremely important in quantum field theory.

8.5 Classical Potentials and Renormalization

8.5.1 *Potential of a Uniformly Charged Plane*

We attempt to understand the concepts of regularization and renormalization on the level of the classical theory. To this end, we consider a uniform plane sheet of charge, with a surface charge density σ_0, lying in the $z = 0$ plane. The goal is to determine the value of the potential at $\vec{r} = x\,\hat{e}_z + y\,\hat{e}_y + z\,\hat{e}_z$ with the condition that the potential is to vanish at the surface. The differential surface area for this problem is $dS' = dx'dy'$. This can be used in Eq. (2.24),

$$\Phi(\vec{r}) = \frac{1}{4\pi\epsilon_0} \int \frac{\rho(\vec{r}')}{|\vec{r} - \vec{r}'|} d^3r' = \frac{1}{4\pi\epsilon_0} \int \frac{\sigma_0}{|\vec{r} - \vec{r}'|} dx'\,dy', \tag{8.176}$$

to obtain the required integral

$$\begin{aligned}
\Phi(x, y, z) &= \Phi(z) \\
&= \frac{1}{4\pi\epsilon_0} \int_{-\infty}^{\infty} \int_{-\infty}^{\infty} \frac{\sigma_0}{\sqrt{(x')^2 + (y')^2 + z^2}} dx'dy' + \Phi_0 \\
&= \frac{\sigma_0}{4\pi\epsilon_0} \int_0^{\infty} \frac{1}{\sqrt{\rho^2 + z^2}} 2\pi\rho\,d\rho + \Phi_0. \tag{8.177}
\end{aligned}$$

This integral is divergent. It even diverges linearly for large R. In order to handle this difficulty, we note that the potential is independent of x and y. We assume that the surface has a finite, but large radius R centered on the z axis. The radius R acts as a cutoff for the xy integration, and it respects the symmetry of the problem. The regularized potential is

$$\begin{aligned}
\Phi(z) &= \frac{\sigma_0}{2\epsilon_0} \int_0^R \frac{\rho}{\sqrt{\rho^2 + z^2}} d\rho + \Phi_0(R) \\
&= \frac{\sigma_0}{2\epsilon_0} \left[\sqrt{R^2 + z^2} - |z| \right] + \Phi_0(R). \tag{8.178}
\end{aligned}$$

Here, we have used the result

$$\frac{\partial}{\partial\rho} \sqrt{\rho^2 + z^2} = \frac{1}{2} \frac{2\rho}{\sqrt{\rho^2 + z^2}} = \frac{\rho}{\sqrt{\rho^2 + z^2}}. \tag{8.179}$$

We have assumed that the counter term $\Phi_0(R)$ is a function of R only, and does not depend on z. It, therefore, is physically unobservable. We can now fix a "renormalization condition" by requiring the potential to acquire a specific value at a specific point, because we know that only potential differences matter. We choose the condition that the the potential be zero at $z = 0$. This requires that $\Phi_0(R) = -\frac{\sigma_0}{2\epsilon_0}R$. We can finally remove the regularization at the end of the calculation, and let R go to infinity,

$$\Phi(z) = \lim_{R \to \infty} \frac{\sigma_0}{2\epsilon_0} \left[R\sqrt{1 + \frac{z^2}{R^2}} - |z| - R \right] = -\frac{\sigma_0}{2\epsilon_0} |z|. \tag{8.180}$$

We have used the fact that

$$\lim_{R \to \infty} \sqrt{R^2 + z^2} - R = \lim_{R \to \infty} R \left(1 + \frac{z^2}{2R^2} \right) - R = \lim_{R \to \infty} \frac{z^2}{2R} = 0 \, . \tag{8.181}$$

The resulting electric field is directed along the z axis, as expected. The electric field is

$$\vec{E}(z) = -\vec{\nabla} \Phi = \begin{cases} \hat{e}_z \dfrac{\sigma_0}{2\epsilon_0} & (z > 0) \\[2mm] -\hat{e}_z \dfrac{\sigma_0}{2\epsilon_0} & (z < 0) \end{cases} \, . \tag{8.182}$$

This behavior is consistent with intuition. We can unify the two above formulas as

$$\vec{E}(z) = \hat{e}_z \frac{\sigma_0}{2\epsilon_0} \, \text{sgn}(z) \, , \tag{8.183}$$

where the sign function is defined as $\text{sgn}(z) = 1$ for $z > 0$ and $\text{sgn}(z) = -1$ for $z < 0$. The value $\text{sgn}(0)$ is fixed to be zero by convention. The derivative of the sign function is

$$\frac{\text{d} \, \text{sgn}(z)}{\text{d}z} = 2 \, \delta(z) \, . \tag{8.184}$$

We can thus verify that

$$\vec{\nabla} \cdot \vec{E}(z) = \frac{\partial}{\partial z} E_z(z) = 2 \, \delta(z) \, \frac{\sigma_0}{2\epsilon_0} = \frac{1}{\epsilon_0} \, \sigma_0 \, \delta(z) = \frac{1}{\epsilon_0} \rho(x, y, z) \, . \tag{8.185}$$

We convince ourselves that the integral for the electric field strength is convergent,

$$E_z(z) = \frac{\sigma_0}{4\pi\epsilon_0} \int_0^\infty \frac{1}{|\vec{r} - \vec{r}'|^2} \frac{z - z'}{|\vec{r} - \vec{r}'|} \, \text{d}x' \, \text{d}y' = \frac{\sigma_0}{4\pi\epsilon_0} \int_0^\infty \frac{1}{\rho^2 + z^2} \frac{z}{\sqrt{\rho^2 + z^2}} \, 2\pi\rho \, \text{d}\rho$$

$$= \frac{\sigma_0 z}{2\epsilon_0} \int_0^\infty \frac{\rho}{(\rho^2 + z^2)^{3/2}} \, \text{d}\rho = \frac{\sigma_0 z}{4\epsilon_0} \left[-\frac{2}{(\rho^2 + z^2)^{1/2}} \right]_{\rho^2 = 0}^{\rho^2 = \infty} = \frac{\sigma_0}{2\epsilon_0} \frac{z}{|z|} \, . \tag{8.186}$$

We thus reproduce the previously obtained result.

8.5.2 *Potential of a Uniformly Charged Long Wire*

Let us consider an infinitely long, charged wire that extends along the x axis, and let us try to evaluate the electrostatic potential a distance z away from the wire. By symmetry, the potential can only be a function of z. We start from Eq. (2.24),

$$\Phi(\vec{r}) = \frac{1}{4\pi\epsilon_0} \int \frac{\rho(\vec{r}')}{|\vec{r} - \vec{r}'|} \, \text{d}^3 r' \, . \tag{8.187}$$

The charge density ρ which corresponds to a long charged rod along the x-axis, reads as

$$\rho(\vec{r}') = \lambda \, \delta(y') \, \delta(z') \, , \tag{8.188}$$

where λ is the charge per unit length along the wire. The potential is to be evaluated at the point $\vec{r} = (0, 0, z)$, where it reads as follows,

$$\Phi(z) = \frac{\lambda}{4\pi\epsilon_0} \int_{-\infty}^{\infty} \frac{1}{\sqrt{z^2 + (x')^2}}\, dx' = \frac{\lambda}{2\pi\epsilon_0} \int_0^{\infty} \frac{1}{\sqrt{z^2 + (x')^2}}\, dx'. \tag{8.189}$$

Observe that the integration limits have changed. This expression is formally infinite, but the divergence at the upper limit of the integration domain is only logarithmic. First we regularize this integral by introducing an upper cutoff for the x'-integration, which, again, respects the symmetry of the problem,

$$\Phi(z) = \frac{\lambda}{2\pi\epsilon_0} \int_0^{R} \frac{1}{\sqrt{z^2 + (x')^2}}\, dx', \tag{8.190}$$

where R is assumed to be large (at the end of the calculation, we let $R \to \infty$). Because the integral diverges, we have to subtract a counter term. We cannot impose the renormalization condition $\Phi(z = 0) = 0$ because the regularized integral would otherwise be divergent at zero. Fixing the potential to be zero at an intermediate point $z = z_0$, we can devise the following counter term,

$$C = \frac{\lambda}{2\pi\epsilon_0} \int_0^{R} \frac{1}{\sqrt{z_0^2 + (x')^2}}\, dx' \tag{8.191}$$

which also is infinite in the limit $R \to \infty$ and independent of z. In view of the result

$$\int dx' \frac{1}{\sqrt{z^2 + x'^2}} = \ln\left(z + \sqrt{z^2 + x'^2}\right), \tag{8.192}$$

the renormalized expression,

$$\begin{aligned}
\Phi(z) &= \frac{\lambda}{2\pi\epsilon_0} \lim_{R \to \infty} \int_0^{R} \left(\frac{1}{\sqrt{z^2 + (x')^2}} - \frac{1}{\sqrt{z_0^2 + (x')^2}} \right) dx' \\
&= \frac{\lambda}{2\pi\epsilon_0} \lim_{R \to \infty} \left(\ln(R + \sqrt{R^2 + z^2}) - \ln(R + \sqrt{R^2 + z_0^2}) - \ln(|z|) + \ln(|z_0|) \right) \\
&= -\frac{\lambda}{2\pi\epsilon_0} \ln\left(\frac{|z|}{|z_0|} \right) \tag{8.193}
\end{aligned}$$

is finite in the limit $R \to \infty$ (removal of the cutoff). By differentiation, we obtain

$$E_z(z) = -\frac{\partial}{\partial z}\Phi(z) = \frac{\lambda}{2\pi\epsilon_0} \frac{\partial}{\partial z} \ln\left(\frac{|z|}{|z_0|} \right) = \frac{\lambda}{2\pi\epsilon_0 z}, \tag{8.194}$$

which is positive for $z > 0$ and negative for $z < 0$. The integral defining the field strength is

$$E_z(z) = \frac{\lambda}{4\pi\epsilon_0} \int_{-\infty}^{\infty} \frac{z}{(z^2 + (x')^2)^{3/2}}\, dx' = \frac{\lambda z}{2\pi\epsilon_0} \frac{1}{|z|^2} = \frac{\lambda}{2\pi\epsilon_0 z}. \tag{8.195}$$

This confirms the result based on the evaluation of the potential. In field theory, a renormalization condition formulated at a finite, but nonvanishing, z_0 is otherwise known as an "intermediate renormalization."

8.5.3   *Charged Structures in* **0.99** *and* **1.99** *Dimensions*

Let us try to generalize the above approach. A charged, long wire is a one-dimensional structure, whereas a charged plane sheet is a two-dimensional structure. We investigate the question of whether it is possible to perform the regularization of the above two examples by investigating charged structures in 0.99 and 1.99 dimensions, i.e., by slightly displacing the dimension of the structure from an integer dimension, using so-called dimensional regularization.

We still operate in three spatial dimensions, i.e., the Coulomb law is unchanged. Only the dimensionality of the charged structure is changed. We first need the volume Ω_D of the D-dimensional unit sphere. The matching of the result for integer dimensions to the result for non-integer dimensions is done as follows. On the one hand, if we integrate all variables which span $\mathbb{R}^D$ one by one, for a "Gaussian bell" function, then we obtain

$$
\int_{\mathbb{R}^D} \frac{\mathrm{d}^D k}{(2\pi)^D}\, \mathrm{e}^{-\vec{k}^2} = \left[\int_{-\infty}^{\infty} \frac{\mathrm{d}k}{2\pi}\, \mathrm{e}^{-k^2} \right]^D = \left(\frac{\sqrt{\pi}}{2\pi} \right)^D = \frac{1}{2^D\, \pi^{D/2}} = \frac{\pi^{D/2}}{(2\pi)^D}. \tag{8.196}
$$

On the other hand, if we single out the integral over the angular variable $\int \mathrm{d}\Omega_d$, then we have

$$
\int_{\mathbb{R}^D} \frac{\mathrm{d}^D r}{(2\pi)^D}\, \mathrm{e}^{-\vec{r}^2} = \frac{1}{(2\pi)^D} \left[\int \mathrm{d}\Omega_D \right] \int_0^{\infty} \mathrm{d}r\, r^{D-1}\, \mathrm{e}^{-r^2} = \frac{\Gamma(D/2)}{2(2\pi)^D} \left[\int \mathrm{d}\Omega_D \right], \tag{8.197}
$$

where we recall that the Gamma function is defined as $\Gamma(x) = \int_0^{\infty} t^{x-1}\, \exp(-t)\, \mathrm{d}t$, for $x > 0$. So, we have

$$
\Omega_D = \int \mathrm{d}\Omega_D = \frac{2\, \pi^{D/2}}{\Gamma(D/2)}. \tag{8.198}
$$

This result takes the following values,

$$
\Omega_D = \begin{cases} 2 & (D = 1) \\ 2\pi & (D = 2) \\ 4\pi & (D = 3) \end{cases}. \tag{8.199}
$$

One may ask about the origin of the result $\Omega_{D=1} = 2$. The reason is that we attempt to replace an integral over the entire D-dimensional space $\int_{\mathbb{R}^D} \mathrm{d}^D r$ with an integral of the form proportional to $\int_0^{\infty} \mathrm{d}r\, r^{D-1}$ over a radially symmetric function. Then, in one dimension, for a function $f(x) = f(-x)$, with $x \in \mathbb{R}$, we have a factor two because the regimes $x \in (-\infty, 0)$ and $x \in (0, \infty)$ have to be treated separately.

For a radially symmetric integrand $f(\vec{r}) = f(r)$, one has

$$
\int_{\mathbb{R}^D} \mathrm{d}^D r\, f(\vec{r}) = \Omega_D \int_0^{\infty} \mathrm{d}r\, r^{D-1}\, f(r), \tag{8.200}
$$

in a general dimension D (where $r = |\vec{r}|$ and the function f is defined accordingly, depending on the scalar or vector character of its argument). In a non-integer dimension D, slightly displaced from the integer n by ε,

$$
D = n - \varepsilon, \tag{8.201}
$$

one has to restore the physical dimensionality of the integrand by the introduction of a reference length scale which we denote by z_0 and which is not necessarily equal to the reference length scale z_0 used for the renormalization of the charged long wire used above. (For $n = 1$ and $n = 2$, and $\varepsilon = 0.01$, we verify the title of the current section.) The appropriate replacement which preserves the physical dimension, is

$$r^D \to r^{n-\varepsilon} z_0^\varepsilon = r^D z_0^\varepsilon. \tag{8.202}$$

We denote the uniform charge density in n dimensions as Ξ. For one dimension, we have $\Xi = \lambda$, while $\Xi = \sigma$ in two dimensions. Finally,

$$\Phi(z) = \frac{1}{4\pi\varepsilon_0} \Omega_D \int_0^\infty dr\, r^{D-1}\, z_0^\varepsilon \frac{\Xi}{\sqrt{r^2 + z^2}}, \qquad D = n - \varepsilon. \tag{8.203}$$

We now consider the integral $(z_0 > 0)$

$$J(D) = \Omega_D \int_0^\infty dr\, r^{D-1}\, z_0^\varepsilon \frac{1}{\sqrt{r^2 + z^2}} = \pi^{\frac{1}{2}(D-1)} \Gamma\left(\tfrac{1}{2}(1 - D)\right) z_0^\varepsilon |z|^{D-1}, \tag{8.204}$$

which is strictly valid only for $D < 1$, because we otherwise have a divergence at the upper limit of integration. However, we courageously assume the validity of Eq. (8.204) even for general $D \geq 1$ by analytic continuation.

For $D = 1 - \varepsilon$, an expansion for "almost integer $D = 1$" results in

$$J(D = 1 - \varepsilon) = \pi^{-\varepsilon/2} (z_0)^\varepsilon |z|^{-\varepsilon} \Gamma\left(\tfrac{1}{2}\varepsilon\right) = \frac{2}{\varepsilon} - \gamma_E - \ln(\pi) - 2 \ln\left(\frac{|z|}{z_0}\right) + \mathcal{O}(\varepsilon). \tag{8.205}$$

In the expansion for small ε, the "replica trick"

$$x^\varepsilon = \exp\left[\varepsilon \ln(x)\right] = 1 + \varepsilon \ln(x) + \mathcal{O}(\varepsilon^2) \tag{8.206}$$

and the formula

$$\Gamma(\varepsilon) = \frac{1}{\varepsilon} - \gamma_E + \mathcal{O}(\varepsilon) \tag{8.207}$$

prove useful. Here, γ_E is the Euler–Mascheroni constant $\gamma_E = 0.57712\ldots$. The complete potential is thus obtained as $(D = 1 - \varepsilon)$

$$\Phi(z) = \frac{\lambda}{2\pi\varepsilon_0} \frac{1}{\varepsilon} - \frac{\lambda}{4\pi\varepsilon_0} (\gamma_E + \ln(\pi)) - \frac{\lambda}{2\pi\varepsilon_0} \ln\left(\frac{|z|}{z_0}\right) + \mathcal{O}(\varepsilon)$$

$$= \frac{\lambda}{2\pi\varepsilon_0} \frac{1}{\varepsilon} - \frac{\lambda}{2\pi\varepsilon_0} \ln\left(\frac{|z|}{z_0} \sqrt{\pi}\, e^{\gamma_E/2}\right) + \mathcal{O}(\varepsilon). \tag{8.208}$$

We have evaluated the integral (8.208) without having a concrete renormalization condition at hand. In the so-called MS (minimal subtraction) renormalization scheme, one now simply throws away the divergent term of order $1/\varepsilon$ and writes

$$\Phi(0, 0, z) = -\frac{\lambda}{2\pi\varepsilon_0} \ln\left(\frac{|z|}{\tilde{z}_0}\right), \qquad \tilde{z}_0 = \frac{z_0}{\sqrt{\pi}\, e^{\gamma_E/2}}. \tag{8.209}$$

A posteriori, i.e., after evaluating the integral, we see that the MS scheme corresponds to a renormalization condition where the potential is required to be zero at a distance $\tilde{z}_0$ from the wire, where $\tilde{z}_0$ is not equal to the originally introduced parameter z_0, but differs only by a multiplicative constant.

In the modified minimal subtraction $\overline{\text{MS}}$ scheme ("MS-bar scheme"), we would also throw away the constant terms in Eq. (8.208), i.e., those with the Euler–Mascheroni constant and the logarithm of π. In this case, the meaning of z_0 is the same as the one used in the calculation described in Sec. 8.5.2. In any case, if one uses the MS or $\overline{\text{MS}}$ scheme, one has to work out the renormalization condition fulfilled by the calculated potential a posteriori. In field theory, one then has to adjust the renormalization scale z_0 that determines the "coupling constant" to the appropriate scale, using a multiplicative renormalization. This is beyond our scope here; however, the principle is clear.

For $D = 2 - \varepsilon$, the expansion is

$$J(D = 2 - \varepsilon) = \pi^{\frac{1}{2} - \frac{1}{2}\varepsilon} (z_0)^{\varepsilon} (|z|)^{1-\varepsilon} \Gamma\left(\tfrac{1}{2}(\varepsilon - 1)\right) = \sqrt{\pi}\,|z|\left(-2\sqrt{\pi}\right) = -2\pi\,|z| + \mathcal{O}(\varepsilon). \tag{8.210}$$

It is characteristic of dimensional regularization that only logarithmic divergences give rise to divergent terms in the dimensional analytic continuation variable ε. Consequently, expanding the potential (8.204) using the result (8.203) about $\varepsilon = 0$ for $D = 2 - \varepsilon$, one recovers the result

$$\Phi(z) = -\frac{\sigma}{2\varepsilon_0}\,|z| + \mathcal{O}(\varepsilon), \qquad D = 2 - \varepsilon, \tag{8.211}$$

where we have used the fact that Ξ has the meaning of the surface charge density σ in two dimensions. This is the result originally derived in Eq. (8.180).

8.6 Open Problems in Classical Electromagnetic Theory

8.6.1 *Abraham–Minkowski Controversy*

We shall conclude this book with a brief digression on physical problems which constitute (at least partially) open problems and leave room for further investigation. One of the most basic open questions in classical electrodynamics, which still has not been conclusively answered, concerns the proper definition of the Maxwell stress tensor (see Sec. 1.3.2) in a dielectric. It is commonly known as the ABRAHAM[1]–MINKOWSKI[2] controversy. The classic literature references are [33, 34]. Let us try to shine some light onto the discrepancy between the two formulations. We recall from Eq. (1.122) that the force density acting on an electromagnetic medium can be written as [see Eq. (1.122)]

$$\vec{f} = \vec{\nabla} \cdot \mathbb{T} - \frac{\partial}{\partial t}\vec{S}, \qquad \vec{S} = \epsilon_0 \vec{E} \times \vec{B}, \tag{8.212}$$

---

[1] Max Abraham (1875–1922)
[2] Hermann Minkowski (1864–1909)

where we recall that $\vec{f}$ denotes the force density, and the components of the stress tensor are [see Eq. (1.119)]

$$\mathbb{T}_{ij} = \epsilon_0 \left(E_i E_j - \frac{1}{2} \delta_{ij} \vec{E}^2 \right) + \frac{1}{\mu_0} \left(B_i B_j - \frac{1}{2} \delta_{ij} \vec{B}^2 \right). \tag{8.213}$$

The scaled Poynting vector $\vec{S}$ differs from $\vec{S}$ used in Eq. (1.122) by a prefactor $1/c^2$; the scaled form allows for simpler prefactors in the generalization to the case of dielectric. Indeed, one might imagine two possible generalizations of the Poynting vector,

$$\vec{S} \to \vec{S}^{\mathrm{M}} = \vec{D} \times \vec{B}, \qquad \vec{S} \to \vec{S}^{\mathrm{A}} = \frac{1}{c^2} \vec{E} \times \vec{H}, \tag{8.214}$$

both of which coincide with the original form $\vec{S} = \epsilon_0 \vec{E} \times \vec{B}$ in vacuum, i.e., for $\epsilon_r = \mu_r = 1$. Of course, the superscripts M and A refer to Minkowski and Abraham. The Minkowski form of the stress tensor is

$$\mathbb{T}_{ij}^{\mathrm{M}} = E_i D_j + H_i B_j - \frac{1}{2} \left(\vec{E} \cdot \vec{D} + \vec{H} \cdot \vec{B} \right) \delta_{ij}. \tag{8.215}$$

It is not symmetric under the interchange $i \leftrightarrow j$, but one can show that

$$\vec{f} = \vec{\nabla} \cdot \mathbb{T}^{\mathrm{M}} - \frac{\partial}{\partial t} \vec{S}^{\mathrm{M}}, \tag{8.216}$$

which carries a formal resemblance with Eq. (8.212). The formal resemblance of Eqs. (8.216) and (8.212) is a primary argument in favor of the Minkowski formulation.

Let us now dwell on Abraham's alternative formalism. First, we mention that, based on the phenomenological Maxwell equations (6.86), one may derive the formula,

$$\vec{\nabla} \cdot \vec{S}^{\mathrm{A}} = -\vec{E} \cdot \vec{J}_0 - \vec{E} \cdot \frac{\partial}{\partial t} \vec{D} - \vec{H} \cdot \frac{\partial}{\partial t} \vec{B}, \tag{8.217}$$

which carries a formal resemblance with Eq. (1.106) and would suggest that the Abraham form of the field momentum should be used. Note, however, that we cannot simply define the energy density as

$$u = \vec{H} \cdot \vec{B} + \vec{E} \cdot \vec{D}, \tag{8.218}$$

because the time derivative contains mixed terms,

$$\partial_t u = \partial_t \vec{H} \cdot \vec{B} + \vec{H} \cdot \partial_t \vec{B} + \partial_t \vec{E} \cdot \vec{D} + \vec{E} \cdot \partial_t \vec{D}. \tag{8.219}$$

In this case, the Abraham form of the force law is

$$\vec{f}^{\mathrm{A}} = \vec{\nabla} \cdot \mathbb{T}^{\mathrm{A}} - \frac{\partial}{\partial t} \vec{S}^{\mathrm{A}}, \tag{8.220}$$

where $\vec{f}^{\mathrm{A}}$ is the Abraham force density, which differs from the Lorentz force density $\vec{f}$. For an isotropic and dispersionless medium, one can show that [35]

$$\vec{f}^{\mathrm{A}} = \vec{f} + \frac{1}{c^2} \left(\epsilon_r - 1 \right) \frac{\partial}{\partial t} \vec{E} \times \vec{H}. \tag{8.221}$$

The additional term in comparison to the Lorentz force is sometimes called the Abraham force. Despite considerable effort, a final word on the subject matter seems to be lacking.

Quantum mechanically, for a refractive index $n = \sqrt{\epsilon_r}$, one can show that the Minkowski momentum density corresponds to a photon momentum

$$p^{\mathrm{M}} = \frac{n\,\hbar\,\omega}{c}\,, \tag{8.222}$$

while the Abraham version reads as

$$p^{\mathrm{A}} = \frac{\hbar\,\omega}{n\,c}\,. \tag{8.223}$$

A relatively recent proposal concerning the detection of the Abraham force with a succession of short optical pulses has been given in Ref. [36]. The controversy still is an active field of research, even if the Abraham formulation is generally preferred in the literature [19, 37].

8.6.2 *Relativistic Dynamics with Radiative Reaction*

When a charged particle is accelerated, it radiates off electromagnetic fields and, therefore, field energy. The backreaction of the electromagnetic field onto the emitting particle is known as the radiative reaction. One might think that it would be very difficult to find a formula that connects the backreaction of the radiation onto the emitting particle. However, by equating the energy loss of the particle with the radiated power, one is actually led to a very intuitive *ansatz* for the backreaction force $\vec{F}_{\mathrm{rad}}$. In the nonrelativistic approximation, the derivation is quite straightforward. Indeed, the problem continues to attract considerable attention, even after a number of physicists have already spent considerable time and effort analyzing its intricacies (for a necessarily incomplete list of literature references, see Refs. [38–48]).

In the nonrelativistic approximation, the power dissipated by an accelerated particle of elementary charge e, dissipated into the radiation, is given by the Larmor formula

$$P_{\mathrm{rad}} = m\,\tau_{\mathrm{rad}}\,\dot{\vec{v}}^{\,2}\,, \qquad \tau_{\mathrm{rad}} = \frac{2\,\alpha\,\hbar}{3\,m\,c^2} = \frac{2\,\alpha\,\lambda_e}{3\,c} = 6.26 \times 10^{-24}\,\mathrm{s}\,. \tag{8.224}$$

Here, $\vec{a} = \dot{\vec{v}}$ is the acceleration, and τ_{rad} is the characteristic emission time for the radiation. Note that the radiation damping time τ_{rad} depends on the mass of the radiating particle. In Eq. (8.224), the numerical result for τ_{rad} is valid provided m denotes the electron mass. In the nonrelativistic limit, Eq. (8.224) was originally derived by Lorentz [49].

We do not derive Eq. (8.224) here, but we can motivate the equation by matching the time-averaged (over an oscillation period), radiated power of a radiating dipole given in Eq. (5.109) with the momentous kinetic energy of the oscillating dipole,

$$\langle P_{\text{rad}} \rangle = \frac{c}{3} \frac{k^4 \vec{p}_0^2}{4\pi\epsilon_0} \rightarrow \frac{2c}{3} \frac{k^4 \langle \vec{p}_0^2 \rangle}{4\pi\epsilon_0} = \frac{2}{3c^3} \frac{\omega^4 \langle \vec{p}_0^2 \rangle}{4\pi\epsilon_0}$$

$$\rightarrow \frac{2 e^2}{3 c^3} \frac{\omega^4 \langle \vec{r}^2 \rangle}{4\pi\epsilon_0} \rightarrow \frac{2 e^2}{3 c^3} \frac{\dot{\vec{v}}^2}{4\pi\epsilon_0} = m \frac{2\alpha \lambda_e}{3c} \dot{\vec{v}}^2 = m \tau_{\text{rad}} \dot{\vec{v}}^2 . \tag{8.225}$$

The justification for the replacements is as follows: *(i)* The maximum dipole moment $\vec{p}_0^2$ is equal to twice the time-averaged dipole moment $\langle \vec{p}_0^2 \rangle$, *(ii)* the average dipole moment is related to position as $\langle \vec{p}_0^2 \rangle = e^2 \langle \vec{r}^2 \rangle$. In the sense of a derivative in Fourier space, we can then replace in *(iii)* the expression $\omega \rightarrow i\partial_t$, to obtain the replacement $\omega^4 \langle \vec{r}^2 \rangle \rightarrow \vec{a}^2$, where $\vec{a} = \dot{\vec{v}} = \ddot{\vec{r}}$ is the acceleration, whose modulus does not depend on time for a harmonic oscillator. Assuming, then, that the relation is valid at any given time, i.e., for a momentous acceleration, we arrive at the formula for the radiated power P. The fine-structure constant is $\alpha = e^2/(4\pi\hbar\epsilon_0 c) \approx 1/137.036$.

The expression (8.224) for P_{rad} is valid for the energy radiated off from the particle. By energy conservation, the radiated power has to be compensated by the action of the radiative-reaction force $\vec{F}_{\text{rad}}$, so that

$$\int_{t_1}^{t_2} \vec{F}_{\text{rad}} \cdot \vec{v}\, dt = - \int_{t_1}^{t_2} P_{\text{rad}}\, dt = - \int_{t_1}^{t_2} m \tau_{\text{rad}} \dot{\vec{v}} \cdot \vec{v}\, dt . \tag{8.226}$$

Provided boundary terms can be neglected, one obtains the result

$$\int_{t_1}^{t_2} \left(\vec{F}_{\text{rad}} - m \tau_{\text{rad}} \ddot{\vec{v}} \right) \cdot \vec{v}\, dt = 0, \qquad \vec{F}_{\text{rad}} = m \tau_{\text{rad}} \ddot{\vec{v}} . \tag{8.227}$$

The radiative backreaction force $\vec{F}_{\text{rad}}$ depends on the time derivative of the acceleration, and the solution of the corresponding equation of motion therefore requires an additional boundary condition. Indeed, the equation of motion

$$\vec{F} = m \dot{\vec{v}} = \vec{F}_{\text{ext}} + \vec{F}_{\text{rad}} = \vec{F}_{\text{ext}} + m \tau_{\text{rad}} \ddot{\vec{v}} , \tag{8.228}$$

for the particle involves the external force $\vec{F}_{\text{ext}}$ and the backreaction force $\vec{F}_{\text{rad}}$. For a vanishing force $\vec{F}_{\text{ext}} = \vec{0}$, one has a runaway solution,

$$\vec{F} = m \vec{a}(t) = m \tau_{\text{rad}} \dot{\vec{a}}(t) , \tag{8.229a}$$

$$\vec{a}(t) = \vec{a}_0 \exp(t/\tau_{\text{rad}}) , \qquad \vec{v}(t) = \vec{a}_0 \tau_{\text{rad}} \exp(t/\tau_{\text{rad}}) + \vec{v}_0 , \tag{8.229b}$$

$$\vec{v}(t) = \vec{a}_0 \tau_{\text{rad}}^2 \exp(t/\tau_{\text{rad}}) + \vec{v}_0 t + \vec{r}_0 . \tag{8.229c}$$

If we postulate that the kinetic energy of the particle needs to remain finite forever, then we can eliminate the runaway solution, but this comes at a price: One can show [47] that the solutions which preserve the finiteness of the energy suffer from problems of causality, namely, the particle would accelerate before the external force is turned on. One can eliminate the runaway solution in a different manner, by replacing $\vec{F}_{\text{rad}} \rightarrow m\,\vec{a}(t)$, to write

$$\vec{F} = m\,\vec{a} = \vec{F}_{\text{ext}} + \tau_{\text{rad}} \left(\frac{\mathrm{d}}{\mathrm{d}t} \vec{F}_{\text{ext}} \right), \tag{8.230}$$

in which case the right-hand side remains zero in the case of a vanishing external force $\vec{F}_{\text{ext}}$. This is equivalent, within the nonrelativistic approximation, to the transition from the Abraham–Lorentz to the Landau–Lifshitz formulation of the backreaction force, which will be discussed in the following.

The first steps toward a relativistic formulation of backreaction were performed by Dirac [40]. One uses the four-velocity β^μ and the proper time τ,

$$\beta^\mu = \left(\gamma, \gamma \frac{\vec{v}}{c} \right), \qquad \mathrm{d}\tau = \left(1 - \frac{v^2}{c^2} \right)^{1/2} \mathrm{d}t. \tag{8.231}$$

The Lorentz–Abraham–Dirac equation is obtained [48] as a natural generalization of Eq. (8.227) to the relativistic domain,

$$m \frac{\mathrm{d}\beta_\mu}{\mathrm{d}\tau} = q F_{\mu\nu} \beta^\nu + \tau_{\text{rad}} \left(g_{\mu\nu} - \beta_\mu \beta_\nu \right) \frac{\mathrm{d}^2 \beta^\nu}{\mathrm{d}\tau^2}. \tag{8.232}$$

The replacement which leads to the "Landau–Lifshitz" form (8.230) corresponds to the replacement

$$\frac{\mathrm{d}^2 \beta^\nu}{\mathrm{d}\tau^2} \approx \frac{\mathrm{d}}{\mathrm{d}\tau} \frac{\mathrm{d}\beta^\nu}{\mathrm{d}\tau} \rightarrow \frac{\mathrm{d}}{\mathrm{d}\tau} \frac{q}{m} F^{\nu\rho} \beta_\rho, \tag{8.233}$$

where the latter form corresponds to the zeroth-order approximation (without the radiative reaction term). Indeed, in Chap. 76 of Ref. [50], one finds the Landau–Lifshitz equation

$$m \frac{\mathrm{d}\beta_\mu}{\mathrm{d}\tau} = q\,F_{\mu\nu} \beta^\nu + \tau_{\text{rad}} \left(g_{\mu\nu} - \beta_\mu \beta_\nu \right) \frac{\mathrm{d}}{\mathrm{d}\tau} \left[q\,F^{\mu\rho} \beta_\rho \right]. \tag{8.234}$$

This equation has no runaway solutions, in contrast to Eq. (8.232).

The Lorentz–Abraham–Dirac equation (8.232) is sometimes written as follows,

$$m \frac{\mathrm{d}\beta_\mu}{\mathrm{d}\tau} = q F_{\mu\nu} \beta^\nu + \tau_{\text{rad}} \left(\frac{\mathrm{d}^2 \beta_\mu}{\mathrm{d}\tau^2} + \frac{\mathrm{d}\beta^\nu}{\mathrm{d}\tau} \frac{\mathrm{d}\beta_\nu}{\mathrm{d}\tau} \beta^\mu \right), \tag{8.235}$$

which corresponds to an integration by parts of the terms in Eq. (8.232). The "Landau–Lifshitz" version of this equation [44, 46, 51] reads as follows,

$$m \frac{\mathrm{d}\beta_\mu}{\mathrm{d}\tau} = q\,F_{\mu\nu} \beta^\nu + \tau_{\text{rad}} \left[\frac{q}{m} \beta^\alpha \left(\partial_\alpha F_{\mu\nu} \right) \beta^\nu + \frac{q^2}{m^2} F_{\mu\nu}\,F^{\nu\rho} \beta_\rho \right.$$

$$\left. + \frac{q^2}{m^2} \left(F^{\nu\alpha} \beta_\alpha \right) \left(F_{\nu\beta} \beta^\beta \right) \beta_\mu \right]. \tag{8.236}$$

This equation is sometimes given in a slightly different form in the literature. E.g., in comparison to Eq. (2) of Ref. [46], one replaces $q \to -e$ and takes notice of the fact that in the second term in square brackets in Eq. (8.236), the author of Ref. [46] interchanges the order of indices and uses the antisymmetry of the field strength tensor. In Refs. [44, 51], it is argued that on the critical manifold, i.e., on the manifold of solutions of the Landau–Lifshitz equation which fulfill "reasonable" boundary conditions and which are, therefore, unstable against any small perturbations which would otherwise lead to runaway solution, the Landau–Lifshitz version of the radiative reaction describes the physical trajectories correctly. Furthermore, it is argued that the trajectory described by the Landau–Lifshitz equation represents an "unstable fixed point" of the manifold described by the trajectories under a renormalization group analysis, and that any modifications beyond the Landau–Lifshitz equation are small for any realistic value of the field strength. However, in Ref. [47], it is argued that the "truncation" of the Abraham–Lorentz–Dirac equation in the sense of Landau and Lifshitz can lead to physically nonsensical trajectories near the onset of time-dependent perturbations, and this statement is illustrated on a number of example cases.

Alternatively, one can point out that, for reasons of principle, the problem cannot be solved on the level of quantum mechanics and requires the added power of quantum electrodynamics, where the "photon recoil", i.e., the backreaction of radiation emission onto the energy and the momentum of the emitting particle, is being taken into account at all interaction vertices without further approximations. However, for an electron moving in a strong, external, classical field, it would be impossible to describe all interactions to all orders, using the formalism of QED, and therefore, the classical models are still useful. Lorentz [49] traced the mechanism of the backreaction to the fact that a spatially extended object emitting radiation actually cannot fulfill Newton's third law, but one has to take into account the backreaction of the emitted radiation onto itself. Within the nonrelativistic approximation, for a spherical radiation emitting object of radius R, the result for the self-reaction force $\vec{F}_{\text{self}}$ can be expressed as follows,

$$\vec{F}_{\text{self}} = \frac{m\tau_{\text{rad}}}{2R^2} \left[\vec{v}\left(t - \frac{2R}{c}\right) - \vec{v}(t) \right],$$

$$\vec{v}\left(t - \frac{2R}{c}\right) = \vec{v}(t) - \frac{2R}{c}\,\dot{\vec{v}}(t) + \frac{1}{2}\left(\frac{2R}{c}\right)^2 \ddot{\vec{v}}(t) + \dots,$$

$$F_{\text{self}} = -\frac{mc\tau_{\text{rad}}}{R}\,\vec{a} + m\tau_{\text{rad}}\,\dot{\vec{a}},$$

$$m\,\vec{a} = \vec{F}_{\text{self}} + \vec{F}_{\text{ext}} = -\frac{m\,c\,\tau_{\text{rad}}}{R}\,\vec{a} + m\tau_{\text{rad}}\,\dot{\vec{a}} + \vec{F}_{\text{ext}},$$

$$\left(m + \frac{m\,c\,\tau_{\text{rad}}}{R}\right)\vec{a} = \vec{F}_{\text{ext}} + m\tau_{\text{rad}}\,\dot{\vec{a}} = \vec{F}_{\text{ext}} + \vec{F}_{\text{rad}}. \tag{8.237}$$

This equation looks suspiciously similar to Eq. (8.228), but with a "renormalized" mass

$$m \to m + \frac{m\,c\,\tau_{\text{rad}}}{R}. \tag{8.238}$$

In some sense, taking the backreaction into account, we find a "dynamical" correction to the electron's mass, which anticipates the mass renormalization inherent to the field-theoretical formulation of QED.

One can also attempt to include the spin precession of the electron in the external field, which results in the Bargmann–Michel–Telegdi equation described in Ref. [52]. A further development is the Barut–Zanghi equation derived in Ref. [53]. By applying the Landau–Lifshitz prescription to the Barut–Zanghi equation, which eliminates runaway solutions to the Lorentz–Dirac equation, the authors of Ref. [48] arrive at a variant of the Bargmann–Michel–Telegdi equation which does not have runaway solutions. Indeed, the authors of Ref. [48] argue that, at least for spin-polarized orbits in a constant magnetic field, the orbits of their magnetically inter-acting electron theory (with spin) decay as physically expected. They also point out that it should be possible to test the equations using a small (few MeV) electron synchrotron or a muon storage ring.

We should point out some open problems. Indeed, in Ref. [48], no attempt is made to rigorously derive the proposed "Landau–Lifshitz form" of the Bargmann–Michel–Telegdi equation, or, to analyze its properties on the critical manifold, as done for the "Landau–Lifshitz form" of the Lorentz–Abraham–Dirac equation (without spin) in Refs. [44, 51]. This constitutes an open problem. The next step would then be to include the "classical spin" in the equation of motion of an electron accel-erated by a laser field; this would otherwise generalize the considerations reported in Ref. [46].

8.6.3 *Electrodynamics in General Relativity*

According to classical nonrelativistic physics, the coordinates assigned to an event by different observers are connected by Galilei transformations. Time, in particular, is absolute. By contrast, according to special relativity, time and space transform into each other, but still, in flat space, once the relative velocity of two observers is known, no matter how far they are apart, they can uniformly transform the space and time coordinates assigned to an event by a global Lorentz transformation. In special relativity, time no longer is an absolute quantity. In general relativity, the situation again is different: Time and space, locally, have a Lorentzian geometry, i.e., locally, it is always possible to choose the space-time metric as $g^{\mu\nu} = \text{diag}(1, -1, -1, -1)$. However, it is not possible to extend this local reference frame throughout the Universe. Space-time is curved, and one gives up, in particular, the Euclidean postulate that two (locally) parallel curves never can cross each other at infinity.

From differential geometry, we know that Christoffel symbols enter the definition of the covariant derivative, which is a generalized derivative adapted to curved geometries. For a position-dependent metric $\bar{g}_{\mu\nu}(x)$, the covariant components $A_\mu(x)$ are connected to the contravariant components $A^\mu(x)$ of a vector field $A^\mu(x)$ by the relations

$$x_\mu = \bar{g}_{\mu\nu}(x)\, x^\nu\,, \qquad\qquad x^\mu = \bar{g}^{\mu\nu}(x)\, x_\nu\,,$$
$$A_\mu(x) = \bar{g}_{\mu\nu}(x)\, A^\nu(x)\,, \qquad A^\mu(x) = \bar{g}^{\mu\nu}(x)\, A_\nu(x)\,. \tag{8.239}$$

From now on, we suppress the dependence of the metric on the coordinates and just write $\bar{g}_{\mu\nu}$ for $\bar{g}_{\mu\nu}(x)$ for the curved-space metric, in order to distinguish it from the flat-space metric given in Eq. (8.30). We also employ units with $\hbar = c = \epsilon_0 = 1$, anticipating the natural unit system used in high-energy physics.

Let us briefly motivate the definition of the covariant derivative. First of all, we recall that a vector $\vec{A}$ actually is independent of a coordinate system. A local basis consists of the set of unit vectors $\vec{e}_\mu$,

$$\vec{A} = A^\mu\, \vec{e}_\mu\,, \qquad \vec{e}_\mu \cdot \vec{e}_\nu = \bar{g}_{\mu\nu}\,, \qquad \vec{e}^{\,\mu} \cdot \vec{e}_\nu = \bar{g}^\mu{}_\nu = \delta^\mu{}_\nu\,. \tag{8.240}$$

Because the basis vectors $\vec{e}_\mu$ are non-constant, an additional term emerges as one calculates the derivative of the vector $\vec{A}$ with respect to a coordinate,

$$\frac{\partial}{\partial x^\nu}\vec{A} = \partial_\nu \vec{A} = (\partial_\nu A^\mu)\,\vec{e}_\mu + A^\mu\,(\partial_\nu \vec{e}_\mu) = (\partial_\nu A^\mu)\,\vec{e}_\mu + A^\mu\,\vec{e}_\lambda\, g^{\lambda\rho}\,(\vec{e}_\rho \cdot \partial_\nu \vec{e}_\mu)$$

$$= (\partial_\nu A^\mu)\,\vec{e}_\mu + A^\mu\,\vec{e}_\lambda\, g^{\lambda\rho}\, \Gamma_{\rho\mu\nu} = (\partial_\nu A^\mu)\,\vec{e}_\mu + A^\mu\,\vec{e}_\lambda\, \Gamma^\lambda{}_{\mu\nu}\,, \tag{8.241}$$

where $\Gamma_{\rho\mu\nu} = \vec{e}_\rho \cdot \partial_\nu \vec{e}_\mu$ is a Christoffel symbol, and the Einstein summation convention has been used (all the Greek indices are summed from zero to three). We now swap the dummy indices (summation) in the second term according to $(\lambda \leftrightarrow \mu)$ and identify the components of the covariant derivative,

$$\partial_\nu \vec{A} = \vec{e}_\mu \left[(\partial_\nu A^\mu) + \Gamma^\mu{}_{\lambda\nu}\, A^\lambda \right]\,, \qquad \nabla_\nu A^\mu = \partial_\nu A^\mu + \Gamma^\mu{}_{\lambda\nu}\, A^\lambda\,. \tag{8.242}$$

While the Christoffel symbols $\Gamma^\mu{}_{\lambda\nu}$ are not tensors, one can still raise the first index with the metric,

$$\Gamma_{\rho\mu\nu} = \vec{e}_\rho \cdot \frac{\partial \vec{e}_\mu}{\partial x^\nu}\,, \qquad \Gamma^\beta{}_{\mu\nu} = \bar{g}^{\beta\rho}\, \Gamma_{\rho\mu\nu}\,. \tag{8.243}$$

A rather straightforward calculation shows that

$$\Gamma_{\beta\mu\nu} = \frac{1}{2}\left(\frac{\partial \bar{g}_{\beta\mu}}{\partial x^\nu} + \frac{\partial \bar{g}_{\beta\nu}}{\partial x^\mu} - \frac{\partial \bar{g}_{\mu\nu}}{\partial x^\beta} \right)\,. \tag{8.244}$$

By inspection, one can easily derive the symmetry properties

$$\Gamma_{\beta\mu\nu} = \Gamma_{\beta\nu\mu}\,, \qquad \Gamma^\beta{}_{\mu\nu} = \Gamma^\beta{}_{\nu\mu}\,. \tag{8.245}$$

The covariant derivative of a vector field $A^\mu(x)$ has the properties,

$$\nabla_\mu A^\nu = \partial_\mu A^\nu + \Gamma^\nu{}_{\mu\beta} A^\beta, \qquad \partial_\mu \equiv \frac{\partial}{\partial x^\mu}, \tag{8.246a}$$

$$\nabla_\mu A_\nu = \partial_\mu A_\nu - \Gamma^\beta{}_{\mu\nu} A_\beta, \tag{8.246b}$$

$$\overline{g}^{\beta\nu} \nabla_\mu A_\nu = \nabla_\mu \left(\overline{g}^{\beta\nu} A_\nu \right) = \nabla^\beta A_\nu, \tag{8.246c}$$

i.e., the covariant derivative transforms as a tensor (unlike the partial derivative), and its indices can therefore be raised and lowered with the metric, both before as well as after the covariant differentiation. The Riemann curvature tensor $R^\rho{}_{\sigma\mu\nu}$ is of rank four and can be expressed in terms of the Christoffel symbols,

$$R^\rho{}_{\sigma\mu\nu} = \partial_\mu \Gamma^\rho{}_{\nu\sigma} - \partial_\nu \Gamma^\rho{}_{\mu\sigma} + \Gamma^\rho{}_{\mu\lambda} \Gamma^\lambda{}_{\nu\sigma} - \Gamma^\rho{}_{\nu\lambda} \Gamma^\lambda{}_{\mu\sigma}. \tag{8.247}$$

For the Ricci tensor $R_{\mu\nu}$, one contracts the first and the third index of the Riemann curvature tensor,

$$R_{\mu\nu} = R^\alpha{}_{\sigma\alpha\nu}, \qquad R = \overline{g}^{\mu\nu} R_{\mu\nu} = R^\mu{}_\mu. \tag{8.248}$$

The scalar curvature R of space-time is obtained as the contraction of the Ricci tensor. The Einstein–Hilbert action S_{EH} is given by

$$S_{EH} = \int d^4x \sqrt{-\det g}\, R, \qquad \frac{\delta S_{EH}}{\delta \overline{g}^{\alpha\beta}} = R_{\alpha\beta} - \frac{1}{2} \overline{g}_{\alpha\beta} R, \tag{8.249}$$

and the variational derivative of the action with respect to the metric tensor leads to the condition $R_{\alpha\beta} - \frac{1}{2} \overline{g}_{\alpha\beta} R = 0$, which is the Einstein–Hilbert equation for a region of the Universe, in which there is no matter and no other fields which give rise to a nonvanishing energy-momentum tensor. The variational condition $\delta S_{EH} = 0$ is equivalent to saying that the global integral over the curvature of space-time R acts like a soap film whose surface tension integral is being minimized by the shape assumed by space-time geometry, which is four-dimensional but might be embedded in a higher-dimensional manifold which would defy our intuitive understanding. The integration measure $\int d^4x \sqrt{-\det g}$ is Lorentz-invariant. The coupling of the space-time curvature to the matter density in the Universe can be written in terms of a sum of the Einstein–Hilbert action S_{EH} and the matter action S_M, where the latter can be expressed in terms of the matter Lagrangian L_M,

$$S_{GR} = \frac{1}{16\pi G} S_{EH} + S_M, \qquad S_M = \int d^4x \sqrt{-\det g}\, L_M, \tag{8.250a}$$

$$T_{\mu\nu} = -\frac{2}{\sqrt{-\det g}} \frac{\delta S_M}{\delta \overline{g}^{\mu\nu}} = -\frac{2}{\sqrt{-\det g}} \frac{\partial(\sqrt{-\det g}\, L_M)}{\partial \overline{g}^{\mu\nu}} = -2 \frac{\partial L_M}{\partial \overline{g}^{\mu\nu}} + \overline{g}_{\mu\nu} L_M. \tag{8.250b}$$

The latter functional derivative acts as a definition of the energy-momentum tensor $T_{\mu\nu}$. The form given in Eq. (8.250b) is valid if the matter Lagrangian does not depend on derivatives of the metric. Examples of the energy-momentum tensor include a collection of particles,

$$T^{\mu\nu} = \sum_a \int \mathrm{d}\tau_a \frac{\delta^{(4)}(x - x(\tau_a))}{\sqrt{-\det g}} \, m_a \, v_a^\mu \, v_a^\nu \, . \tag{8.251}$$

Here, the v_a^μ are the four-vector components of the velocity v_a of particle a, and τ_a is the proper time of the trajectory of particle a. For an electromagnetic field, the stress–energy tensor is given by Eq. (8.102),

$$T^{\mu\nu} = -\left(F^{\mu\alpha} \, F^\nu{}_\alpha - \frac{1}{4} \overline{g}^{\mu\nu} \, F^{\alpha\beta} \, F_{\alpha\beta}\right) , \tag{8.252}$$

where we recall that $\epsilon_0 = \mu_0 = 1$ for the current discussion. Under the inclusion of the energy-momentum tensor, the variational equations read as follows,

$$R^{\alpha\beta} - \frac{1}{2} R \overline{g}^{\alpha\beta} = 8\pi G \, T^{\alpha\beta} \, . \tag{8.253}$$

These equations have all been derived under a variation of the action with respect to the metric. They still do not clarify how the Maxwell equations need to be generalized to curved space-time, and this will be explored next.

It is not hard to guess that for a free electromagnetic field, one simply has to use the Lorentz-invariant measure $\int \mathrm{d}^4x \sqrt{-\det g}$ in the action integral for the electromagnetic field,

$$S_{EM} = \int \mathrm{d}^4x \sqrt{-\det g} \left(-\frac{1}{4} F^{\mu\nu} F_{\mu\nu}\right) . \tag{8.254}$$

The field-strength tensor, in the curved space-time, is given as follows,

$$\nabla_\mu A_\nu - \nabla_\nu A_\mu = \partial_\mu A_\nu - \partial_\nu A_\mu \, . \tag{8.255}$$

Here, we have used the definition (8.246) of the covariant derivative and the symmetry properties (8.245) of the Christoffel symbols. Variation with respect to the metric leads to the energy-momentum tensor already given in Eq. (8.252), modulo some total divergence. However, the variational principle also dictates that the action S_{EM} should be invariant under a variation of the four-vector field A^μ. The corresponding Euler–Lagrange equation gives rise to the condition

$$\frac{\partial}{\partial x_\beta} \left(F^{\mu\nu} \sqrt{-\det g}\right) = 0 \, , \qquad \nabla^\beta F^{\alpha\beta} = 0 \, , \tag{8.256}$$

where ∇^β is the covariant derivative. The covariant derivative of the field-strength tensor can be expressed in terms of Christoffel symbols,

$$\nabla_\mu F_{\alpha\beta} = \partial_\mu F_{\alpha\beta} - \Gamma^\rho{}_{\alpha\mu} F_{\rho\beta} - \Gamma^\rho{}_{\beta\mu} F_{\alpha\rho} \, . \tag{8.257}$$

If we consider a source term, then Eq. (8.254) gets modified to read

$$S_{EM} = \delta \int \mathrm{d}^4x \sqrt{-\det g} \left(-\frac{1}{4} F^{\mu\nu} F_{\mu\nu} - J^\mu \, \delta A_\mu\right)$$

$$\frac{\delta S_{EM}}{\delta A^\nu} = \frac{\partial}{\partial x_\nu} \left(F^{\mu\nu} \sqrt{-\det g}\right) - J^\mu \sqrt{-\det g} = 0 \, . \tag{8.258}$$

In component form, we have

$$\frac{1}{\sqrt{-\det g}} \frac{\partial}{\partial x_\beta} \left(F^{\mu\nu} \sqrt{-\det g} \right) = J^\beta , \qquad \nabla^\beta F^{\mu\nu} = J^\beta , \qquad (8.259)$$

where ∇^β is the covariant derivative. The equations (8.259) are the generalizations of the inhomogeneous Maxwell equations, $\partial_\mu F^{\mu\nu} = J^\mu$. For the homogeneous Maxwell equations, we had defined in Eq. (8.89), for a flat space-time, the tensor $\tilde{F}^{\mu\nu} = \frac{1}{2} \epsilon^{\mu\nu\rho\delta} F_{\rho\delta}$, where $\epsilon^{\mu\nu\rho\delta}$ is the totally antisymmetric tensor. The homogeneous equations were obtained as $\partial_\mu \tilde{F}^{\mu\nu} = 0$.

Obviously, in curved space-time, one would replace partial derivatives ∂_β with covariant derivatives ∇_β to obtain the following equation, which is the appropriate generalization to curved space-time,

$$\epsilon^{\alpha\beta\gamma\delta} \nabla_\beta F_{\gamma\delta} = 0 . \qquad (8.260)$$

Any cyclic permutation of the indices $\beta\gamma\delta$ leaves the result invariant, and therefore we can reformulate the condition (8.260) as follows,

$$\epsilon^{\alpha\beta\gamma\delta} \nabla_\beta F_{\gamma\delta} = \frac{1}{3} \epsilon^{\alpha\beta\gamma\delta} \left(\nabla_\beta F_{\gamma\delta} + \nabla_\gamma F_{\delta\beta} + \nabla_\delta F_{\beta\gamma} \right) = 0 . \qquad (8.261)$$

The value of the index α is arbitrary, and so we have

$$\nabla_\beta F_{\gamma\delta} + \nabla_\gamma F_{\delta\beta} + \nabla_\delta F_{\beta\gamma} = \partial_\beta F_{\gamma\delta} + \partial_\gamma F_{\delta\beta} + \partial_\delta F_{\beta\gamma} = 0 , \qquad (8.262)$$

which is otherwise known as the Bianchi identity. We recall that the covariant derivative of the Maxwell tensor has been defined in Eq. (8.257).

A very interesting problem concerns the question whether a freely falling electron (in the gravitational field of a star, say) emits radiation or not. On the one hand, one can argue that any accelerated electron should radiate, a statement which would in principle pertain to gravitational interactions. On the other hand, one can argue that gravitational interactions are special, in the sense that any massive particle follows a geodesic, described by the equation

$$\frac{\mathrm{d}^2 x^\mu}{\mathrm{d}^2 \tau} + \Gamma^\mu{}_{\rho\sigma} \frac{\mathrm{d}x^\rho}{\mathrm{d}\tau} \frac{\mathrm{d}x^\sigma}{\mathrm{d}\tau} = 0 , \qquad (8.263)$$

i.e., it follows a "straight line" in the curved space-time, and thus, it is highly questionable if a straightforward generalization of the concept of the "radiating electron" is applicable to gravitational interactions. In the literature, it is sometimes said that a "particle on a geodesic does not radiate", keeping in mind that a massive body follows a geodesic trajectory described by Eq. (8.263).

The last word on this issue still needs to be spoken. In order to treat this problem systematically, one has to formulate the problem in such a way that the Liénard–Wiechert potentials are calculated, using the formalism of curved space-time, for a particle moving along a specific trajectory. One thus needs to solve the equations that couple the potentials to the sources in curved space-time, as given in Eq. (8.259). Or, expressed differently, one has to calculate the Green functions of the electromagnetic theory in curved space-time [54]. A particularly interesting case is given by the stationary (time-independent) Schwarzschild metric,

$$\bar{g}_{\mu\nu} = \text{diag}\left(e^{\nu}, -e^{\lambda}, -r^2, -r^2 \sin^2\theta\right), \tag{8.264}$$

$$e^{\nu} = e^{-\lambda} = 1 - \frac{2GM_P}{r}, \tag{8.265}$$

where the Schwarzschild radius is $r_s = 2G M_P/c^2$, and G is Newton's gravitational constant. Furthermore, M_P is the mass of the planet under consideration. If M_P is the mass of the Earth, then we have $r_s \approx 0.0089\,\text{m}$. For a gravitational center and a given trajectory of a freely falling body in the Schwarzschild metric, the scalar and vector potentials for the pointlike particle should be expressed as a function of the space-time coordinates along the geodesic. Using the curved-space Green functions mentioned above, one should be able to calculate the scalar and vector potentials, and calculate the Poynting vector which describes the dissipation of energy through radiation. A systematic way of treating this problem has been outlined in Ref. [54]; the conclusion seems to be that a particle on an (almost) stable orbit (almost) does not radiate. We recall that, in view of the perihelion precession (of Mercury and other planets), there actually are no stable orbits in general relativity in the Schwarzschild metric. Also, in Ref. [54], one reaches the conclusion that a particle on a "crash" orbit toward the center of the gravitating object (almost) does not radiate. However, a particle on a hyperbolic orbit seems to give off a substantial amount of radiation. In general, this is a very difficult problem, both conceptually and technically, and it may well be worth the effort to recalculate the formulas presented in Ref. [54].

8.7 Exercises

• **Exercise 8.1**: Consider the Lagrangian (in classical mechanics)

$$\mathcal{L} = \frac{1}{2}m\vec{v}^{\,2} - V(|\vec{r}|), \tag{8.266}$$

for a single particle, and its variation under rotations characterized by a vector of rotation angles $\vec{\varphi}$. Under the assumption that $\mathcal{L}$ is independent of the rotation angles $\vec{\varphi}$, show that the conserved quantity, in the sense of Noether's theorem, is the angular momentum.

• **Exercise 8.2**: Show the commutation relations for the orbital angular momentum operators (8.43),

$$\left[L^i, L^j\right] = i\epsilon^{ijk}L^k, \tag{8.267}$$

by acting on a general test function $f = f(\vec{r})$.

• **Exercise 8.3**: Show the identity (8.48),

$$(\vec{\varphi} \times \vec{u}) \times \vec{v} + \vec{u} \times (\vec{\varphi} \times \vec{v}) = \vec{\varphi} \times (\vec{u} \times \vec{v}), \tag{8.268}$$

with the help of the ϵ tensor.

• **Exercise 8.4**: Derive the Biot–Savart law from the Lorenz-gauge coupling of the vector potential to the current, in steady state,

$$\left(\frac{\partial^2}{\partial t^2} - \vec{\nabla}^2\right)\vec{A}(\vec{r},t) = \mu_0\,\vec{J}(\vec{r},t) \quad \Rightarrow \quad \vec{\nabla}^2\vec{A}(\vec{r}) = -\mu_0\,\vec{J}(\vec{r}). \tag{8.269}$$

Hint: Write an "action-at-a-distance" solution to Eq. (8.269) (which is not problematic because we considering a steady state) and take the curl of the vector potential.

• **Exercise 8.5**: Start from the relativistic transformation of the magnetic field,

$$\vec{B}' = \gamma\left(\vec{B} - \frac{1}{c^2}(\vec{v}\times\vec{E})\right) + (1-\gamma)\frac{\vec{B}\cdot\vec{v}}{v^2}\vec{v}. \tag{8.270}$$

These formulas are valid for the transformation from the rest frame to the moving (primed) frame. Recall that, for a Lorentz boost along the z axis, the backtransformation of z and t reads as

$$x = x'. \qquad y = y', \qquad z = \gamma z' + \gamma\,v_x\,t', \qquad t = \gamma t' + \gamma\frac{v}{c^2}z'. \tag{8.271}$$

Derive an expression for the following partial derivative, keeping t' constant,

$$\left.\frac{\partial}{\partial z'}\right|_{t'} = \left.\frac{\partial z}{\partial z'}\right|_{t'}\frac{\partial}{\partial z} + \left.\frac{\partial t}{\partial z'}\right|_{t'}\frac{\partial}{\partial t} = F\frac{\partial}{\partial z} + G\frac{\partial}{\partial t}, \tag{8.272}$$

where the variable which is being kept constant is indicated as a subscript and you calculate the terms F and G. Show that for $\vec{v} = v\,\hat{e}_z$, i.e., under a boost in the z direction,

$$B_x' = \gamma\left(B_x + \frac{v}{c^2}E_y\right), \qquad B_y' = \gamma\left(B_y + \frac{v}{c^2}E_x\right), \qquad B_z' = B_z. \tag{8.273}$$

Furthermore, show that

$$\frac{\partial B_x'}{\partial x'} + \frac{\partial B_y'}{\partial y'} + \frac{\partial B_z'}{\partial z'} = \frac{\partial B_x}{\partial x} + \frac{\partial B_y}{\partial y} + \frac{\partial B_z}{\partial z} + \frac{v\gamma}{c^2}\left[\frac{\partial B_z}{\partial t} + \left(\vec{\nabla}\times\vec{E}\right)_z\right] \tag{8.274}$$

where $\left(\vec{\nabla}\times\vec{E}\right)_z = \partial E_x/\partial y - \partial E_y/\partial x$ is the z component of the curl of the electric field. Show that the expression given in Eq. (8.274) vanishes and interpret your result in terms of the absence of magnetic monopoles in the primed and unprimed coordinate systems, and in terms of (perhaps, if applicable) the Faraday law.

• **Exercise 8.6**: Start from the relativistic transformation of the electric field and charge density,

$$\vec{E}' = \gamma\left(\vec{E} + \vec{v}\times\vec{B}\right) + (1-\gamma)\frac{\vec{E}\cdot\vec{v}}{v^2}\vec{v}, \qquad \rho' = \gamma\left(\rho - \frac{1}{c^2}\vec{J}\cdot\vec{v}\right). \tag{8.275}$$

These formulas are valid for the transformation from the rest frame to the moving (primed) frame.

Set $\vec{v} = v_x\,\hat{e}_x$ and write explicitly the components E_x', E_y' and E_z' as a function of the unprimed (rest-frame) components of the fields E_x, E_y and E_z and B_x, B_y and B_z. Recall that, for a Lorentz boost along the x axis, the backtransformation of x and t reads as

$$x = \gamma x' + \gamma\,v_x\,t', \qquad y = y', \qquad z = z', \qquad t = \gamma t' + \gamma\frac{v_x}{c^2}x'. \tag{8.276}$$

Derive an expression for the following partial derivative, keeping t' constant,

$$\frac{\partial}{\partial x'}\bigg|_{t'} = \frac{\partial x}{\partial x'}\bigg|_{t'}\frac{\partial}{\partial x} + \frac{\partial t}{\partial x'}\bigg|_{t'}\frac{\partial}{\partial t} = K\frac{\partial}{\partial x} + L\frac{\partial}{\partial t}, \tag{8.277}$$

where the variable which is being kept constant is indicated as a subscript and you calculate the terms K and L.

In the moving frame, Gauss's law reads as follows,

$$\frac{\partial}{\partial x'}E_x' + \frac{\partial}{\partial y'}E_y' + \frac{\partial}{\partial z'}E_z' = \frac{1}{\epsilon_0}\rho'. \tag{8.278}$$

Again, assuming $\vec{v} = v_x\,\hat{e}_x$, use the results for the components of $\vec{E}'$ derived previously, and Eq. (8.277), to rewrite the left-hand side of Eq. (8.278) in terms of the unprimed components of the fields. Then, transform ρ' on the right-hand side of Eq. (8.278) according to Eq. (8.275), again assuming $\vec{v} = v_x\,\hat{e}_x$. Show that, in the unprimed (rest) frame, Eq. (8.278) is equivalent to a linear superposition of the "unprimed" Gauss's law in the rest frame, and of the x-component of the Ampere–Maxwell law in the rest frame.

• **Exercise 8.7:** We had encountered in Eq. (8.203) the following integral for potential generated by a uniformly charged, $D = n - \varepsilon$ dimensional structure,

$$\Phi(z) = \frac{1}{4\pi\varepsilon_0}\Omega_D\int_0^\infty dr\,r^{D-1}z_0^\varepsilon\frac{\Xi}{\sqrt{r^2 + z^2}} = \frac{\Xi}{4\pi\varepsilon_0}J(D), \qquad D = n - \varepsilon, \tag{8.279}$$

Here, $J(D)$ is defined as

$$J(D) = \Omega_D\int_0^\infty dr\,r^{D-1}z_0^\varepsilon\frac{1}{\sqrt{r^2 + z^2}}, \qquad \Omega_D = \frac{2\,\pi^{D/2}}{\Gamma(D/2)}. \tag{8.280}$$

(a.) Show that

$$J(D) = \pi^{\frac{1}{2}(D-1)}\,\Gamma\left(\tfrac{1}{2}(1 - D)\right)z_0^\varepsilon\,|z|^{D-1}. \tag{8.281}$$

You may use the result

$$\int_0^\infty dr\,\frac{r^{D-1}}{\sqrt{1 + r^2}} = \frac{\Gamma(\frac{1}{2}(1 - D))\Gamma(\frac{1}{2}D)}{2\sqrt{\pi}}. \tag{8.282}$$

(b.) Investigate a "zero-dimensional uniformly charged structure" (a point charge). To this end, first show that, for $D = 0 - \varepsilon$,

$$J(D = 0 - \varepsilon) = \pi^{-\frac{1}{2}(1+\varepsilon)}z_0^\varepsilon\,z^{-1-\varepsilon}\Gamma\left(\frac{1 + \varepsilon}{2}\right) = \frac{1}{z} + \mathcal{O}(\varepsilon). \tag{8.283}$$

Give a heuristic argument why you identify the uniform charge density $\Xi \equiv q$ in zero dimensions, where q is the value of the point charge. Finally, show that

$$\Phi(z) = \frac{q}{4\pi\varepsilon_0\,z}, \qquad D = 0 - \varepsilon, \tag{8.284}$$

and interpret your result in terms of the well-known potential generated by a point charge.

• **Exercise 8.8**: Show Eq. (8.217).

• **Exercise 8.9**: Investigate the Abraham–Minkowski controversy discussed in Sec. 8.6.1 on the basis of the microscopic model introduced in Sec. 6.2.

• **Exercise 8.10**: Form your own opinion regarding the material presented in Sec. 8.6.2. Study the literature and attempt to derive an equation for the description of radiative reaction free from the problems of runaway solutions. Attempt to reconcile the approach with quantum electrodynamics.

• **Exercise 8.11**: Study the literature on Christoffel symbols. Show Eq. (8.244). As a next step, attempt to generalize the approach to the gradient, curl and Laplacian in spherical and cylindrical coordinates, with reference to the discussion in Sec. 2.3.2 [see Ref. [16]]. The covariant derivatives in Eq. (8.260) act on vectors according to

$$\nabla_\alpha V^\beta = \partial_\alpha V^\beta + \Gamma^\beta{}_{\alpha\mu} V^\mu \qquad (8.285)$$

and on tensors as follows,

$$\nabla_\alpha F^{\beta\gamma} = \partial_\alpha F^{\beta\gamma} + \Gamma^\beta{}_{\alpha\mu} F^{\mu\gamma} - \Gamma^\gamma{}_{\alpha\rho} F^{\beta\rho}. \qquad (8.286)$$

How would the Coulomb law change around a massive black hole? Use Eq. (8.259), specialized to the Schwarzschild geometry.

Bibliography

[1] M. Spivak, *Calculus on Manifolds: A modern approach to classical theorems of advanced calculus*. Addison-Wesley, Reading, MA (1992).

[2] L. Parker and S. M. Christensen, *MathTensor: A System for Doing Tensor Analysis by Computer*. Addison-Wesley, Reading, MA (1994).

[3] B. O'Neill, *Elementary Differential Geometry*, 2nd edn. Academic Press, New York (1997).

[4] M. Spivak, *A Comprehensive Introduction to Differential Geometry, Vols. 1–5*, 3rd edn. Publish or Perish, Houston, TX (1999).

[5] F. W. J. Olver, D. W. Lozier, R. F. Boisvert and C. W. Clark, *NIST Handbook of Mathematical Functions*. Cambridge University Press, Cambridge (2010).

[6] H. Bateman, *Higher Transcendental Functions*, Vol. 1. McGraw-Hill, New York (1953).

[7] M. Abramowitz and I. A. Stegun, *Handbook of Mathematical Functions*, 10th edn. National Bureau of Standards, Washington, D. C. (1972).

[8] G. N. Watson, *A Treatise on the Theory of Bessel Functions*. Cambridge University Press, Cambridge, UK (1922).

[9] F. W. J. Olver, *Asymptotics and Special Functions*. Academic Press, New York, NY (1974).

[10] S. Wolfram, *Mathematica-A System for Doing Mathematics by Computer*. Addison-Wesley, Reading, MA (1988).

[11] U. D. Jentschura and E. Lötstedt, Numerical Calculation of Bessel, Hankel and Airy Functions, *Comput. Phys. Commun.* **183**, pp. 506–519 (2012).

[12] S. Hildebrandt and A. J. Tromba, *Mathematics of Optimal Form*. W. H. Freeman, New York (1984).

[13] R. Courant and D. Hilbert, *Methods of Mathematical Physics—Volume 1*. Interscience, New York (1966).

[14] G. Strang, *Introduction to Linear Algebra*, 4th edn. Wellesley–Cambridge Press, Wellesley, MA (2009).

[15] C. D. Meyer, *Matrix Analysis and Applied Linear Algebra*. Society for Industrial and Applied Mathematics, Philadelphia, PA (2001).

[16] D. A. Clarke, *A Primer on Tensor Calculus*, Saint Mary's University, Halifax, Nova Scotia, Canada (unpublished) (2011).

[17] C. Cohen-Tannoudji, B. Diu and F. Laloë, *Quantum Mechanics (Volume 1)*, 1st edn. J. Wiley & Sons, New York (1978a).

[18] C. Cohen-Tannoudji, B. Diu and F. Laloë, *Quantum Mechanics (Volume 2)*, 1st edn. J. Wiley & Sons, New York (1978b).

[19] J. D. Jackson, *Classical Electrodynamics*, 3rd edn. J. Wiley & Sons, New York, NY (1998).

[20] C. Itzykson and J. B. Zuber, *Quantum Field Theory*. McGraw-Hill, New York (1980).

[21] J. Schwinger, L. L. DeRaad, K. A. Milton and W.-Y. Tsai, *Classical Electrodynamics*. Perseus, Reading, MA (1998).

[22] B. J. Wundt and U. D. Jentschura, Sources, Potentials and Fields in Lorenz and Coulomb Gauge: Cancellation of Instantaneous Interactions for Moving Point Charges, *Ann. Phys. (N.Y.)* **327**, pp. 1217–1230 (2012).

[23] U. D. Jentschura, *Precision theory of atoms: Quantum electrodynamics at work*. World Scientific, Singapore (scheduled for 2017).

[24] F. Rohrlich, Causality, the Coulomb field, and Newtons law of gravitation, *Am. J. Phys.* **70**, pp. 411–414 (2002).

[25] J.-J. Labarthe, The vector potential of a moving charge in the Coulomb gauge, *Eur. J. Phys.* **70**, pp. L31–L32 (1999).

[26] G. Russakoff, A Derivation of the Macroscopic Maxwell Equations, *Am. J. Phys.* **38**, pp. 1188–1195 (1970).

[27] G. Łach, M. DeKieviet and U. D. Jentschura, Multipole Effects in Atom–Surface Interactions: A Theoretical Study with an Application to He–α-quartz, *Phys. Rev. A* **81**, p. 052507 (2010).

[28] W. Sellmeier, Zur Erklärung der abnormen Farbenfolge im Spectrum einiger Substanzen, *Annalen der Physik und Chemie* **219**, pp. 272–282 (1871).

[29] E. D. Palik, *Handbook of Optical Constants of Solids*. Academic Press, San Diego (1985).

[30] B. Bertotti, L. Iess and P. Tortura, A test of general relativity using radio links with the Cassini spacecraft, *Nature (London)* **425**, pp. 374–376 (2003).

[31] A. Wolski, *Theory of Electromagnetic Fields, Part II: Standing Waves*, CERN Accelerator School, Specialised Course on Radiofrequency for Accelerators (unpublished) (2010).

[32] H. B. G. Casimir and D. Polder, The Influence of Radiation on the London-van-der-Waals Forces, *Phys. Rev.* **73**, pp. 360–372 (1948).

[33] H. Minkowski, *Die Grundgleichungen für die elektromagnetischen Vorgänge in bewegten Körpern*, Nachr. Ges. Wiss. Göttingen, Math.–Phys. Kl. **1**, 53–111 (1908).

[34] M. Abraham, *Zur Elektrodynamik bewegter Körper*, Rend. Circ. Math. Palermo **28**, 1–28; *Sull'Elettrodinamica di Minkowski, ibid.* **30**, 33–46 (1910).

[35] P. W. Milonni and R. W. Boyd, Momentum of Light in a Dielectric Medium, *Adv. Opt. Photonics* **2**, pp. 519–533 (2010).

[36] I. Brevik and S. A. Ellingsen, Detection of the Abraham force with a succession of short optical pulses, *Phys. Rev. A* **86**, p. 025801 (2012).

[37] L. D. Landau and E. M. Lifshitz, *The Classical Theory of Fields, Volume 2 of the Course on Theoretical Physics*. Pergamon Press, Oxford, UK (1960).

[38] M. Abraham, Prinzipien der Dynamik des Elektrons, *Ann. Phys. (Leipzig)* **315**, pp. 105–179 (1902).

[39] M. Abraham, Zur Theorie der Strahlung und des Strahlungsdruckes, *Ann. Phys. (Leipzig)* **319**, pp. 236–287 (1904).

[40] P. A. M. Dirac, Classical theory of radiating electrons, *Proc. Roy. Soc. London, Ser. A* **167**, p. 148 (1938).

[41] B. S. DeWitt and R. W. Brehme, Radiation Damping in a Gravitational Field, *Ann. Phys. (N.Y.)* **9**, p. 220259 (1960).

[42] J. M. Hobbs, A Vierbein Formalim of Radiation Damping, *Ann. Phys. (N.Y.)* **47**, pp. 141–165 (1968).

[43] H. Levine, E. J. Moniz and D. H. Sharp, Motion of extended charges in classical electrodynamics, *Am. J. Phys.* **45**, pp. 75–78 (1977).

[44] H. Spohn, The critical manifold of the Lorentz-Dirac equation, *Europhys. Lett.* **50**, pp. 287–292 (2000).

[45] R. Medina, Radiation reaction of a classical quasi-rigid extended particle, *J. Phys. A* **39**, pp. 3801–3816 (2006).

[46] A. di Piazza, Exact Solution of the Landau–Lifshitz Equation in a Plane Wave, *Lett. Math. Phys.* **83**, pp. 305–313 (2008).

[47] D. J. Griffiths, T. C. Proctor and F. D. Schroeter, Abraham–Lorentz versus Landau–Lifshitz, *Am. J. Phys.* **78**, pp. 391–402 (2010).

[48] A. Kar and S. G. Rajeev, On the relativistic classical motion of a radiating spinning particle in a magnetic field, *Ann. Phys. (N.Y.)* **326**, pp. 958–967 (2011).

[49] H. A. Lorentz, Electromagnetic phenomena in a system moving with any velocity smaller than that of light, *Kon. Akad. Weten. Amsterdam. Proc.* **6**, pp. 809–831 (1904).

[50] L. D. Landau and E. M. Lifshitz, *Quantum Mechanics, Volume 3 of the Course on Theoretical Physics.* Pergamon Press, Oxford, UK (1958).

[51] H. Spohn, *Dynamics of Charged Particles and their Radiation Field.* Cambridge University Press, Cambridge, England (2004).

[52] V. Bargmann, L. Michel and V. I. Telegdi, Precession of the Polarization of Particles Moving in a Homogeneous Magnetic Field, *Phys. Rev. Lett.* **2**, pp. 435–436 (1959).

[53] A. O. Barut and N. Zanghi, Classical Model of the Dirac Electron, *Phys. Rev. Lett.* **52**, pp. 2009–2012 (1984).

[54] R. A. Breuer, *Gravitational Perturbation Theory and Synchrotron Radiation (Lecture Notes in Physics Vol. 44).* Springer, Heidelberg (1975).

Index